PHOTOCHEMICAL PROCESSES
IN ORGANIZED MOLECULAR SYSTEMS

North-Holland
Delta Series

NORTH-HOLLAND
AMSTERDAM • LONDON • NEW YORK • TOKYO

Photochemical Processes in Organized Molecular Systems

Proceedings of the Memorial Conference for
the late Professor Shigeo Tazuke
Yokohama, Japan
September 22 -24, 1990

Editor-in-Chief

Kenichi Honda

Tokyo Institute of Polytechnics

Executive Editors

Noboru Kitamura
Research Development Corporation of Japan

Hiroshi Masuhara
Osaka University

Co-Editors

Tomiki Ikeda
Masahiko Sisido
Tokyo Institute of Technology

Mitchell A. Winnik
University of Toronto

1991

NORTH-HOLLAND
AMSTERDAM • LONDON • NEW YORK • TOKYO

North-Holland
Elsevier Science Publishers B.V.
P.O. Box 211
1000 AE Amsterdam
The Netherlands

Sole distributors for the U.S.A. and Canada:

Elsevier Science Publishing Company, Inc.
655 Avenue of the Americas
New York, N.Y. 10010
U.S.A.

Library of Congress Cataloging-in-Publication Data

Photochemical processes in organized molecular systems : proceedings
   of the memorial conference for the late professor Shigeo Tazuke,
   Yokohama, Japan, September 22-24, 1990 / editor-in-chief, Kenichi
   Honda ... [et al.].
        p.   cm. -- (North-Holland delta series)
     Papers from the International Conference on Photochemical
   Processes in Organized Molecular Systems.
     Includes bibliographical references and indexes.
     ISBN 0-444-88878-0
     1. Photochemistry--Congresses.   I. Tazuke, Shigeo, 1935-   .
   II. Honda, Kenichi.   III. International Conference on Photochemical
   Processes in Organized Molecular Systems (1990 : Yokohama-shi,
   Japan)   IV. Series.
   QD701.P455 1991
   541.3'5--dc20   91-25550
                                                               CIP

ISBN: 0 444 88878 0

Printed in The Netherlands

PREFACE

It was indeed a great discouragement for the chemistry community in Japan in 1989 to lose Professor Shigeo Tazuke of the Tokyo Institute of Technology. His colleagues, friends, and students will not forget his genuine enthusiasm for chemistry and the pioneering spirit of his research work.

The International Conference on "Photochemical Processes in Organized Molecular Systems" held at Yokohama International Convention Center on September 22-24, 1990 was dedicated to the late Professor Shigeo Tazuke. The scientific program consisted of 21 invited lectures, 8 invited posters, and 73 contributed posters. There were over 270 participants including 12 scientists from abroad. Professor Tazuke had wide interests in science. To reflect the breadth of his interests, the Conference covered four separate areas in which he was active: I. photoinduced electron and energy transfer processes, II. photoredox reactions in solution, III. photochemistry in organized molecular systems, and IV. toward integrated molecular systems.

It was our hope that the high quality of the talks and the active discussion would enhance progress in both fundamental and technological aspects of photochemistry and photophysics. Our memory of the enthusiasm of the late Professor Shigeo Tazuke in science and his life contributed to the success of this Conference.

On behalf of the Organizing Committee, I would like to express our hearty thanks to all the authors for their contributions.

Kenichi Honda
Tokyo, Japan

THE MEMORIAL CONFERENCE FOR THE LATE PROFESSOR SHIGEO TAZUKE

The Memorial Conference for the late Prof. Shigeo Tazuke was sponsored and financially supported by the Prof. Shigeo Tazuke Memorial Committee, to which many personal and institutional donations were made. The Conference was co-sponsored by Japanese Photochemical Association, The Chemical Society of Japan, Kanto Branch, The Society of Polymer Science, Japan, and the Research Laboratory of Resources Utilization, Tokyo Institute of Technology.

For the organization of the Conference, Miss Shoko Hitomi, Microphotoconversion Project, JRDC, helped us to prepare the circulars and the book of abstracts of the Conference. The students of the late Prof. Shigeo Tazuke, Photochemical Process Division, Tokyo Institute of Technology, served as the organizing staff during the Conference. We would like to express our sincere thanks to all the persons who contributed in many ways to the Conference.

Organizing Committee

Chairman: K. Honda (Tokyo)
Member: T. Endo (Yokohama)   Y. Hayashi (Yokohama)
T. Ikeda (Yokohama)   M. Irie (Fukuoka)
H.-B. Kim (Yokohama)   N. Kitamura (Kyoto)
H. Masuhara (Kyoto)   M. Murakami (Yokohama)
M. Sisido (Yokohama)

PHOTOCHEMICAL PROCESSES IN ORGANIZED MOLECULAR SYSTEMS
Memorial Symposium for the late Professor SHIGEO TAZUKE

1990. 9. 22 ~ 24   YOKOHAMA

Professor Shigeo Tazuke
(February 27, 1935 - July 11, 1989)

THE LATE PROFESSOR SHIGEO TAZUKE

Professor Shigeo Tazuke was born in Osaka on February 27, 1935. He received bachelor's degree in 1957 and master's degree in 1959, both from Kyoto University. He received Ph.D in 1962 from Leeds University, England, under the supervision of Lord Professor F.S. Dainton. In 1962-1968, he was a research fellow at Research Institute for Production Development, Kyoto, and in 1968-1972 was a research associate at Kyoto University. He joined the faculty at Research Laboratory of Resources Utilization, Tokyo Institute of Technology in 1972 as an associate professor and was promoted to a full professor in 1980. Professor Tazuke had been a summer faculty fellow at IBM Research Laboratory, San Jose in 1979.

Professor Shigeo Tazuke received the Award of the Society of Polymer Science, Japan, in 1981.

Professor Shigeo Tazuke served as a chairman or a member of organizing committee for Materials Research Society International Meeting, NATO workshop, and IUPAC Symposiums. He had been a plenary or a keynote lecturer at numerous international conferences.

Professor Shigeo Tazuke authored over 270 publications and contributed numbers of chapters to over 40 books.

Professor Shigeo Tazuke's research accomplishments encompass photochemistry and photophysics of both organic and inorganic compounds, polymer science, and photoresponsive materials. Besides his scientific activities, he took a leading role in the Society of Polymer Science, Japan and Japanese Photochemical Association. He made major contributions to research and development programs of the Ministry of Education, Science, and Culture, the Ministry of International Trade and Industry, as well as of the Science and Technology Agency, Japan.

ACKNOWLEDGEMENTS

On behalf of the Editorial Board, we would like to express our heartfelt thanks to all the authors who contributed papers to this volume. Special thanks are also due to our friends for their kind courtesy to review several papers collected in this volume, and to Miss Shoko Hitomi, Microphotoconversion Project, who helped us to edit this volume. We thank Dr. Hiroaki Misawa, Microphotoconversion Project, who provided technical advice for preparing off-set manuscripts.

Finally, we would like to express our sincere thanks to Mrs. Hiroko Tazuke for the partial support to publish this volume.

Hiroshi Masuhara and Noboru Kitamura
Executive Editors
Kyoto, Japan

TABLE OF CONTENTS

Chapter I: PHOTOINDUCED ELECTRON AND ENERGY TRANSFER PROCESSES

Chapter III: PHOTOCHEMISTRY IN ORGANIZED MOLECULAR SYSTEMS

Chapter IV: TOWARD INTEGRATED MOLECULAR SYSTEMS

Chapter I:
Photoinduced Electron and
Energy Transfer Processes

Photochemical Processes in Organized Molecular Systems
K. Honda (Editor-in-Chief)
© Elsevier Science Publishers B.V., 1991

ULTRAFAST LASER PHOTOLYSIS INVESTIGATIONS ON THE PHOTOINDUCED ELECTRON
TRANSFER AND DYNAMICS OF TRANSIENT ION PAIRS

Noboru MATAGA

Department of Chemistry, Faculty of Engineering Science,
Osaka University, Toyonaka, Osaka 560, Japan

In order to elucidate the mechanisms and dynamics regulating the
photoinduced charge separation (CS) and charge recombination (CR) of
produced transient charge transfer (CT) or ion pair (IP) states as well as
related reaction processes, we have made femtosecond–picosecond laser
photolysis studies on various donor (D)–acceptor (A) systems combined by
spacer or directly by single bond, uncombined fluorescer–quencher pairs
including exciplexes, and CT complexes of aromatic hydrocarbons with
various A's such as cyanocompounds and acid anhydrides.

On the basis of the results obtained by comparative studies on the above
systems concerning the effect of the magnitude of electronic interaction
responsible for the electron transfer (ET) between D and A, energy gap and
reorganization energy associated with ET, solvation of transient CT or IP
state and solvent dynamics on the ET processes, we have obtained a deep
insight into the nature of the photoinduced CT or ET phenomena.

1. INTRODUCTION

Depending on the strength of electronic interaction responsible for
electron transfer (ET) between electron donor (D) and acceptor (A), ET
mechanism can be nonadiabatic and adiabatic. In the limit of the strong
interaction in rigidly held D and A systems, the excited state can be regarded
as a very polar single molecule for which the first theoretical formula of
fluorescence Stokes shift due to the interaction with polar solvent was given
in 1955 by Lippert[1] and the present author[2], and has been extended recently by
Bagchi et al.[3] and others to take into account its dynamical aspects.

$$h(\nu_a - \nu_f) = \text{Const.} + \frac{2(\vec{\mu}_e - \vec{\mu}_g)^2}{a^3} \left[\frac{\varepsilon_s - 1}{2\varepsilon_s + 1} - \frac{\varepsilon_\infty - 1}{2\varepsilon_\infty + 1} \right] \tag{1}$$

The relation between the Stokes shift of fluorescence and the stabilization
energies from the Franck–Condon (FC) state to the equilibrium state due to the

reorientation interactions of solute–solvent dipoles in the excited state and
the ground state is given by eq 1.

It should be noted here that the nonequilibrium polarization with respect
to the solute–solvent interaction in the FC state in the light absorption and
emission involved in the calculation of eq 1 is essentially similar to that
involved in the calculation of ET rate in polar solution as developed by
Marcus[4] in 1956. Namely, the solvent reorganization parameter λ_0 derived
first by Marcus for the one electron transfer between charged spheres of
radius a_1 and a_2 with center to center distance R is given by,

$$\lambda_0 = (e)^2 [\frac{1}{2a_1} + \frac{1}{2a_2} - \frac{1}{R}][\frac{1}{\varepsilon_\infty} - \frac{1}{\varepsilon_s}] \ . \tag{2}$$

The equivalent quantity in the case of the solvation of solute dipole is given
from eq 1 as,

$$\lambda_0' = \frac{(\vec{\mu}_e - \vec{\mu}_g)^2}{a^3} [\frac{\varepsilon_s - 1}{2\varepsilon_s + 1} - \frac{\varepsilon_\infty - 1}{2\varepsilon_\infty + 1}] \tag{3}$$

The above treatment are concerned with rather ideal case of one electron
transfer between two charged spheres interacting only weakly and that of the
change of the dipole moment due to the intramolecular charge transfer (CT) of
solute molecule composed of strongly interacting donor and acceptor groups
combined rigidly.   Moreover, in both extreme cases, the solute–solvent
interaction is approximated by linear response, assuming the dielectric
continuum model for solvent.

However, in many cases of actual systems of strongly interacting D and A,
the electronic structure of the (D, A) system seems to change gradually
accompanied with some geometrical change in the course of solvation.   For
example, the ET process in the strongly interacting (D, A) systems combined
directly by single bond are vigorously investigated at present by ultrafast
laser spectroscopy, and actually some of them seems to show such complex CS
processes[5,6].   Moreover, such behavior in the charge separation process may
not be limited to the system combined by single bond but also exciplexes and
excited CT complexes will show similar phenomena.   In relation to this
problem, the present author proposed such mechanism of photoinduced CS process
in 1960's in the case of strongly interacting (D, A) systems of exciplexes[7]
and excited CT complexes[8] in polar solvents. Namely, in order to interpret
the much larger decrease of the exciplex fluorescence yield ϕ_f than lifetime
τ_f with increase of the solvent polarity, the following mechanism of the
solvent polarity dependent electronic and geometrical structure of the
exciplex was proposed[7,8].

With increase of solvent polarity, the energy of ET configuration $(A^- D^+)$

is lowered and coefficient a increases, and b and c decrease in the exciplex wavefunction Ψ_e.

$$\Psi_e \sim a\Phi(A^-D^+) + b\Phi(A^*D) + c\Phi(AD^*) \qquad (4)$$

Since the fluorescence transition moment of the exciplex and the excited CT complex is determined mainly by those between LE (locally excited) configurations $\Phi(A^*D)$, $\Phi(AD^*)$ and ground state $\Psi_g \sim \Phi(AD)$, radiative transition

$$<\Psi_e \mid \vec{\mu}_{op} \mid \Psi_g> \sim a<\Phi(A^-D^+) \mid \vec{\mu}_{op} \mid \Phi(AD)> + b<\Phi(A^{*g}D) \mid \vec{\mu}_{op} \mid \Phi(AD)>$$
$$+ c<\Phi(AD^*) \mid \vec{\mu}_{op} \mid \Phi(AD)> \qquad (5)$$

probability k_f will decreases significantly with increase of solvent polarity while nonradiative transition probability k_{nr} will increase because of the decrease of the energy gap between fluorescent state and the ground state. Furthermore, strong solvation of polar exciplex will induce some geometrical structural change accompanied with the change of exciplex electronic structure, which will further make the exciplex electronic structure more ionic. These solvent effects will lead to the much larger decrease of ϕ_f than τ_f with increase of solvent polarity.

$$\phi_f = k_f/(k_f+k_{nr}), \quad \tau_f = 1/(k_f+k_{nr}) \qquad (6)$$

The above idea proposed by the present author was further examined theoretically by means of nonlinear Schrödinger equation with the solvent polarization dependent electronic wavefunction and vice versa[9,10], although the solvation induced geometrical structural change of exciplex was not taken into account in these treatments.

In any way, since the electronic structure of the (D, A) system is affected not only by the solvent polarity but also by the strengths of D and A (i.e. the oxidation potential of D and reduction potential of A, etc.), the above idea suggests also that, depending on the solvent polarity as well as strengths of D and A, several or many different kinds of exciplexes or excited CT complexes, and ion pairs produced from them or directly by ET due to weak interaction, can exist. This fact will be very important in elucidating the reaction mechanisms of the systems in condensed phase undergoing the photoinduced charge separation, charge recombination as well as chemical reaction product formation.

In this article, we discuss the above problem of the varieties of (D, A) interactions on the basis of the results obtained by femtosecond–picosecond laser photolysis investigations mainly on the strongly interacting (D, A) systems, including the case of some organic photochemical product formation which proceeds via ion pair states.

2. PHOTOINDUCED CS IN D AND A GROUPS COMBINED DIRECTLY BY SINGLE BOND

The intramolecular exciplexes are very suitable model systems which can be used to examine the above problems of the effect of the strength of electronic interaction between D and A on the CS process. We compare here the results of our femtosecond laser photolysis measurements on $p-(CH_3)_2N-\phi-(9-anthryl)(A_0)$ and $p-(CH_3)_2N-\phi-CH_2-(9-anthryl)(A_1)$ in alkanenitrile solutions[5]. In these systems, photoinduced charge transfer occurs from N,N-dimethylaniline (DMA) moiety to excited anthryl group. By femtosecond time-resolved absorption spectral measurement, it is possible to observe directly the time-dependent change of the electronic structure of these systems.

In the case of A_1, it has been clearly shown that the photoexcitation is initially localized in anthracene part and then ET takes place showing clearly the rise of absorption band at 400 nm characteristic to DMA cation in the intramolecular ion pair state, $D^+-CH_2-A^-$. This spectral change can be reproduced by assuming the simple two state model, $D-CH_2-A^* \rightarrow D^+-CH_2-A^-$. The rise time ($\tau_r$) of $D^+-CH_2-A^-$ state in n-butyronitrile (BCN) and n-hexanenitrile (HexCN) solutions at room temperature has been obtained to be 1.0 ps and 1.4 ps, respectively. These values of rise time are rather near to the solvation time of dipole τ_S in these solvent determined by measurements of dynamic fluorescence Stokes shift of probe molecule[11].

In the case of A_0 with much stronger electronic interaction between D and A, however, the photoinduced CS process cannot be described by such simple two state model but it is necessary to assume a gradual change of electronic structure from less polar one to a more polar one due to an extensive solvation and accompanied with some geometrical change. We have estimated approximate rise time of the CT state of A_0 to be $\tau_r \sim 5.0$ ps in HexCN and $\tau_r \sim 2.7$ ps in BuCN. These values are considerably longer compared with the values of A_1 in the same solvent. These results indicate that for the photoinduced CS in such strongly interacting D-A system as A_0, more extensive solvation than in weakly interacting $D-CH_2-A$ system and some geometrical change are necessary in order to prevent electronic delocalization interaction in CS process. We can see analogous photoinduced CS processes in the case of another strongly interacting (D, A) system of some CT complexes as discussed in the next section.

3. PHOTOINDUCED CS PROCESSES OF CT COMPLEXES

From the viewpoints described in sections 1. and 2. concerning the strongly interacting (D, A) systems, we have examined the photoinduced CS processes (IP formation processes) of various CT complexes and CR decay as well as dissociation processes of produced IP's by means of femtosecond-picosecond laser photolysis and time-resolved absorption spectral measurements[12-17].

In the case of the 1,2,4,5-tetracyanobenzene (TCNB)–toluene (Tol) complex in acetonitrile, the solvent reorientation can induce the CS with time constant shorter than 1 ps to a considerable extent but not completely and, for the complete CS leading to the IP formation, further relaxation process with time constant (τ_{CS}) of ca. 20 ps is necessary[12,15]. This τ_{CS} value is much longer than the solvent reorientation relaxation time in acetonitrile, which indicates that further intracomplex structural change assisted by solvation is necessary for the IP formation by complete CS.

The τ_{CS} value in acetonitrile solution has been confirmed to become shorter by lowering the oxidation potential of donor; τ_{CS}=41 ps, 20 ps, 13 ps, 12 ps, 7∿8 ps, 5∿6 ps for the TCNB complexes with benzene, toluene, mesitylene, p-xylene, durene and hexamethylbenzene donors, respectively[15]. Presumably, the extent of the intracomplex structural change necessary for the complete CS will become smaller in the case of the donor with lower ionization potential. The IP formed in this way by the relaxation process from the FC excited state of the complex will be the contact IP (CIP) without intervening solvent between D^+ and A^- ions in the pair.

We have confirmed also that, when the electron affinity of electron acceptor becomes larger the formation of CIP from the excited CT complex becomes faster. For example, our direct observation[17] of CIP formation in the excited state of PMDA (pyromellitic dianhydride, which is a stronger electron acceptor than TCNB) complex with Tol in acetonitrile is shown in Figure 1. We can see clearly in Figure 1(A) that the absorption band at 0.5 ps delay time is considerably broader and a little red shifted compared with that at 15 ps, which indicates that it takes several ps to ten ps for the formation of CIP from the excited state of the CT complex also in the case of PMDA–Tol system. By comparing the simulation of the time profile of absorbance with the observed one in Figure 1(B), the rise time of the CIP state was estimated to be ca. 7 ps in addition to its lifetime of ca. 30 ps. Similar analysis of the results of PMDA–hexamethylbenzene complex in acetonitrile indicated a few ps as the rise time of CIP in addition to its lifetime of ca. 5 ps.

The above results suggests that, for stronger donor or acceptor, the CIP formation becomes faster, which indicates that for such complexes with stronger D or A, the extent of the structural change is smaller, in general. Actually, in the case of the complexes of aromatic hydrocarbons with very strong electron acceptors such as TCNQ (tetracyanoquinodimethane) and TCNE (tetracyanoethylene), the CIP formation is almost instantaneous, and the structure of CIP seems to be very close to the excited CT complex itself[17].

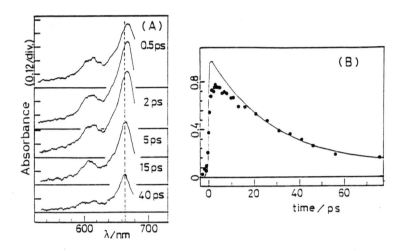

Figure 1

(A) Time-resolved transient absorption spectra of PMDA-Tol CT complex in acetonitrile solution measured with femtosecond laser photolysis system. [PMDA]=0.1 M, [Tol]=0.3 M

(B) Time profile of PMDA⁻ absorbance of PMDA-Tol IP. ●: observed value, ── : simulation taking into account the exciting pulse width and assuming the instantaneous formation of CIP immediately after excitation.

4. CR DECAY OF THE IP PRODUCED BY CT COMPLEX EXCITATION IN COMPARISON WITH THAT OF THE IP FORMED BY CS AT ENCOUNTER BETWEEN FLUORESCER AND QUENCHER

4.1. CR decay of the IP formed by encounter collisional quenching of fluorescence

We have made systematic studies on the energy gap dependence of CR rate of geminate IP produced by CS at encounter between fluorescer and quencher in acetonitrile solution by directly observing the CR deactivation process competing with the dissociation[18] as indicated in eq 7, by means of the picosecond spectroscopy monitoring the time dependence of the absorbance of geminate IP.

$$^1A^* + D \text{ or } A + {}^1D^* \rightarrow {}^1(A_S^- ... D_S^+) \xrightarrow{k_{diss}} A_S^- + D_S^+ \quad (7)$$

$$\uparrow h\nu \qquad\qquad \downarrow k_{CR}$$

$$A + D \qquad\qquad A...D$$

The geminate IP in this case is believed to be mainly the solvent separated IP (SSIP) with intervening solvent molecules between the ions in the pair. As an example, time-resolved transient absorption spectra of 1.12-benzperylene (BPer)-phthalic anhydride (PA) system in acetonitrile and time profile of BPer$^+$ absorbance in SSIP as well as dissociated state in accordance with the reaction scheme of eq 7 are indicated in Figure 2. By conducting similar

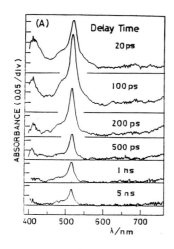
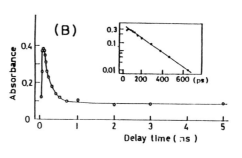

Figure 2

(A) Picosecond time-resolved absorption spectra of BPer-PA system in acetonitrile. [BPer]=2×10^{-4} M, [PA]=0.5 M

(B) Time profile of the BPer$^+$ absorbance at 513 nm in BPer-PA system. Insert: semilogarithmic plot of the absorbance obtained by subtracting the constant value of absorbance at long delay time.

measurements on various (D, A) systems including wide range of free energy gap, $-\Delta G^{\circ}_{ip}$, between SSIP and ground states, we have given first the experimental confirmation of the bell-shaped energy gap dependence of k_{CR} of SSIP, including all of the inverted, top and normal regions[18]. The observed bell-shaped energy gap dependence of k_{CR} of SSIP is in an approximate agreement with the conventional ET theories assuming weak electronic interaction between D and A.

4.2. CR decay of the IP formed by CT complex excitation

Previously we recognized some relations between ionization potential of electron donor and relative ionic dissociation yield obtained by exciting TCNB complexes as well as PMDA complexes in acetonitrile solution with nanosecond laser pulse[19]. That is, the results indicated the tendency of the decrease of relative dissociation yield with decrease of the ionization potential of the donor[19]. However, no such detailed systematic study as described above in the case of the geminate IP formation by diffusional encounter in the fluorescence quenching reaction has been made on the CR processes of the IP's formed by excitation of the CT complexes in acetonitrile solutions.

As we have already discussed in Section 3., the CS process immediately after excitation of the CT complex in acetonitrile leads to the formation of the CIP. It is believed in general that, when the ionic photodissociation of the CT complex occurs in strongly polar solvent, the CIP undergoes change to SSIP, from which the dissociation takes place, as indicated in eq 8.

$$^1(a^{-\delta'} \cdot D^{+\delta'})_S^{FC} \rightarrow {}^1(A^- ... D^+)_S \xrightarrow{k_{solv}} {}^1(A_S^- ... D_S^+) \xrightarrow{k_{diss}} A_S^- + D_S^+ \qquad (8)$$

$$\uparrow h\nu \qquad k_{CR}^{CIP} \qquad k_{CR}^{SSIP}$$

$$^1(A^{-\delta} \cdot D^{+\delta})_S$$

Moreover, it is believed in general that the SSIP in eq 8 is the same or very similar to that formed by encounter collisional quenching of fluorescence between the same (A, D) pair in the same solvent.

However, in most of the systems we have studied by direct quantitative observations of the time dependence of the IP absorbance in acetonitrile solutions, the existence of the definite SSIP in the course of the ionic dissociation was not clear except the TCNB-Tol and -xylene systems[16]. In many other systems, the observed results can be well reproduced by the simple scheme of eq 9.

$$^1(A^{-\delta'} \cdot D^{+\delta'})_S^{FC} \rightarrow {}^1(A^- ... D^+)_S \xrightarrow{k_{diss}} A_S^- + D_S^+ \qquad (9)$$

$$\uparrow h\nu \qquad k_{CR}^{CIP}$$

$$^1(A^{-\delta} \cdot D^{+\delta})_S$$

We have made comparative studies on the behaviors of IP's of the same D and A system, where one is produced by CT complex excitation and the other is formed by encounter collisional quenching of fluorescence. In general, the CIP formed by the CT complex excitation shows a much faster decay and smaller dissociation yield compared with the case of the SSIP formed by CS at diffusional encounter.

For example, in the case of the pyrene (D)-PA (A) system in acetonitrile[13,17,18], $k_{CR}^{CIP} = 7.1 \times 10^9$ s^{-1} in comparison with $k_{CR}^{SSIP} = 1.5 \times 10^9$ s^{-1} and, for the anthracene (D)-PA (A) system in acetonitrile, $k_{CR}^{CIP} = 1.3 \times 10^{10}$ s^{-1} in comparison with $k_{CR}^{SSIP} = 2.2 \times 10^9$ s^{-1}. An extreme example of the very large difference between k_{CR}^{CIP} and k_{CR}^{SSIP} is the case of the pyrene(D)–TCNE (A) system in acetonitrile solution, where $k_{CR}^{SSIP}=2.6\times10^9$ s^{-1} and $k_{diss}=2.5\times10^9$ s^{-1} as obtained by examining the IP formed by CS at diffusional encounter[18] while $k_{CR}^{CIP}=2\times10^{12}$ s^{-1} as determined by femtosecond laser spectroscopy[13,17]. These differences between the k_{CR}^{SSIP} and k_{CR}^{CIP} values of the same D and A systems must be ascribed to the difference of the structure between CIP and SSIP.

4.3. Comparison of the energy gap dependence of the CR decay of CIP with that of the SSIP

By the direct observation of the CR decay of CIP's as described above, we have examined the energy gap dependence of k_{CR}^{CIP} over a wide free energy gap range, and compared with that of k_{CR}^{SSIP}, as shown in Figure 3. We can see

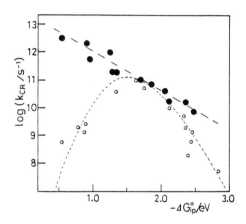

Figure 3
Energy gap dependence of the CR rate constant of the IP produced by CT complex excitation (●) compared with that formed by CS at diffusional encounter in fluorescence quenching reaction (○).

clearly that the energy gap dependence of the CR rate of the CIP studied on similar or the same (D, A) pair in the same solvent, acetonitrile, is quite different from the bell–shaped one observed for the SSIP, and the relation between the k_{CR}^{CIP} and $-\Delta G_{ip}^{\circ}$ (the free energy gap between the IP and the ground state) is given by,

$$k_{CR}^{CIP} = \alpha\exp[-\gamma|\Delta G_{ip}^{\circ}|] \qquad (10)$$

where α and γ are constants independent of ΔG_{ip}°. This energy gap dependence of the CR rate constant of the CIP state is qualitatively analogous to that of the radiationless transition probability in the "weak coupling" limit[20].

Although the exact and quantitative interpretation of the relation in eq 10 is not clear at the present stage of the investigation, we give here a qualitative interpretation for the strong difference between k_{CR}^{CIP} and k_{CR}^{SSIP}, especially in the region of small $|\Delta G_{ip}^{\circ}|$ values, that is, why the normal region is not observed in the case of the CIP. The interpretation is based on the conclusion derived in Section 3. concerning the CIP formation process from the excited CT complex and the dependence of the structure of CIP on the nature of D and A. That is, with increase of the strength of D and A (decrease of ionization potential of D and increase of electron affinity of A), the formation of CIP from the excited CT complex becomes faster and the electronic and geometrical structure of the CIP including surrounding solvent becomes closer to the excited CT complex itself.

Schematically, the stronger the D and A, the closer the position of the potential minimum of CIP on the reaction coordinate (including D, A configuration and solvation) to that of the CT complex itself as shown in Figure 4. This makes difficult the observation of normal region, and the CR decay rate becomes very large with decrease of the energy gap between the CIP and the ground state owing to the increase of the mixing of the ET configuration with the ground configuration.

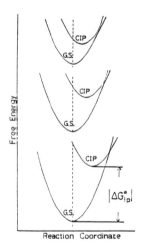

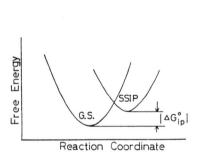

Figure 4
Free energy curves for the CIP states and the ground state (G.S.) of D, A systems vs. reaction coordinate. Change of the positioin of the potential minimum depending on the change of the $-\Delta G_{ip}^{\circ}$ value illustrating that the CR reaction of CIP is in the inverted region for all $-\Delta G_{ip}^{\circ}$ values.

Figure 5
Relation between the free energy curves of the ground state and the SSIP state corresponding to the small $-\Delta G_{ip}^{\circ}$ value, where the large horizontal shift of the potential minimum of SSIP against the ground state brings the CR rate constant of SSIP to the normal region.

On the other hand, the horizontal shift of the potential minimum of the SSIP state on the reaction coordinate may be much larger than that of the CIP. Therefore, for sufficiently strong (D, A) systems with small energy gap, $-\Delta G_{ip}^{\circ}$, the CR rate is in the normal region as indicated in Figure 5.

5. ROLES PLAYED BY DIFFERENT KINDS OF IP'S IN BENZOPHENONE–AMINE PHOTOCHEMICAL REACTIONS

Finally, we discuss an example of the femtosecond–picosecond laser photolysis studies on the role of various IP's in the organic photochemical reaction of benzophenone(BP)–amine system. The investigation on the mechanisms of this hydrogen abstraction reaction is a rather old problem. However, there are still many unclear points concerning the mechanism of this well-known reaction. Quantitative femtosecond–picosecond laser photolysis and direct observation of reaction processes by quantitative time-resolved absorption spectral measurements are of crucial importance. By means of such investigations, we have elucidated detailed mechanisms of this reaction[21-24]. In this article, we discuss the reaction mechanisms of BP-N,N-dimethylaniline (DMA) system in acetonitrile solution on the basis of our results of measurements and we will show that the reaction mechanisms proposed by Peters, et al.[25-28] on this system are completely wrong.

Results of our investigations on BP–DMA system in acetonitrile with relatively high concentration (0.3–1.0 M) of DMA (which is frequently used in the photochemical studies of this system) indicate the formation of (BP$^-$ ...DMA$^+$) IP's in the very early stage immediately after the excitation with hundreds femtosecond laser pulse. Detailed analysis of the results shows clearly that the singlet CIP is formed by excitation of weak BP–DMA CT complex formed in the ground state and also singlet SSIP is formed by ET reaction between ^{1}BP* and DMA in competition with the intersystem crossing. (We have determined the time constant of the intersystem crossing of the free ^{1}BP* in acetonitrile solution to be 9+2 ps.) Moreover, it should be noted here that the rapid ET reaction between ^{1}BP* and DMA is mainly due to the transient effect in the quenching reaction.

According to the above analysis of observed results the ET processes of the excited BP–DMA system in the early stage can be summarized as follows.

$$
\begin{array}{ccc}
& \sim 5 \text{ ps} & \\
^1(\text{BP}^-...\text{DMA}^+)_{\text{com}} & ^1\text{BP}^* + \text{DMA} \xrightarrow{} & ^1(\text{BP}^-...\text{DMA}^+)_{\text{enc}} \quad (11) \\
h\nu \uparrow \; \{85 \text{ ps} & \uparrow h\nu & \searrow \; ^3\text{BP}^* + \text{DMA} \\
(\text{BP}...\text{DMA}) \xrightarrow{} & \text{BP} + \text{DMA} & \\
\quad k_g = 0.1 \sim 0.5 & &
\end{array}
$$

where the decay time 85 ps of the ^{1}CIP, ^{1}(BP$^-$...DMA$^+$)$_{com}$, was determined by exciting exclusively the CT complex at 397 nm with Raman shifted laser pulse and by observing the decay of the absorption band with peak at 740 nm due to the BP$^-$ in the ^{1}(BP$^-$...DMA$^+$)$_{com}$. We have confirmed that, during the decay of ^{1}CIP, the spectral band shape does not show any change and no ketyl radical formation as well as ionic dissociation are observed. This behavior of ^{1}(BP$^-$...DMA$^+$)$_{com}$ can explain the well-known fact that the benzopinacol product decreases with increase of the amine concentration.

We have confirmed that the ^{3}BP* formed by intersystem crossing competing with ET reaction in the ^{1}BP* state undergoes the ET reaction with DMA leading to the formation of SSIP, ^{3}(BP$^-$...DMA$^+$)$_{enc}$. Moreover, by means of the detailed analysis of the time profiles of absorbances at several different wavelengths into the contributions of such species as ^{1}BP*, ^{1}(BP$^-$...DMA$^+$)$_{com}$, ^{1}(BP$^-$...DMA$^+$)$_{enc}$, ^{3}BP*, ^{3}(BP$^-$...DMA$^+$)$_{enc}$, BPH$^\cdot$, etc, with computer simulation, and examining several solutions of different DMA concentrations, we have established the reaction schemes of eq 12 and eq 13 for ^{3}(BP$^-$...DMA$^+$)$_{enc}$ and ^{1}(BP$^-$...DMA$^+$)$_{enc}$, respectively, and determined the rate constants k_{PT}, k_{ID} and k_{CR}.

$$^3BP^* + DMA \rightarrow {}^3(BP^-...DMA^+)_{enc} \begin{array}{l} \xrightarrow{k_{ID}} BP_S^- + DMA_S^+ \\ \xrightarrow{k_{PT}} (BPH^\cdot...DMA^\cdot) \rightarrow BPH^\cdot + DMA^\cdot \end{array} \qquad (12)$$

$$^1BP^* + DMA \rightarrow {}^1(BP^-...DMA^+)_{enc} \begin{array}{l} \xrightarrow{k_{IP}} BP_S^- + DMA_S^+ \\ \longrightarrow (BPH^\cdot...DMA^\cdot) \rightarrow BPH^\cdot + DMA^\cdot \\ \xrightarrow{k_{PT}} \\ \searrow_{k_{CR}} BP + DMA \end{array} \qquad (13)$$

The obtained rate constants for these SSIP's together with those of ^{1}CIP are given in Table 1. We have confirmed that the rate constants obtained here are independent of [DMA] over wide range of its change.

Table 1. Dependence of the Reaction Rate Constant of the Ion Pairs upon the Mode of its Produciton

Ion Pair	k_{PT}/s^{-1}	k_{ID}/s^{-1}	k_{CR}/s^{-1}
^{3}(BP$^-$...DMA$^+$)$_{enc}$	5.4×10^9	1.4×10^9	——
^{1}(BP$^-$...DMA$^+$)$_{enc}$	6.6×10^8	9.5×10^8	5.8×10^8
^{1}(BP$^-$...DMA$^+$)$_{com}$	$\ll 10^8$	$\leq 4 \times 10^8$	1.1×10^{10}

The quite different behaviors of these three kinds of ion pairs are closely related to the problem we have discussed in sections 3. and 4. For example, the much larger k_{CR} value of $^1(BP^-...DMA^+)_{com}$ compared with that of $^1(BP^-...DMA^+)_{enc}$ is in a good agreement with the general tendency observed for CIP and SSIP in sections 3. and 4. and $k_{CR}=1.1\times10^{10}$ s^{-1} is rather close to the k_{CR} value of other CIP's observed for the similar $-\Delta G_{ip}^\circ$ values. With respect to k_{PT}, our tentative interpretations are as follows.

$^1(BP^-...DMA^+)_{com}$ seems to have a rather rigid structure where the components ions can not reorient to take the structure suitable for the proton transfer before recombination, while $^1(BP^-...DMA^+)_{enc}$ will have a more loose structure and is more long-lived to change the structure to the more suitable one for the proton transfer. The difference of k_{PT} between $^1(BP^-...DMA^+)_{enc}$ and $^3(BP^-...DMA^+)_{enc}$ may be also ascribed to the difference of the structure. Since the $^1(BP^-...DMA^+)_{enc}$ is produced by ET between the near-by BP* and DMA in the transient effect of the quenching reaction, it will have a more specific and tight structure favorable for ET but not suitable for proton transfer.

Concerning the structure of the ion pair and its relation to the proton transfer within the ion pair, Peters, et al.[25-28] concluded that the ET reaction leading to the formation of SSIP was followed by change of structure to CIP where BPH' formation by proton transfer took place. This conclusion is based on the experimental results that the absorption peak of BP$^-$ in the ion pair blue-shifted with increase of the delay time. They assigned the band with peak at 720 nm to BP$^-$ in SSIP and that with peak at 690 nm to CIP. Their conclusion was based on the assumption that ET took place exclusively at encounter between $^3BP^*$ and DMA even in such concentrated solution of 1~5 M.

However, at such high concentration of DMA, our results show unambiguously that ^{1}CIP produced by excitation of the ground state weak CT complex, ^{1}SSIP formed by encounter between $^1BP^*$ and DMA, and ^{3} SSIP formed by encounter between $^3BP^*$ and DMA, exist, and ^{1}CIP shows the BP$^-$ absorption band at the longest wavelength and SSIP as well as the dissociated ions show BP$^-$ band at much shorter wavelength. ^{1}CIP undergoes rapid CR decay without reaction and without dissociation, which causes apparent blue shift of the spectra. We have confirmed this by the fact that no spectral shift can be observed when we subtract the contribution of ^{1}CIP band from the observed spectra. On the other hand, ^{3}SSIP and ^{1}SSIP give ketyl radical by proton transfer within the ion pair, and ^{3}SSIP which has the most loose structure gives the much higher yield of ketyl radical.

ACKNOWLEDGEMENT

The present article is a discussion concerning the nature of photoinduced charge separation process and charge recombination of ion pairs as well as chemical reactions proceeding via ion pairs, on the basis of the results of our ultrafast laser photolysis studies on strongly interacting D, A systems supported by a Grant-in-Aid (No. 6265006) from the Japanese Ministry of Education, Science and Culture to the author.

REFERENCES

1) E. Lippert, Z. Naturforsch. 10a (1955) 541.

2) N. Mataga, Y. Kaifu and M. Koizumi, Bull. Chem. Soc. Jpn. 28 (1955) 690.

3) B. Bagchi, D. W. Oxtoby and G. Fleming, Chem. Phys. 86 (1984) 257.

4) R. A. Marcus, J. Chem. Phys. 24 (1956) 966.

5) N. Mataga, S. Nishikawa, T. Asahi and T. Okada, J. Phys. Chem. 94 (1990) 1443.

6) P. F. Barbara, T. J. Kang. W. Jarzeba and T. Fonseca, Solvation Dynamics and Ultrafast Electron Transfer, in: Perspectives in Photosynthesis, eds. J. Jortner and B. Pullman (Kluwer-Academic, Dordrecht, 1990) pp. 273-292.

7) N. Mataga, T. Okada and N. Yamamoto, Chem. Phys. Lett. 1 (1967) 119.

8) N. Mataga and Y. Murata, J. Am. Chem. Soc. 91 (1969) 3144.

9) H. Beens and A. Weller, Excited Molecular π-Complexes in Solution, in: Organic Molecular Photophysics, vol. 2, ed. J. B. Birks (John-Wiley & Sons, London, 1975) pp. 159-215.

10) N. Mataga, Electronic Structures and Dynamical Behavior of Some Exciplex Sytstems, in: The Exciplex, eds. M. Gordon and R. Ware (Academic Press, New York, 1975) pp. 113-144.

11) M. A. Kahlow, T. J. Kang and P. F. Barbara, J. Phys. Chem. 91 (1987) 6452.

12) H. Miyasaka, S. Ojima and N. Mataga, ibid. 93 (1989) 3380.

13) T. Asahi and N. Mataga, ibid. 93 (1989) 6575.

14) S. Ojima, H. Miyasaka and N. Mataga, ibid. 94 (1990) 4147.

15) S. Ojima, H. Miyasaka and N. Mataga, ibid. 94 (1990) 5834.

16) S. Ojima, H. Miyasaka and N. Mataga, ibid. 94 (1990) 7534.

17) T. Asahi and N. Mataga, ibid., in print.

18) N. Mataga, T. Asahi, Y. Kanda, T. Okada and T. Kakitani, Chem. Phys. 127 (1988) 249.

19) J. Hinatu, F. Yoshida, H. Masuhara and N. Mataga, Chem. Phys. Lett. 59 (1978) 80.

20) R. Englman and J. Jortner, Mol. Phys. 18 (1970) 145.

21) H. Miyasaka and N. Mataga, Bull. Chem. Soc. Jpn. 63 (1990) 131.

22) H. Miyasaka, K. Morita, M. Kiri and N. Mataga, Ultrafast Phenomena VII, 1990, p.498.

23) H. Miyasaka, K. Morita, K. Kamada and N. Mataga, Bull. Chem. Soc. Jpn. 63 (1990) 3385.

24) H. Miyasaka, N. Mataga, et al., to be published.

25) K. S. Peters, S. C. Felich and C. G. Shaefer, J. Am. Chem. Soc. 102 (1980) 5701.

26) C. G. Shaefer and K. S. Peters, ibid. 102 (1980) 7567.

27) J. D. Simon and K. S. Peters, ibid. 103 (1981) 6403; ibid. 105 (1983) 4875.

28) L. E. Manring and K. S. Peters, ibid. 107 (1985) 6452.

Photochemical Processes in Organized Molecular Systems
K. Honda (Editor-in-Chief)
© Elsevier Science Publishers B.V., 1991

DYNAMICS OF PHOTOINDUCED ELECTRON TRANSFER IN SOLUTION

Ian R. GOULD,* Ralph H. YOUNG,† and Samir FARID*

*Corporate Research Laboratories, Eastman Kodak Company, Rochester, New York 14650-2109, USA
†Research and Technology Development, Copy Products Division, Eastman Kodak Company, Rochester, New York 14650-2021, USA

Three key intermediates can be identified in photoinduced bimolecular electron transfer reactions. These are the encounter pair ($A^*(S)D$), the contact radical-ion pair (CRIP, $A^{\bullet-}D^{\bullet+}$) or exciplex/excited CT complex ($A^*D \leftrightarrow A^{\bullet-}D^{\bullet+}$), and the solvent-separated radical-ion pair (SSRIP, $A^{\bullet-}(S)D^{\bullet+}$). The efficiency with which such reactions result in the formation of free radical ions in solution ($A^{\bullet-} + D^{\bullet+}$) depends upon the competition between an electron transfer reaction, and a reaction involving a change in solvation within each of these three intermediates. In the encounter pair, the competition is between solvent displacement to form contact species, and direct formation of the SSRIP via a "long distance" electron transfer (ca. 6 - 8 Å) reaction. For the present systems, CRIP formation occurs with near unit efficiency in low polarity solvents, but direct formation of $A^{\bullet-}(S)D^{\bullet+}$ is observed in the medium polarity o-dichlorobenzene, in which charge-transfer interactions between the excited acceptor (A^*) and the solvent appear to enhance the long distance electron-transfer rate via a superexchange type mechanism. In acetonitrile the probability of direct formation of $A^{\bullet-}(S)D^{\bullet+}$ depends upon the energetics of the long distance electron transfer reaction, consistent with Marcus "normal" region behavior. In the contact and solvent separated radical-ion pairs the competition is between return electron transfer to reform the neutral starting materials and solvation and separation processes. The rate constants for the return electron transfer reactions are in the Marcus "inverted" region. The electronic coupling matrix element which characterizes these reactions is ca. 2 orders of magnitude lower in the solvent-separated pair compared to the contact pair. Conversely, in acetonitrile the solvent reorganization energy for the SSRIP is ca. 1 eV larger than that of the corresponding CRIP. Estimates for the reorganization energies for the CRIP reactions are obtained independently by using CRIP emission spectroscopy.

1. INTRODUCTION

Photoinduced electron transfer reactions play a central role in many diverse processes including photosynthesis,[1] silver halide photography,[2] electrophotography,[3] and radical and cationic polymerizations.[4] In addition, many chemical reactions have been identified as occurring via photoinduced electron transfer, including isomerization, dimerization, rearrangement, nucleophilic addition, oxygenation, and fragmentation reactions.[5]

In all of these processes a common challenge is to understand the factors which control the efficiencies of charge separation, and to minimize the energy wasting return electron transfer reactions. In this regard, the rate constants for forward and return electron transfer in a variety of media have been under intensive investigation for a number of years. Systems which have been studied include proteins,[6] rigidly[7] and flexibly[8] linked donor/acceptor molecules, donors and acceptors in fluid[9] and rigid[10] homogeneous solutions, and reactions at heterogeneous interfaces.[11]

One of the issues which is commonly addressed in such studies is the effect of the distance between the acceptor and the donor on the rate of electron transfer. Normally an exponential distance dependence is assumed.[12] Studies of proteins, of rigidly linked molecules and of reactions in glassy matrices have demonstrated that this distance dependence is a function of the intervening medium.[6,7,10] The influence of donor/acceptor separation on electron transfer rates is also an important issue in bimolecular reactions in fluid media, since this separation distance is different for the key intermediates in these processes.

A convenient method for evaluating the efficiencies of these reactions in polar solvents involves the determination of the yields of free radical ions.[13] These yields can be understood in terms of a reaction scheme in which an electron transfer step in each of three intermediates competes with a process involving a change in solvation. In this report we describe the roles of these intermediates in determining the overall efficiencies of photoinduced bimolecular electron transfer reactions in fluid media, with an emphasis on the factors controlling their competing reactions.

2. GENERAL SCHEME

A general outline of the intermediates and individual reaction steps required to understand the present data is shown in Scheme I. The intermediates can be distinguished by means of their energies and the approximate center-to-center distance of the acceptor and donor. In order of increasing energy, the intermediates are characterized as ion-pair states and locally excited states. For two molecules with aromatic π systems, such as those used in this work, the smallest center-to-center distance is achieved in a face-to-face configuration, i.e. at the van der Waals contact distance of ca. 3.5 Å. The next larger center-to-center distance to be considered occurs in the solvent separated or "loose" configuration. The interaction between the donor and acceptor is obviously significantly weaker than when in the contact configuration, yet can be strong enough to allow electron transfer to compete with solvent displacement to form the contact species, or further solvation to

SCHEME I.   Intermediates in Bimolecular Photoinduced Electron Transfer Reactions in Fluid Media (the relevant rate constants are given in. eqs 1 - 3).

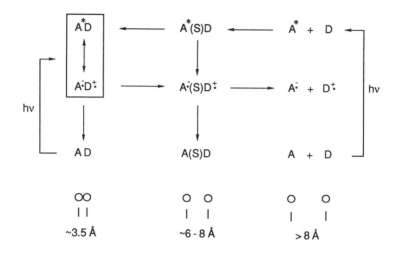

form free species in solution. The solvent-separated pair, whether an ion pair or a locally excited state, is not a static entity with a fixed configuration, but has a loose, constantly changing geometry in which the two π systems are not necessarily parallel.   In a conventional polar solvent such as acetonitrile, the average center-to-center separation in the solvent-separated radical-ion pair is often considered to be ~ 6 - 8 Å,[9a,b] i.e. larger than the contact separation by an amount which is roughly equal to the dimension of the intervening solvent. At larger separation distances in which there are more solvent layers between the acceptor and donor, the interaction between the reactants is sufficiently weak that the rates of electron transfer are too slow to compete with further solvation and desolvation processes, although, in rigid media and proteins, where there are no other competing reactions, electron transfer can still be detected over large distances.[6,10]

Scheme I includes all of the important intermediates which are required to understand the data discussed here, although more complete schemes could be written, for example to take into account intersystem crossing.[9]   In addition, many of the reactions of Scheme I may be reversible, but these steps are omitted for clarity.

This discussion of the intermediates has not yet taken into account the possibility of mixing between the various states. Such mixing increases with orbital overlap, and is thus likely to be strong at contact distances, and also increases with decreasing energy gap between the mixing states. These considerations suggest that the two states in Scheme I which are most likely to be strongly mixed are the contact locally excited and ion-pair states (i.e. $A^*D \leftrightarrow A^{\bullet-}D^{\bullet+}$, Scheme I). The term exciplex in organic photochemistry is used to describe excited contact charge-separated species, which are characterized by a wide range of such mixing between locally excited and ion pair states.[9] Species which can be described as exciplexes range from an excited acceptor (or donor) which is slightly perturbed by interaction with a donor (or acceptor), to an essentially pure contact radical-ion pair.

The excited contact charge-separated states are defined as exciplexes when the ground state in the contact configuration is repulsive, i.e. when no stable CT complex is formed, and as excited CT complexes when a bound complex is formed in the ground state.[9,14] The transition from exciplexes to excited CT complexes is a gradual one. In this work, the extent of charge-separation in the contact pair is high, and therefore we will use the term contact radical-ion pair (CRIP, $A^{\bullet-}D^{\bullet+}$) to refer to the exciplex or excited CT complex, recognizing that a more accurate description for some of the systems should include a small admixture of other locally excited states, i.e. $A^*D \leftrightarrow A^{\bullet-}D^{\bullet+}$.

According to Scheme I, when an excited free acceptor approaches a donor via diffusive processes, an encounter pair is formed ($A^*(S)D$). This species is the first of three intermediates in which competitive processes occur. In this case the competition is between electron transfer (k_{et}) to form a solvent-separated radical-ion pair (SSRIP, $A^{\bullet-}(S)D^{\bullet+}$), and solvent displacement (k_{-s}) to form a contact pair (A^*D), eq 1.

$$A^*(S)D \underset{k_{-s}}{\overset{k_{et}}{\rightleftarrows}} \begin{array}{l} A^{\bullet-}(S)D^{\bullet+} \quad \text{electron transfer in solvent separated pair} \\ A^*D \qquad\qquad \text{solvent displacement} \end{array} \tag{1}$$

Electron transfer in A^*D, followed by solvent reorganization and thermalization, leads to a contact radical-ion pair (CRIP, $A^{\bullet-}D^{\bullet+}$). If a ground state CT complex (AD) is formed, the CRIP can be formed directly by excitation of this species (Scheme I).

Once formed, the CRIP can undergo deactivation via several paths, including nonradiative and radiative decay to the neutral ground state and intersystem crossing. In polar solvents such as acetonitrile the most important reactions are solvation to form the SSRIP (k_{solv}), and nonradiative decay via return electron transfer, $(k_{-et})_{cp}$, which occur in competition, eq 2.

$$A^{\bullet -}D^{\bullet +} \underset{k_{solv}}{\overset{(k_{-et})_{cp}}{\rightleftarrows}} \begin{array}{l} AD \\ \\ A^{\bullet -}(S)D^{\bullet +} \end{array}$$

return electron transfer
in contact pair

solvent penetration
and solvation

(2)

The third competition involves the SSRIP, which can be formed directly from $A^*(S)D$, as mentioned above, or via solvation of the CRIP (Scheme I). In polar solvents the important reactions of the SSRIP are deactivation by return electron transfer, $(k_{-et})_{ss}$, and separation to form free radical ions, k_{sep}, eq 3.

$$A^{\bullet -}(S)D^{\bullet +} \underset{k_{sep}}{\overset{(k_{-et})_{ss}}{\rightleftarrows}} \begin{array}{l} A(S)D \\ \\ A^{\bullet -} + D^{\bullet +} \end{array}$$

return electron transfer
in solvent separated pair

solvent separation
(cage escape)

(3)

In each intermediate the competition occurs between an electron transfer reaction and a reaction involving movement (displacement and rearrangement) of solvent molecules. All three of the solvent rearrangement reactions should depend on the solvent polarity and viscosity, and to some extent upon the molecular structures of A and D. Electron-transfer theories consistently relate reaction rates and thermodynamics. Therefore, in this work we have studied a series of A and D molecules with varying driving force for electron transfer so that the partitions between the electron transfer and solvent rearrangement processes can be varied in a systematic manner.

3. RESULTS AND DISCUSSION

In this work the electron acceptors are the excited states of 9,10-dicyanoanthracene (DCA) and 2,6,9,10-tetracyanoanthracene (TCA).[13] The electron donors are simple alkyl-substituted benzenes.[13] The acceptors and donors are designed to have structures which are as similar as possible (for example substituents with steric bulk or heteroatoms are not used), but to have differing redox potentials so that the driving force for electron transfer can be varied.

3.1. Reactions of the Excited Encounter Pair, $A^*(S)D$

Excitation of the free acceptor results in the formation of an excited encounter pair $A^*(S)D$. From the encounter pair a SSRIP can be formed via electron transfer (k_{et}) while the donor and acceptor remain separated, in competition with solvent displacement (k_{-s}), which ultimately leads to the

CRIP. The efficiency, α, of CRIP formation is given by eq 4.

$$\alpha \quad = \quad \frac{k_{-s}}{k_{-s} + k_{et}} \tag{4}$$

Contact and solvent-separated radical-ion pairs are most readily distinguished by the fact that emission can often be observed from the former and not the latter.[9] Indeed, the emission of the CRIP can be used to obtain an estimate for α, using two different methods. The overall quantum yield for CRIP emission, Φ_f, is given in eq 5, in which k_f and τ are the radiative rate constant and the lifetime of the CRIP. Φ_f and τ are experimentally measurable quantities, and so α can be determined if k_f is known. The radiative rate constant depends upon the emission frequency cubed, v_f^3, which is also measurable experimentally, and an electronic transition moment squared, M^2, eq 6.[15] Therefore, to estimate α by measurement of the radiative rate, a method for the determination of M for the CRIP is required. Alternatively, if α is known then M can be determined.

$$\Phi_f \quad = \quad \alpha \; k_f \; \tau \tag{5}$$

$$k_f \quad = \quad \frac{64 \; \pi^4}{3 \; h} \; n^3 \; \tilde{v}_f^3 \; M^2 \tag{6}$$

The second method for the estimation of α relies upon the formation of ground-state CT complexes. CT complex formation can be observed between TCA and several of the alkylbenzene donors, such as durene, pentamethyl-benzene and hexamethylbenzene, as an increase in a broad absorption above 450 nm, and as a decrease in the vibrational structure in the 350 - 420 nm region, as illustrated in Figure 1 for the TCA/pentamethylbenzene system in o-dichlorobenzene. The perturbation in the absorption spectrum of DCA in the presence of these donors is only minor. Indeed, the changes in this case at high [D] are due largely to the macroscopic change in the medium. The absence of significant complex formation means that for this acceptor only the radiative rate method, described above, can be used for estimating α.

Whereas the efficiency of CRIP formation from the encounter pair is given by eq 4, the corresponding efficiency for excitation of a ground-state CT complex is unity. Thus, in principle, α can be determined by measuring the ratio of the CRIP emission quantum yield for excitation of the monomer and the CT complex. This experiment is most conveniently performed by comparing the absorption spectrum of an equilibrium mixture (with equilibrium constant K) of the free acceptor, A, and the complex, AD (eq 7),

$$A + D \xrightleftharpoons{K} AD \qquad (7)$$

with the CRIP excitation spectrum. When α is unity, the excitation spectrum is identical to the absorption spectrum of the same solution, since, by definition, excitation of either A or AD leads to the CRIP with unit efficiency. However, if α is less than unity, excitation of A does not always lead to the CRIP emission. In this case the ratio of the absorbance of the CT complex to that of the monomer will be higher in the excitation spectrum than in the corresponding absorption spectrum. The probability α can thus be determined from a comparison of these spectra.

The ratio $(R)_{abs}$ of the optical density at two different wavelengths, λ_1 and λ_2, in the absorption spectrum of an equilibrium mixture of A and AD is given by eq 8, in which ε_1 and ε_2 are the extinction coefficients of A or AD at λ_1 and

$$(R)_{abs} = \frac{\varepsilon_1{}^A + \varepsilon_1{}^{AD} K[D]}{\varepsilon_2{}^A + \varepsilon_2{}^{AD} K[D]} \qquad (8)$$

$$(R)_{exc} = \frac{\varepsilon_1{}^A + \varepsilon_1{}^{AD} K[D]/\alpha}{\varepsilon_2{}^A + \varepsilon_2{}^{AD} K[D]/\alpha} \qquad (9)$$

λ_2. The ratio in the corresponding excitation spectrum is given by eq 9. Accordingly, α can be obtained as a ratio of donor concentrations [D]/[D'], in which [D'] is the concentration for which the absorption spectrum matches the excitation spectrum at concentration [D] (Figure 1).

Alternatively, if λ_2 is a wavelength where A does not absorb, then eqs 8 and 9 reduce to 10 and 11, respectively. In this case α is given simply by the ratio of the slopes of the linear plots of $(R)_{exc}$ and $(R)_{abs}$ vs. $1/[D]$.

$$(R)_{abs} = \frac{\varepsilon_1{}^{AD}}{\varepsilon_2{}^{AD}} + \frac{\varepsilon_1{}^A}{\varepsilon_2{}^{AD} K} \frac{1}{[D]} \qquad (10)$$

$$(R)_{exc} = \frac{\varepsilon_1{}^{AD}}{\varepsilon_2{}^{AD}} + \alpha \frac{\varepsilon_1{}^A}{\varepsilon_2{}^{AD} K} \frac{1}{[D]} \qquad (11)$$

It is important to note that these experiments clearly identify those cases in which CRIP formation is 100% efficient for quenching of the free acceptor. However, these experiments do not distinguish whether this is the result of α being unity, or whether a SSRIP is formed which subsequently "collapses" to give the CRIP with unit efficiency (not shown in Scheme I). However, if the experiments show that the CRIP is not formed with 100% efficiency, then α can not be unity.

Estimates for α were obtained using both methods in solvents of low polarity, including cyclohexane, dioxane, carbon tetrachloride, trichloroethylene and fluorobenzene. In these solvents it is easy to show that CRIP formation is essentially 100% efficient for those systems in which TCA forms CT complexes, since the absorption and excitation spectra are indistinguishable.[16] In these low polarity systems, α is almost certainly unity and the direct formation of a SSRIP from the encounter pair is unlikely to occur since these reactions would be endothermic.[9b] The various combinations of donor and solvent result in CRIP emission over a range of frequencies, v_f. In those systems for which α was unity, the transition moment M could be obtained using eqs 5 and 6. It was found that M was a function of v_f, becoming smaller as the emission frequency decreased.[16] This change in M was attributed to differences in the electronic structures of the CRIP due to varying degrees of mixing of the pure ion-pair state with the acceptor locally excited state.[16] Nevertheless, M is a smooth function of v_f, and in fact a value for M could be predicted from the v_f value for any CRIP emission. Thus, k_f values could be calculated for the CRIP of the DCA/alkylbenzene systems in nonpolar and polar solvents from eq 6, using a value for M from the TCA data at the appropriate v_f. Comparison with experimental k_f values obtained from the emission quantum yields and lifetimes of these CRIP (eq 5) showed that α was unity for all of the DCA systems in these solvents. Thus we obtain the unsurprising result that the CRIP are formed with unit efficiency in the bimolecular quenching reactions of both TCA and DCA in these low polarity solvents (Table 1).

Indeed, using the radiative rate method it can be shown that α is unity for essentially all of the DCA reactions, even in solvents as polar as acetonitrile (Table 1). However, as the solvent polarity increases there are clear indications that this is not the case for the TCA reactions. For example, in o-dichlorobenzene (which has a dielectric constant, $\varepsilon = 9.9$, similar to that of methylene chloride) the excitation spectrum for the CRIP emission of TCA/pentamethylbenzene is clearly not the same as the absorption spectrum of the same solution, as shown in Figure 1. Based on a quantitative analysis of these spectra as described above, a value for α of ca. 0.5 was obtained for this system. Furthermore, from measurements of the CRIP radiative rates in this solvent, eq 5 indicates a value for α of ca. 0.5 - 0.55 for TCA with p-xylene, 1,2,4-trimethylbenzene, durene, and pentamethylbenzene. Thus, we estimate that quenching of the excited TCA by the alkylbenzenes in this solvent leads to contact ion pairs with an efficiency of only ~0.5.

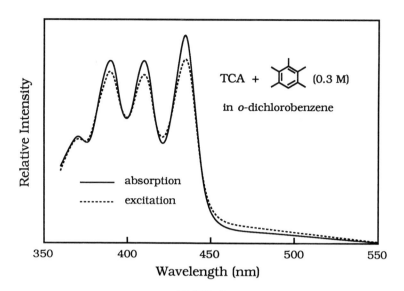

FIGURE 1
Absorption and excitation spectra (for CT emission at 650 nm) of TCA/pentamethylbenzene (0.3 M) in o-dichloro-benzene at 10°C. The absorption spectrum at twice the donor concentration (0.6 M, not shown) is very similar to the excitation spectrum shown in the Figure.

Electron transfer to form a SSRIP while in the encounter pair A*(S)D is the reaction most likely to compete with CRIP formation, as indicated in Scheme I. An attractive rationalization for this behavior is that excited TCA, which is easier to reduce than excited DCA by 0.47 V,[13] undergoes weak CT interaction with the solvent, o-dichlorobenzene. In the encounter pair such CT interactions may enhance the rate of electron transfer from the donor (k_{et}) via a "superexchange" type mechanism.[17] Indeed, from the transition moment of $^1TCA^*$ in o-dichlorobenzene, there is an indication of considerable CT interaction of $^1TCA^*$ with the solvent.[16] This behavior contrasts with that in methylene chloride, in which radiative rate measurements indicate a larger transition moment for $^1TCA^*$, and a value for α of unity for the CRIP of TCA. Presumably the better donating ability of the o-dichlorobenzene is responsible for the increased CT interaction with $^1TCA^*$.

Further evidence for the importance of solvent polarity as a factor in determining the value of α comes from studies in acetonitrile, in which the SSRIP are considerably stabilized compared to less polar solvents. In this solvent it is likely that α is unity for those systems in which CRIP formation appears to be 100% efficient since once a SSRIP is formed, the "desolvation" to form a CRIP has to compete with return electron transfer within the SSRIP and separation to form free radical ions, which occur with rates greater than 10^9 s^{-1}.[13] Radiative rate measurements of the CRIP of DCA with durene and pentamethylbenzene are consistent with a high CRIP formation efficiency for these donors. With hexamethylbenzene, however, the measured Φ_f/τ suggests a somewhat smaller α (~ 0.7) in this case. With TCA as the acceptor, the CRIP emissions are very weak (emission quantum yields < 10^{-4}), and therefore it is difficult to determine α accurately using either method. Using the higher oxidation potential donors such as m-xylene and mesitylene, measurements of the emission intensities as a function of donor concentration are consistent with a large value for α (greater than 0.7 and probably close to 1.0). Using the lower oxidation potential donors including durene, pentamethylbenzene and hexamethylbenzene the CRIP emissions increase with increasing donor concentration in a manner which indicates that CRIP formation occurs almost exclusively via excitation of the ground state CT complex. Intermediate behaviour is observed using 1,2,4-trimethylbenzene, which has an oxidation potential between those of durene and mesitylene, with CRIP formation being 25% efficient in this case. Thus the efficiency for direct formation of SSRIP appears to depend strongly upon the driving force for this reaction. The reaction becomes faster as the exothermicity increases (the oxidation potential of the donor decreases), which corresponds to a Marcus "normal" region for reactions with exothermicities less than the reaction reorganization energy, λ.[12] We previously estimated λ to be ca. 1.9 eV for electron transfer reactions in solvent-separated radical-

Table 1. Efficiencies of CRIP formation $(\alpha)^a$ as a function of solvent for two cyanoanthracene acceptors with alkylbenzene donors

| | | α | |
solvent	ε^b	DCA	TCA
cyclohexane dioxane carbon tetrachloride trichloroethylene fluorobenzene	2 - 6	1.0	1.0
methylene chloride o-dichlorobenzene	9 - 10	1.0 1.0	1.0 0.5
acetonitrile	36	≤ 1	~0 - 1^c

^{a} defined in eq 4. ^{b} dielectric constant of the solvent.
^{c} varies with electron transfer driving force (see text).

ion pairs (i.e. for $(k_{-et})_{ss}$),[13] which is significantly larger than the exothermicities of the SSRIP forming reactions discussed here ($-\Delta G_{et}$ from 0.3 to 0.65 eV).[13] Presumably both k_{et} and $(k_{-et})_{ss}$ will have similar reorganization energies and thus the strong dependence of the CRIP formation efficiency on ΔG_{et} is not surprising.

The results in the solvents studied so far seem to indicate that a combination of at least two solvent parameters, the *donor ability* and the *polarity*, in addition to the *free energy of the SSRIP formation reaction* can play an important role in determining the competition between electron transfer and solvent displacement in the encounter pair. The solvent viscosity should also affect the rate of the displacement process (k_{-s}), and hence the partitioning between these reactions.

3.2. Reactions of the Solvent Separated Radical Ion Pair, $A^{\bullet-}(S)D^{\bullet+}$

Bimolecular quenching of $^1TCA^*$ by the low oxidation potential donors leads directly to SSRIP. Bimolecular quenching of $^1TCA^*$ by the high oxidation potential donors, and of $^1DCA^*$ by all of the donors results in the formation of CRIP. As described below, however, in these latter cases the return electron transfer reactions are sufficiently exothermic that the rates of the return electron transfer reactions within the CRIP are smaller than the rates of solvation to yield SSRIP. Thus, for the bimolecular quenching reactions in acetonitrile, the yields of free radical ions are determined by the dynamics of the SSRIP only. The SSRIP undergo return electron transfer to reform the neutral starting materials, $(k_{-et})_{ss}$, or separation to form free radical ions in solution, k_{sep}, eq 3, Scheme I. The quantum yield for free radical ion formation under these conditions, $(\Phi_{ions})_{ss}$, is given by eq 12.

$$(\Phi_{ions})_{ss} = \frac{k_{sep}}{k_{sep} + (k_{-et})_{ss}} \tag{12}$$

If k_{sep} is known, then $(k_{-et})_{ss}$ can be obtained directly from $(\Phi_{ions})_{ss}$. As indicated previously, the donor/acceptor systems were designed so that the exothermicity of the return electron transfer could be varied by changing the donor and acceptor redox potentials. The redox energy of a donor/acceptor pair in acetonitrile is equal to the negative of the free energy for the return electron transfer reaction of the SSRIP in that solvent ($-\Delta G_{-et}$, eq 13).[13] Thus, from measurements of $(\Phi_{ions})_{ss}$ for different donor/acceptor pairs, $(k_{-et})_{ss}$ can be obtained as a function of ΔG_{-et}.

$$-\Delta G_{-et} = E^{ox}_{D} - E^{red}_{A} \tag{13}$$

The separation rate constant, k_{sep}, has not been measured directly, but Weller et al have estimated a value of 5×10^8 s^{-1} for the pyrene/dimethylaniline radical-ion pair in acetonitrile, from studies of the effects of magnetic fields on these reactions.[9a] In addition, from quantitative product analysis studies, a value of ca. 5×10^8 s^{-1} was also obtained for the TCA/phenylacetylene and DCA/1,1-diphenylethylene pairs.[18] This rate constant does not appear to be very dependent upon the chemical structure of the radical ions in acetonitrile, and it seems especially reasonable to assume a constant value for the present, structurally related systems.

Using TCA and DCA with various alkylbenzene donors, values for $(\Phi_{ions})_{ss}$ were obtained for 22 radical-ion pairs using a transient absorption technique described in detail elsewhere.[13] Values for $(k_{-et})_{ss}$ were thus obtained using eq 12, with a value of 5×10^8 s^{-1} for k_{sep}. These data are plotted in Figure 2.

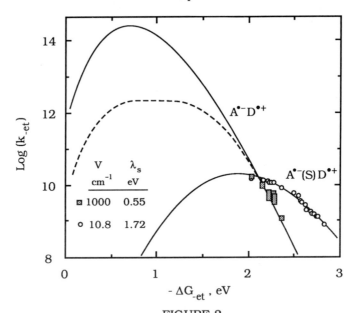

FIGURE 2

Plots of log (k_{-et}) for contact and solvent separated radical ion pairs of cyanoanthracenes/alkylbenzenes in acetonitrile vs. the driving force $(-\Delta G_{-et})$ (data from refs 13 and 26). The solid curves represent fits to the data using the golden rule based theory described in the text, with the values of the parameters shown. The dashed curve is a fit to the contact radical ion pair data using a non-perturbation theory in which the maximum rate is given by a characteristic solvent relaxation time.

The most significant aspect of the data is that the reaction rates decrease as the return electron transfer reaction becomes more exothermic (i.e. with increasing $-\Delta G_{-et}$). The most reasonable explanation of this behavior is that these reactions are in the Marcus "inverted" region for electron transfer.[19]

Theoretical treatments of the rates of electron transfer reactions were described in the 1950's by Marcus.[19] Marcus' original theory was derived using classical mechanics, although more recent theories take quantum effects into account.[12] The reactions are described using a golden rule type formula in which the rate is given as the product of an electronic coupling matrix element squared (V^2) and a Franck-Condon weighted density of states, which contains the dependence of the rate on the reaction free energy (ΔG_{-et}), and two reorganization energies, λ_s and λ_v. The former is related to rearranged low frequency (mainly solvent) modes which may be treated classically. The latter is related to rearranged high frequency internal vibrational modes of the donor and acceptor, and is typically associated with a single averaged frequency, v_v. A schematic representation of the theoretical model is given in Figure 3. The electron transfer process from the initial (ion pair) state is considered to be composed of a sum of individual electron transfer rates to various high frequency (v_v) vibrational sublevels of the final

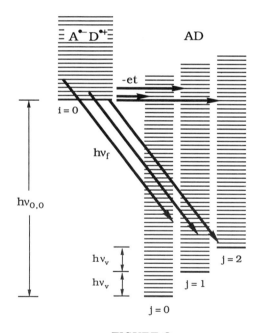

FIGURE 3
Schematic representation of thermal and optical return electron transfer from the lowest vibrational level of a contact radical-ion pair (i = 0), to different vibrational levels, j, of the neutral ground state.

(neutral) state (indicated by j in Figure 3). Each high frequency vibrational sublevel is further composed of a quasicontinuum of low frequency states (associated with λ_s). When the reaction free energy is varied, the rate of the return electron transfer changes because the neutral products are formed in different sets of vibrationally excited states, as indicated by the horizontal arrows in Figure 3.

In order to fit the SSRIP data in Figure 2, values have to be obtained for the parameters V, λ_s, λ_v, and v_v. We have recently obtained estimates of 0.2 eV and 1400 cm^{-1} for λ_v, and $\tilde{v}_v$ by fitting CRIP emission spectra as described in the next section. Taking these two values as fixed, the curve fitting procedure consists of finding the best values for V and λ_s which fit the data. The solid curve through the SSRIP data in Figure 2 represents the best fit using values for V and λ_s of 10.8 cm^{-1} and 1.72 eV.

The value for λ_s obtained in these studies is considerably larger than those reported in the literature for self-exchange reactions in acetonitrile, which are on the order of 0.25 - 0.5 eV.[20] However, it is important to note that the reactions in the present case are for solvent-separated radical ions, whereas the self-exchange reactions presumably occur when the species are in contact. As described in the following section, we expect considerable differences between the extent of solvation for solvent separated and contact pairs, and indeed the reorganization energy obtained for reaction in the contact radical pairs is similar to those which characterize the exchange reactions. The solvent separated nature of the ion pair in the present case also accounts for the rather small value for V, since this is related to orbital overlap,[21] which must be small in the solvent separated configuration.

The inverted region has now been observed for many electron transfer reactions,[10a,22] including return electron transfer in several other geminate ion pair systems.[23] A key requirement for observing the inverted region is that the electron transfer reaction must be first order, and that diffusion of the donor and the acceptor is not necessary for reaction. For photoinduced bimolecular systems in homogeneous solution the initial electron transfer to form the geminate pair is second order, whereas the return electron transfer reaction is first order. In a famous study of electron transfer fluorescence quenching in acetonitrile, Rehm and Weller demonstrated that the inverted region could not be observed for the initial electron transfer reaction.[24] Since this is a second order process, when the intrinsic electron transfer reaction is very fast the rate determining step for ion pair formation becomes the rate of diffusion, as observed experimentally.[24] The present results on these and other geminate ion pair systems are important because they clearly demonstrate the role of the inverted region in bimolecular electron transfer reactions in solution.   The variation in return electron-transfer rates measured so far using the cyanoanthracene acceptors corresponds to a wide range in the quantum yields for free radical ion formation, from as high as ca. 0.75 to as low as 0.01.[13] In the absence of chain amplification mechanisms and chemical reactions in the geminate pairs, these values represent the limiting quantum yields for product formation.   In most of the other systems in which the inverted region has been observed the acceptor and donor are held at a fixed distance in rigid media or using rigid spacer groups.[10a,22] As model systems for the study of electron-transfer rates, although the relative geometry and donor/acceptor separation distance are not fixed in the ion pairs, different donors and acceptors with different chemical properties can easily be studied in such systems.   In this manner we have been able to investigate the effects of molecular dimension and isotopic substitution on the rates of electron transfer,[25a,b] and to compare charge recombination and charge shift reactions.[25c]

3.3. Reactions of the Contact Radical Ion Pair, $A^{\bullet-}D^{\bullet+}$

As indicated above, bimolecular quenching of $^1TCA^*$ by the low oxidation potential donors leads directly to the SSRIP, bypassing the CRIP. The CRIP can, however, be formed by excitation of the corresponding ground-state CT complexes. Ground-state CT complexes of TCA are formed in the presence of high concentrations of the alkylbenzene donors, and absorb at longer wavelengths than the free acceptor.   Thus the SSRIP and the CRIP of the same donor and acceptor, can be generated independently by changing the excitation wavelength and donor concentration, and hence the dynamics of their electron transfer reactions can be compared directly.

The different properties of the two types of geminate radical-ion pair should have a significant influence on the rates of return electron transfer in these species. For example, both ion pairs can be considered to be excited states of the corresponding donor/acceptor "supermolecule".   However, emission is only ever observed from the CRIP and not the SSRIP, because of the weak electronic coupling which characterizes the larger center-to-center distance within the SSRIP.[9] The rates of electron transfer depend on the extent of electronic coupling and solvation via the quantities V and λ_s,[12] and, as a consequence, the rates of return electron transfer could be quite different for the SSRIP and CRIP.

The quantum yield for formation of free radical ions for excitation of a CT complex $((\Phi_{ions})_{cp})$ is given by eq 14, in which Φ_{solv} is the yield of SSRIP which are formed from the CRIP. Values for $(\Phi_{ions})_{cp}$ were obtained directly by excitation at 460 nm, since only the CT complexes absorb at wavelengths longer than ca. 450 nm. At 410 nm both the free TCA and the complexes absorb. Values for $(\Phi_{ions})_{ss}$ were obtained for 410 nm excitation by extrapolation to zero donor concentration. Additional estimates for $(\Phi_{ions})_{cp}$ were obtained from the 410 nm experiments from an analysis of the dependence of the free radical-ion yield on the donor concentration.[26] $(\Phi_{ions})_{cp}$ will always be less than $(\Phi_{ions})_{ss}$ if $(k_{-et})_{cp}$ can compete effectively with k_{solv}, and indeed for all of the complexes studied this was found to be the case. Thus Φ_{solv} can be evaluated using eq 14a, and $(k_{-et})_{cp}$ obtained using eq 14b, if k_{solv} is known.

$$(\Phi_{ions})_{cp} \quad = \quad \Phi_{solv} \ (\Phi_{ions})_{ss} \tag{14a}$$

$$\Phi_{solv} \quad = \quad \frac{k_{solv}}{k_{solv} \ + \ (k_{-et})_{cp}} \tag{14b}$$

Extremely weak emissions can be observed from the CRIP of TCA. The lifetimes of these emissions, τ, were determined to be 55, 88, 96, 95, 107, and 84 ps for hexamethylbenzene, pentamethylbenzene, durene, 1,2,3,5-tetramethylbenzene, 1,2,3,4-tetramethylbenzene and 1,2,4-trimethylbenzene respectively.[26] The lifetimes are given by $1/(k_{solv} + (k_{-et})_{cp})$, and thus $(k_{-et})_{cp}$ can be determined by using eq 14b. These data, calculated using the Φ_{solv} from both the 410 nm and the 460 nm experiments, are plotted as a function of ΔG_{-et} in Figure 2.

Compared to the SSRIP, the CRIP electron-transfer rates cover a smaller range of ΔG_{-et}, which is insufficient to allow accurate estimates of the variable parameters V and λ_s, using curve fitting. Furthermore, it is not clear that eq 13 gives accurate values for ΔG_{-et} for the CRIP's since, compared to the SSRIP, the ion-pair energy will tend to be raised because of decreased solvation, and lowered because of the coulombic stabilization which occurs when the ions are in contact. Therefore we sought a method to obtain independent estimates of the reorganization energies and for ΔG_{-et} in this case.

It is known that thermal return electron transfer in ion pairs $((k_{-et})_{cp})$ is closely related to the CT (CRIP) emission in these species,[27] as illustrated in Figure 3. Just as in the return electron transfer reaction, the CT emission process occurs to give a vibrationally-excited neutral state. The rearranged modes associated with the return electron transfer and the CT emission are identical. In the return electron transfer process (represented by the horizontal arrows in Figure 3), when the reaction free energy ΔG_{-et} is varied, the reaction rate changes because the neutral products are formed in different sets of vibrationally excited states. The CT emission process (represented by the diagonal arrows) occurs to give neutral products in different vibrational states depending upon the energy of the emitted light. The energy dependence of the CT emission process can be obtained from a single donor/acceptor pair by determining the relative probabilities of light emission at different emission energies, as obtained from a single emission spectrum. The CT emission process can be described using a formula which is closely related to that for k_{-et}, in which the radiative rate is given as a function of emission energy $(h\nu_f)$, and depends upon λ_s, λ_v and ν_v and on $(\Delta G_{-et} + h\nu_f)$, in an identical manner to the dependence of k_{-et} on these parameters and on ΔG_{-et}.[26]

Shown in Figure 4 is the CT emission spectrum for the DCA/pentamethylbenzene system in acetonitrile together with a fit to the spectrum obtained using the values of the parameters indicated in the figure. In this

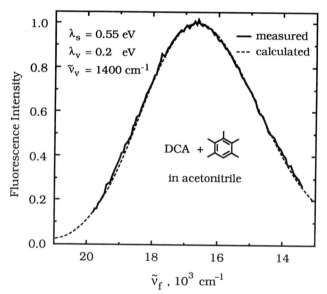

FIGURE 4

Measured and calculated contact radical ion pair (exciplex) emission spectra for DCA/pentamethylbenzene in acetonitrile.

case, ΔG_{-et} was also used as a fitting parameter. Unlike the thermal return electron transfer data, almost the complete energy dependence of the optical electron transfer is known from the emission spectrum. Furthermore, the emission occurs in an important region of the energy spectrum since the maximum occurs when the Franck-Condon factors are optimized, which corresponds to the energy at which the maximum rate would occur for the thermal return electron transfer (i.e. the rate for which $-\Delta G_{-et} \approx \lambda_s + \lambda_v$). *The decrease in emission intensity at higher and lower emission energies corresponds to the Marcus normal and inverted regions.* Thus, compared to the k_{-et} data, much more accurate estimates for the reorganization parameters can be obtained from fitting the emission spectra. Furthermore, the ability to fit spectra in more than one solvent allows the total reorganization energy to be divided into λ_s and λ_v more accurately.[16]

The best values for λ_v and $\tilde{v}_v$ obtained in this way are 0.20 eV and 1400 cm^{-1} respectively. The values for λ_s and $-\Delta G_{-et}$ which give the best fit to the spectrum in acetonitrile are 0.55 eV and 2.67 eV. The value for $-\Delta G_{-et}$ is larger than the energy obtained from the redox potentials of this pair (eq 13), but by only 0.05 eV, which suggests that the solvation and coulombic effects discussed earlier are almost exactly offset. Therefore the $-\Delta G_{-et}$ for all of the CRIP in Figure 2, are taken to be the same as for the SSRIP, eq 13. For the $(k_{-et})_{cp}$ data, therefore, the only fitting parameter is V. A good fit to the data is obtained when a value of 1000 cm^{-1} is used (Figure 2). It is encouraging that the results of the Franck-Condon analysis of the CT emission give a good fit to the $(k_{-et})_{cp}$ data.

The large value of V for the CRIP data suggests that the golden rule type analysis of the electron transfer data might not be appropriate.[12] Additionally, the predicted maximum rate of ca. 10^{14} s^{-1} might not be realistic because the reactions could become controlled by solvent dynamical effects at these rates.[28] The dashed curve in Figure 2 is a fit to the CRIP data using the non-perturbative approach described by Jortner and Bixon,[28] in which the maximum rate is related to a characteristic solvent relaxation time.[26,28] Although this curve probably represents more accurately the dependence of rate on ΔG_{-et} at low exothermicities, the fact that the theories agree both with each other and with the data in the high exothermicity region supports the use of the golden rule approach for the determination of V with this data.[26]

The solvent reorganization energies and electronic matrix elements are quite different for the two ion pairs. These effects can be understood by reference to the schematic model shown in Figure 5. There are two major factors which contribute to the different solvation of the two sets of ions. First, there are no solvent molecules between the ions in the CRIP. Second, because of the proximity of the ions in the CRIP, there is considerable cancellation of their orienting influence on the solvent molecules in the outer solvent shells, and to a lesser extent on the inner shell.   Both effects act to

Contact Radical Ion Pair
CRIP

Solvent Separated Radical Ion Pair
SSRIP

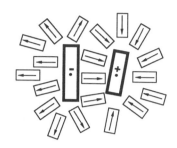

High Electronic Coupling
Weak Solvation

Low Electronic Coupling
Strong Solvation

FIGURE 5
Schematic representation of contact and solvent separated
radical ion pairs in a polar solvent (e.g. acetonitrile). The
arrows represent the solvent dipole orientation.

reduce the overall solvation of the CRIP compared to the SSRIP.[19] The
electronic coupling is understandably larger in the CRIP compared to the
SSRIP since V is related to orbital overlap, which will be large when the ions
are in contact and small when separated by a layer of solvent molecules.[21]

4. CONCLUSION

The combination of quantitative measurements of CRIP emissions
(quantum yields, lifetimes and spectral distributions) and measurements of
radical-ion yields has proved to be extremely useful in studies of the
intermediates in photoinduced bimolecular electron transfer reactions in
fluid solution. The results of these studies have led to a number of
conclusions concerning the competing reactions of the intermediates which
control the overall dynamics of these processes. In solvents of low or
moderate polarity, direct formation of a SSRIP in the encounter pair (k_{et})
which can not compete with solvent displacement (k_{-s}) leading to CRIP
formation, unless specific CT interactions of the excited state with the
solvent can increase the ratio of k_{et} to k_{-s}. Although this particular effect is
probably due mainly to an increase in k_{et}, a decrease in k_{-s} may also occur,
since the structure of the encounter pair might be significantly influenced by
the CT interactions. In acetonitrile the ratio of k_{et} to k_{-s} is observed to
increase with the exothermicity of the SSRIP forming reaction, suggesting
Marcus "normal" region behavior. Diffusive quenching of $^1TCA^*$ in this

solvent with the low oxidation potential donors and excitation of their corresponding ground state CT complexes lead to different yields for formation of free radical ions. These results are interpreted as being due to the formation of CRIP for excitation of the ground state complexes, and the formation of SSRIP in the diffusive quenching reaction, and therefore provide support for the conclusion that α is less than unity in these systems. For the CRIP and SSRIP intermediates, return electron transfer reactions occur in competition with solvation and separation processes. The partitions in these cases are influenced by variations in the rates of the electron transfer reactions, which occur in the Marcus "inverted" region, and which can be understood in terms of current theories for these processes. The differences in the rates of electron transfer in the two ion pair intermediates are due to the larger center-to-center separation distance in the SSRIP compared to the CRIP, which results in an electronic coupling matrix element which is ca. two orders of magnitude smaller in the SSRIP compared to the CRIP, and a solvent reorganization energy which is ca. 1 eV larger.

REFERENCES

1) R. K. Clayton, Photosynthesis: Physical Mechanisms and Chemical Patterns (Cambridge University, Cambridge, 1980).

2) (a) P. B. Gilman, Pure Appl. Chem. 49 (1977) 357. (b) T. H. James, Chemical sensitization, spectral sensitization, and latent image formation in silver halide photography, in: Advances in Photochemistry, Vol. 13, eds. D. H. Volman, K. Gollnick, and G. S. Hammond (Wiley, New York, 1986) p. 329.

3) J. H. Perlstein and P. M. Borsenberger, Photoconductive properties of organic assemblies and a comparison with dark conductors, in: Extended Linear Chain Compounds, Vol. 2, ed. J. F. Miller (Plenum, New York, 1982) Chap. 8.

4) (a) D. F. Eaton, Dye sensitized photopolymerization, in: Advances in Photochemistry, Vol. 13, eds. D. H. Volman, K. Gollnick, and G. S. Hammond (Wiley, New York, 1986) p. 427. (b) J. L. R. Williams, D. P. Specht and S. Farid, Polymer Engineering and Science 23 (1983) 1022. (c) F. D. Saeva, Photoinduced electron transfer (PET) bond cleavage reactions, in: Topics in Current Chemistry, Vol. 156, ed. J. Mattay (Springer-Verlag, Berlin, 1990) p. 59.

5) (a) S. L. Mattes and S. Farid, Photochemical electron-transfer reactions of olefins and related compounds, in: Organic Photochemistry, Vol. 6, ed. A. Padwa (Marcel Dekker, New York, 1983) p. 233. (b) G. J. Kavarnos and N. J. Turro, Chem. Rev. 86 (1986) 401. (c) M. A. Fox, and M. A. Channon eds., Photoinduced Electron Transfer, Part C. Photoinduced Electron Transfer Reactions: Organic Substrates (Elsevier, Amsterdam, 1988).

6) (a) G. McLendon, Acc. Chem. Res. 21 (1988) 160. (b) J. Winkler, D. Nocera, K. Yocum, E. Bordignon and H. B. Gray, J. Am. Chem. Soc. 104

(1982) 5798. (c) S. Isied, C. Kuehn and G. Worosilia, J. Am. Chem. Soc. 104 (1982) 7659. (d) J. McGourty, N. Blough and B. Hoffman, J. Am. Chem. Soc. 105 (1983) 4470. (e) B. E. Bowler, A. L. Raphael and H. B. Gray, Prog. Inorg. Chem. in press.

7) For a recent review see, M. R. Wasielewski, Distance dependencies of electron transfer reactions, in: Photoinduced Electron Transfer, Part A. Conceptual Basis. eds. M. A. Fox, and M. A. Channon (Elsevier, Amsterdam, 1988) p. 161.

8) (a) H. Staerk, W. Kuehnle, R. Treichel and A. Weller, Chem. Phys. Lett. 118 (1985) 19. (b) P. Vanderauwera, F. C. DeSchryver, A. Weller, M. A. Winnik and K. A. Zachariasse, J. Phys. Chem. 88 (1984) 2964. (c) N. Mataga, S. Nishikawa, T. Asahi and T. Okada, J. Phys. Chem. 94 (1990) 1443.

9) (a) A. Weller, Z. Phys. Chem. (Wiesbaden) 130 (1982) 129. (b) A. Weller, Z. Phys. Chem. (Wiesbaden) 133 (1982) 93. (c) H. Masuhara and N. Mataga, Acc. Chem. Res. 14 (1981) 312. (d) N. Mataga, Pure Appl. Chem. 56 (1984) 1255.

10) (a) J. R. Miller, J. V. Beitz and R. K. Huddleston, J. Am. Chem. Soc. 106 (1984) 5057. (b) T. Guarr, M. Maguire and G. McLendon, J. Am. Chem. Soc. 107 (1985) 5104.

11) (a) M. A. Fox, Acc. Chem. Res. 16 (1983) 314. (b) M. Grätzel, Acc. Chem. Res. 14 (1981) 376. (c) A. Fujishima and K. Honda, Nature 238 (1972) 37.

12) (a) J. J. Hopfield, Proc. Natl. Acad. Sci. USA 71 (1974) 3640. (b) J. Ulstrup and J. Jortner, J. Chem. Phys. 63 (1975) 4368. (c) R. A. Marcus and N. Sutin, Biochem. Biophys. Acta 811 (1985) 265. (d) R. P. Van Duyne and S. F. Fischer, Chem. Phys. 5 (1974) 183. (e) R. A. Marcus, J. Chem. Phys. 81 (1984) 4494.

13) I. R. Gould, D. Ege, J. E. Moser and S. Farid, J. Am. Chem. Soc. 112 (1990) 4290.

14) R. S. Mulliken and W. B. Pearson, Molecular Complexes: A Lecture and Reprint Volume (Wiley, New York, 1969).

15) J. B. Birks, Photophysics of Aromatic Molecules (Wiley, New York, 1970).

16) L. Mueller, R. H. Young, I. R. Gould, S. Farid and A. C. Albrecht, unpublished results. The full details of this work will be published elsewhere.

17) (a) H. M. McConnel, J. Chem. Phys. 35 (1961) 508. (b) J. R. Miller and J. V. Beitz, J. Chem. Phys. 74 (1981) 6746.

18) (a) S. L. Mattes and S. Farid, J. Chem. Soc. Chem. Commun. (1980) 126. (b) S. L. Mattes and S. Farid, J. Am. Chem. Soc. 105 (1983) 1386. (c) S. L. Mattes and S. Farid, J. Am. Chem. Soc. 108 (1986) 7356.

19) (a) R. A. Marcus, J. Chem. Phys. 24 (1956) 966. (b) R. A. Marcus, Annu. Rev. Phys. Chem. 15 (1964) 155.

20) L. Eberson, Electron-transfer reactions in organic chemistry, in: Advances in Physical Organic Chemistry, Vol. 18, eds. V. Gold and D.

Bethell (Academic, London, 1982) p. 79.

21) (a) M. D. Newton, J. Phys. Chem. 90 (1986) 3734. (b) M. D. Newton, ACS Symp. Ser. 198 (1982) 255.

22) (a) G. L. Closs and J. R. Miller, Science 240 (1988) 440. (b) M. R. Wasielewski, M. P. Niemczyk, M. P. Svec and E. B. Pewitt, J. Am. Chem. Soc. 107 (1985) 1080. (c) M. P. Irvine, R. J. Harrison, G. S. Beddard, P. Leighton and J. K. M. Sanders, Chem. Phys. 104 (1986) 315.

23) (a) T. Ohno, A. Yoshimura and N. Mataga, J. Phys. Chem. 90 (1986) 3296. (b) N. Mataga, T. Asahi, Y. Kanda, T. Okada and T. Kakitani, Chem. Phys. 127 (1988) 249. (c) E. Vauthey, P. Suppan and E. Haselbach, Helv. Chim. Acta 71 (1988) 93. (d) K. Kemnitz, Chem. Phys. Lett. 152 (1988) 305. (e) P. P. Levin, P. F. Pluzhnikov and V. A. Kuzmin, Chem. Phys. Lett. 147 (1988) 283.

24) D. Rehm and A. Weller, Isr. J. Chem. 8 (1970) 259.

25) (a) I. R. Gould, D. Ege, J. E. Moser and S. Farid, J. Am. Chem. Soc. 110 (1988) 1991. (b) I. R. Gould and S. Farid, J. Am. Chem. Soc. 110 (1988) 7883. (c) I. R. Gould, J. E. Moser, B. Armitage, S. Farid, J. L. Goodman and M. S. Herman, J. Am. Chem. Soc. 111 (1989) 1917.

26) I. R. Gould, R. H. Young, R. E. Moody and S. Farid, J. Phys. Chem. submitted.

27) R. A. Marcus, J. Phys. Chem. 93 (1989) 3078.

28) J. Jortner and M. Bixon, J. Chem. Phys. 88 (1988) 167.

Photochemical Processes in Organized Molecular Systems
K. Honda (Editor-in-Chief)
© Elsevier Science Publishers B.V., 1991

CONTROL OF ELECTRON TRANSFER RATES BY MOTIONS OF ATOMS AND MOLECULES

John R. Miller

Chemistry Division, Argonne National Laboratory, Argonne, IL   60439

Recently, measurements have been made of the temperature dependence of long distance intramolecular electron transfer reactions in molecules in which donor and acceptor groups are held at a fixed distance apart.   For very highly exoergic electron transfer reactions, the rates show a remarkable independence of temperature.   This temperature independence can be predicted quite precisely from the results of previous experiments which measured only the dependence of rates on free energy change.   The temperature dependence can be ascribed to "nuclear tunneling" in high frequency vibrational modes which are coupled to the electron transfer reaction.   While this is not surprising, the dominance of even a small coupling to high frequency modes in determining the temperature dependence is surprising.   The importance of the high frequency modes is likely to be apparent in other contexts, particularly cases in which "dynamic solvent" effects occur for electron transfer reactions in highly viscous media.

1. INTRODUCTION

This paper discusses some issues in electron transfer (ET).   The paper will present interpretation of recent data from our laboratory[1,2] on temperature dependence of electron transfer rates and a somewhat speculative connection of that analysis to questions of adiabaticity/nonabiabaticity and solvent-dynamic control of ET.   Temperature dependence[3] and the question of adiabaticity[4] are subjects of considerable interest to Professor Shigeo Tazuke, who had a strong interest in the subject of electron transfer[3-12].   The point of view presented herein on the question of control, or sometimes lack of control, of ET rates by sluggish solvent dynamics is unconventional.   It may also be justifiably open to criticism because it is only intuitive.   It is offered in an adventuresome spirit that I think Shigeo Tazuke would have liked.

2. TEMPERATURE-INDEPENDENT RATES OF ELECTRON TRANSFER

The classical Marcus theory[13-15] (eq 1) made the remarkable prediction that rate constants, k_{ET}, of highly exoergic (inverted region) electron transfer (ET) reactions would decrease if the thermodynamic driving force, $-\Delta G°$, became greater than λ, the reorganization energy of the reactants and the surrounding solvent:

$$k_{ET} = \kappa \frac{k_B T}{h} \exp \frac{-(-\Delta G°-\lambda)^2}{4\lambda kT} \cdot \qquad (1)$$

Although that prediction was controversial for almost two decades, work in this laboratory confirmed that electron transfer rates decreased in the inverted region[16-19]. According to classical Marcus theory, the inverted region is due to the reappearance of an activation energy. Therefore, inverted-region ET reactions are predicted by the classical Marcus theory to be strongly temperature-dependent. We have measured the temperature dependence of two highly exoergic intramolecular electron transfer reactions in order to provide a direct, critical test of this theory[2]. Although these reactions are relatively slow, they show almost no activation energy, in disagreement with the expectations of Marcus theory, but in accord with theories that include quantum-mechanical treatments of high-frequency molecular vibrations of the donor and acceptor groups[20-26].

This important feature of ET chemistry was elaborated by pulse radiolysis measurements of intramolecular electron transfer rates. Specifically, we studied the intramolecular charge-shift ET from biphenylyl anion (B$^-$) to a neutral electron acceptor A, where A = benzoquinonyl (Q, $\Delta G° = -2.10\,eV$) or 5-chlorobenzoquinonyl (ClQ, $\Delta G° = -2.29$ eV) in the rigid molecule, ASB.

A = Cl Q   A = Q

The intramolecular and intermolecular ET rate constants were measured in the temperature range -94 °C to 100 °C by pulse radiolysis, which was used to add an electron to the bifunctional molecule. Time-dependent concentrations of the radical anions were measured by their optical absorptions. The results of these experiments with QSB and ClQSB are shown in Figure 1.

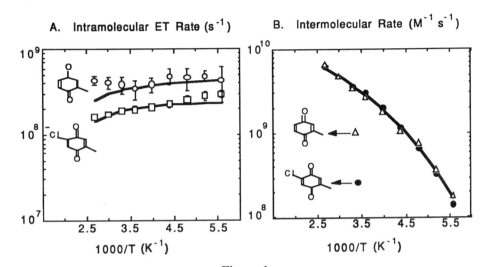

Figure 1

(A) Temperature dependence of the intramolecular ET rate constants for QSB⁻ and ClQSB⁻ in 2-methyltetrahydrofuran (MTHF). The solid lines were calculated by using eq 3 with the experimentally determined parameters. The theoretical $k_{intra}(T)$ of ClQSB has been multiplied by a factor of 1.69. (B) Temperature dependence of the intermolecular ET rate constants for QSB⁻ and for QSB and ClQSB⁻ and ClQSB. The solid line in the lower portion represents the best fit to the VTF equation as discussed in the text.

The observed rate constants, k_{intra} and k_{inter}, are plotted against reciprocal temperature in Figure 1. The values of k_{intra} for both QSB and ClQSB are almost temperature independent with a smaller value for the more exoergic ClQSB. For ClQSB, k_{intra} even increases slightly upon lowering temperature. These results are most remarkable. Their novelty prompted us to examine the intermolecular rate constants for ET under the same conditions by using mixtures of the monofunctional model compounds BS and QS or ClQS.

For the same two molecules, the intermolecular ET rate constants, k_{inter}, (Figure 1B) behave oppositely: k_{inter} is quite sensitive to temperature but is insensitive to $\Delta G°$, which may demonstrate that unique features of the inverted region are being masked by the rate-limiting transport processes for bimolecular reactions. This is because k_{inter} is determined primarily by the rate of diffusion. For both molecules, k_{inter} could be fit to the phenomenological VTF equation[27-29]:

$$k = k_0 exp[E_0/(T-T_0)] , \qquad (2)$$

with T_0 taken to be 105 K to achieve the best correlation. It is not surprising that T_0 is close to the glass-transition temperature of MTHF. Thus, the intermolecular rate constants can be rationalized on the basis of conventional concepts. The rate constants for the intramolecular ET reactions of QSB⁻ and ClQSB⁻ are completely at variance with conventional Marcus theory as expressed in eq 1. However, the results can be readily understood on the basis of the same concepts that we have used[18,19] to account for the rate constants of other ET reactions of compounds ASB (see structure above) with varied A groups such as

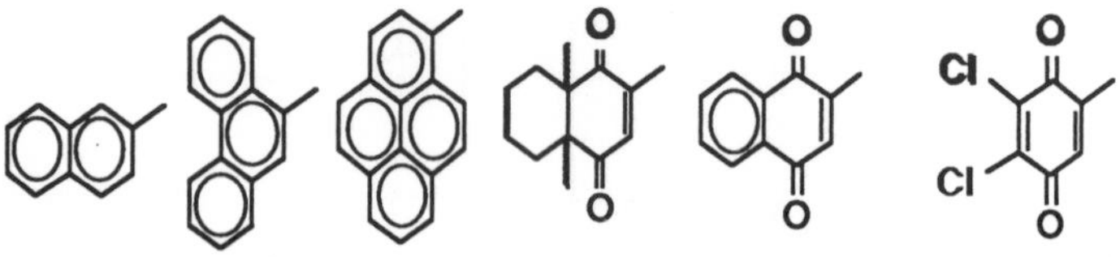

Rate data for electron transfer to these acceptors (and to Q and ClQ) for which $\Delta G°$ ranges from -0.06 to -2.5 eV are well described by the nonadiabatic ET theory[20-26] of eq 3, in which Franck-Condon weighted density of states (FCWD) is responsible for temperature dependence:

$$k_{intra} = (2\pi/\hbar) |V^2|FCWD;$$

$$FCWD = (2\hbar\lambda_s k_B T)^{-1/2} \sum_{w=0}^{\infty} \frac{e^{-S}S^w}{w!} exp\{-[(\lambda_s+\Delta G°+whv)\frac{2}{4}\lambda_s k_B T]\}; \quad S = \lambda_v/hv. \quad (3)$$

The solvent reorganization energy ($\lambda_s = 0.75$ eV for MTHF), the inner reorganization energy ($\lambda_v = 0.45$ eV) of high frequency modes, represented by a single average frequency (hv = 1500 cm⁻¹) of skeletal vibrations of the donor and acceptor groups and electronic coupling matrix element (V = 6 cm⁻¹) were determined from dependence of k_{intra} on $\Delta G°$ (see Figure 2).

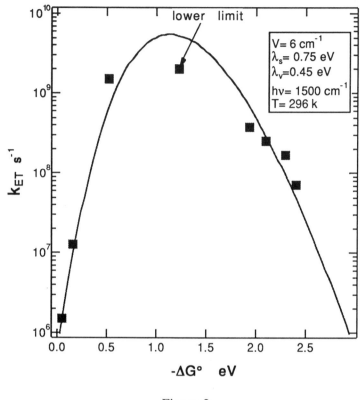

Figure 2

Experimental determination of intramolecular electron transfer rate constants (k_{ET}) of as a function of free energy change as measured by pulse radiolysis in 2-methyltetrahydrofuran. The curve is a best fit to the experimental points using eq 3.

According to eq 3 the rates are very weakly dependent on temperature because of the quantum-mechanical nature of the high-frequency modes. These modes are "frozen" in the temperature range examined in this work, thus their Franck-Condon factors attenuate the rates and are not improved by increasing temperature. Quantum effects, which are absent in classical Marcus theory, make the high-frequency modes efficient at disposing of the excess energy in these highly exoergic reactions. The classical Marcus theory correctly predicts a barrier to reaction in the inverted region, but does not account for the quantum effects (nuclear tunneling) that enable the nuclei to pass through the barrier rather than over it. Thus, the rates exhibit only a very weak temperature dependence.

In addition, λ_s of MTHF increases[1,2] by 20% from 100 °C to -94 °C, which is responsible for the slightly *negative* activation energies (-0.18 kcal/mol for ClQSB). If λ_s were independent of temperature, a more-normal, weakly positive activation energy (~0.5 kcal/mol) would have been expected.

3. HOW CAN WE ACCOUNT FOR THE REMARKABLE TEMPERATURE INDEPENDENCE OF THE RATES?

The rates of the highly exoergic intramolecular electron transfer reactions are slowed by the Franck-Condon factors, but they are remarkably independent of temperature. How can we account for this temperature independence of the rates? According to the classical Marcus theory (eq 1) the rates of these reactions would be extremely sensitive to temperature. If λ in eq 1 is set equal to 1.3 eV to match the maximum the observed rate vs. free energy curves, then over our temperature range the electron transfer rates in QSB would have been expected to change by a factor of 200 and those in ClQSB by a factor of 3,000. The actual observed temperature dependence, which is extremely weak, is described very well by eq 3 which includes high-frequency quantum mechanical vibrations of the donor and acceptor groups. These quantum mechanical vibrations, which are coupled to the electron transfer process, are almost entirely due to changes in the bond lengths for carbon-carbon and carbon-oxygen double bonds of the aromatic and ketone groups involved.

The Marcus theory of eq 1 has the quadratic form, $(-\Delta G° - \lambda)^2$, because of the quadratic dependence of the potential energy of free energy curves on displacement shown on Figure 3.

In the right panel of Figure 3 we see that there is a barrier to reaction because that reaction is so exoergic. The high-frequency modes can penetrate this barrier by nuclear tunneling, that is, the vibrational wave function of the reactant's curve can penetrate through the barrier region due to its quantum mechanical nature and overlap with the vibrational wave function of the product's potential energy curve. While the vibrational overlap (Franck-Condon factors) could be improved in excited vibrational levels, the improvement cannot compensate for the large Boltzman factor required to excite to a higher vibrational level. Therefore, in the total Franck-Condon factor for the high-frequency vibrations

FC (high frequency) = FC (v=0) + FC (v=1) $e^{-hv/kT}$ + FC (v=2) $e^{-2hv/kT}$ + ... (4)

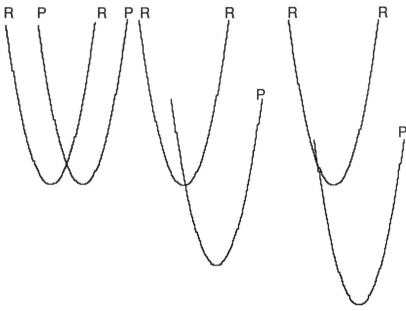

Figure 3
Potential energy curves for weakly exoergic (left), moderately exoergic (middle), and highly exoergic (right) electron transfer reactions. In each case the reactants' potential energy surface is labelled R and the product potential energy surface is labelled P. Activation barriers occur in both the weakly exoergic and strongly exoergic cases. Splitting by the electronic coupling between the reactants and products, V, is not shown.

At moderate temperatures, only the first term is significant for high-frequency molecular vibrations. That is true because the Boltzman factor for thermally exciting one quantum of a high-frequency vibration of 1500 cm^{-1} is 6×10^{-4} at room temperature. It is therefore clear that if all of the reorganization were due to such high-frequency modes, for example the electron transfer occurred in the gas phase or in a solvent with no polar molecules, then the electron transfer rates would be independent of temperature.

The puzzle is, however, when the majority (0.75 eV) of the reorganization energy is due to low-frequency modes from the solvent, why is the electron transfer rate nearly independent of temperature? We can examine why eq 3 makes this prediction. In Figure 4 we plot the predictions of eq 3 for rate as a function of free energy at three different temperatures. We can see that at the low temperatures, the rates are dramatically different in the weakly exoergic region. In the strongly exoergic region, however, the shape of the curves are basically the same except that at low temperatures deep "valleys" develop in the curve, which begins to look like a vibrational spectrum in the gas phase. When the temperature is low, the classical solvent modes can no longer fill in the valleys between the peaks due to the various vibrational states of the products. If our reaction occurs at a free energy change that is near one of the peaks, eq 3 would predict almost no temperature dependence, whereas a strong temperature dependence would be predicted at a free energy change corresponding to one of the "valleys". But the deep valleys themselves predicted by eq 3 are not realistic.

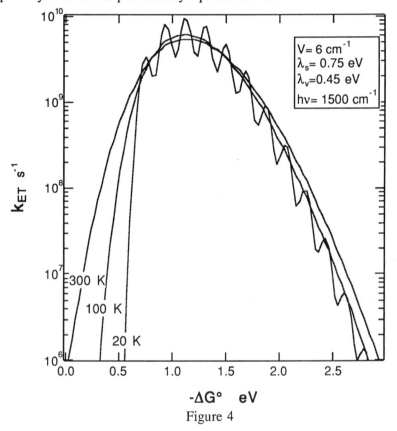

Figure 4

Plots of ET rate as a function of free energy change according to eq 3 at T = 300, 100 and 20 K.

Eq 3 assumes that all of the vibrations have the same frequency, 1500 cm^{-1}. In real molecules, however, there is a range of frequencies, most of them between 1350 and 1700 cm^{-1}, because there are vibrations[24] of different frequencies which are displaced as we change from the reactants to the products. There is even some reorganization of modes in frequencies of just a few hundred cm^{-1}. Therefore, in highly exoergic reactions where several vibrational quanta are dissipated in the products, these quanta will have a number of different frequencies so that the vibrational spectra will be quite congested: the peaks will become slightly smaller and the valleys will be filled in. Furthermore, particularly as the temperature is reduced, some rotational motions of polar molecules will change into librational motions, which have reasonably high frequencies. The frequencies of these librations are even likely to increase at low temperatures, exceeding 100 cm^{-1} in many cases. Therefore, we expect that in the highly exoergic region the rate vs. free energy curve will not have substantial valleys, but instead will be a reasonably smooth curve to very low temperatures. Experimental evidence that this can be so has been found in photosynthetic reaction centers by Gunner and Dutton[30].

The ET rates as a function of temperature presented in Figure 1A, however, were well described by eq 3 which did not yet include the fact that the high-frequency modes have a range of frequencies, or that there are solvent librational motions coupled to the electron transfer reaction. That is because, as we can see in Figure 4, even classical solvent modes with a reorganization energy of 0.75 eV still provide enough width to the rate vs. free energy "spectrum" that they fill in the valleys at the vibrational levels, even at 100 K. In eq 3, the solvent vibrations are assumed to behave classically, and to obey the Marcus expression, like that in eq 1. Figure 5 illustrates how the width of a spectrum of such classical modes changes with temperature down to very low temperatures. We can see that the width remains wide enough to reasonably well fill in the valleys between vibrational lines until the temperature drops below 100 K.

We will therefore conclude that for highly exoergic electron transfer reactions, accompanied by substantial high frequency reorganization energies, that the rates will be almost completely independent of temperature. For the weakly exoergic reactions, though, as we can see from Figure 4, it is quite another story. Looking at Figure 4 we can see that there are four regions:

$\Delta G° < \lambda_s$: The rates are strongly dependent upon motion of the solvent and therefore strongly dependent upon temperature.

$\lambda_s <- \Delta G° < \lambda_s + \lambda_v$: The rate is slowed by the solvent, but is not strongly dependent upon temperature.

$-\Delta G° \cong \lambda_s + \lambda_v$: The rate is fast, and does not depend upon motion of the solvent and therefore does not depend on temperature.

$-\Delta G° >> \lambda_s + \lambda_v$: The rate is slow, but is nearly independent of solvent motions and therefore of temperature.

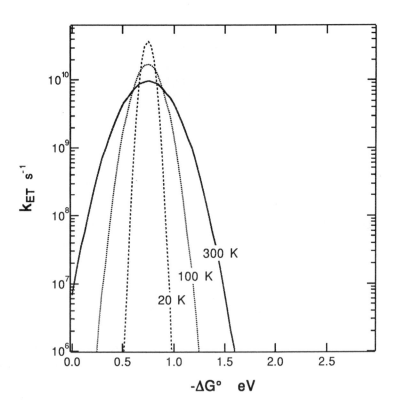

Figure 5

Classical Rate vs. $\Delta G°$ "spectra" which illustrate how the width from the classical solvent modes changes with temperature. Parameters are the same as in Figure 4 except that $\lambda_v = 0.0$.

We can understand in an intuitive way the reason for these different dependencies on the solvent motions and therefore on temperature. For weakly exoergic reactions ($-\Delta G° < \lambda_s, \lambda_v$) both the solvent reorganization energy and the reorganization energy due to high-frequency vibrations present barriers to the reaction. The free energy change for the reaction is not sufficient to surmount these barriers, so that the two barriers must be surmounted either by thermal energy or, as is likely in the case of high-frequency modes, by zero point motions ("nuclear tunneling"). If nuclear tunneling helps to transmit the barrier due to the high-frequency reorganization energy, that will not help in transmitting the barrier formed by solvent reorganization. We can think of this situation that the two barrier lie in series with each other, and both must be somehow transmitted or surmounted to reach a reactive configuration.

In the very highly exoergic region, however, the situation is quite different. Now the reason that there is a barrier to reaction is because there is too much energy driving the reaction and there is a problem dissipating the energy. In this case, the solvent reorganizations are not providing a barrier to reaction. Increased reorganization energy would, in fact, lower the barrier to reaction which is opposite to the situation in the weakly exoergic region. Because the high frequency vibrations are more effective at dissipating energy, they can and do dissipate a much larger fraction of the energy than would be expected from their smaller (0.4 eV) reorganization energy. Since the problem is not to surmount two energy barriers, but is to dissipate energy, the energy can be dissipated by whichever modes are most effective. That is, in highly exoergic reactions, the two reorganizations function in parallel rather than in series. This is a substantial part of the reason why the high-frequency modes can dominate the shape of the free-energy curve in the highly exoergic region but not in the weakly exoergic region.

Of course, the mere presence of reorganization in high-frequency modes is not sufficient to make rates of highly exoergic ET reactions independent of temperature. The amount of coupling (the reorganization energy) of these modes is important. Yet only an amazingly small amount of coupling is needed. Figure 6 illustrates that remarkably little coupling to high frequency modes is needed to dramatically alter the temperature dependence.

Temperature dependence of highly exoergic ($\Delta G° = -2.29$ eV as in ClQSB) electron transfer rates for varying amounts of reorganization energy λ_v of high frequency molecular vibrations according to eq 3. The solvent reorganization energy is constant at 0.75 eV and the other parameters are the same as in Figure 3.

Strong temperature dependence is seen only when there is no reorganization of high frequency vibrational modes.

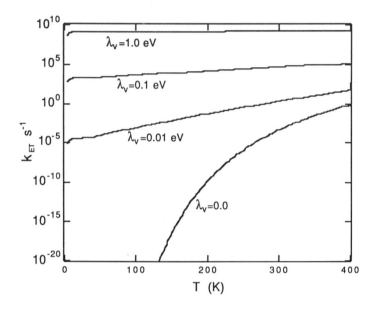

Figure 6

ET rate calculated by eq 3 as a function of temperature with a large, constant classical reorganization energy, $\lambda_s = 0.75$ eV, and varied high-frequency reorganization energy λ_v, with a vibrational frequency of 1500 cm^{-1}.

In Figure 6 we see that even a reorganization energy of 0.01 eV (about 1% of the total reorganization energy) can dramatically alter the temperature dependence of the ET rates. Probably no ET reaction involving molecules can have such a small reorganization of molecular vibrations, so we may conclude weak temperature-dependence will be a general feature of highly-exoergic ET reactions. Jortner pointed out the dominant effect of high frequency modes on the rates of highly exoergic ET reactions several years ago. Now we see that this dominance is even more profound when temperature dependence is examined in detail.

ET rates will depend weakly on temperature for highly exoergic reactions, but some additional temperature dependence can result if $\Delta G°$ of λ depends on temperature. For example, in charge separation or recombination reactions

$D + A \rightleftharpoons D^+ \cdot + A\dot{}$ or in many reactions of metal ions $\Delta G°$ can change substantially with temperature, which can make the rates temperature-dependent. In the present experiments on charge shift reactions, $\Delta G°$ is nearly independent of temperature.

4. SPECULATIONS ON CONTROL OF ET RATES BY SOLVENT DYNAMICS.

In recent years, a number of authors have, both experimentally[31-45] and theoretically[20-58], investigated situations in which the motions of solvent molecules becomes comparable to or slower than the rate of an electron transfer process. In these situations, both the classical Marcus theory (eq 1) and nonadiabatic expressions like eq 3 could fail because they do not include the control of the rate by solvent dynamics. This occurs when thermal energy is sufficient, according to eqs 1 or 3, to allow the solvent to reach the transition state needed for electron transfer, but when this inherent rate of electron transfer is so fast or motions of the solvent so slow that the solvent simply does not have time to move fast enough to reach that transition state. Evidence for this situation was first obtained by Kosower Huppert and coworkers[31-36]. The theories give the result that the solvent relaxation time, usually taken as the longitudinal relaxation time, τ_L, becomes an upper limit for the ET rate.

On the other hand we have measured ET rates in rigid glasses[16,18,59,60,61] where the longitudinal relaxation times are surely longer than the time scale of the experiment (10^{-7} - 10^2 s^{-1}). The results of those experiments did not indicate that the rates were limited by solvent relaxation times and appeared, in fact, to be well-described by eq 3[16,18]. The proposition that solvent relaxation times give an upper limit to ET rates seems questionable. The concepts raised above may provide some insight into this enigma of solvent dynamic control of ET rates.

There are strong parallels between the solvent dynamic control of electron transfer rates, and the control by temperature, described above. For electron transfer to occur, it is required that the nuclei adjust to "transition" configuration in which the energies of the reactants and product states are equal. We saw above that in the case of temperature dependence this transition configuration can be reached by tunneling motions of high frequency molecular vibrations which, in highly exoergic reactions, can make up for the inability of the solvent to be thermally activated to a transition configuration. Reaching a transition configuration is a requirement for reaction. But if the solvent is unable to move toward the transition configuration, it does not matter whether the reason is low temperature or sluggish solvent relaxation. The roles of high frequency modes (molecular vibrations and librations) in providing alternate routes to the transition configuration will be the same, and therefore the conditions for control of electron transfer rates by solvent dynamics should be very much like those controlled by the effect of temperature on solvent motion:

$\Delta G° < \lambda_s$: The rates are strongly dependent upon motion of the solvent and therefore strongly dependent upon rates of solvent relaxation

$\lambda_s <- \Delta G° < \lambda_s + \lambda_v$: The rate is slowed by the solvent, but is not strongly dependent upon solvent relaxation.

$-\Delta G° \cong \lambda_s + \lambda_v$: The rate is fast, and does not depend upon motion of the solvent.

$-\Delta G° >> \lambda_s + \lambda_v$: The rate is slow, but is nearly independent the rate of solvent relaxation.

What may we conclude if these four conditions correctly describe the control of electron transfer rates by solvent dynamics? Firstly, like temperature dependence, solvent dynamic control will definitely be possible for weakly exoergic reactions. But for optimally exoergic reactions (condition 3), or highly exoergic reactions (condition 4), when the solvent is unable to adjust to reach a configuration, high frequency molecular vibrations will adjust to reach nuclear configurations for which the energy of reactants and products are equal. Just as for the case of temperature dependence, energy matching could be a problem, that is, the "valleys" in the rate vs. free energy curve, but just as with the case of temperature dependence, these valleys will in reality tend to disappear in highly exoergic reactions because of the involvement of multiple frequencies in the skeletal vibrations of donor/acceptor groups and of librational motions of the solvent. If we assume, for example, that the solvent is equilibrated for the reactants, but is completely unable to move (an extreme solvent dynamic effect), then an adequate description of the classical solvent motion is that it is just like a solvent which is frozen at absolute zero. Its effect will simply be to reduce the energy gap (that is, to reduce the free energy change) by an amount equal to solvent reorganization energy, and electron transfer rates could be given by eq 3 with the temperature set equal to zero. We only need to make the proviso that energy matching will be achieved by coupling of to high frequency modes including librational motions.

A theory has been constructed to include high frequency molecular vibrations along with sluggish solvent modes[57]. However in this case the solvent is very simple, possessing only a slow relaxation time and no librational motions. Therefore the ET rate "spectrum" would be, except for rare, exact coincidences with vibrational lines, entirely "valleys" for times shorter than the relaxation time. The rates are indeed expected to be controlled by relaxation times, even in the highly-exoergic region. A more realistic theory would include a large variety of vibrational and librational motions. It would, I believe, fail to be dominated by solvent relaxation time except in the weakly-exoergic region. Such a theory would be difficult to construct and is not presently available.

The same restrictions that limit control of ET rates should also apply to the possibility of the reaction becoming "adiabatic". Transition state based theories such as the Marcus theory (eq 1), which predict that the rates of electron transfer will cease to increase with increasing electronic coupling when the coupling becomes large enough that the rate exceeds $\approx 10^{13}$ s^{-1}. This is the so-called adiabatic limit of electron transfer rates. It would be expected to occur because, when the electronic coupling is large enough, it is sufficient to guarantee electron transfer every time the reactants and products reach an intersection of reactants and products potential energy surfaces. The highly exoergic electron transfer reactions, however, the reactants would frequently reach energy coincidence with products by high-frequency vibrational motions, and the reactants will be removed from this intersection region by the same high-frequency motions. Therefore, larger values of the electronic coupling will produce larger electron transfer rates. Electron transfer rates of 10^{14} s^{-1}, and perhaps somewhat larger, should be possible. As expressed in the above inequalities for solvent dynamic control, weakly exoergic electron transfer reactions should still be subject to the adiabatic limit imposed by the solvent. This is because the reaction, when it is weakly exothermic, cannot reach a transition configuration without substantial participation of the solvent (the solvent barrier and high-frequency "barriers" are in series). When the solvent has reached a favorable configuration, it is true that high-frequency modes might take the reaction in the reactants and products in and out of energy coincidence a number of times. But it is only the mean probability averaged over those high-frequency vibrational coincidences, multiplied by the number of these coincidences that occur while the solvent is in a favorable configuration, that determines the net probability for the reaction. It is still a requirement that the solvent reach a configuration favorable for reaction. In the weakly exoergic regime the adiabatic limit will be determined primarily by solvent motion. Evidence for the occurrence of the adiabatic limit is rare, but has been obtained by Richardson and Taube some years ago[62,63] for weakly exoergic (in fact, isoenergetic) reactions.

The conditions for achieving solvent dynamic control of electron transfer rates, then, are very difficult. We need to have rates that are faster than the rates of solvent motions. Because typical solvent motions are on the order of 10^{-12} to 10^{-11} s^{-1}, this means we must consider very fast electron transfer reactions. However, the electron transfer reactions must be weakly exoergic ($\Delta G° < \lambda_s$). Therefore, to observe solvent dynamic effects it is necessary to have a very large electronic coupling to get such a fast rate for a weakly exoergic reaction. Typically, examples where solvent dynamic control has been reported are with conjugated molecules, where there are no "insulating" groups between the donor and acceptor portions of the molecules so that the electronic coupling is very large.

5. CONCLUSIONS

We've seen that high-frequency vibrations drastically modify the behavior of highly exoergic electron transfer reactions. The Marcus theory is no longer of aid, in particular, the strong temperature dependence it predicts is almost totally absent. This is due to the overwhelming role of high-frequency vibrational motions in highly exoergic reactions. The high-frequency vibrations would also play a similar dramatic role in enhancing rates of electron transfer in highly exoergic reactions where solvent dynamic control and adiabaticity might otherwise limit the rate of reaction.

REFERENCES

1) N. Liang, J.R. Miller and G.L. Closs, J. Am. Chem. Soc. 111 (1989) 8740.

2) N. Liang, J.R. Miller and G.L. Closs, J. Am. Chem. Soc. 112 (1990) 5353.

3) H.-B. Kim, N. Kitamura, Y. Kawanishi and S. Tazuke, J. Am. Chem. Soc. 109 (1987) 2506.

4) N. Kitamura, R. Obata, H.-B. Kim and S. Tazuke, J. Phys. Chem. 91 (1987) 2033.

5) R.K. Guo, N. Kitamura and S. Tazuke, J. Phys. Chem. 94 (1990) 1404.

6) R. Hayashi, S. Tazuke and C.W. Frank, ACS Symp. Ser. 358(Photophys. Polym.): (1987) 135.

7) R. Hayashi, S. Tazuke and C.W. Frank, Macromolecules 20 (1987) 983.

8) Y. Kawanishi, N. Kitamura and S. Tazuke, J. Phys. Chem. 90 (1986) 2469.

9) H.-B. Kim, N. Kitamura, Y. Kawanishi and S. Tazuke, J. Phys. Chem. 93 (1989) 5757.

10) N. Kitamura, R. Obata, H.-B. Kim and S. Tazuke, J. Phys. Chem. 93 (1989) 5764.

11) N. Kitamura, H.-B. Kim, S. Okano and S. Tazuke, J. Phys. Chem. 93 (1989) 5750.

12) T. Ohno, A. Yoshimura, N. Mataga, S. Tazuke, Y. Kawanishi and N. Kitamura, J. Phys. Chem. 93 (1989) 3546.

13) R.A. Marcus, J. Chem. Phys. 24 (1956) 966.

14) R.A. Marcus, Can. J. Chem. 37 (1959) 155.

15) R.A. Marcus, J. Chem. Phys. 43 (1965) 58.

16) J.V. Beitz and J.R. Miller, J. Chem. Phys. 71 (1979) 4579.

17) J.R. Miller, L.T. Calcaterra and G.L. Closs, J. Am. Chem. Soc. 106 (1984) 3047.

18) J.M. Miller, J.V. Beitz and R.K. Huddleston, J. Am. Chem. Soc. 106 (1984) 5057.

19) G.L. Closs, L.T. Calcaterra, N.J. Green, K.W. Penfield and J.R. Miller, J. Phys. Chem. 90 (1986) 3673.

20) N.R. Kestner, J. Logan and J. Jortner, J. Phys. Chem. 78 (1974) 2148.

21) J. Ulstrup and J. Jortner, J. Chem. Phys. 63 (1975) 4358.

22) J. Jortner, J. Chem. Phys. 64 (1976) 4860.

23) R.P. Van Duyne and S.F. Fischer, Chem. Phys. 5 (1974) 183.

24) S.F. Fischer and R.P. Van Duyne, Chem. Phys. 26 (1977) 9.

25) R.R. Dogonadze, A.M. Kuznetsov and M.A. Vorotyntsev, Z. Phys. Chem. N. F. 100 (1976) 1.

26) R.R. Dogonadze, A.M. Kuznetsov, M.A. Vorotyntsev and M.G. Zakaroya, J. Electroanal. Chem. 75 (1977) 315.

27) G.H. Fredrickson, Annu. Rev. Phys. Chem. 39 (1988) 149.

28) G.S. Fulcher, J. Am. Chem. Soc. 8 (1925) 339.

29) H. Vogel, Phys. Z. 22 (1921) 645.

30) M.R. Gunner and P.L. Dutton, J. Am. Chem. Soc. 111 (1989) 3400.

31) D. Huppert, H. Kanety and E.M. Kosower, Faraday Discuss. Chem. Soc. 74: (1982) 161.

32) E.M. Kosower, H. Kanety, H. Dodiuk, G. Striker, T. Jovin, H. Boni and D. Huppert, J. Phys. Chem. 87 (1983) 2479.

33) E.M. Kosower and D. Huppert, Chem. Phys. Lett. 96 (1983) 433.

34) D. Huppert, V. Ittah and E.M. Kosower, Chem. Phys. Lett. 144 (1988) 15.

35) D. Huppert, V. Ittah, A. Masad and E.M. Kosower, Chem. Phys. Lett. 150 (1988) 349.

36) D. Huppert, V. Ittah and E.M. Kosower, Chem. Phys. Lett. 159 (1989) 267.

37) A. Masad, D. Huppert and E.M. Kosower, Chem. Phys. 144 (1990) 391.

38) J.D. Simon and S.-G. Su, J. Chem. Phys. 87 (1987) 7016.

39) J.D. Simon and S.-G. Su, J. Phys. Chem. 92 (1988) 2395.

40) J.D. Simon and S.-G. Su, J. Phys. Chem. 94 (1990) 3656.

41) M.A. Kahlow, T.J. Kang and P.F. Barbara, J. Phys. Chem. 91 (1987) 6452.

42) T.J. Kang, M.A. Kahlow, D. Giser, S. Swallen, V. Nagarajan, W. Jarzeba and P.F. Barbara, J. Phys. Chem. 92 (1988) 6800.

43) M.J. Weaver, G.E. McManis, W. Jarzeba and P.F. Barbara, J. Phys. Chem. 94 (1990) 1715.

44) M. McGuire and G. McLendon, J. Phys. Chem. 90 (1986) 2549.

45) J.S. Bashkin, G. McLendon, S. Mukamel and J. Marohn, J. Phys. Chem. 94 (1990) 4757.

46) L.D. Zusman, Chem. Phys. 49 (1980) 295.

47) G. van der Zwan and J.T. Hynes, J. Chem Phys. 76 (1982) 2993.

48) J.T. Hynes, Ann. Rev. Phys. Chem. 36 (1985) 573.

49) J.T. Hynes, J. Phys. Chem. 90 (1986) 3701.

50) H.J. Kim and J.T. Hynes, J. Phys. Chem. 94 (1990) 2736.

51) R.A. Marcus and H. Sumi, J. Electroanal. Chem. Interfacial Electrochem. 204 (1986) 59.

52) H. Sumi and R.A. Marcus, J. Chem. Phys. 84 (1986) 4894.

53) H. Sumi and R.A. Marcus, J. Chem. Phys. 84 (1986) 4272.

54) W. Nadler and R.A. Marcus, J. Chem. Phys. 86 (1987) 3906.

55) I. Rips and J. Jortner, J. Chem. Phys. 87 (1987) 2090.

56) I. Rips and J. Jortner, J. Chem. Phys. 88 (1988) 818.

57) J. Jortner and M. Bixon, J. Chem. Phys. 88 (1988) 167.

58) I. Rips, J. Klafter and J. Jortner, Solvation dynamics and solvent-controlled electron transfer., in: Photochemical Energy Conversion, Proc. 7th Int. Conf. Photochem. Convers. Storage Solar Energy, Evanston, IL, 31 Jul-5 Aug, 1988, Vol. , Ed. Jr. and D. Meisel J.R. Norris (Elsevier, New York, 1989) 1.

59) J.R. Miller and J.V. Beitz, J. Chem. Phys. 74 (1981) 6746.

60) J.R. Miller, J.A. Peeples, M.J. Schmitt and G.L. Closs, J. Am. Chem. Soc. 104 (1982) 6488.

61) J.R. Miller, K.W. Hartman and S. Abrash, J. Am. Chem. Soc. 104 (1982) 4296.

62) D.E. Richardson and H. Taube, J. Am. Chem. Soc. 105 (1983) 40.

63) D.E. Richardson and H. Taube, Coord. Chem. Rev. 60 (1984) 107.

Photochemical Processes in Organized Molecular Systems
K. Honda (Editor-in-Chief)
© Elsevier Science Publishers B.V., 1991

FLUORESCENCE PROBES BASED ON ADIABATIC PHOTOCHEMICAL REACTIONS

W. RETTIG[a], R. FRITZ[a,b] and J. SPRINGER[b]

I.N. Stranski-Institut f. Phys. & Theor. Chemie[a] and Institut
f. Techn. Chemie, Fachgebiet Makromol. Chemie[b],
Techn. Univ. Berlin, Strasse des 17. Juni 112,   1000 Berlin
12, FRG

Four types of fluorescence probes are shown based on TICT,
excimer and Dewar isomer forming compounds. TICT and excimer
probes are usable in liquid surroundings, and quantitative
free volume information can be obtained by applying current
theories. This information also reflects the size of the
necessary reaction volume of the probe.
TICT and Dewar isomer forming compounds are shown to be probes
usable even for glassy polymers. In the latter case, decay
time distributions can be recovered which reflect the
distribution function of the polymer free volume.

1. INTRODUCTION

Adiabatic photochemical reactions, such as the formation of
fluorescent Twisted Intramolecular Charge Transfer (TICT) states
through intramolecular twisting, can be used as indicators to
show whether the surrounding allows for this reaction, either by
making possible sufficiently rapid diffusion (site exchange of
solute and solvent) or by furnishing reaction space whithin the
microscopic cavities of a glassy polymer (free volume).

In the case of the well-studied TICT compounds (for reviews
see reference[1-3]), many molecules with dual fluorescence are
known. This is caused by the possibility of forming, through
intramolecular twisting and charge separation, a fluorescent
product state (TICT or A* state, twisted conformation, longer
wavelength or "A" fluorescence) in addition to the emission from
the "normal" or B* state (planar molecule, short wavelength or
"B" fluorescence). There exists, therefore, a precursor-successor
relationship between B* and A* states, the latter being
characterized by a growing-in (risetime) of its fluorescence
which, in medium polar solvents, matches the decay time of the B*
state[2]. The excited state potential energy surface, in the case

of dimethylaminobenzonitrile (DMABN) can be shown to be highly dependent on the polarity of the surrounding. This is reflected in polarity-dependent activation energies for the reaction B* → A* [4]. In highly polar environments such as alcohols, barrierless relaxation B* → A* is thus possible, and this leads to a more complicated time profile of the B* decay[3,5-8] and to the possibility of A* risetimes shorter than B* decay times[3,5].

On the one hand, this strong solvent polarity dependence has advantages for the use of these molecules as fluorescent probes, because the red shift of the TICT emission yields a measure for the effective micropolarity of the surrounding (for example polar solvent mixtures or polymer glasses[9]). On the other hand, this complicates their use as free volume probes because media which differ in both viscosity and polarity exert their influence in two ways i) through changes of their viscous drag on the twisting motion leading from B* to A* (viscosity influence) ii) through changes of the potential energy profile due to a preferential lowering of the TICT state with increased solvent polarity (polarity influence). In principle, these two factors can be separated, and two approaches of achieving this are shown below.

The first involves the development of a TICT fluorescence probe which works even in nonpolar environments. For different alkanes, for example, the changes in polarity are negligible, and the viscosity influence becomes rate-determining. The same applies to the comparison of polymer melts (or synthetic oils) where macroscopic viscosity changes are brought about by a variation of the polymer chainlength.

The second approach involves the comparison of two TICT probes necessitating different free volumes for reaction. This latter approach can even be used for glassy media (polymers far below the glass transition temperature T_g).

TICT probes lend themselves to the functional incorporation into a polymer, for example as side chains where they can be separated from the main polymer chain by spacers of varying length. Several such examples have been studied by Tazuke and coworkers[10-13], and informations regarding the side chain mobility in the neighbourhood of the main chain as well as changes of the effective micropolarity as a function of spacer length have been derived.

These TICT-copolymers are highly specialized, and an approach using unlinked TICT fluorescent probes gives the possibility for an easy comparison of different solvents and polymers, even those of technical importance (see below), and enables the study of highly refined polymers which have been fractionated according to the degree of polymerization and characterized previously.

In addition to TICT-probes, we also compare in the present study other classes of fluorescent probes, characterized by different adiabatic photoreactions. A general kinetic scheme applicable to all these cases is the following one

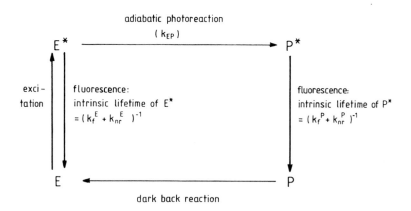

KINETIC SCHEME

where E^* stands for the educt or precursor excited state, and P^* for the product state. The adiabatic photoreaction k_{EP} leads to a quenching of the E^* fluorescence (its lifetime being shortened to $(k_f^E + k_{nr}^E + k_{EP})^{-1}$), and therefrom, the rate constant k_{EP} can be extracted and studied as a function of the parameters of the surrounding. In the case of TICT systems, E^* and P^* correspond to the fluorescent B^* and TICT (A^*) states. A possible excited state reversibility (reaction $P^* \rightarrow E^*$, omitted from the scheme) can complicate the observed fluorescence decay behaviour. Other examples studied are intramolecular excimers where E^* corresponds to the species where the excitation is localized on one of the aromatic units, and P^* is the fluorescing excimer state.

There are also numerous cases, where P^* is nonfluorescent (or so weakly that it cannot be observed), and a well-characterized example are the triphenylmethane (TPM) dyes[3,14,15]. Comparison of TPM dyes with substituents of different size and flexibility have been used to study microviscosity (free volume) effects in mono- and polyvalent alcohols[16]. The malonitrile fluorescence probes used by Loutfy and coworkers to monitor the time dependence of methacrylate bulk polymerization and to characterize the microviscosity properties of these polymers[17-23] work on a similar basis with a nonfluorescent P* state (twisted conformation). Rigidification of the polymeric network leading to a suppression of the photoreaction k_{EP} therefore leads to a strong increase of the fluorescence quantum yield from E*. These probes are closely related to the TICT probes because the adiabatic photoreaction consists of a twisting of single and/or double bonds. There is no principle difference between the twisting of a single or a double bond in the excited state, and both processes can be understood and described in a unified model involving biradicaloid states[3,24-26].

Finally, a third type of photoreaction, that of valence bond isomerization towards a Dewar structure, will be shown to be very useful in conjunction with the properties of a fluorescence probe. Similar to the last type of probe, it leads to a nonfluorescent product (a funnel in the excited state, halfway between the aromatic and the Dewar form), but the butterfly folding reaction needs comparatively little reaction volume, such that it partially occurs even in rigid polymer glasses and can be detected by the shortening of the fluorescence lifetimes.

2. EXPERIMENTAL

Corrected fluorescence spectra have been measured on a Perkin-Elmer 650/60 spectrofluorimeter. Fluorescence decays were determined with the single photon counting technique using an experimental setup described previously[16] and synchrotron radiation from BESSY, (Berliner Elektronenspeicherring-Gesellschaft für Synchrotronstrahlung mbH), as excitation source.

The TICT compounds were the same samples as used previously[2,27]. The excimer and "Dewar" probes (for structures see below) were gifts of H. Bouas-Laurent and H. Dreeskamp, respectively. The compounds were introduced into the solid polymers (polymethylmethacrylate (PMMA) and homologues, and polystyrene (PS)) by preparing solutions of polymer and probe in chloroform, and casting them onto a quartz plate such that, after solvent evaporation, the film thickness was about 200 μm, and the optical density of the probe remained small (<1). The films were then dried in high vacuum for 4 days under heating at a temperature close to T_g. For liquid solutions, similarly small optical density conditions wered used. The alkane solvents were of the highest optical quality commercially available (Merck Uvasol or Fluka puriss.), polydimethylsiloxane oils were commercial products from Rhône Poulenc. Solutions were not degassed. For the oils and for some of the less well purified alkanes, a fluorescent impurity was present which showed a lifetime of several nanoseconds. In the time resolved measurements, it was straightforward to distinguish this impurity component from the (subnanosecond) component of the TICT probe. This shows the validity of this approach even to unpurified commercial polymers.

3. RESULTS AND DISCUSSION

3.1. Fluorescence Probes for Liquids and for Polymer Melts.

Excimer forming probes have been used since several years to extract free volume parameters for polymers above the glass transition temperature T_g[28-31]. This was achieved by treating the temperature dependent excimer formation kinetics with the theory of Williams, Landel and Ferry (WLF theory)[32,33] as described below.

Here, we used the same procedure and applied it to both excimer and TICT forming probes, with the goal of comparison between the two. Whereas TICT formation is a pure intramolecular rotational process, excimer formation additionally necessitates translational components and is therefore more difficult to achieve in highly viscous or glassy media (Figure 1).

As TICT probe, PYREST, a derivative of the dual fluorescing dimethylaminobenzonitrile (DMABN) was used (Probe A) which shows the TICT fluorescence band even in nonpolar hydrocarbon solvents and which was developed as shown below. As excimer probe,

DIPHANT[28-31], was used (Probe B). The chemical structures of these probes and a schematic representation of the underlying photochemical processes are given in Figure 1.

Probe A: (TICT formation)

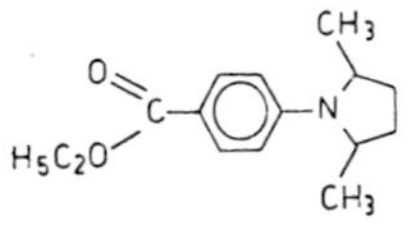

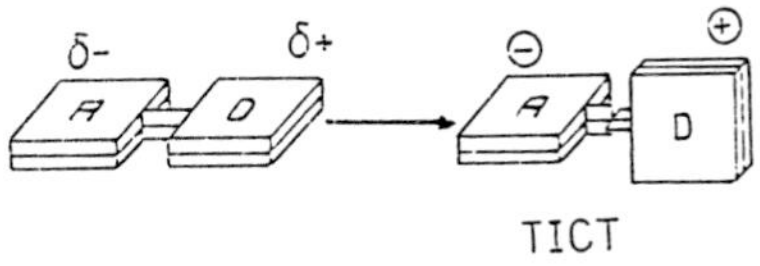

PYREST

Probe B: (Excimer formation)

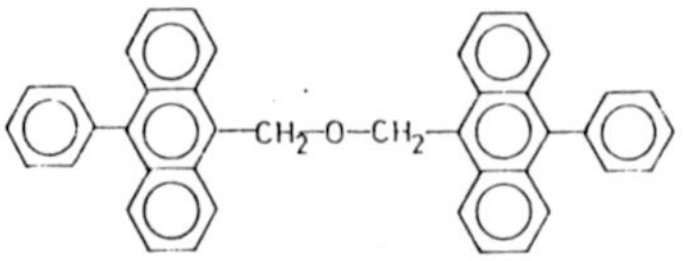

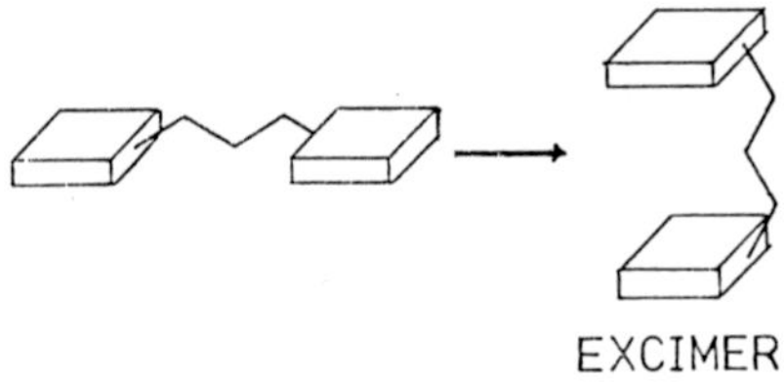

DIPHANT

EXCIMER

FIGURE 1

Chemical structure of the TICT and excimer probe and schematic representation of the adiabatic photochemical processes connected with these probes.

Figure 2 shows a comparison of the fluorescence behaviour of DMABN, the derived ester DMABEE, and of PYREST in medium polar n-butyl chloride, and in an alkane solvent. DMABEE shows more TICT fluorescence than DMABN, because the "normal" state (B*, ^1L_b-type in Platt's nomenclature) is raised energetically in an ester whereas the energy of the ^1L_a state (and of the TICT state correlating with it[34]) is lowered relative to that of the nitrile[35,36]. On the other hand, it is known that "pretwisting" of the amino group (i.e. a nonplanar ground state conformation) enhances TICT formation[2,37], and it is therefore understandable, that PYREST (pretwisted by some 10-15°)[37] shows even more TICT fluorescence than DMABEE. Indeed, TICT formation is so preferable in this compound that the TICT fluorescence band even occurs in inert solvents, without the help of a polar solvent stabilization (Figure 2).

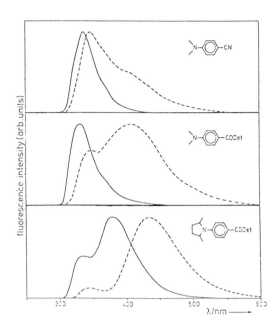

FIGURE 2

Fluorescence spectra of DMABN (upper), DMABEE (middle) and PYREST (lower) in n-butyl chloride (---) and saturated hydrocarbon solvent (——) at room temperature. Note also the strong solvatochromic red shift of the TICT band of PYREST between the two solvents.

Excimer probes, on the other hand, are more difficult to predict and to develop. Figure 3 shows, as an example, the spectra of DIPHANT and of the structurally closely related 9-anthryl intramolecular excimer A3A. Although the latter shows clear signs of excimer formation in time resolved measurements[38], no distinct excimer band is seen, in contrast to DIPHANT.

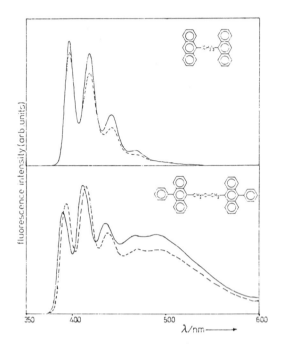

FIGURE 3

Fluorescence   spectra   of   A3A   (1,3-di(9-anthryl)propane)   and
DIPHANT in cyclohexane (————) and acetonitrile (---) at room
temperature. Note, that the excimer band of DIPHANT is not
redshifted with increasing solvent polarity.

The solvatochromic shift of the TICT band in PYREST (and other
TICT forming dyes) can be used to probe the micropolarity of the
medium, in addition to its free volume properties. This is not
possible with the excimer probe DIPHANT which lacks this redshift
(Figure 3).

Figure 4 shows the temperature dependence of the fluorescence
decay times of the short wavelength (E*) bands for probes A and B
dissolved in commercial polydimethylsiloxanes (PDMS) possessing a
macroscopic viscosity of 100 and 1000 mPas at room temperature
(S100 and S1000, respectively).

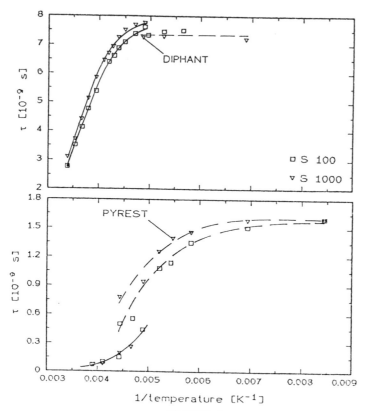

FIGURE 4

Short wavelength fluorescence decay times of PYREST (probe A) and
DIPHANT (probe B) in polydimethylsiloxane S100 (□) and S1000 (▽)
as a function of temperature. Near the crystallisation
temperature T_k (around -80 °C), PYREST shows a discontinuity due
to partial crystallization, and hysteresis effects (cooling part
(——), heating part (---)). Partial relaxation still occurs below
T_k. This is not so for DIPHANT where the excimer forming
photoreaction is completely stopped below T_k. Note that there is
very little difference for the decay times measured in S100 and
S1000.

From Figure 4, several conclusions can be drawn: i) The
excimer formation process is stopped more easily than the TICT
formation process. ii) Some TICT relaxation can be observed even
below the crystallization temperature T_k. iii) Although the
macroscopic viscosity differs by a factor of 10, the rates of the
microscopic photochemical processes are barely affected.

The temperature dependent relaxation times τ can be converted
into the rates $k(T)$ of the corresponding photochemical reaction
by the aid of eq. (1),

$$k(T) = \tau^{-1}(T) - \tau_{ref}^{-1} \tag{1}$$

where the decay time τ_{ref} is taken at a temperature where photoreaction has stopped (plateau region of Figure 4).

$k(T)$ can then be treated within the WLF-theory by using eq.2

$$\left(\log \frac{k_o}{k(T)} \right)^{-1} = -\frac{1}{c_1^o} - \frac{c_2^o}{c_1^o(T-T_o)} \tag{2}$$

where c_1^o and c_2^o are the WLF parameters, determined from least squares fitting of the experimental points to eq. (2), and are related to the free volume fraction $f_o = v_{free}/v_{total}$ at a reference temperature T_o by eq. (3a) and (3b). B is a constant and α_f is the thermal expansion coefficient of the free volume.

$$c_1^o = \frac{B}{f_o \ln 10} \tag{3a}$$

$$c_2^o = \frac{f_o}{\alpha_f} \tag{3b}$$

Figure 5 shows how values for $1/c_1$ can be derived by extrapolation of a straight line[39]. For pobes A and B, significantly different c_1 values are determined, whereas the difference in c_1 is negligible when comparing the two polymeric solvents S100 and S1000.

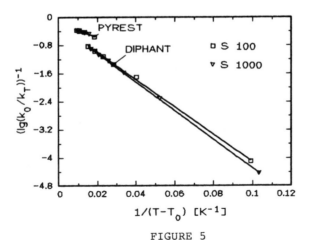

FIGURE 5

WLF plot for probes A and B in the PDMS solvents S100   (□) and S1000 (▽).

The results for the free volume fraction f_o, including those for the monomeric alkane solvents[39] are given in Table 1.

Table 1: Free volume fractions f_g = v_{free}/v_{total} at T_g derived from the intercept of WLF plots[39], for the TICT probe PYREST and the excimer probe DIPHANT in polydimethylsiloxane oils of different macroscopic viscosity and in acyclic and cyclic alkane solvents.

	PDMS S100	PDMS S1000	3MP [a)]	MCH/MCP [b)]
TICT probe	0.08	0.08	0.016	0.06
excimer probe	0.05	0.05	-	-

a) 3-methylpentane; b) methylcyclohexane/methylcyclopentane

An obvious conclusion from table 1 is that neither the TICT nor the excimer probe reflects the tenfold viscosity difference for the two PDMS oils. Thus, macroscopic viscosity, in this case, is not directly related to the microscopic viscous properties of the medium, as seen by the probe, but probably by more large-scale interactions such as different size and/or entanglement of the polymer coils.

For the alkane solvents, cyclic structures seem to possess a larger free volume than the acyclic ones.

Finally, the free volume fractions determined with the TICT probe are sizeably larger than those determined with the excimer probe. This probably reflects the different need of the two probes for reaction volume. A possible model accounting for these probe-dependent free volume effects is given in Figure 6.

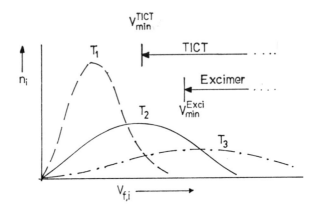

FIGURE 6

Free volume model relating to the relaxation of fluorescence probes at different temperatures ($T_1 < T_2 < T_3$). The abscissa corresponds to the size of a given microscopic free volume void $v_{f,i}$ being part of the total free volume $v_f(T)$. The ordinate gives the abundance n of $v_{f,i}$. The curves thus represent a hypothetical distribution of v_f with

$$v_f(T) = \sum_{i=1}^{\infty} n_i \cdot v_{f,i}$$

For a lowering of the temperature (---), the importance of the larger free volume sizes is reduced, and the distribution function shifts to the left. The microscopic reaction volume of the probes, however, (indicated by v_{min}, the minimal volume required) is not temperature dependent. Therefore, at lower temperature, a smaller fraction of the total free volume can be sensed. If a probe requires a larger reaction volume (excimer with respect to TICT probe), the free volume fraction sensed is reduced for similar reasons. For temperatures above T_g and/or T_k (-.-.), relaxation of the medium during the excited state reaction can become important, and sizeably larger reaction amplitudes become possible. This is indicated by the strongly broadend distribution function. Probes with different reaction volumes still sense different fractions of the total available effective free volume.

3.2. Probes usable in Polymer Glasses

The comparative results shown in Figure 4 make clear that, whereas DIPHANT as excimer probe is totally unable to sense free volume effects in the temperature region below T_k, the TICT probe shows some effect but its interpretation in a WLF treatment is doubtful because WLF theory is valid only for amorphorus phases. For an alternative approach a TICT probe with still smaller reaction volume is desirable. This is outlined in Figure 7.

Probe C: (TICT formation with pretwisting)

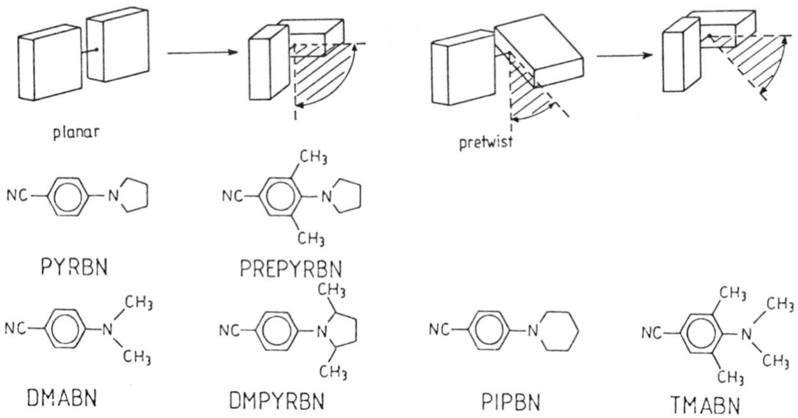

FIGURE 7
Schematic effect of pretwisting on the necessary reaction volume
of TICT formation. Some examples of TICT compounds with differing
ground state twist angle are also given.

This approach uses steric hindrance to planarity which leads
to pretwisted compounds. As an example, table 2 lists the ground
state twist angles for some of the compounds in Figure 7, as
determined by photoelectron spectroscopy. PREPYRBN, a highly
desirable compound with a predicted twist angle between 40 and
60°, has not yet been synthesized.

Table 2: Approximate mean spectroscopic ground state twist angles
<φ> as determined from photoelectron spectroscopy[37,40] for some
derivatives of DMABN.

compound	PYRBN	DMABN	DMPYRBN	PIPBN	TMABN
<φ>	0	5	10	30	60

Due to the negligible relaxation of the surrounding in glasses
during the excited state lifetime, only a small fraction of the
total number of excited molecules can undergo the TICT reaction,
namely those which happen to have an adjacent free volume void
larger than v_{min}. This small fraction can, however, easily be
detected due to its long wavelength TICT emission. Inspection of
the dual fluorescence properties of the dyes therefore yields
information on the free volumes. A pretwisted dye (with reduced

v_{min}^{TICT}) will have a better chance of relaxing to the TICT state and therefore should show an increased relative TICT contribution ϕ_A/ϕ_B in the fluorescence spectrum.

The strong dependence of ϕ_A/ϕ_B on solvent polarity (see Figure 2) necessitates the use of pairs of probes to investigate and compare polymer glasses with different micropolarity. Ideally, a pair with exactly the same size of the rotors would be desirable, and these should be as rigid as possible.

PYRBN, with a somewhat strained 5-ring system, seems to possess less degrees of internal flexibility than DMABN (Table 2), and therefore the pair PYRBN/PREPYRBN should constitute a very good pair of probes.

As PREPYRBN is not yet available, we will take DMABN ($<\phi>$=5°) and PIPBN ($<\phi>$ = 30°) instead. Figure 8 shows spectra of these two compounds in polymer glasses of different polarity, the less polar poly(methylmethacrylate) PMMA and its more polar hydroxylated counterpart poly(hydroxyethylmethacrylate) PHEMA.

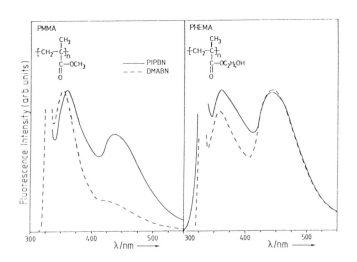

Figure 8
Dual fluorescence spectra of PIPBN (———) and DMABN (———) embedded in glassy polymers. Excitation wavelength was 330 nm [41].

Due to the higher medium polarity, the relative TICT contribution for both compounds is larger in PHEMA. But there are subtle changes:

In PMMA, DMABN shows a smaller relative TICT contribution than PIPBN, in PHEMA a larger one. This is an indication that the matrix effect hindering TICT formation is stronger for PIPBN when comparing PHEMA to PMMA. With the assumption, that the reaction volume is somewhat larger in PIPBN (reduced angular gap but considerably larger rotor size), this leads to the conclusion that PHEMA possesses a smaller free volume than PMMA.

Still more information can be gained from a probe with a different type of photoreaction, that of formation of the Dewar isomer [42]. Figure 9 shows the relevant structures.

Probe D: (Dewar formation with butterfly folding)

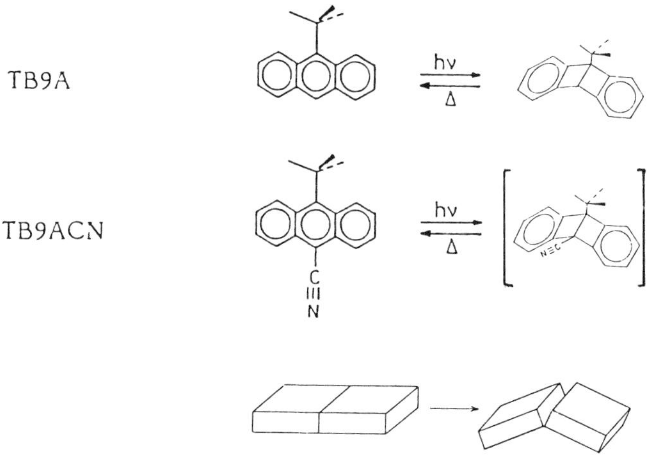

TB9A

TB9ACN

FIGURE 9
Structures of probes with excited state Dewar formation, and the relevant butterfly folding process. For TB9ACN, the Dewar product has not yet been isolated, but the excited state reaction is even faster than for TB9A.

Figure 10 displays the reaction profile based on quantum chemical calculations[45].

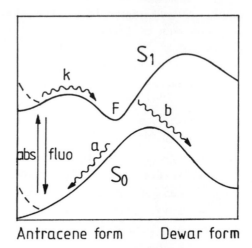

Antracene form   Dewar form

FIGURE 10

Schematic representation of S_0 and S_1 energy profiles for Dewar formation in TB9A and TB9ACN. The excited state funnel F is very close to the ground state surface and leads to fluorescence quenching (rate constant k). Most of the molecules return to the anthracene form via pathway a, only few to the Dewar form (pathway b), because F is placed to the left of the ground state barrier. The steric effect of the tert-butyl substituent is indicated by the broken line. Without this "prefolding", Dewar formation is not observed.

Relaxation to the funnel F halfway between the anthracene and Dewar anthracene structure leads to fluorescence quenching of the anthracene form (E* in the kinetic scheme), and the corresponding rate constant k can be extracted similarly as shown in the cases of probes A and B. In liquid solution, k is so fast, that virtually all fluorescence is quenched in the case of TB9ACN. In a polymer glass at low temperature, the lifetime approaches that of an anthracene derivative without Dewar formation. But a net shortening of the lifetimes is seen as the temperature is increased. This is exemplified in Figure 11.

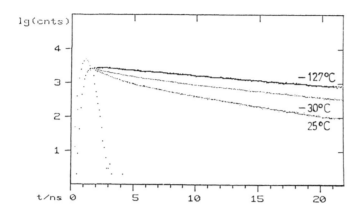

FIGURE 11

Fluorescence decay curves of TB9ACN in PMMA at different temperatures. Only at the lowest temperature, the decay is monoexponential (τ = 14.9 ns). For room temperature, the nonexponential decay can be well described by a triexponential model (τ_1 = 0.08 ns (A_1=0.83), τ_2 = 1.9 ns (A_2 = 0.1), τ_3 = 9.9 ns (A_3 = 0.07)).

The clearly visible nonexponential decay behaviour in polymers at increased temperatures[44,45] can be explained by the inhomogeneous distribution of free volume v_f around the probe molecules TB9ACN (see also the model in Figure 6). For molecules with a very small neighbouring free volume void v_f, k is reduced to such an extent that only the intrinsic E* decay (14.9 ns at 146 K) remains. Other molecules possess surroundings with a larger available reaction space, and k is increased. It can be seen that in this way the distribution of free volume (Figure 6) is translated into a distribution of decay times yielding the observed nonexponential decay behaviour. In principle, the underlying decay time distributions can be recovered from the nonexponential decays[46-48] but very high quality data are needed.

In Figure 12, we display our preliminary results obtained by fitting a symmetrical Gaussian lifetime distribution which is truncated at the longest observed decay time (15 ns).

W. Rettig, R. Fritz and J. Springer

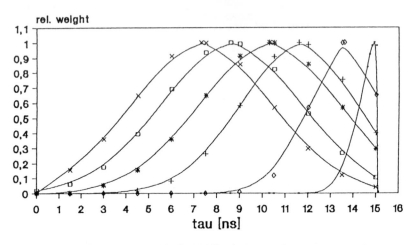

FIGURE 12
Decay time distribution for different temperatures recovered by
fitting Gaussian lifetime distributions to the observed
nonexponential decays of TB9ACN in PMMA.

It can be seen that the most probable lifetime $<\tau>$ becomes
shorter and the width of the distribution broader as the
temperature is increased. For similar temperatures in liquid
solution, the lifetime is unmeasurably short ($\tau <$ 0.1   ns).
Therefore, $<\tau>$ and the width of the distribution reflect the
effect of the surrounding in slowing down k and not a classical
activation barrier effect.

One obvious and important conclusion from these results is
that there is indeed still a reduction of free volume (a
narrowing of the free volume distribution function in Figure 6)
on lowering the temperature, even far below T_g. This is an
addition to the assumptions of the WLF model (frozen segmental
motion below T_g) and points to the possibility for further
refinements.

ACKNOWLEDGEMENT
    The authors wish to thank Professors H. Bouas-Laurent
(Bordeaux), K. Muellen (Mainz) and H. Dreeskamp (Braunschweig)
for the gift of the compounds DIPHANT, A3A and TB9ACN,

respectively. A supply of the PDMS oils S100 and S1000 from Rhône-Poulenc is also gratefully acknowledged, as well as the support by the Bundesministerium für Forschung und Technologie (projects 05 314 FAI5 and 05 414 FAB1) and by the Techn. Univ. Berlin (IFP 4).

REFERENCES

1) Z.R. Grabowski, K. Rotkiewicz, A. Siemiarczuk, D.J. Cowley and W. Baumann, Nouv. J. Chim. 3 (1979) 443.

2) W. Rettig, Angew. Chem. Int. Edit. Engl. 25 (1986) 971.

3) E. Lippert, W. Rettig, V. Bonačić-Koutecký, F. Heisel and J.A. Miehé, Advan. Chem. Phys. 68 (1987) 1.

4) J.M. Hicks, M.T. Vandersall, Z. Babarogic and K.B. Eisenthal, Chem. Phys. Lett. 116 (1985) 18; J.M. Hicks, T.T. Vandersall, E.V. Sitzmann, and K.B. Eisenthal, Chem. Phys. Lett. 135 (1987) 413.

5) F. Heisel and J.A. Miehé, Chem. Phys. Lett. 100 (1983) 183; Chem. Phys. 98 (1985) 233.

6) W. Rettig, M. Vogel, E. Lippert and H. Otto, Chem. Phys. 103 (1986) 381.

7) S.R. Meech and D. Phillips, Chem. Phys. Lett. 116 (1985) 262.

8) B. Bagchi, G.R. Fleming and D.W. Oxtoby, J. Chem. Phys. 78 (1983) 7375.

9) E. Butty and P. Suppan, Polymer Photochemistry 5 (1984) 171; P. Suppan, Chimia 42 (1988) 320; K.A. Al-Hassan and T. Azumi, Chem. Phys. Lett. 145 (1988) 49 and 146 (1988) 121.

10) R. Hayashi, S. Tazuke, and C.W. Frank, Macromolecules 20 (1987) 983 and Chem. Phys. Lett. 135 (1987) 123.

11) S. Tazuke, R.K. Guo, and R. Hayashi, Macromolecules 21 (1988) 1046.

12) S. Tazuke, R.K. Guo, and T. Ikeda, J. Phys. Chem. 94 (1990) 1408.

13) S. Tazuke, R.K. Guo, T. Ikeda, and T. Ikeda, Macromolecules 23 (1990) 1208.

14) M. Vogel and W. Rettig, Ber. Bunsenges. Phys. Chem. 89 (1985) 962.

15) W. Rettig, Appl. Phys. B45 (1988) 145.

16) M. Vogel and W. Rettig, Ber. Bunsenges. Phys. Chem. 91 (1987) 1241.

17) R.O. Loutfy and K.Y. Law, J. Phys. Chem. 84 (1980) 2803.

18) R.O. Loutfy, Macromolecules 14 (1981) 270.

19) K.Y. Law, Chem. Phys. lett. 75 (1980) 545.

20) K.Y. Law, Photochem. and Photobiol. 33 (1981) 799.

21) R.O. Loutfy and B.A. Arnold, J. Phys. Chem. 86 (1982) 4205.

22) R.O. Loutfy, Macromolecules 16 (1983) 678.

23) R.O. Loutfy, Pure & Appl. Chem. 58 (1986) 1239.

24) V. Bonačić-Koutecký, J. Koutecký and J. Michl, Angew. Chem. Int. Ed. Engl. 26 (1987) 170.

25) V. Bonačić-Koutecký and J. Michl, J. Am. Chem. Soc. 107 (1985) 1765.

26) W. Rettig in "Modern Models of Bonding and Delocalization" (Molecular structure and energetics, vol. 6), J. Liebman and A. Greenberg, ed., VCH Publ., New York, 1988, chapter 5, p. 229.

27) W. Rettig and G. Wermuth, J. Photochem. 28 (1985) 351.

28) E. Pajot-Augy, L. Bokobza, L. Monnerie, A. Castellan and H. Bouas-Laurent, Macromolecules 17 (1984) 1490.

29) L. Bokobza, C. Pham-Van-Cang, C. Giordano, L. Monnerie, J. Vandendriessche, F.C. De Schryver and E.G. Kontos, Polymer 28 (1987) 1867.

30) L. Bokobza, C. Pham-Van-Cang, C. Giordano, L. Monnerie, J. Vandendriessche and F.C. DeSchryver, Polymer 29 (1988) 251.

31) L. Bokobza, C. Pham-Van-Cang, L. Monnerie, J. Vandendriessche and F.C. De Schryver, Polymer 30 (1989) 45.

32) M.L. Williams, R.F. Landel, J.D. Ferry, J.Am.Chem.Soc. 77, (1955) 3701

33) I.D. Ferry, Viscoelastic properties of polymers, 3th Ed. Wiley & Sons, Inc., New York (1976)

34) W. Rettig and V. Bonačić-Koutecký, Chem. Phys. Lett. 62 (1979) 115.

35) G. Wermuth, Z. Naturforsch. 38a (1983) 368.

36) G. Wermuth, Z. Naturforsch. 38a (1983) 641.

37) W. Rettig and R. Gleiter, J. Phys. Chem. 89 (1985) 4676.

38) W. Rettig, unpublished.

39) R. Fritz, W. Rettig and J. Springer, to be published.

40) W. Rettig, D. Braun, P. Suppan, E. Vauthey and K. Rotkiewicz, to be published.

41) K.A. Al-Hassan and W. Rettig, Chem. Phys. Lett. 126 (1986) 273.

42) B. Jahn and H. Dreeskamp, Ber. Bunsenges. Phys. Chem. 88 (1984) 42.

43) W. Rettig, D. Braun, W. Carl and P. Kapahnke, to be published.

44) S. Hirayama and Y. Shimono, J. Chem. Soc., Faraday Trans. 2, 80 (1984) 941.

45) S. Hirayama, T. Inoue, Y. Ito, and T. Matsura, Chem. Phys. Lett. 115 (1985) 79.

46) D.R. James and W.R. Ware, Chem. Phys. Lett. 120 (1985) 455.

47) D.R. James and W.R. Ware, Chem. Phys. Lett. 126 (1986) 7.

48) D.R. James, Y.-S. Liu, P. DeMayo and W.R. Ware, Chem. Phys. Lett. 120 (1985) 460.

Photochemical Processes in Organized Molecular Systems
K. Honda (Editor-in-Chief)
© Elsevier Science Publishers B.V., 1991

MULTICOMPONENT FLUORESCENCE DECAY ANALYSIS IN INTRAMOLECULAR EXCIMER FORMATION WITH DIPYRENYLALKANES

Klaas A. ZACHARIASSE, Gerd DUVENECK, Wolfgang KÜHNLE, Uwe LEINHOS, and Peter REYNDERS

Max-Planck-Institut für biophysikalische Chemie, Abteilung Spektroskopie, Am Fassberg, Postfach 2841, D-3400 Göttingen, Federal Republic of Germany

Intramolecular excimer formation with a series of 1,n-di(1-pyrenyl)alkanes 1Py(n)1Py for n = 1–16, 22, and 32 is discussed. Next to the chain-length dependence of the fluorescence behaviour, the monomer and excimer fluorescence decays are treated, with special emphasis on the triple-exponential decays of 1Py(3)1Py. The various kinetic schemes comprising three excited state species, either two monomers and one excimer or one monomer and two excimers, are discussed in detail. It is concluded that excimer formation of 1Py(3)1Py can be completely described with the scheme DMD: a single group of excited state monomers $^1M^*$ in connection with two structurally different excimers $^1D_1^*$ and $^1D_2^*$. From the temperature dependence of the fluorescence decays (decay times and amplitudes) all rate constants in scheme DMD can be determined, giving values for the excimer binding enthalpies ΔH and the changes in entropy ΔS. For 1,3-di(2-pyrenyl)propane, 2Py(3)2Py, only one excimer is formed, as concluded from the double-exponential monomer and excimer fluorescence decays and supported by singlet excited state absorption spectra.

1. INTRODUCTION

Intramolecular excimer formation with diarylalkanes in dilute solution was first reported by Hirayama[1] for 1,3-diphenylpropane, in an investigation of a series of 1,n-diphenylalkanes with $n = 1 - 6$. Excimer emission was only detected for $n = 3$, an observation that led to the formulation of the Hirayama-rule. This rule states that intramolecular excimer formation in molecules A(n)A, where the end groups are linked with n CH_2-groups, only is possible when a trimethylene-chain is present. Later, Chandross[2] studied various dinaphthylalkanes: 1Na(n)1Na with $n = 2$, 3, 4, and 2Na(3)2Na as well as 1Na(3)2Na. Strong excimer emission was only observed for the symmetric dinaphthylpropanes, whereas for 1Na(4)1Na a weak excimer fluorescence was found at low temperatures. These studies, and similar experiments on 1,3-di(N-carbazolyl)propane by Klöpffer[3] and on 1,n-di(N-carbazolyl)alkanes with $n = 1 - 6$ by Johnson,[4] in fact supported the validity of the $n = 3$ rule.[5] However, for the series of 1,n-di(1-pyrenyl)alkanes, 1Py(n)1Py, with $n = 1 - 16$, 22, and 32, intramolecular excimer emission was found for all n except $n = 7$ (see below).[6,7] Similar observations were reported on 1,n-di(9-anthryl)alkanes with $n = 2 - 10$ by Bouas-Laurent[8] and on 1,n-di(dimethylamino)alkanes with $n = 1 - 13$, 16, 18, and 20, by Halpern.[9] This then led to the conclusion that the $n = 3$ rule had to be formulated in a less exclusive sense,[6] indicating that intramolecular excimer formation in diarylalkanes, although optimal for $n = 3$, is very well possible for other n. The specific situation will depend on the physical properties of the aromatic hydrocarbons at the ends of the chain, which determine the stability of the excimers under the energetic restrictions posed by the different alkane chains.[10] It is not surprising that the dipyrenylalkanes proved to be the compounds of choice to observe excimer formation over a broad range of values for n, as pyrene forms the by far most stable intermolecular excimer.[11,12] This molecule, moreover, possesses an exceptionally long fluorescence lifetime (450 ns in cyclohexane[12]), facilitating fluorescence quenching.

It is of interest to point out here, that the $n = 3$ rule of optimal excimer formation and fluorescence quenching for the diarylalkanes, cannot be automatically carried over to compounds with different chains. For example, for the 1,n-di(1-pyrenecarboxy)alkanes, $1PyCOO(CH_2)_n$-$OOC1Py$, with $n = 1 - 16$, 22, and 32, the largest value for the excimer-to-monomer fluorescence intensity ratio is observed for $n = 5$, in methylcyclohexane (MCH) over a large temperature range.[13] In comparable studies on the 1,n-di(1-pyrenemethyl)alkanedioic esters $(1PyCH_2OCO(CH_2)_nOCOCH_21Py)$, in contrast, the most efficient excimer formation was found for $n = 2$.[14]

In general, the following three features are significant in investigations with linked systems A(n)A, as compared to studies involving intermolecular reactions: (a) low concentrations (less than $10^{-5}M$) are possible, (b) the relatively small values of ΔS,[12,15] the entropy difference involved in excimer formation, facilitates the intramolecular reaction. This can lead to efficient excimer formation even when unfavourable stabilization energies ($-\Delta H < 25$ kJ/mol) make the intermolecular process difficult to observe,[3,15,16] and (c) the relative configurations of the end groups A are directed, but also restricted, by the chain.[17]

2. RESULTS AND DISCUSSION

2.1. Intramolecular Excimer Formation with Dipyrenylalkanes

In the series $1Py(n)1Py$ with $n = 1 - 16$, 22, and 32, intramolecular excimer formation occurs for all n except $n = 7$.[7] The fluorescence spectra of these compounds consist of a structureless excimer emission band and a residual monomer fluorescence, see Figure 1 for $1Py(22)1Py$ in MCH.

For $n = 7$, an excimer emission cannot be ascertained from an inspection of its fluorescence spectrum, which is the same as that of the model substance 1-methylpyrene (1MePy).[7] Further, the decay time of the monomer fluorescence of $1Py(7)1Py$ is identical to the lifetime of 1MePy over the entire temperature range (Figure 2a). Nevertheless, a growing-in is observed at 500 nm, a wavelength where excimer emission is found for the other dipyrenylalkanes, see Figure 2b. This can, however, be attributed to intermolecular excimer formation of $1Py(7)1Py$, similar to experiments with 1MePy at 5×10^{-6} M.[18]

In Figure 3a the excimer-to-monomer fluorescence quantum yield ratio Φ'/Φ for $1Py(n)1Py$ in MCH at 25 °C is plotted as a function of chain length n. As can be seen from this plot, a weak excimer emission is observed for $n = 6$ and 8.[7] The chain length dependence of the rate constant of excimer formation k_a (see below) is depicted in Figure 3b.[19]

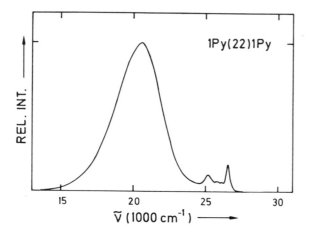

FIGURE 1

Fluorescence spectrum of 1,22-di(1-pyrenyl)docosane (1Py(22)1Py) in methylcyclohexane (5×10^{-6} M) at 25°C.

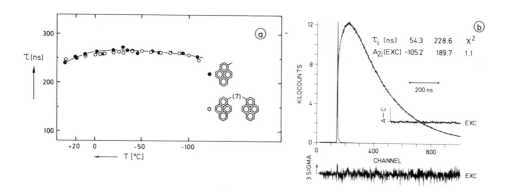

FIGURE 2

(a) Fluorescence decay times (at 376 nm) of 1-methylpyrene (1MePy) and of 1,7-di(1-pyrenyl)-heptane (1Py(7)1Py) in methylcyclohexane (1×10^{-5} M) as a function of temperature.
(b) Fluorescence response function (at 500 nm) of 1Py(7)1Py in methylcyclohexane at 25 °C. Note the growing-in of the curve, which is due to intermolecular excimer formation, see text.

The pattern of the plots in Figure 3 is reminiscent, in its general shape, of plots representing the chemical yield of ring-closure reactions, such as, for example, the cyclic ketone formation from dinitriles.[20] These chemical reaction-yields display a sharp minimum around $n = 10$, and present an alternating pattern for larger n. This behaviour is analogous to that observed for the cycloalkanes, when the difference in the heats of combustion between a ring-methylene and an aliphatic CH_2-group is plotted against the number of methylene groups,[10,21] see Figure 4.

The discrepancy between the minimum occurring at $n = 7$ for the intramolecular excimer formation with 1Py(n)1Py, and the minimum at $n = 10$ in the case of the ring closure reactions,

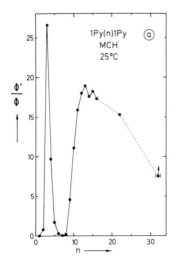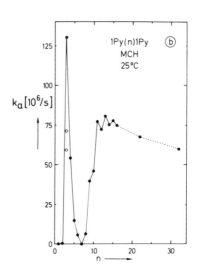

FIGURE 3

(a) Excimer-to-monomer fluorescence quantum yield ratio Φ'/Φ, and (b) the rate constant of excimer formation k_a, for the 1,n-di(1-pyrenyl)alkanes (1Py(n)1Py) in methylcyclohexane at 25 °C as a function of chain length n.

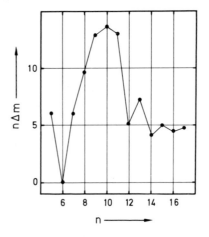

FIGURE 4

Thermochemical strain in cycloalkanes with n CH_2-groups. Δm is the difference in heat of combustion between a methylene group in the cycloalkanes and an aliphatic CH_2-group. See refs 10 and 21.

has been ascribed[7] to the fact that for a chemical reaction the atoms have to approach to a distance of one C-C bond length. In excimer formation, on the other hand, the end groups only have to reach a configuration with an interplanar distance of about 0.35 nm,[12] requiring three CH_2-groups less. These conditions are schematically illustrated by the following pair of molecular structures, showing the difference between a cyclohexane ring and 1Py(3)1Py in the symmetric excimer configuration:

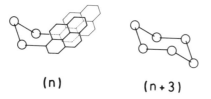

(n) $(n+3)$

The appearance of the minimum in the chemical yield, as well as in the efficiency of excimer formation, is due to the thermochemical strain associated with medium-sized rings (Figure 4).[10] It is seen that the highest strain in the cycloalkanes is observed for $n = 10$. Further, a higher strain is observed for the odd-membered rings. This explains, taking into account the difference of three CH_2-groups as described above, that intramolecular excimer formation is more efficient for odd- rather than even-membered 1Py(n)1Py systems with n larger than 9 (see Figure 3).

2.2. Presence of Ground State Dimers

An important point in the discussion of the kinetics of excimer formation with the di(1-pyrenyl)alkanes and other linked compounds, is the question whether this process is completely dynamic, occurring without the intermediacy of inter- or intramolecular dimers in the ground state. As shown for 1Py(4)1Py (Figure 5), it appears to be possible completely to freeze out excimer emission and therefore excimer formation. This means that excimer formation pathways not requiring the diffusion of the two pyrene moieties do not have to be taken into account. Similar observations were made for all other members of the 1Py(n)1Py family.[22] This conclusion is supported by ^{1}H NMR measurements and by an analysis of the double- and triple-exponential excimer fluorescence response functions of the dipyrenylalkanes.[23] These findings are the basis for the assumption that the excimer concentration is equal to zero at the moment of excitation, a boundary condition that will be used in a further section when solving the differential equations describing excimer formation.

2.3. Analysis of Multicomponent Fluorescence Response Functions

The kinetics of intramolecular excimer formation with the dipyrenylalkanes can best be studied employing time-dependent measurements, such as those utilizing the method of time-correlated single-photon counting (SPC).[19,23] The analysis of the fluorescence response functions (monomer and excimer) of 1Py(3)1Py and other dipyrenylalkanes will now be discussed. Multicomponent decay analysis in polymeric systems[24] will not be considered here.

2.4. Triple-Exponential Decay of the Fluorescence of 1,3-Di(1-pyrenyl)propane

The fluorescence decays of the monomer, $i_M(t)$, and the excimer, $i_D(t)$, of 1Py(3)1Py cannot be fitted with two exponentials. Only with three exponentials acceptable fits are obtained (Figure 6).[19a]

This has been observed to be the case, over an extended temperature range, for a variety

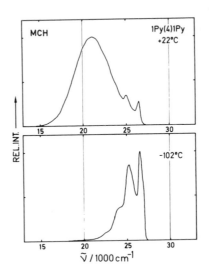

FIGURE 5

Fluorescence spectra of 1,4-di(1-pyrenyl)butane (1Py(4)1Py) in methylcyclohexane (1×10^{-5} M) at 22 and $-102\ ^{\circ}$C. Note the complete absence of excimer emission at the lower temperature.

of solvents of different viscosities: alkanes (from n-pentane to liquid paraffin), acetonitrile, ethanol, tetrahydrofuran, and toluene.[19a,25]

$$i_M(t) = A_{11}e^{-t/\tau_1} + A_{12}e^{-t/\tau_2} + A_{13}e^{-t/\tau_3} \qquad (\tau_1 > \tau_2 > \tau_3) \tag{1}$$

$$i_D(t) = A_{21}e^{-t/\tau_1} + A_{22}e^{-t/\tau_2} + A_{23}e^{-t/\tau_3} \tag{2}$$

It is important to note that the decay parameters $1/\tau_i$ obtained for the monomer are practically identical to those observed for the excimer.[19,27] As will be elaborated below, it is only under this condition that the excited state behaviour of Py(3)Py can be completely described with a kinetic scheme comprising three, and not more than three, excited state species. As an example, the three decay time parameters $1/\tau_1$, $1/\tau_2$, and $1/\tau_3$, of 1Py(3)1Py in n-heptane are plotted as a function of temperature in Figure 7.

It is seen that two of the decay parameters, $1/\tau_1$ and $1/\tau_2$, have a relatively small temperature dependence, whereas the largest decay constant $1/\tau_3$ strongly increases with temperature. This general trend of the temperature dependence of the decay parameters has been observed for 1Py(3)1Py in a variety of solvents, irrespective of solvent nature and viscosity.[19,25]

2.5. Kinetic Schemes Comprising Three Excited State Species

In intermolecular excimer formation[12,19b] the presence of two excited state species, a monomer and an excimer, leads to the observation of double-exponential decays with identical time constants τ_i for excimer and monomer. Similarly, when in the intramolecular case triple-exponential decay is observed featuring identical decay times (τ_1, τ_2, τ_3) for excimer and monomer, it can be concluded that three, and only three, excited state species are involved in

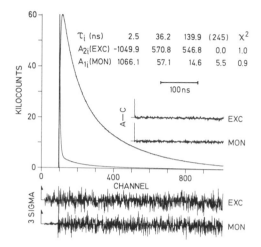

FIGURE 6

Monomer (at 376 nm) and excimer (at 520 nm) fluorescence response functions of 1,3-di(1-pyrenyl)propane (1Py(3)1Py) in n-heptane at 60.5 °C. The monomer and excimer decays are analyzed simultaneously (global analysis), see ref 34. The decays were obtained using picosecond laser excitation (298 nm), with 6×10^4 peak counts in 900 effective channels and 0.486 ns/channel. The values for the decay times (τ_3, τ_2, τ_1) and their preexponential factors A_{2i} (excimer) and A_{1i} (monomer) are given, see eq 1 and 2. The fixed time in parentheses is equal to the lifetime τ_o of 1-ethylpyrene, measured under the same conditions. The weighted deviations, expressed in σ (expected deviations), the autocorrelation functions A-C and the values for χ^2 are also indicated.

the excimer formation process. For a system such as 1Py(3)1Py two possibilities then exist: either two excited state monomers (M^*) and one excimer (D^*), or one monomer and two excimers.[19] In both cases, the two monomers as well as the two excimers will necessarily be kinetically-distinct species. The monomers would differ in the chain conformation, requiring a different number of bond rotations to reach the excimer configuration, whereas the excimers will have distinct relative geometric orientations of the aromatic end groups:

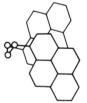

1Py(3)1Py

2.5.1. Two Monomers

In the case of two different monomers, M_1^* and M_2^*, the following kinetic scheme can be set up for the most general situation, where all reactions (back and forth) between the three excited state species have been taken into account:

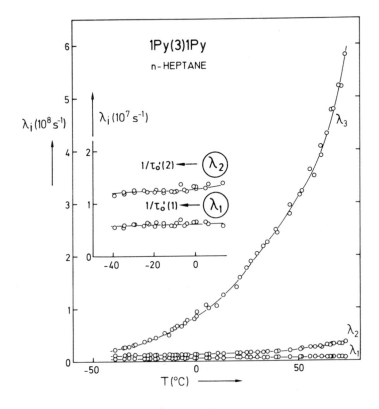

FIGURE 7

Decay parameters λ_i ($= 1/\tau_i$), describing the triple-exponential excimer fluorescence decay $i_D(t)$ of 1,3-di(1-pyrenyl)propane (1Py(3)1Py) in n-heptane (5×10^{-6} M) as a function of temperature.

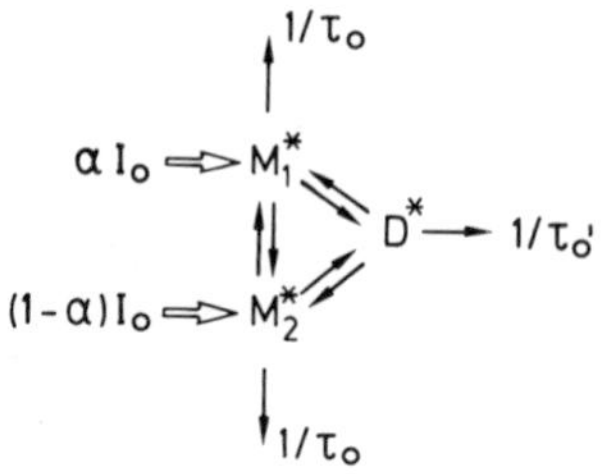

Scheme M2D

In this scheme, τ_o (monomer) and τ_o' (excimer) are the excited state lifetimes, whereas I_o is the intensity of the excitation light of which a fraction α produces M_1^*. The other rate constants will be specified below. For molecules such as 1Py(3)1Py, M_1^* can be identified as the chain conformer needing one bond rotation to reach the excimer structure D^*, whereas two such rotations are required for M_2^*. It can then be inferred[19,25] that the direct pathway from M_2^* to D^* in Scheme M$_2$D can be neglected, as it would involve two simultaneous bond rotations.[26]

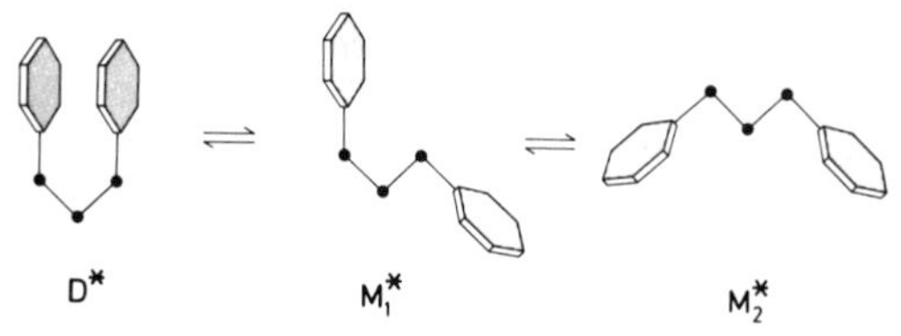

This breaks up the triangular Scheme M$_2$D into Scheme MMD:

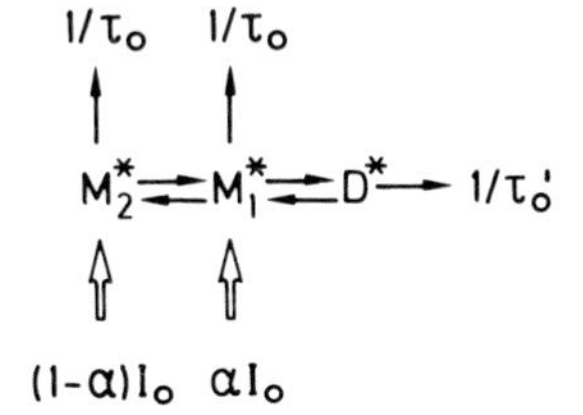

Scheme MMD

Another possible variation of Scheme M$_2$D, one in which the connection between the two excited state monomers is blocked, is Scheme MDM:

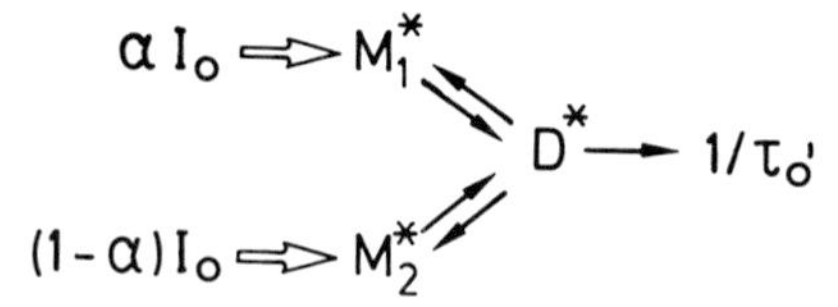

Scheme MDM

It is difficult to find an explanation for the absence, required by Scheme MDM, of a reversible reaction between the two monomers M_1^* and M_2^*, as for a propane chain these conformers can differ only in the number of bond rotations needed to reach the excimer structure, as discussed above.

2.5.2. Two Excimers

When two structurally, and hence kinetically, distinct excimers (D_1^* and D_2^*) are formed from a single group of excited state monomers, Scheme MD$_2$ is applicable, in which all reactions between D_1^*, D_2^* and M^* are allowed:

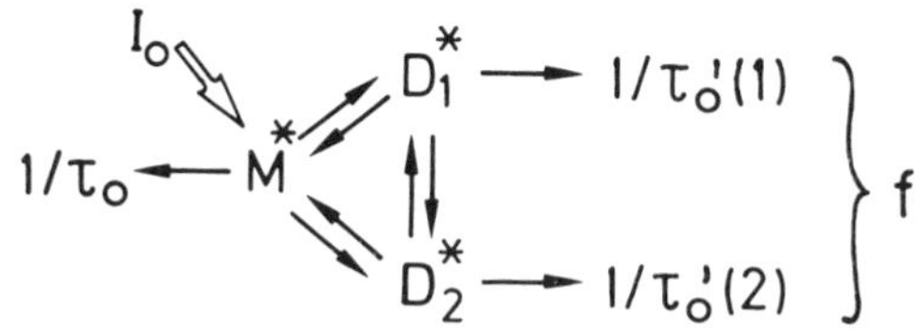

Scheme MD$_2$

This general Scheme MD_2 can be simplified in a manner similar to that presented in the previous section for Scheme M_2D, resulting in the two schemes DMD and MDD. In Scheme DMD it is assumed that the interconversion of the two excimers D_1^* and D_2^* necessarily involves the complete breaking up of the excimer structure. This is equivalent to stating that this process has to pass through the momomer M^*:

$$
\left.
\begin{array}{l}
I_o \searrow \quad D_1^* \longrightarrow 1/\tau_o'(1) \\
1/\tau_o \longleftarrow M^* \\
\qquad\qquad D_2^* \longrightarrow 1/\tau_o'(2)
\end{array}
\right\} f \qquad\qquad \text{Scheme DMD}
$$

For the alternative Scheme MDD, a justification for only allowing the reaction from the kinetically uniform group of monomer conformers M^* to one of the excimers (e.g. D_2^*) is not obvious. The scheme is presented here for the sake of completeness (cf. Scheme MDM):

$$
\left.
\begin{array}{l}
I_o \searrow \quad D_1^* \longrightarrow 1/\tau_o'(1) \\
1/\tau_o \longleftarrow M^* \quad \updownarrow \\
\qquad\qquad D_2^* \longrightarrow 1/\tau_o'(2)
\end{array}
\right\} f \qquad\qquad \text{Scheme MDD}
$$

2.6. Calculation of Decay Parameters and Amplitudes for Scheme DMD

The methodology by which the decay parameters $1/\tau_i$ and their amplitudes A_{ij} (eqs. 1 and 2) can be derived for one of the four kinetic schemes of reduced complexity introduced above, will be presented. The treatment given here for DMD, leads to basically similar results, *mutatis mutandis*, for the three other schemes.[13] In Scheme DMD, repeated below, k_a and k_d are the rate constants for excimer formation and dissociation, respectively.

$$
\left.
\begin{array}{l}
M \\
\quad \searrow I_o \quad k_a(1) \quad D_1^* \longrightarrow 1/\tau_o'(1) \\
\qquad\qquad\qquad k_d(1) \\
1/\tau_o \longleftarrow {}^1M^* \\
\qquad\qquad\qquad k_d(2) \\
\qquad k_a(2) \quad D_2^* \longrightarrow 1/\tau_o'(2)
\end{array}
\right\} f \qquad\qquad \text{Scheme DMD}
$$

When all (intramolecular, i.e., concentration independent) pathways for formation and disappearance of M^*, D_1^*, and D_2^*, are taken into account in a kind of book-keeping method, the following coupled differential equations are obtained for the time dependence of the concentrations of the three species after flash excitation[27]:

$$d[M^*]/dt = -\{k_a(1) + k_a(2) + 1/\tau_o\}[M^*] \quad\quad +k_d(1) \quad\quad [D_1^*] \quad\quad +k_d(2) \quad\quad [D_2^*] \quad (3)$$

$$d[D_1^*]/dt = \quad\quad k_a(1) \quad\quad [M^*] - \{k_d(1) + 1/\tau_o'(1)\}[D_1^*] \quad\quad +0 \quad\quad [D_2^*] \quad (4)$$

$$d[D_2^*]/dt = \quad\quad k_a(2) \quad\quad [M^*] \quad\quad +0 \quad\quad [D_1^*] - \{k_d(2) + 1/\tau_o'(2)\}[D_2^*] \quad (5)$$

These three equations, with the boundary conditions $[M^*] = [M_o^*]$, and $[D_1^*] = [D_2^*] = 0$ at the moment of excitation ($t = 0$), result in the following equations for the time dependence of the concentrations of M^*, D_1^*, and D_2^*:

$$[M^*] = A_{11}e^{-t/\tau_1} + A_{12}e^{-t/\tau_2} + A_{13}e^{-t/\tau_3} = i_{M^*}(t)/k_f \quad (6)$$

$$[D_1^*] = A_{21}e^{-t/\tau_1} + A_{22}e^{-t/\tau_2} + A_{23}e^{-t/\tau_3} = i_{D_1^*}(t)/k_f'(1) \quad (7)$$

$$[D_2^*] = A_{31}e^{-t/\tau_1} + A_{32}e^{-t/\tau_2} + A_{33}e^{-t/\tau_3} = i_{D_2^*}(t)/k_f'(2) \quad (8)$$

The set of the three differential equations (eq. 3-5) leads, with the substitutions

$X = k_a(1) + k_a(2) + 1/\tau_o$,

$Y = k_d(1) + 1/\tau_o'(1)$, and

$Z = k_d(2) + 1/\tau_o'(2)$,

to a (3,3)-determinant:

$$\begin{vmatrix} (-X + 1/\tau) & k_d(1) & k_d(2) \\ k_a(1) & (-Y + 1/\tau) & 0 \\ k_a(2) & 0 & (-Z + 1/\tau) \end{vmatrix} = 0 \quad (9)$$

This determinant gives rise to the cubic equation[28]:

$$\lambda^3 + p\lambda^2 + q\lambda + r = 0 \quad (10)$$

with $p = -(X + Y + Z)$,

$q = (XY + XZ + YZ) - k_a(2)k_d(2) - k_a(1)k_d(1)$, and

$r = XYZ - k_a(2)k_d(2)Y - k_a(1)k_d(1)Z$.

The eigenvalues $1/\tau_i$ and the corresponding eigenvectors (the amplitudes A_{ij}, see eqs.1 and 2) are then determined either analytically by matrix multiplication[13] or employing a computer library program.[29]

2.7. Available Experimental Information

In Scheme DMD eight unknown parameters are involved. Seven of these have already been introduced above: the rate constants of excimer formation $k_a(1)$ and $k_a(2)$, the rate constants of thermal excimer dissociation $k_d(1)$ and $k_d(2)$, two excimer lifetimes $\tau_o'(1)$ and $\tau_o'(2)$, and the monomer lifetime τ_o. In addition, there is still a further unknown. This originates from the fact that the emission bands of the two excimers D_1^* and D_2^* cannot be separated spectroscopically.[19,27] The total excimer fluorescence response function $i_{DMD}(t)$ is therefore a superposition of the two functions for the excimers D_1^* and D_2^* (eqs. 7 and 8):

$$i_{DMD}(t) = k_f'(1)[D_1^*] + k_f'(2)[D_2^*] \quad (11)$$

$$i_{DMD}(t)/k_f'(1) = (A_{21} + f \cdot A_{31})e^{-t/\tau_1} + (A_{22} + f \cdot A_{32})e^{-t/\tau_2} + (A_{23} + f \cdot A_{33})e^{-t/\tau_3} \quad (12)$$

The parameter f in eq. 12 stands for the ratio of the radiative rate constants of D_1^* and D_2^*:

$f = k'_f(2)/k'_f(1)$.[27]

From the experimental point of view, on the other hand, seven independent pieces of information can be obtained from the triple-exponential decay curves: three decay times (τ_1, τ_2, τ_3), two monomer amplitude ratios, one excimer amplitude ratio,[30] and τ_o (the monomer lifetime, determined in dilute solution for a model compound such as 1MePy[27,31]). Clearly, one piece of information is lacking before one can determine the eight unknowns involved in Scheme DMD.

This missing piece of information can be obtained by inspection of a plot of the reciprocal decay parameters, i.e. the decay times τ_i, as a function of temperature. In the case of intermolecular excimer formation, one of these decay times becomes equal to the excimer lifetime τ'_o when the thermal back reaction becomes negligible with respect to the reciprocal excimer lifetime: i.e., when $k_d << 1/\tau'_o$.[19b] The same phenomenon takes place in the more complicated case of triple-exponential decay associated with the intermolecular excimer formation of 1Py(3)1Py (Figure 7). It is seen that for 1Py(3)1Py in n-heptane, $1/\tau_1$ and $1/\tau_2$ become practically constant at low enough temperatures, below 20 and $-10°C$, respectively. In this manner one can obtain an additional parameter δ which is equal to the difference the reciprocal excimer lifetimes: $\delta = 1/\tau'_o(2) - 1/\tau'_o(1)$. The eight unknowns forming part of Scheme DMD can now be calculated from the information contained in the monomer and excimer fluorescence response functions. When these response functions are measured as a function of temperature, the activation energies E_a and E_d as well as the thermodynamic parameters (ΔH and ΔS) of excimer formation can be determined. Details of this procedure will be published elsewhere.[32]

2.8. Fitting the Experimental Data to the Parameters of the Kinetic Schemes

It will now be discussed how to determine which one of the four Schemes MMD, MDM, MDD, or DMD, correctly describes the intramolecular excimer formation with 1Py(3)1Py. To this end, the experimental data obtained from the fluorescence response functions are fitted to the values for these parameters that can be calculated for each of the four schemes.[32] As a first step in this fitting procedure,[29] guesses for the various rate constants and reciprocal lifetimes are given as input data. From these data, values for the decay times τ_i, and for their amplitude ratios are calculated. These calculated values are then compared with the experimental data obtained for these quantities (τ_1, τ_2, τ_3; two monomer amplitude ratios; δ and τ_o, in the case of Scheme DMD). After this first step, the values for the input data are changed until an adequate fit is obtained, using the method of steepest descent. The factor f, introduced in the previous section, is calculated from the excimer amplitude ratio.[32]

Using this procedure, it appeared to be impossible to fit the experimental data-sets for 1Py(3)1Py obtained under a variety of experimental conditions, to the Schemes MMD, MDM, and MDD. Only with DMD, a fitting of experimental and calculated data was found to be possible. For Scheme MMD,[33] for example, two negative amplitudes for the excimer rise and decay curve $i_D(t)$ resulted from the calculations, whereas experimentally only one negative amplitude was observed without exception. Obviously, the experimental data of 1Py(3)1Py

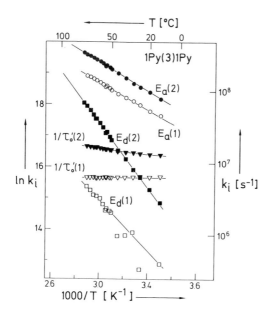

FIGURE 8

Arrhenius plot of the rate constants $k_a(1)$, $k_a(2)$, $k_d(1)$, $k_d(2)$, and of the reciprocal excimer lifetimes $1/\tau'_o(1)$ and $1/\tau'_o(2)$, for intramolecular excimer formation with 1,3-di(1-pyrenyl)propane (1Py(3)1Py) in n-heptane.

cannot be fitted to such a scheme. It is thus concluded that the triple-exponential decays observed with 1Py(3)1Py arise from the presence of two structurally distinct excimers, having identical fluorescence emission bands.[27,31] In one of the following sections, further independent experimental support for the existence of two excimers will be presented.

2.9. Rate Constants as a Function of Temperature

In the manner outlined above, the various rate constants, the excimer lifetimes and the factor f have been extracted from the monomer and excimer fluorescence response functions of 1Py(3)1Py within the context of Scheme DMD. This procedure has been carried out for various solvents, as a function of temperature. The results obtained for 1Py(3)1Py in n-heptane are presented as an Arrhenius plot (Figure 8).

It is seen that straight lines can be drawn through the data points representing the rate constants $k_a(1)$, $k_a(2)$, and $k_d(2)$. From the slopes of these lines the respective activation energies can be obtained (Table I). In contrast, plots of the reciprocal excimer lifetimes, $1/\tau'_o(1)$ and $1/\tau'_o(2)$, are only slightly temperature dependent. A similar behaviour was observed for intermolecular excimer formation with pyrene.[19b] Evidently, the Arrhenius-diagram (Figure 8) in fact consists of two separate diagrams, as a kind of superposition of the two connected processes $M^* \rightleftharpoons D_1^*$ and $M^* \rightleftharpoons D_2^*$ in Scheme DMD.

From the activation energies derived from Figure 8, see Table I, a potential energy scheme

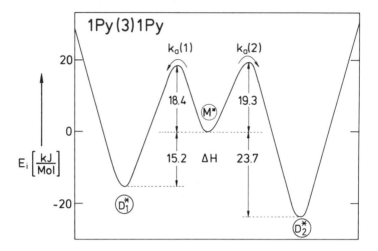

FIGURE 9

Potential energy diagram for the excited state monomer M^* and the two excimers D_1^* and D_2^* in intramolecular excimer formation with 1,3-di(1-pyrenyl)propane (1Py(3)1Py) in n-heptane. The activation energies of excimer formation $E_a(1) = 18$ kJ/mol and $E_a(2) = 19$ kJ/mol, as well as the binding enthalpies $\Delta H(1) = -15$ kJ/mol and $\Delta H(2) = -24$ kJ/mol have been indicated in the diagram. See Table I and Figure 8.

can be deduced, picturing the energetics of the various processes that can take place with 1Py(3)1Py, see Figure 9. The excimer binding enthalpies $(-\Delta H)$ have clearly different values for the two excimers D_1^* and D_2^*. As the emission bands for the two excimers are practically indistinguishable,[27] it then follows that the repulsion energies, i.e., the energies of the Franck-Condon ground state of 1Py(3)1Py reached immediately after excimer emission, with respect to the energy of the fully stretched 1Py(3)1Py molecule, show a difference similar to that observed for the binding enthalpies of D_1^* and D_2^*. It is of interest to note (Table I) that the ΔS values are different for D_1^* and D_2^*, the excimer (D_2^*) with the largest value for the binding enthalpy having the most negative value for ΔS. This is an example of a $\Delta H/\Delta S$ compensation effect, i.e. a linear free energy relationship, which has been observed for a variety of inter- and intramolecular excimers.[16] Further, it is concluded that excimer D_1^*, having a lifetime comparable to that of the symmetric 2Py(3)2Py[27,31] and of [3.3](2,7)-pyrenophane,[18] has a symmetric arrangement of the two pyrene moieties different from the other excimer D_2^*. This second excimer will be similar to the excimer structure for 1Py(22)1Py and pyrene.[18,23c]

2.10. Double-Exponential Decays with 1,3-Di(2-pyrenyl)propane

From the preceding sections in this paper it has to be concluded that the complexity of the fluorescence decays observed with 1Py(3)1Py is not caused by the presence of kinetically different conformers of the propane chain linking the pyrene end groups. This conclusion is sup-

TABLE I: Thermodynamic and Kinetic Data (Scheme DMD) for Intramolecular Excimer Formation with 1,3-Di(1-pyrenyl)propane in n-Heptane.

$E_a(1)^a$	[kJ/mol]	18.4
$E_a(2)$	[kJ/mol]	19.3
$E_d(1)$	[kJ/mol]	33.6
$E_d(2)$	[kJ/mol]	43.0
$\Delta H(1)^b$	[kJ/mol]	−15.2
$\Delta H(2)^b$	[kJ/mol]	−23.7
$\Delta S(1)^c$	[JK^{-1}mol^{-1}]	−14
$\Delta S(2)^c$	[JK^{-1}mol^{-1}]	−55
$\tau_0'(1)$	[ns]	156
$\tau_0'(2)$	[ns]	75
$f=k_f'(2)/k_f'(1)$, (52°C)		2.1

[a] The activation energy E_{diff} for diffusion in n-heptane determined from the solvent viscosity η in an Arrhenius plot of T/η is 10.6 kJ/mol.
[b] $-\Delta H = E_d - E_a$.
[c] $\Delta S = R \ln (k_a^0/k_d^0)$

ported by results obtained with a different dipyrenylpropane isomer, 1,3-di(2-pyrenyl)propane (2Py(3)2Py). For this molecule double-exponential fluorescence response functions with decay times (τ_1,τ_2) that have identical values for monomer and excimer were observed in MCH over an extended

2Py(3)2Py

temperature range.[27,31] This means that only two excited species are present in the case of 2Py(3)2Py, obviously one monomer and one excimer. It is important in this respect that identical τ's are obtained for the monomer and excimer irrespective of temperature, as double-exponential decays might otherwise have been observed as a limiting case (either in the low- or high-temperature limit) of in itself triple-exponential decays.[18,19b]

2.11. Fluorescence Spectra of the Two Dipyrenylpropanes

The excimer emission band of 2Py(3)2Py is similar in its general shape to that observed for 1Py(3)1Py. The width-at-half-maximum $\Delta_{1/2}$ of this band, is even slightly larger for the former molecule. This similarity of the excimer fluorescence supports our finding that the two excimers D_1^* and D_2^* of 1Py(3)1Py have nearly identical excimer stabilization energies ($-\Delta H$).[27,31] As a further support, measurements of the fluorescence response functions at different wavelengths spanning the excimer emission band of 1Py(3)1Py in acetonitrile, resulted in the same values for the amplitude ratios of the decay times τ_1, τ_2, and τ_3, independent of emission wavelength.[25]

2.12. Absorption Spectra of the Two Excimers in 1,3-Di(1-pyrenyl)propane

From the analysis of the fluorescence response functions $i_M(t)$ and $i_D(t)$ discussed above, it was concluded that the two excimers D_1^* and D_2^* should have different relative arrangements of the two pyrene moieties in 1Py(3)1Py. It would then be expected that the singlet excited state absorption spectra would be different for the two excimers. Such a spectrum of 1Py(3)1Py in MCH indeed consists of a superposition of two structureless bands with maxima at 475 nm and 500 nm.[25] Although the decay times associated with these absorption bands cannot be determined as accurately as those from the single-photon-counting experiments described before, the band with the maximum at 500 nm (with a decay time of around 130 ns) can evidently be associated with the excimer D_1^* (156 ns), the other band being connected with D_2^* (75 ns, Table I). For 2Py(3)2Py, in contrast, only one structureless absorption band is observed with a maximum at approximately 495 nm, supporting the conclusion that only one excimer is present for this molecule. The excimer decay time of 160 ns obtained from the time-dependence of absorption curves agrees very well with the excimer lifetime of $\tau_o' = 150$ ns, determined from the single-photon-counting data.[27,31] It is worth noting that the excimer absorption maximum of 2Py(3)2Py is similar to that of the excimer D_1^* of 1Py(3)1Py, in accord with our assumption of a symmetric structure for this species.[25]

2.13. Double- or Triple-Exponential Decay for Diarylalkanes

The conclusion that the observation of triple-exponential decays in the diarylalkanes is not associated with the kinetic anisotropy of the different chain conformers, is supported by the results obtained for other diarylalkanes.[25,27,31] Double-exponential decays are observed with the following compounds: 1,3-diphenylpropane, 1,3-di(4-biphenylyl)propane,[15] 1,3-di(1-naphthyl)propane, the 1,3-di(m-phenanthryl)propanes[17,23] with m = 1, 2, 3, and 9, and for 1,3-di(2-pyrenyl)propane[25,27,31] as well with 1,3-di(N-carbazolyl)propane.[4,23] Triple-exponential decay, on the other hand, is only detected with: 1,3-di(2-naphthyl)propane, 1,3-di(1-pyrenyl)-propane as discussed in detail above, and with 1,3-di(4-pyrenyl)propane.[25,27,31] With 1,3-di(9-anthryl)propane excimer formation was not observed at all, probably because of excessive steric hindrance exerted by the hydrogens at the 1- and 8-positions of the anthracene moiety.[8,23]

Apparently, the observation of triple-exponential decays, which in the case of 1Py(3)1Py has been shown to be due to the presence of two structurally different intramolecular excimers, is not a general phenomenon in excimer formation with diarylalkanes. This statement also applies to linked electron donor/acceptor systems $A(CH_2)_nD$, as will be discussed elsewhere.[23] Higher than triple-exponential decays have been observed for meso- and racemo-2,4-di(1-pyrenyl)pentane. These decays could not be analyzed, except in the low-temperature limit.[23a,23c]

3. CONCLUSION

The fluorescence response functions of 1Py(3)1Py can be fitted to three exponentials with identical decay time constants for the excimer and the monomer. This points to the involvement of three excited state species in intramolecular excimer formation with this compound. It could be shown that only one of the four possible kinetic schemes is in accord with the experimental data, a scheme comprising one monomer and two structurally different excimers. This conclusion was supported by the detection of two different excimer absorption bands for 1Py(3)1Py, having decay times similar to those determined from the fluorescence decays. Higher than double-exponential decay was found to be the exception rather than the rule with the diarylalkanes, indicating that the propane chain conformers do act as a single, kinetically uniform, group in intramolecular excimer formation. Higher than double-exponental decays have only been found when two different excimers are possible. Whether this is the case will depend on the nature of the connecting chain and on the properties of the excimer-forming molecules.

ACKNOWLEDGEMENT

Many thanks are due to G. Striker for making his deconvolution program (modulating functions)[34] available to us. This work has been supported by the Deutsche Forschungsgemeinschaft (Sonderforschungsbereich 93,"Photochemie mit Lasern").

REFERENCES

1) F. Hirayama, J. Chem. Phys. 42 (1965) 3165.

2) E. A. Chandross and C. J. Dempster, J. Am. Chem. Soc. 92 (1970) 3586.

3) W. Klöpffer, Chem. Phys. Lett. 4 (1969) 193.

4) G. E. Johnson, J. Chem. Phys. 61 (1974) 3002.

5) (a) W. Klöpffer, Intramolecular excimers, in: Organic Molecular Photophysics, Vol. 1, ed. J. B. Birks (Wiley, London, 1973) pp. 357-402. (b) F. C. De Schryver, N. Boens, and J. Put, Advan. Photochem. 10 (1977) 359.

6) (a) W. Kühnle and K. A. Zachariasse, Ber. Bunsenges. Physik. Chem. 78 (1974) 1254. (b) W. Kühnle and K. A. Zachariasse, J. Photochem. 5 (1976) 165.

7) K. A. Zachariasse and W. Kühnle, Z. Phys. Chem. NF 101 (1976) 267.

8) A. Castellan, J.-P. Desvergne, and H. Bouas-Laurent, Chem. Phys. Lett. 76 (1980) 390.

9) A. M. Halpern, M. W. Legenza, and B. R. Ramachandran, J. Am. Chem. Soc. 101 (1979) 5736.

10) M. A. Winnik, Chem. Rev. 81 (1981) 491.

11) Th. Förster and H. P. Seidel, Z. Phys. Chem. NF 48 (1965) 58.

12) (a) J. B. Birks, Photophysics of Aromatic Molecules (Wiley, London, 1970). (b) J. B. Birks, ed., Organic Molecular Photophysics, Vol. 1 (Wiley, London, 1973). (c) J. B. Birks, ed., Organic Molecular Photophysics, Vol. 2 (Wiley, London, 1975).

13) K. A. Zachariasse, A. Maçanita, and W. Kühnle, to be published.

14) T. Kanaya, K. Goshiki, M. Yamamoto, and Y. Nishijima, J. Am. Chem. Soc. 104 (1982) 3580.

15) K. A. Zachariasse, W. Kühnle, and A. Weller, Chem. Phys. Lett. 59 (1978) 357.

16) K. A. Zachariasse and G. Duveneck, J. Am. Chem. Soc. 109 (1987) 3790.

17) K. A. Zachariasse, R. Busse, U. Schrader, and W. Kühnle, Chem. Phys. Lett. 89 (1982) 303.

18) P. Reynders, PhD Thesis, Göttingen University, 1988.

19) K. A. Zachariasse, G. Duveneck, and R. Busse, J. Am. Chem. Soc. 106 (1984) 1045. (b) K. A. Zachariasse, Excimer formation with pyrenes on silica surfaces, in: Photochemistry on Solid Surfaces, eds. T. Matsuura and M. Anpo (Elsevier, Amsterdam, 1989) pp. 48-78.

20) (a) M. Sisido, Macromolecules 4 (1971) 737. (b) H. Morawetz and N. Goodman, Macromolecules 3 (1970) 699.

21) J. Coops, H. van Kamp, W. A. Lambregts, B. J. Visser, and H. Dekker, Recl. Trav. Chim. Pays-Bas 79 (1960) 1226.

22) K. A. Zachariasse, unpublished results.

23) (a) P. Reynders, H. Dreeskamp, W. Kühnle, and K. A. Zachariasse, J. Phys. Chem. 91 (1987) 3982. (b) P. Reynders, W. Kühnle, and K. A. Zachariasse, J. Am. Chem. Soc. 112 (1990) 3929. (c) P. Reynders, W. Kühnle, and K. A. Zachariasse, J. Phys. Chem. 94 (1990) 4073.

24) (a) A. J. Roberts, D. V. O'Connor, and D. Phillips, N. Y. Acad. Sci. 366 (1981) 93. (b) S. W. Beavan, J. S. Hargreaves, and D. Phillips, Advan. Photochem. 11 (1979) 207. (c) S. N. Semerak and C. W. Frank, Advan. Polym. Sci. 54 (1984) 31.

25) G. Duveneck, PhD Thesis, Göttingen University, 1986.

26) S. Ito, M. Yamamoto, and Y. Nishijima, Bull. Chem. Soc. Japan 55 (1982) 363.

27) K. A. Zachariasse, R. Busse, G. Duveneck, and W. Kühnle, J. Photochem. 28 (1985) 235.

28) I. N. Bronstein and K. A. Semendjajew, Taschenbuch der Mathematik (Deutsch, Thun, 1981).

29) (a) Eigensystem Subroutine Package (Eispack), 1975. (b) B. S. Garbow and J. J. Dongarra, Eispack Path Chart, 1974.

30) Only one independent excimer amplitude ratio is obtained, as the sum of the excimer amplitudes is equal to zero.

31) K. A. Zachariasse, G. Duveneck, and W. Kühnle, Chem. Phys. Lett. 113 (1985) 337.

32) K. A. Zachariasse and A. Maçanita, to be published.

33) M. Goldenberg, J. Emert, and H. Morawetz, J. Am. Chem. Soc. 100 (1978) 7171.

34) (a) G. Striker, Effective implementation of modulating functions, in: Deconvolution and Reconvolution of Analytical Signals, ed. M. Bouchy (University Press, Nancy, France, 1982) pp. 329-357. (b) J. R. Knutson, J. M. Beechem, and L. Brand, Chem. Phys. Lett. 102 (1983) 501.

Photochemical Processes in Organized Molecular Systems
K. Honda (Editor-in-Chief)
© Elsevier Science Publishers B.V., 1991

EXCITATION HOPPING BETWEEN TWO IDENTICAL CHROMOPHORES ATTACHED TO BOTH ENDS OF ALKANES

Tomiki IKEDA, Bong LEE and Shigeo TAZUKE+

Photochemical Process Division, Research Laboratory of Resources Utilization, Tokyo Institute of Technology, 4259, Nagatsuta, Midori-ku, Yokohama 227, Japan

Excitation hopping behavior between two identical chromophores has been directly observed by time-resolved fluorescence anisotropy (r(t)) measurements with the aid of a picosecond time-correlated single photon counting system. In order to explore the excitation hopping behavior in purely isolated two identical chromophoric systems, we used α , ω -bis(2-naphthyl)-n-alkanes, Nap$(CH_2)_n$Nap [Nn, where n = 3, 5, 7 and 12], and α , ω -bis(9-anthryl)-n-alkanes, An$(CH_2)_m$An [Am, where m = 3, 5, 7, 10, 12, 14 and 18], as well as their model compounds with a single chromophore, 2-ethylnaphthalene (EN) and 9-ethylanthracene (EA). It was demonstrated that r(t) for the model compounds did not change with time during their lifetimes while in Nn and Am with m $\geq$ 7 the initial decay of r(t) was clearly observed, indicating that excitation hopping takes place between the two chromophores attached to both ends of alkanes. Conformational analysis was performed on the basis of potential energy calculations and the results were used to analyze the decay profiles of r(t).

1. INTRODUCTION

Nonradiative energy migration is a primary process occurring in molecular aggregates such as micelles, liquid crystals (lyotropic and thermotropic), monolayers, Langmuir-Blodgett's films and polymers, where identical molecules are packed in close proximity of each other. Organic crystals may be classified into this category, but "molecular aggregate" is usually used for less ordered system.

Nonradiative energy transfer in the molecular aggregate systems was first recognized by Gaviola et al.,[1] Weigert et al.,[2] and Levshin[3] in concentrated dye solutions. They realized that emission from the concentrated dye solution was depolarized while emission from the viscous dilute dye solution retained the polarization. Since the discovery of the "concentration depolarization", the energy migration process has been recognized as an important primary photophysical process in molecular aggregate systems.

+Deceased July 11, 1989.

In polymers with identical side-groups, for example, excitation formed by absorption of photons at one chromophore site moves among the chromophores and is
captured at a specific site called a "trap", where subsequent processes like excimer formation take place.   Under this situation, incorporation of acceptor
molecules may result in efficient collection of the energy of photons, and ideas
of "antenna polymers"[4] and "photon-harvesting polymers"[5] were proposed for the
efficient collection of the energy of light.

   However, it is very difficult to analyze strictly the excitation migration behavior in the molecular aggregate systems.   Difficulties arise from the fact
that in the molecular aggregate systems the location of the initially excited sites
is not clearly defined.   For example, in the polymers, the probability of excitation at the chain end is equal to that of the center of the polymer chain.[6]
However, if the excitation migration operates, efficiency is definitely different
between the two sites.   Thus, for the analysis of the experimental results, account must be taken of distribution of the initially excited sites.[6]   Another
difficulty, much more serious for the exact analysis, is that excitation hops
around in the system like a random walk; thus, to follow the locus of the excitation hopping is quite difficult even if we assume the dipole-dipole transfer a
priori for the excitation hopping mechanism. If we assume that excitation hops
in a polymer coil, we can draw many pathways for the excitation movements:
between the nearest-neighbors and between a pair of chromophores which are far
apart along the polymer chain.   For such complicated systems, the exact solution
of the problem becomes very difficult, requiring multiple averages over 1) initially excited sites, 2) local conformation, 3) distribution of acceptors, and so on,
but approximation may be made.   The most promising theoretical approaches to
this problem were developed by Fayer et al.,[7] Perlstein et al.,[8] Haan and
Zwanzig,[9] Blumen et al.,[10] and Frank et al.[11].

   For the analysis of the excitation migration behavior, *purely isolated bichromophoric systems* are superior to other molecular aggregate systems
like polymers, since in the bichromophoric systems there are no complications associated with the distributions of the initially excited sites and the hopping
pathway. A bichromophoric system with nonidentical donor and acceptor was already investigated in full detail by Stryer and Haughland[12] using oligomers of
poly-L-proline, labeled at one end with a naphthyl (donor) and at the other end
with a dansyl (acceptor) group. They used the sensitized fluorescence from the
acceptor and obtained $R^{-5.9 \pm 0.3}$ dependence (R is the interchromophoric
distance), which is in excellent agreement with the Förster's theory.   However,
very few studies have been performed so far on the excitation migration in a
bichromophoric system with *identical* chromophores.[13]

   In this paper, we explore the excitation hopping behavior between two chromophores attached to both ends of alkanes by means of the picosecond time-resolved fluorescence anisotropy measurements.   Although the bichromophoric compounds possess a great advantage, in that analysis of the results of the time-resolved measurements is free from statistical considerations on the originally excited site and the hopping pathway, they have a disadvantage in that the distribution must be taken into account for the interchromophore distance and the orientation of the two chromophores because of the flexible nature of the links between the two chromophores.   To overcome this problem, a conformational analysis of the bichromophoric compounds was performed.   Calculation of potential energy of nonbonded atoms (van der Waals potential) enabled us to estimate the distribution functions of the interchromophore distance and of the orientation of the two chromophores in the compounds.   In every conformation, the interchromophore distance, the orientation of the two relevant chromophores and the orientation factor were calculated, and the hopping rate constant in that conformation was calculated by applying the Förster's equation.   On the assumption that the distribution of the conformations obeys the Boltzmann distribution, we performed the computer simulation of the expected decays of the fluorescence anisotropy.

2. BICHROMOPHORIC COMPOUNDS

The structures of the bichromophoric compounds, α , ω -bis(2-naphthyl)-n-alkanes, $Nap(CH_2)_nNap$ [Nn, where n = 3, 5, 7 and 12], and α , ω-bis(9-anthryl)-n-alkanes, $An(CH_2)_mAn$ [Am, where m = 3, 5, 7, 10, 12, 14 and 18], as well as their model compounds with a single chromophore, 2-ethylnaphthalene (EN) and 9-ethyl-anthracene (EA) are shown in Figure 1.

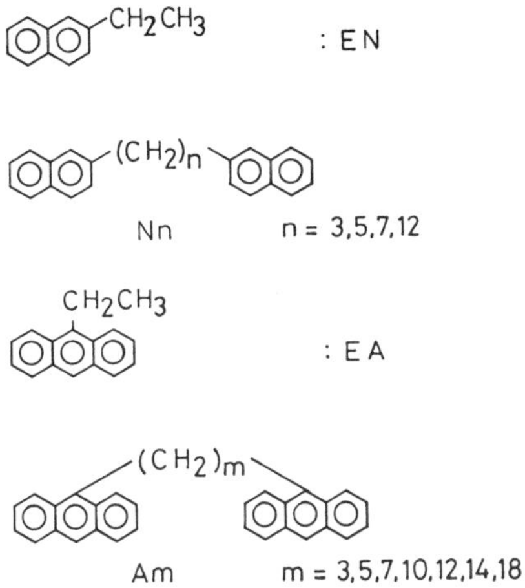

CH_2CH_3 : EN

$(CH_2)_n$ Nn   n = 3,5,7,12

CH_2CH_3 : EA

$(CH_2)_m$ Am   m = 3,5,7,10,12,14,18

FIGURE 1
Structures of bichromophoric compounds used in this study.

3. TIME-RESOLVED MEASUREMENTS

A time-correlated, single-photon counting method was used to measure the fluorescence and the fluorescence anisotropy decays throughout (Figure 2).[14-16] A synchronously pumped, cavity-dumped dye laser operated with a mode-locked Nd:YAG laser was the excitation pulse source with a pulse width of 6 ps fwhm. For excitation of samples, frequency doubling was achieved with a KDP crystal. Fluorescence was monitored at right angles to the excitation path through a monochromator with a microchannel-plate photomultiplier.   Signals from the photomultiplier were amplified with a 1.3 GHz preamplifier, discriminated with a constant fraction discriminator, and used as a stop pulse for a time-to-amplitude converter (TAC).   A start pulse was provided from a fast photodiode monitoring a laser pulse through a discriminator.   Data from the TAC were stored in a multichannel analyzer and then transferred to a microcomputer where decay analysis was performed by an iterative nonlinear least-squares method.

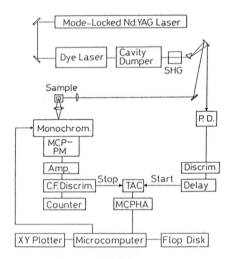

FIGURE 2

Schematic diagram for picosecond time-resolved fluorescence measurements: P.D., photodiode; MCP-PM, microchannel-plate photomultiplier; MCPHA, multichannel pulse-height analyzer.

The decays of the fluorescence anisotropy were measured using pulse excitation.   A vertically polarized excitation light pulse was obtained through a Babine-Soleil compensator, and $I_{\parallel}$ (t) and $I_{\perp}$ (t) were measured alternatively through a polarizer, which was placed in front of the monochromator.   Here, $I_{\parallel}$ (t) and $I_{\perp}$ (t) represent the fluorescence intensities observed with parallel and perpendicular orientation of the polarizer to the excitation pulse. $I_{\parallel}$ (t) and $I_{\perp}$ (t) were measured alternatively by rotating the polarizer periodically in order to eliminate any artifacts due to the long-term laser instability. A depolarizer was also placed just in front of the monochromator to remove anisotropic characteristics of the monochromator against $I_{\parallel}(t)$ and $I_{\perp}(t)$.

The fluorescence anisotropy, $r(t)$, was calculated by eq 1 where deconvolution of the instrument response function was performed.

$$r(t) = [I_{\parallel}(t) - I_{\perp}(t)] / [I_{\parallel}(t) + 2I_{\perp}(t)] \qquad (1)$$

4. CONFORMATIONAL ANALYSIS[14,15]

Conformational analysis was performed on the basis of potential energy calculation including the van der Waals interaction between nonbonded atoms and the intrinsic torsional potential energies associated with rotation about C-C bonds.[17-19] For calculation of the van der Waals interaction, a Lennard-Jones potential function was employed in which the potential energy V_{ij} between the i'th and j'th nonbonded atoms is expressed by eq 2 where d_{ij} is the distance between the i'th and j'th nonbonded atoms and A_{ij} and B_{ij} are coefficients.

$$V_{ij} = B_{ij}/d_{ij}^{12} - A_{ij}/d_{ij}^{6} \qquad (2)$$

The attractive term coefficient A_{ij} in eq 2 can be determined by applying the Slater-Kirkwood equation modified by Scott and Scheraga,[20] but in this study we adopted the A_{ij} and B_{ij} values reported in the literature.[18] The potential function was modified by considering the molecule-solvent interaction according to the procedure employed by Flory.[21] For molecules in a solvent, intramolecular contact with an alteration in the molecular conformation occurs at the expense of intermolecular interaction with the solvent. The molecule-solvent interaction is attractive so that inclusion of this effect would result in suppression of the attractive term in eq 2 at larger distances, leading to effective elimination of the minimum in the function. This means that calculations can be performed by the potential functions given in eq 3 where V_{ij}^{0} is the value of V_{ij} at $d_{ij} = d_{ij}^{0}$ (the sum of the van der Waals radii of the interacting atoms).

$$V_{ij}^{*} = V_{ij} - V_{ij}^{0} \qquad d_{ij} < d_{ij}^{0}$$

$$V_{ij}^{*} = 0 \qquad d_{ij} \geq d_{ij}^{0} \qquad (3)$$

The total potential energy due to the van der Waals interaction was calculated as the sum over all pairs of nonbonded atoms. The intrinsic torsional potential energy for the skeletal alkyl chain, $E_{tor}(\phi)$, was calculated for the rotational angle ϕ by eq 4 where E_0 is the torsional barrier energy (E_0 = 2.8 kcal/mol).[22]

$$E_{tor}(\phi) = (E_0/2)(1 + \cos 3\phi) \qquad (4)$$

The total potential energy was obtained as the sum of the potential energies due to the van der Waals interaction and the intrinsic torsional potentials.

The fraction of the i'th conformation, f_i, was calculated by eq 5 on the assumption that the distribution of conformations obeys the Boltzmann distribution where E_i is the calculated potential energy for the i'th conformation.

$$f_i = \exp(-E_i/RT) / \Sigma \exp(-E_i/RT) \qquad (5)$$

On every conformation, the interchromophore distance ($\underline{R}$) and the angles between two vectors along the long axis of the chromophore rings (θ_L) and along the short axis of the chromophore rings (θ_S) were calculated, and the distribution functions for $\underline{R}$, θ_L and θ_S were evaluated on the basis of eq 5. Here, $\underline{R}$ was defined as the distance between the centers of the two aromatic rings.

Typical conformational energy maps are shown in Figures 3 – 6 for N3 and A3. These energy maps were constructed from the results of the potential energy calculation at intervals of 5°. Contours are drawn at intervals of 0.5 kcal/mol relative to the energy minima. Figure 5 demonstrates that the most stable conformation in A3 is situated at $\underline{tt}$ ($\phi_2 = \phi_3 \simeq 180°$), and is surrounded by four equivalent $\underline{tg}$ conformations. It is of particular interest to compare the stable conformations of A3 and N3. In the latter, the $\underline{gg}$ conformations ($\phi_2 = \phi_3 \simeq 60°$ and $\phi_2 = \phi_3 \simeq 300°$) were also involved in the stable conformations (Figure 3) while in A3 such conformations are excluded from the stable conformation. This result clearly indicates that the 9-anthryl moiety is much more bulky than the 2-naphthyl group, thus the number of the stable conformations is limited in Am compounds.

The conformational energy maps for A3 as a function of ϕ_3 and ϕ_4 are shown in Figure 6, where ϕ_1 and ϕ_2 were fixed at 90° and 180° (a), and 90° and 60° (b), respectively, and calculation was performed at 5° intervals. Figure 6 shows a remarkable feature of 9-anthryl alkanes that the stable conformation is limited only in the small range of the rotational angles of ϕ_3 and ϕ_4 due to the steric hindrance of the bulky 9-anthryl moieties. This is in sharp contrast to the conformation of 2-naphthyl alkanes in which the potential energies are insensitive to the rotation of the naphthalene ring (Figure 4).

Distribution functions of the interchromophore distance ($\underline{R}$) for Nn and Am at 77 K are shown in Figures 7 and 8, where calculation of the distribution was

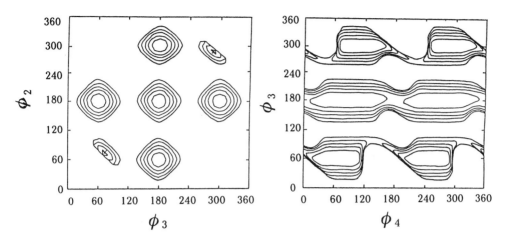

FIGURE 3
Conformational energy map for N3 as a function of ϕ_2 and ϕ_3. ϕ_1 and ϕ_4 were fixed at 90°. Contours are shown at intervals of 0.5 kcal/mol relative to the energy minima.

FIGURE 4
Conformational energy map for N3 as a function of ϕ_3 and ϕ_4. ϕ_1 and ϕ_2 were fixed at 90° and 180°. Contours are shown at intervals of 0.5 kcal/mol relative to the energy minima.

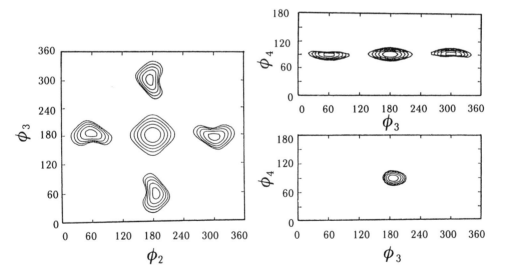

FIGURE 5
Conformational energy map for A3 as a function of ϕ_2 and ϕ_3. ϕ_1 and ϕ_4 were fixed at 90°. Contours are shown at intervals of 0.5 kcal/mol relative to the energy minima.

FIGURE 6
Conformational energy map for A3 as a function of ϕ_3 and ϕ_4. ϕ_1 and ϕ_2 were fixed at 90° and 180° (a) and at 90° and 60°(b), respectively. Contours are shown at intervals of 0.5 kcal/mol relative to the energy minima.

performed at a 0.2 increment.

It is clearly seen that in N3 and A3, the distribution of $\underline{R}$ is quite narrow at 77 K and bimodal with two maxima. In Nn and Am with n, m ≥ 5, the distribution is broader as expected from the increased number of the skeletal atoms. An ensemble-averaged interchromophore distance, $\underline{R}$, was calculated by eq 6 where $\underline{R}_i$ is the interchromophore distance of the i th conformer and $f_i(\underline{R}_i)$ is the fraction of the i th conformers.

$$\underline{R} = \Sigma \ \underline{R}_i f_i(\underline{R}_i) \ / \ \Sigma \ f_i(\underline{R}_i) \qquad (6)$$

The $\underline{R}$ values thus calculated are 8.41 Å (N3), 9.15 Å (N5), 10.79 Å (N7), 7.05 Å (A3), 9.45 Å (A5), 11.28 Å (A7), 13.84 Å (A10) and 15.45 (A12).

Angular distributions of θ_L and θ_S in Am at 77 K are shown in Figures 9 and 10, respectively. These angular distribution functions were computed with fixed rotational angles of $\phi_1 = \phi_{m+1} = 90°$.

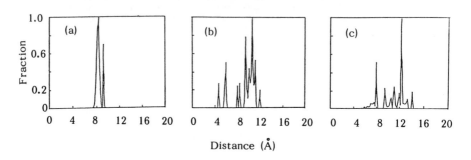

FIGURE 7
Distribution functions of the interchromophore distance for N3(a), N5(b), and N7(c) at 77 K.

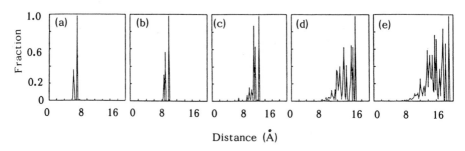

FIGURE 8
Distribution functions of the interchromophore distance for A3(a), A5(b), A7(c), A10(d), and A12(e) at 77 K.

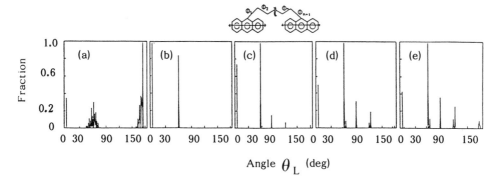

FIGURE 9

Angular distributions of the two vectors along the long axis (θ_L) of the anthracene ring.   (a)A3; (b)A5; (c)A7; (d)A10; (e)A12.

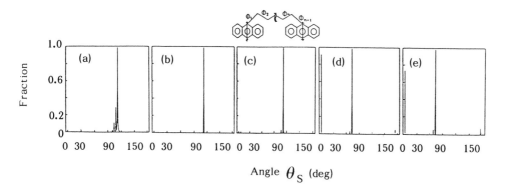

FIGURE 10

Angular distributions of the two vectors along the short axis (θ_S) of the anthracene ring.   (a)A3; (b)A5; (c)A7; (d)A10; (e)A12.

5. FLUORESCENCE ANISOTROPY DECAYS[14,15]

In order to exclude the contribution of rotational diffusion of the chromophores to the fluorescence depolarization, the time-resolved fluorescence anisotropy measurements were performed at 77 K in glass matrix of 2-methyltetrahydrofuran (MTHF) at the concentration of 10^{-5} M.   In the rigid glass matrix of MTHF at 77 K, no excimer emission was observed in any bichromophoric compounds and no photoreaction proceeded to any appreciable extent even in A3. Furthermore, at 77 K the absorption and fluorescence spectra of Nn and Am were similar in each homolog, and in addition the fluorescence decay curves were similarly analyzed by a single-exponential function with the lifetime of about 80 ns for Nn (also EN) and of about 11 ns for Am including EA, indicat-

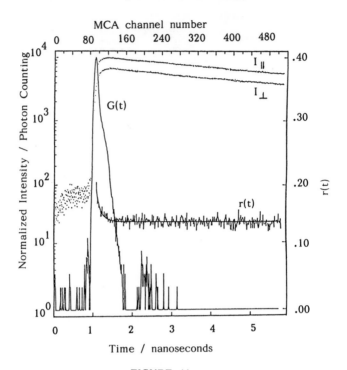

FIGURE 11

Polarized fluorescence decays $I_\parallel$ and $I_\perp$, instrument response function G(t), and anisotropy decay r(t) of A12 in MTHF at 77 K: λ_{ex} = 370 nm, λ_{em} = 417 nm; time division, 11.36 ps/channel. The solid line shows the best fit curve based on r(t) = A exp(-2 ω t) + B.

ing that the properties of the excited states are quite similar in each homolog.

In Figure 11 are shown $I_\parallel$ (t) and $I_\perp$ (t), the instrument response function, G(t), and the anisotropy decay, r(t), of A12 measured in MTHF at 77 K. Figure 12 shows the anisotropy decays of all anthracene compounds. It is clearly seen that r(t) for EA does not change with time, which indicates that excitation remains at the originally excited site during its lifetime. In Am with m ≥ 7, the decay of r(t) is clearly observed. This means that emission from a transition moment, which is different from that of the originally excited site, contributes to the observed emission, and is a clear piece of evidence that excitation hopping takes place between the two 9-anthryl moieties attached to both ends of alkanes. In A3 and A5, no decay of r(t) was observed even in the shortest time range available (10.73 ps/ch) for our time-resolved measurement system. However, in both compounds, residual polarization r_∞, which is the r(t) value at t = ∞, is different from the r(t) value observed in EA, thus it is reasonabe to assume that in these compounds the decay of r(t) is too fast to be detectable with our apparatus.

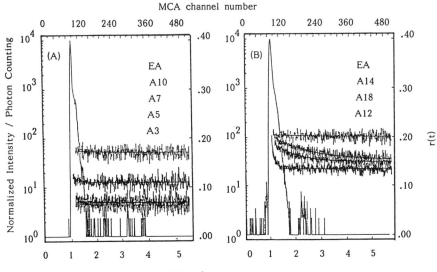

FIGURE 12

Anisotropy decays of Am's and EA measured in MTHF at 77 K: (A) EA, A3, A5, A7, and A10, λ_{ex} = 347 nm, λ_{em} = 393 nm, time division 10.73 ps/channel. (B) EA, A12, A14, and A18, λ_{ex} = 370 nm, λ_{em} = 417 nm, time division 11.36 ps/channel. The solid lines show the best fit curves based on r(t) = A exp(-2 ωt) + B.

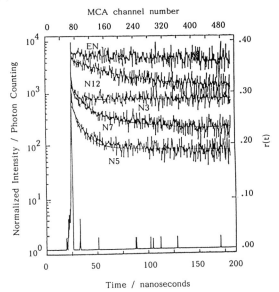

FIGURE 13

Anisotropy decays of Nn's and EN measured in MTHF at 77 K: EN, N3, N5, N7, and N12, λ_{ex} = 318 nm, λ_{em} = 335 nm, time division 364.58 ps/channel. The solid lines show the best fit curves based on r(t) = A exp(-2 ωt) + B.

In the naphthalene compounds, the same behaviors of the fluorescence anisotropy decays were observed as shown in Figure 13 where the initial decay of the
anisotropy was clearly observed for N3 - N12 while r(t) of EN remained un-
changed during its lifetime.

6. SIMULATION OF FLUORESCENCE ANISOTROPY DECAY CURVES[15]

Because of the known direction of the transition moment of the 9-anthryl
moiety, an exact analysis based on Förster's equation is possible.   Simulation of
the fluorescence anisotropy decay curves was performed on the basis of the conformational analysis.   As described in Section 4, for each conformation the interchromophore distance ($\underline{R}_i$), the angle between two vectors along the long   axis
(θ_{Li}) and the short axis (θ_{Si}) of the anthracene ring, and the orientation factor defined by eq 7 were calculated:[23]

$$\kappa_i = \cos \theta_{DAi} - 3\cos \theta_{Di}\cos \theta_{Ai} \qquad (7)$$

Here, θ_{DAi} is the angle between the transition moment vectors of the donor
and the acceptor, θ_{Di} and θ_{Ai} are the angles between these transition moments and the direction of $\underline{R}_i$, respectively.   It is well known that the ^1L_a band
in the anthracene derivatives corresponds to the $S_0 \rightarrow S_1$ transition and its transition moment lies along the short axis of the anthracene ring.[24]   Thus, in Am
and EA, θ_{DAi} in eq 7 can be reasonably replaced by θ_{Si}.

In the incoherent excitation hopping between two identical chromophores, the
following equation is applicable to describe the fluorescence anisotropy decay
profiles:[13-15]

$$r(t) = (1/2)(r_0 - r_1)\exp(-2 \omega t) + (1/2)(r_0 + r_1) \qquad (8)$$

Here, r_0 is the limiting value of r and can be estimated from a case where no
fluorescence depolarization occurs.   Thus, the r_0 value for the present case can
be determined from the r_0 value of EA. r_1 is the anisotropy of fluorescence
emitted entirely by the chromophore which is not originally excited.   In the
simulation procedure, we calculated the r_1 value of the i th conformer (r_{1i}) by
eq 9.[23]

$$r_{1i} = (r_0/2)(3\cos^2 \theta_{Si} - 1) \qquad (9)$$

In eq 8, ω is the hopping rate constant.   If we assume the Förster's mechanism for the excitation migration, ω of the i th conformer with $\underline{R}_i$ and κ_i
can be written in the form of eq 10.[23]

$$\omega_i = \frac{9(\ln 10)\,\Phi_D}{128\,\pi^5 N n^4\,\tau_D} \cdot \frac{\kappa_i^2}{\underline{R}_i^6} \cdot J(v) \tag{10}$$

Here, N is Avogadro's number, Φ_D and τ_D are the donor emission quantum yield and lifetime in the absence of the acceptor, $J(v)$ is the spectral overlap integral of the donor emission and the acceptor absorption, n is the refractive index of the medium. With respect to Φ_D and τ_D, we can use the values of Φ_F and τ of EA. $J(v)$ was determined from the absorption and the fluorescence spectra of EA as 1.5×10^{15} cm^6/mol at 77 K. This value is somewhat larger than the value reported for 9-methylanthracene at room temperature.[25] With respect to n of MTHF at 77 K, it has not been reported so far, thus we used the value at 20°C (1.405). Substituting these values in eq 10, we obtain a simpler expression for ω_i:

$$\omega_i = K \cdot \frac{\kappa_i^2}{\underline{R}_i^6} \tag{11}$$

where,

$$K = \frac{9(\ln 10)\,\Phi_D}{128\,\pi^5 N n^4\,\tau_D} \cdot J(v) \tag{12}$$

By substituting eqs 9 and 11 in eq 8, we obtain the expected decay of the fluorescence anisotropy, $r_i(t)$, for one particular conformation i. The observable decay of the fluorescence anisotropy can be obtained by ensemble-averaging over all conformations with the assumption that the distribution of the conformations obeys the Boltzmann distribution.

$$r(t) = \Sigma\, r_i(t)\exp(-E_i/RT)\,/\,\Sigma \exp(-E_i/RT) \tag{13}$$
$$= \Sigma\, r_i(t) f_i$$

Our program of the fluorescence anisotropy simulation enabled us to draw the expected decay profiles in any time region. Figure 14 shows the expected decays of $r(t)$ in three time regions. In Figure 14(a), the time division is 0.1 ps/ch which is not obtainable in our apparatus, and Figure 14(b) and (c) show the

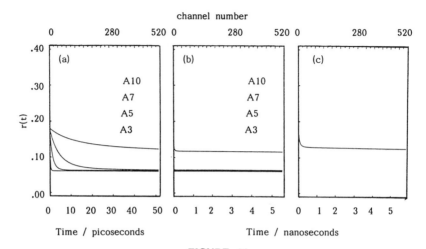

FIGURE 14

Expected decays of fluorescence anisotropy obtained by computer simulation based on the results of conformational analysis: (a)A3, A5, A7, and A10, time division 0.1 ps/channel, r_0 = 0.18; (b)A3, A5, A7, and A10, time division 10.73 ps/channel, r_0 = 0.18; (c)A12, time division 11.36 ps/channel, r_0 = 0.206.

results of the simulation with the same time division (10.73 ps/ch and 11.36 ps/ch) as employed in our time-resolved anisotropy measurements (Figure 12). In our simulation procedure, no variables (fitting parameters) are involved. Only experimentally determined parameters are used: Φ_D, τ_D and J(v) which can be unequivocally determined from the model compound, EA. Simulation based on the strict application of the Förster equation was only possible in the case where each chromophore possesses a definite direction of the transition moment. Comparison of the simulated decay profiles of r(t) (Figure 14(b) and (c)) with the experimentally observed decay profiles (Figure 12A and B) revealed that the experimentally observed decay profiles were well reproduced by the simulated decay curves.

In A3 and A5, the initial decays of r(t) could not be observed (Figure 12) because of higher hopping rates than those detectable with our apparatus, and this behavior was exactly reflected in the simulated curves (Figure 14(b)). Furthermore, the experimentally observed values of the residual polarization, r_∞, were in good agreement with the simulated values of r_∞. In particular, in the experimentally observed curves, r_∞ increased in the order of A3 < A5 < A7, although the differences were very small. This behavior was well reproduced in the simulated curves. In addition, in Figure 12A r_∞ of A10 was much larger than those of A3, A5 and A7, which may be a consequence of the odd-even effect associated with alkyl spacers. In fact, the angular distributions of θ_S in A10 and A12 were clearly different from those of A3, A5 and A7, which exhibited similar distribution functions (Figure 9). The r_∞ values are predominant-ly determined by the relative orientation of the two anthryl groups, thus excel-lent agreement between the experimentally observed curves and the simulated ones demonstrates that the actual orientation of Am is well reflected in the potential energy calculation.

The hopping rate constants observed for Am are larger by approximately two order of magnitude than those observed for Nn when compared with the com-pounds with the same alkyl spacer length. This is mainly attributable to larger values of Φ_D and ε_A, and shorter τ_D of the anthryl moiety than those of the 2-naphthyl moiety. From the lifetimes and the hopping rate constants, we can estimate the hopping frequency during the lifetime. In A7, the hopping frequency is ~ 28, and this value is ~ 18 times larger than that of N7.

REFERENCES

1) E. Gaviola and P.Z. Pringsheim, Z. Phys. 24 (1924) 24.

2) F. Weigert and G.Z. Kapper, Z. Phys. 25 (1924) 99.

3) W.L. Levshin, Z. Phys. 26 (1924) 274.

4) X.-X. Ren and J.E. Guillet, Macromolecules 18 (1985) 2012; J.E. Guillet and W.A. Rendall, Macromolecules 19 (1986) 224.

5) J.S. Hargreaves and S.E. Webber, Macromolecules 18 (1985) 734; F. Bai, C.H. Cheng and S.E. Webber, Macromolecules 19 (1986) 2484.

6) C.W. Frank, G.H. Fredrickson and H.C. Andersen, in: Photophysical and Photo-chemical Tools in Polymer Science, ed. M.A. Winnik (Reidel, Dordrecht, 1986) pp. 495.

7) C.G. Gochanour, H.C. Andersen and M.D. Fayer, J. Chem. Phys. 70 (1979) 4254; R.F. Loring, H.C. Andersen and M.D. Fayer, J. Chem. Phys. 76 (1982) 2015; J. Baumann and M.D. Fayer, J. Chem. Phys. 85 (1986) 4087.

8) R.P. Hemenger and R.M. Perlstein, J. Chem. Phys. 59 (1973) 4064.

9) S.W. Haan and R. Zwanzig, J. Chem. Phys. 68 (1978) 1879.

10) A. Blumen and J. Manz, J. Chem. Phys. 71 (1979) 4694.

11) G.H. Fredrickson, H.C. Andersen and C.W. Frank, Macromolecules 16 (1983) 1456; G.H. Fredrickson, H.C. Andersen and C.W. Frank, Macromolecules 17 (1984) 54; G.H. Fredrickson, H.C. Andersen and C.W. Frank, Macromolecules 17 (1984) 1496.

12) L. Stryer and R. Haughland, Proc. Natl. Acad. Sci. USA 58 (1967) 719.

13) R.S. Moog, A. Kuki, M.D. Fayer and S.G. Boxer, Biochemistry 23 (1984) 1564.

14) T. Ikeda, B. Lee, S. Kurihara, S. Tazuke, S. Itoh and M. Yamamoto, J. Am. Chem. Soc. 110 (1988) 8299.

15) T. Ikeda, B. Lee, S. Tazuke and A. Takenaka, J. Am. Chem. Soc. 112 (1990) 4650.

16) T. Ikeda, S. Kurihara and S. Tazuke, J. Phys. Chem. 94 (1990) 6550. S. Tazuke, R.K. Guo and T. Ikeda, J. Phys. Chem. 94 (1990) 1408.

17) P. Flory, Statistical Mechanics of Chain Molecules (Wiley, New York, 1969).

18) A.J. Hopfinger, Conformational Properties of Macromolecules (Academic, New York, 1973).

19) S. Itoh, M. Yamamoto and Y. Nishijima, Bull. Chem. Soc. Jpn. 55 (1982) 363.

20) R.A. Scott and H.A. Scheraga, J. Chem. Phys. 42 (1965) 2209.

21) D.A. Brant and P.J. Flory, J. Am. Chem. Soc. 87 (1965) 2791.

22) A. Abe, R.L. Jernigan and P.J. Flory, J. Am. Chem. Soc. 88 (1966) 631.

23) R. Dale and J. Eisinger, in: Biochemical Fluorescence: Concept, eds. R.F. Chen and H. Edelhoch (Marcel Dekker, New York, 1975) Vol. 1 pp. 115.

24) J.B. Birks, Photophysics of Aromatic Molecules (Wiley, New York, 1970).

25) I.B. Berlman, Energy Transfer Parameters of Aromatic Compounds (Academic Press, New York, 1973).

Photochemical Processes in Organized Molecular Systems
K. Honda (Editor-in-Chief)
© Elsevier Science Publishers B.V., 1991

PICOSECOND DYNAMICS OF DYES IN HETEROGENEOUS ENVIRONMENTS

K. Yoshihara and K. Kemnitz[*]

Institute for Molecular Science, Myodaiji, Okazaki 444, Japan

Surfaces exert an ordering influence on adsorbed molecules. The macroscopic confinement in two dimensions, in combination with a distribution of adsorption sites of two- and three-dimensional microscopic geometries, and the presence of surface-specific unidirectional forces, results in a rich photophysical and photochemical behaviour of aromatic molecules in the adsorbed state.

1. INTRODUCTION

Excited-state dynamics of organic molecules are widely used to monitor physico-chemical properties of micro-environments immediately adjacent to the probe molecule. In this review we describe the temperature-dependent influence of surfaces on the rate of internal conversion (IC),[1-3] and on the rates of energy[4] and electron[5-11] transfer, and compare some of the results with that obtained in solid matrix.[2,3] Surfaces also can have a profound influence on the state of aggregation and on the kinetics of the adsorbed dimers.[4,12] As probe molecules served rhodamine B (RhB), rhodamine 101 (Rh101), pyronine B (PyB), thiacarbocyanine (TCC), pseudoisocyanine (PIC), and malachite green (MG). Examples of surface-specific physico-chemical behaviour will be given for: (a) 2-dimensional energy transfer,[4] (b) the presence of fluorescent dimers in the adsorbed state at increased surface coverage,[4,12] (c) temperature-dependent equilibrium between slow and fast adsorption sites,[7] (d) solvent effect on adsorbed molecules,[1,5,7] (e) electron transfer between adsorbed monomers and organic[5,6,8,10] and inorganic[7] semiconductor substrates in dry and wet systems, and between J-aggregates and AgBr microcrystals,[9,11] (f) retardation and enhancement of IC by regular and irregular adsorption sites,[1,4] respectively, (g) the effect of surface-adsorbate vibration,[2,7] and (h) a free-volume effect in the adsorbed state.[1-3]

---

[*] present address: Microphotoconversion Project, ERATO Program, 15 Morimoto-cho, Shimogamo, Sakyo-ku, Kyoto 606, Japan.

Figure 1 schematically describes   observed surface phenomena
discussed in the present work.

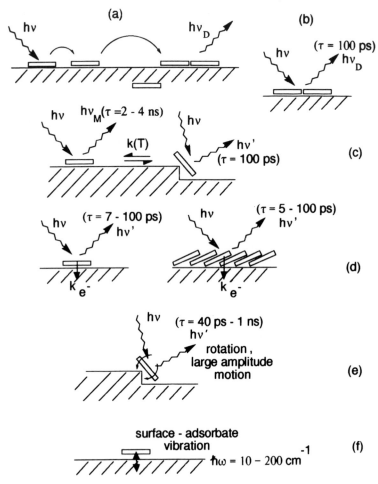

FIGURE 1
Schematic representation of surface-induced physico-chemical
response by excited dye molecules in the adsorbed state: (a)
Energy migration among monomers and transfer of excitation energy
to nonfluorescent dimer: two-dimensional Förster energy transfer.[4]
(b) Emission by fluorescent dimers with short lifetime of about
100 ps.[4,12] (c) Thermal equilibrium between sites of long (planar
site) and short (nonplanar site) fluorescence lifetimes.[7,9] (d)
Electron transfer of monomer[5-7] (left) and J-aggregate[9,11] (right)
in solvent-free adsorption systems with unusually small reorientation energy. (e) Free-Volume effect experienced by large aromatic
molecules, adsorbed at distorted, nonplanar sites: relaxation of
the excited molecule into the free-volume and resultant enhanced
rate of IC.[1-3] (f) Surface-Adsorbate vibration with its potential
influence on rates of internal conversion, energy, and electron
transfer.[2,7]

The above interfacial phenomena were applied to monitor the quality and site-distribution of organic single crystal surfaces and its surface phase transition,[1] as well as the liquid-solid phase transition between melt and microcrystalline phase.[13] The temperature-dependent free-volume on surfaces in relation to that in solid matrix had been investigated by studying isomerization of cyanine dyes and the rate of IC of malachite green.[2-3] Surfaces can produce unexpected and novel effects, so we recently observed a "transient, isoemissive point" in the temperature-dependent fluorescence decays of cyanine dyes adsorbed on silica gel.[14]

2. PHOTOPHYSICS AND CHEMISTRY IN THE ADSORBED STATE

Figure 2 introduced the molecular probes used to investigate the influence of surfaces on internal conversion and other physicochemical properties of adsorbed molecules.

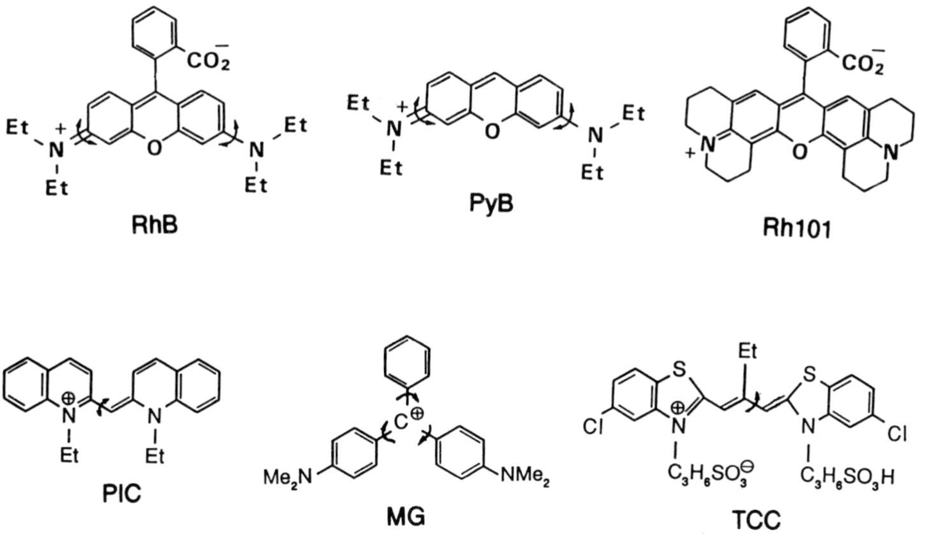

FIGURE 2
Dye molecules used as fluorescent probes in the present surface studies. The arrows indicate molecular motions, leading to enhanced internal conversion.

2.1. Energy Transfer in Two Dimensions

Two-dimensional Förster energy transfer from excited monomer species to nonfluorescent dimers of RhB is observed at increasing surface coverage, as shown in Figures 1a and 3.

2.2. Fluorescent Xanthene Dimers in the Adsorbed State

Dimers, usually considered as nonfluorescent in homogeneous solution, become strongly fluorescent in the adsorbed state (Figure 1b), and are even dominating the fluorescence decays of monolayer systems. Figure 3a shows the exponential contribution of fluorescent in-plane surface dimers to the overall fluorescence decay, which also includes minor contributions from Förster energy transfer.

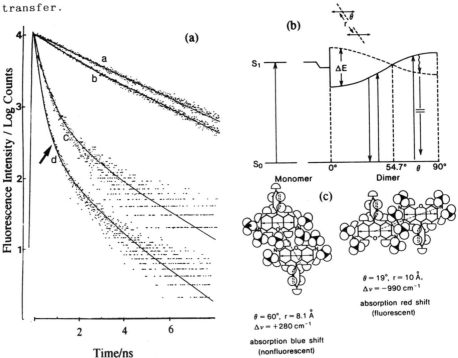

FIGURE 3

(a) Fluorescent xanthene dimers in the adsorbed state: exponential decay (arrow) of dimers (τ_D = 100 ps) in a monolayer, d, of RhB adsorbed on quartz. The total decay can be analyzed by $I(t) = A_1 \exp(-t/\tau_D) + A_2 \exp[-t/\tau_M - 1.354\Gamma(t/\tau_M)^{1/3}]$, where the second term represents two-dimensional Förster-type energy transfer.[4] Decays a, b, and c correspond to coverages of about 0.01, 0.1, and 0.5, respectively. (b) Simple exciton model and (c) two possible geometries of fluorescent (right) and nonfluorescent (left) RhB in-plane dimers, potentially existing in the adsorbed state.[12]

2.3. Electron Transfer in Solvent-Free Adsorption Systems

Solvent-free adsorption systems, comprised of singlet excited dye and substrate electron donor or acceptor (Figure 1d), are mainly distinguished, in the absence of solvent, by their excep-

tionally small total reorientation energy. Values of 0.15 - 0.20
eV had been found for electron transfer in the systems of
RhB/organic single crystals, RhB/semiconductors, and for J-aggre-
gates/AgBr. Such small reorientation energy results in a very slim
parabola of Marcus-type, characterized by a very steep increase
of the rate constant of electron transfer with decreasing free
energy gap (Figure 4a, ref. 11).

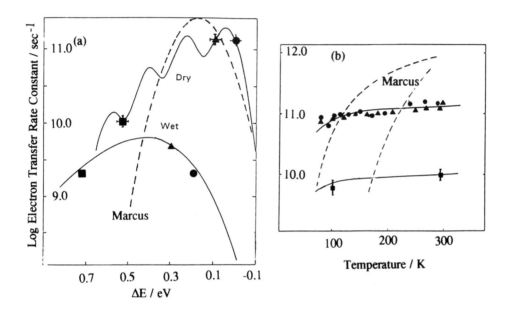

FIGURE 4
Electron transfer in the solvent-free adsorption system of
RhB/organic single crystals:[5] anthracene (circle), pyrene (trian-
gle), and perylene (square). Inverted region-type behaviour and
strong deviation from classical Marcus theory (dashed) of dry and
wet systems (a), and weakly activated electron transfer rates of
dry system (b).

A second prominent feature of the above electron transfer
systems is the nearly temperature-independent rate of electron
transfer (Figures 4b, 5, and 6a), which is in agreement with
quantum-mechanical theory (solid lines in Figure 4b) and in strong
deviation from the classical prediction (dashed lines).

Temperature-independent rate constants of electron transfer in
the range of 300 - 4 K had been observed for xanthene dyes adsorbed on polycrystalline films of electron-donating or electron-
accepting aromatic hydrocarbons.[10]

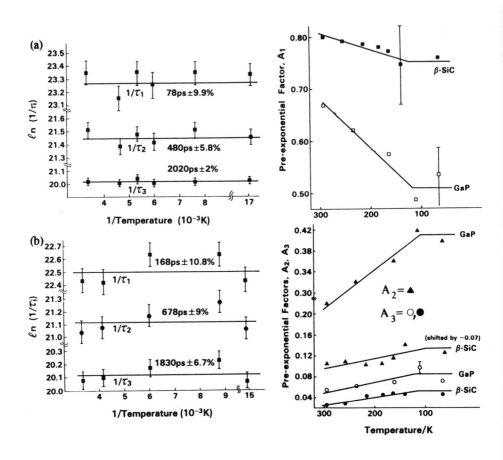

FIGURE 5

Electron transfer in the solvent-free system of RhB adsorbed on single crystals of inorganic semiconductors: temperature-independent fluorescence lifetimes (left) of ß-SiC (a) and GaP (b), and strongly temperature-dependent pre-exponential factors of three-exponential fluorescence decays (right).[7]

2.4. Temperature-Dependent Equilibrium of Populations of Slow and Fast Adsorption Sites

Figure 1c schematically describes an equilibrium between molecules adsorbed at ideal (planar) and nonideal (step) sites. In case of RhB/ß-SiC and RhB/GaP, these sites seem to correspond to slow and fast rate of electron transfer, respectively, and an equilibrium between both species is apparent in temperature-dependent pre-exponential factors of two- or three-exponential fluorescence decays (Figure 5, right).

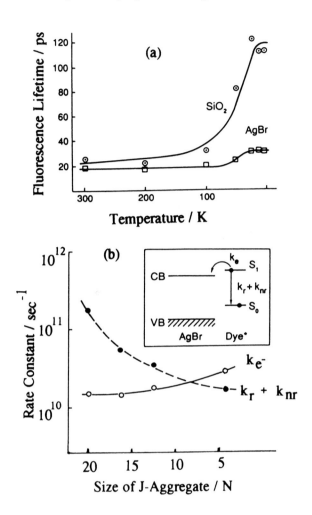

FIGURE 6

Electron transfer in the solvent-free system of J-aggregates of thiacyanine adsorbed on AgBr microcrystals, and comparison with the reference system of J-aggregate/silica gel (no electron transfer). (a) Plot of fluorescence lifetimes vs. temperature. (b) The rate constant of electron transfer and the sum of radiative and nonradiative rate constants as a function of aggregate size are shown for the AgBr system.[9]

2.5. Free-Volume in the Adsorbed State

Figures 7 compares the temperature-dependent fluorescence decays of MG in solid matrix of EPA (a) with that in the adsorbed state (b). A drastic difference in the temperature behaviour of both systems is obvious: strong dependence in solid matrix and almost complete independence in the adsorbed state.

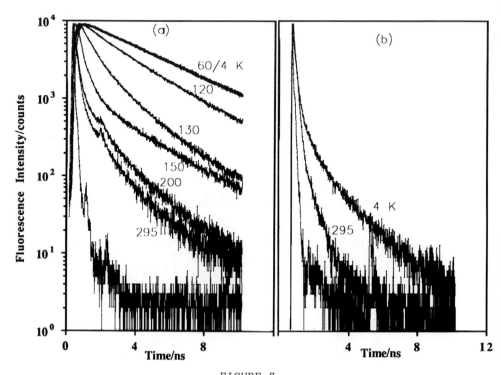

FIGURE 7

Rate of IC as a function of the environment: MG as a sensitive
free-volume probe. Strongly temperature-dependent fluorescence
decays in EPA matrix (a) and almost temperature-independent decays
in the adsorbed state (b).[2-3]

Surface-induced changes in the potential surfaces, especially
in that of the excited state, can explain the temperature-independent internal conversion of MG, and that of cyanine and xanthene dyes in the adsorbed state. Figure 8 compares the potential
surfaces in homogeneous (dashed) and heterogeneous environment.
Unidirectional surface-forces, in the presence of free-volume,
alter the excited state potential, creating an all-downhill reaction path at the surface-adjacent side, which is made responsible
for the observed temperature-independent fluorescence decays.
Figure 8b displays the model compound stilbene, adsorbed at an
idealized step-site, and its "pseudoisomerization" in the excited
state. Note that free-volume, as found at distorted sites, is
necessary for such large-amplitude motion in the adsorbed state.
Similarly in the case of xanthene dyes, free-volume in combination
with unidirectional   surface-forces are thought to allow the

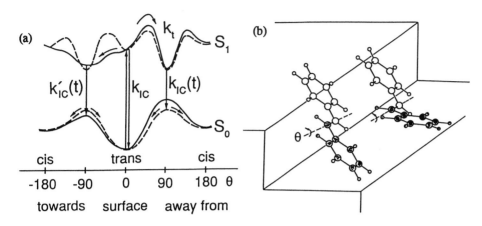

FIGURE 8
Internal conversion and surface-modified potential surfaces (a) of
large organic molecules adsorbed at distorted sites, and relaxation into free-volume in the excited state (b).[2]

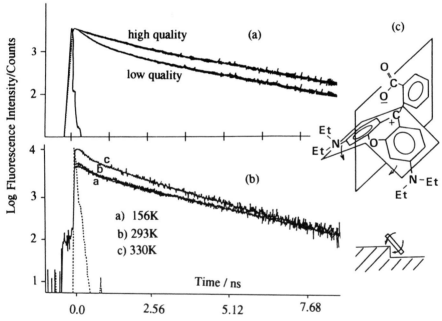

FIGURE 9
(a) RhB as surface probe:[1] low-quality single crystals of phenanthrene are revealed by their large contribution (40%) of short
fluorescence lifetime (τ_1 = 200 ps), the high-quality crystal has
a contribution of 26% only, in contrast. (b) Increase of distorted
sites on surface of phenanthrene single crystal, when passing
through the bulk phase transition, ranging from 318 to 345 K. (c)
Potential mechanism of enhanced IC of RhB, Rh101, and PyB adsorbed
at distorted sites: presumed butterfly-type motion of the xanthene
skeleton, which is similar to the relaxation mechanism of triphenylmethane dyes.

postulated butterfly-type relaxation mechanism, as shown in Figure 9c. RhB is shown as oxygen-bridged analog to triphenylmethane dyes with its xanthene skeleton bending in the excited state.

Figure 9 shows the application of RhB as surface probe, monitoring surface quality and the change of adsorption site distribution during a phase transition.

2.6. Surface-Adsorbate Vibration

A surface-adsorbate vibration, schematically depicted in Figure 1f, can influence the rates of energy and electron transfer and that of internal conversion. Figure 10a shows the mean square displacement in dependence of temperature and as a function of the surface-adsorbate vibrational frequency, and Figure 10b displays the probability of displacement from the equilibrium value. For a frequency of 100 cm^{-1}, the vibration is completely frozen for temperatures below 50 K, and already at 100 K, the probability to find the molecule at a distance of 0.2 Å from the equilibrium position is more than 10 times lower compared to room temperature.[2,7] Energy and electron transfer from an adsorbed molecule to the substrate carry an intrinsic temperature dependence of $1/r^3$ and $exp(-r)$, respectively, and thus are influenced by the probability $p(x)$ of finding the molecule at a given distance x from the equilibrium value x_0 (Figure 10b). Figure 10 uses the harmonic approximation for qualitative visualization of the intrinsically inharmonic situation on the surface. In case of IC at low temperatures, a smaller average separation from the surface in case of an anharmonic potential, will result in enhanced rigidity and reduced IC.

2.7. Transient, Isoemissive Point

Figure 11 shows the peculiar temperature-dependent fluorescence decays of J-aggregate adsorbed on silica gel, intersecting in an unique "transient, isoemissive point". The analysis in terms of three exponentials reveals the surprising constancy of the integrated fluorescence to within 5%.[14] This novel phenomenon is not yet understood in detail and is mentioned here to demonstrate the surprising variety of physicochemical behaviour of aromatic molecules in the adsorbed state.

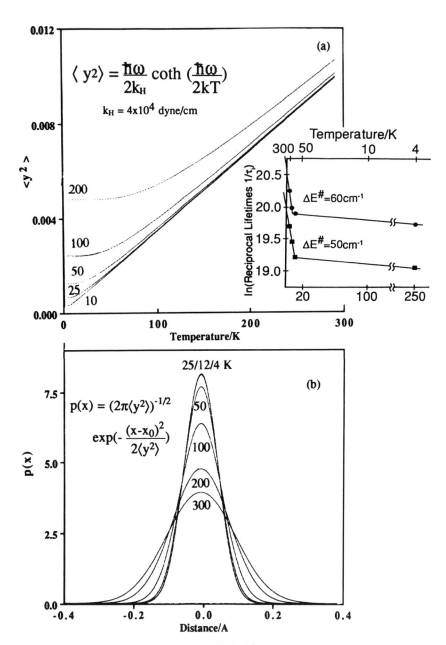

FIGURE 10

Surface-Adsorbate vibration: mean square displacement $\langle y^2 \rangle$ vs. temperature as function of frequency. The inset shows the peculiar temperature dependence of TCC adsorbed on silica gel that might be caused by the above surface-specific vibration.[2] (b) Analogous display of probability of displacement, p(x), from the potential energy minimum at x_0, for $\hbar w = 100$ cm^{-1}.

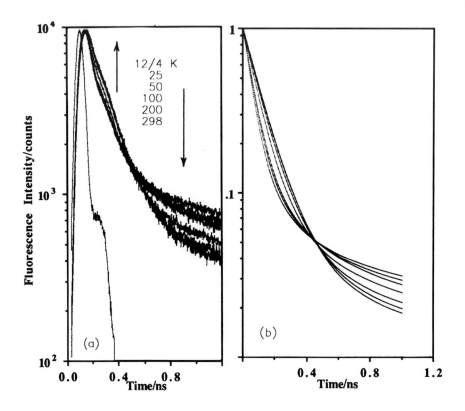

FIGURE 11

(a) Transient, isoemissive point of temperature-dependent fluorescence decays of J-aggregates in the adsorbed state. (b) Recalcula-
tion, using parameters obtained from three-exponential analysis.[14]

3. CONCLUSIONS

The presence of surfaces drastically influences a broad array
of physico-chemical properties of aromatic molecules in the adsorbed state. Sensitive picosecond spectroscopy in combination
with temperature studies in the wide range from 300 to 4 K, accessible in solvent-free systems, has been shown to be a valuable
tool in revealing surface-induced deviations from standard homogeneous behaviour.

References

1) K. Kemnitz, N. Tamai, I. Yamazaki, N. Nakashima, and K. Yoshihara, J. Phys. Chem. 91 (1987) 1423.

2) K. Kemnitz and K. Yoshihara, J.Phys.Chem. (1990), in press.

3) K. Kemnitz and K. Yoshihara, Chem.Lett. (1990) 1789.

4) K. Kemnitz, T. Murao, I. Yamazaki, N. Nakashima, and K. Yoshihara, Chem.Phys.Lett. 101 (1983) 337.

5) K. Kemnitz, N. Nakashima, and K. Yoshihara, J.Phys.Chem. 92 (1988) 3915.

6) K. Kemnitz, N. Nakashima, and K. Yoshihara, in 'Photochemistry on Solid Surfaces' eds. M. Anpo and T. Matsuura, Elsevier Science, Amsterdam, 1989.

7) K. Kemnitz, N. Nakashima, and K. Yoshihara, J.Phys.Chem. 93 (1989) 6704.

8) K. Yoshihara, N. Nakashima, and K. Kemnitz, in 'Lasers in Atomic, Molecular, and Nuclear Physics' ed. V.S. Letokhov, World Sci. Publ. 1989.

9) K. Kemnitz, K. Yoshihara, and T. Tani, J.Phys.Chem. 94 (1990) 3099.

10) K. Kemnitz and K. Yoshihara, to be submitted.

11) T. Tani, T. Suzumoto, K. Kemnitz, K. Yoshihara, in preparation.

12) K. Kemnitz, N. Tamai, I. Yamazaki, N. Nakashima, and K. Yoshihara, J.Phys.Chem. 90 (1986) 5094.

13) K. Kemnitz and K. Yoshihara, to be submitted.

14) K. Kemnitz, K. Yoshihara, T. Tani, Chem.Lett. (1990) 1785.

Chapter II:
Photoredox Reactions in Solution

Photochemical Processes in Organized Molecular Systems
K. Honda (Editor-in-Chief)
© Elsevier Science Publishers B.V., 1991

INTRAMOLECULAR, PHOTOCHEMICAL ELECTRON AND ENERGY TRANSFER

Thomas J. MEYER

Department of Chemistry, University of North Carolina at Chapel Hill, Chapel Hill, North Carolina 27599-3290

A series of chromophore-quencher complexes has been prepared based on polypyridyl complexes of Ru^{II}, Os^{II} or Re^{I}. Following metal-to-ligand-charge-transfer (MLCT) excitation of these complexes, intramolecular electron or energy transfer processes occur rapidly and with high efficiencies. These complexes may provide the building blocks for more complex molecular assemblies that can function as devices.

1. INTRODUCTION

It is a characteristic feature of polypyridyl complexes of Ru^{II}, Os^{II}, and Re^{I} that intense absorption bands appear in their visible and low-energy, ultra-violet spectra. These bands arise from metal-to-ligand-charge-transfer (MLCT) transitions, eq 1. [1-3]

$$Ru(bpy)_3^{2+} \xrightarrow{h\upsilon} Ru^{III}(\overline{bpy})(bpy)_2^{2+*}$$

$$(d\pi)^6 \qquad\qquad (d\pi)^5(\pi^*)^1 \qquad\qquad (1)$$

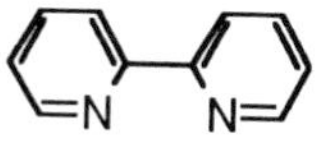

(bpy)

Typically, the excited states that result from these transitions are stable and have been shown to undergo facile oxidative or reductive electron transfer, eq 2, 3. [4,5]

$$Ru(bpy)_3^{2+*} + PQ^{2+} \longrightarrow Ru(bpy)_3^{2+} + PQ^+$$

$$(2)$$

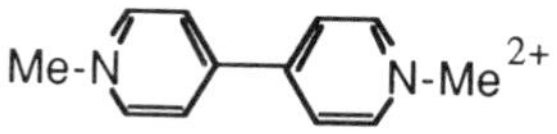

$Me\text{-}N$ ⃝ ⃝ $N\text{-}Me^{2+}$

(PQ^{2+})

$$\text{Ru(bpy)}_3^{2+*} + \text{10-MePTZ} \longrightarrow \text{Ru(bpy)}_3^{2+} + \text{10-MePTZ}^+$$

$$(3)$$

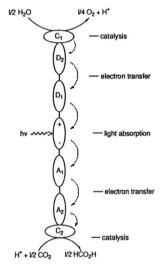

(10-MePTZ)

There is a well-developed synthetic chemistry for modifying these complexes. When combined with their excited state properties, this leads to the possibility of creating complex molecular assemblies in which the directionality and extent of electron or energy transfer can be controlled. If this can be achieved, it may be possible to prepare devices whose properties depend on molecular level events and can be controlled in a systematic way by synthetic modification.

One target is the preparation of artificial systems that mimic the behavior of the reaction center of photosynthesis. A model for such a device is illustrated in Figure 1.

Figure 1. Molecular, Photochemical-Electrochemical Reactor

In this device, single photonic excitation of a chromophore is followed by a sequence of electron transfer events. Directionality in the electron transfer chain is built in by controlling the redox potentials of the individual donor (D) and acceptor (A) sites thus creating a "free energy cascade". The outcome of this sequence is that oxidative equivalents are swept by a "free energy broom" to a catalyst site when the oxidation of water to dioxygen occurs. Reductive equivalents are swept to a second site where carbon dioxide is reduced. It is well-separated spatially in order to avoid back electron transfer. In the net sense, the device provides a way to transfer electrons from water to carbon dioxide upon illumination

A second target is illustrated in Figure 2.

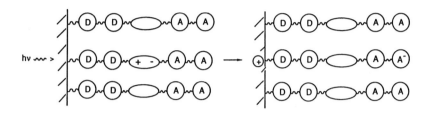

Figure 2. Directed Electron Transfer at a Molecular-Electrode Interface.

In this example the goal is to construct a related, molecular-level apparatus, but attached in a spatially controlled way to an electrode surface. Upon excitation and electron transfer, such a device would provide a basis for creating a photopotential by transferring oxidative equivalents to the molecular-electrode interface and reductive equivalents to the molecular-solution interface. This structure could also provide a basis for device-like applications.

Both targets cited here are, at least for now, only goals. A complete understanding of the complexities involved and the details of the underlying chemistry remain to be developed. The goal of this account is to summarize the progress that has been made in constructing photochemical-electron transfer components that might find a place in such devices. These components are based on MLCT excited states of polypyridyl complexes of Ru^{II}, Os^{II}, or Re^{I}.

2. MLCT EXCITED STATES

The electronic structures of polypyridyl complexes of Ru^{II}, Os^{II}, and Re^{I} are complex.[2,3,6,7] In their excited state manifolds there are series of low-lying excited states whose origins are metal-centered, (dd), ligand-centered, ($\pi\pi^*$), or MLCT in character. Visible absorption spectra are dominated by MLCT transtions to states that are largely singlet in character. Emission occurs from orbitally equivalent states which are largely triplet in character. These are not pure spin states, they are mixed by spin-orbit coupling. The spin-orbit coupling constants for these ions are in the range 1000-3000 cm^{-1}.

Typical examples of complexes of this type are $Os(phen)_3^{2+}$, $Re(bpz)(CO)_3Cl$, or $Ru(bpy)_3^{2+}$. One of their virtues in excited state studies, is the versatility that exists in the synthetic chemistry that is used to prepare them. Systematic changes in properties can be made either by changing the metal or the surrounding ligands. The ability to modify ligands and to exploit the coordination chemistry of the metal ions has provided the basis for the chemistry that is described below. It also provides a basis for preparing more complex structures in the future.

(phen)   (bpz)

Many fundamental studies have been devoted to establishing the photophysical properties of MLCT excited states.[1-4, 6-11] Based on these studies it has been established that: 1) There are typically a series of low-lying MLCT excited states that contribute to excited state properties.[12] There are additional, low-lying MLCT and dd states that can play a role in nonradiative decay. This is especially true near room temperature where population of these states can become significant.[10,13] Population of dd states following MLCT excitation can lead to ligand loss photochemistry.[10] 2) The excited states are not highly distorted. The individual bond displacements in the C-C and C-N bonds of the polypyridyl ligands, which act as electron acceptors, are in the range of 0.01-0.02 Å.[10e,f,14] These estimates have been made based on resonance Raman measurements and by extracting structural parameters from emission spectral profiles by using a Franck-Condon analysis. 3) Because of the relatively small distortions in the excited states, nonradiative decay follows the energy gap law, eq 4.[15,16]

$$k \; \alpha \; e^{-\gamma E_0 / h\upsilon} \qquad (4)$$

$$\gamma = \ln \frac{E_0}{Sh\upsilon} - 1$$

In this equation E_0 is the energy gap, and $h\upsilon$ is the quantum spacing for the average of the several (6-8) ring stretching modes that act as the dominant energy acceptors. The quantity S is the electron-vibrational coupling constant. It is proportional to the square of the difference in equilibrium displacement between the excited and ground states for the average vibration. Low frequency modes and the solvent play a lesser role as energy acceptors. 4) By analyzing emission spectral profiles, it is possible to extract the kinetic parameters that allow relative rate constants for non-radiative decay to be calculated.[10e,f,15] 5) The free energies of the excited states above the ground states can also be calculated by the results of emission spectral fitting. When combined with ground state redox potentials, this quantity allows redox potentials for excited state couples to be estimated, eq. 5. Depending on the metal-ligand combination, these potentials can be varied over ranges that can exceed 1 V.[10d]

$$Ru^{III}(b\bar{p}y)(bpy)^{2+*} + e \longrightarrow Ru^{III}(b\bar{p}y)(bpy)_2^+ \quad (5a)$$

$$Ru^{III}(bpy)_3^{3+} + e \longrightarrow Ru^{II}(b\bar{p}y)(bpy)_2^{2+*} \quad (5b)$$

These excited states are capable of undergoing facile electron or energy transfer. Examples of electron transfer to or from $[Ru(bpy)_3]^{2+*}$ were shown in eq 2 and 3. An example of energy transfer is shown in eq 6.[17]

$$Ru(bpy)_3^{2+*} + An \longrightarrow Ru(bpy)_3^{2+} + An$$

$$(6)$$

(An)

The variation of rate constant k with driving force for these reactions can be accounted for by available theories.[4,5,18] These analyses show that the excited states are quite facile towards either electron or energy transfer. The kinetic facility has two origins. The first is in the relatively minor structural changes that exist between the ground state and the oxidized or reduced products,e.g., $Ru(bpy)_3^{3+}-Ru(bpy)_3^{2+}-Ru(bpy)_3^+$, and between the excited states, $[Ru(bpy)_3]^{2+*}$, and the oxidized or reduced products. These complexes are also relatively large, the molecular radius of $[Ru(bpy)_3^{2+}]$ is ~7 Å. Because of the large molecular volumes, the reorganizational energy that arises from the reorientation of solvent dipoles during electron transfer is relatively small.

In summary, based on polypyridyl complexes of Ru^{II}, Os^{II}, and Re^I, access is available to a family of well-defined excited states. The properties of these excited states can be modified systematically by making synthetic changes and they have the ability to undergo rapid electron or energy transfer.

3.   PREPARING MOLECULAR ASSEMBLIES. THE BUILDING BLOCKS

Well-defined synthetic techniques are available for preparing mixed-ligand, polypyridyl complexs of Ru^{II}, Os^{II}, or Re^I.[19,20] The other ligands that have been included in the coordination spheres include nitriles, phosphines, arsines, pyridines, halides or CO. When combined with the organic chemistry used to modify the polypyridyl ligands, the coordination chemistry has led to the preparation of a series of complexes which contain, at the same time, an MLCT chromophore and attached electron or energy transfer donors or acceptors. The resulting, chromophore-quencher complexes provide useful building blocks for the preparation of more complex structures.

The organic chemistry for preparing derivatized pyridyl or polypyridyl ligands which contain attached energy or electron transfer groups has been developed in some detail.[19d-f] Much of this chemistry relies on initial deprotonation or partial oxidation of the derivative 4,4'-dimethyl-2,2'-bipyridine. With functional groups such as -CH$_2$Br or -CHO at the 4,4' position of bpy, there are a number of linking possibilites.

By exploiting these procedures, it has been possible to prepare several ligands of this type. Examples are illustrated below where there are linked anthryl groups for energy transfer and either pyridinium or phenothiazyl groups for electron transfer.

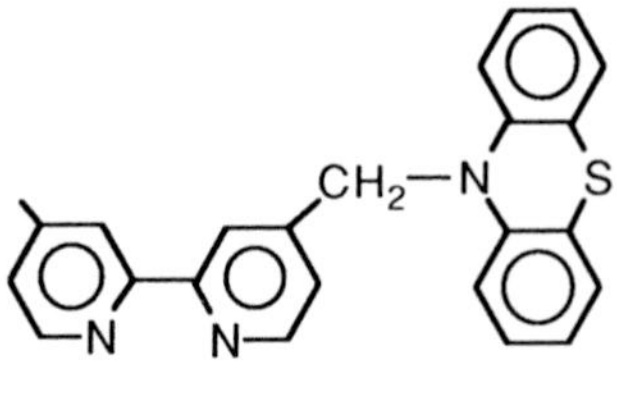

(bpy-PTZ)

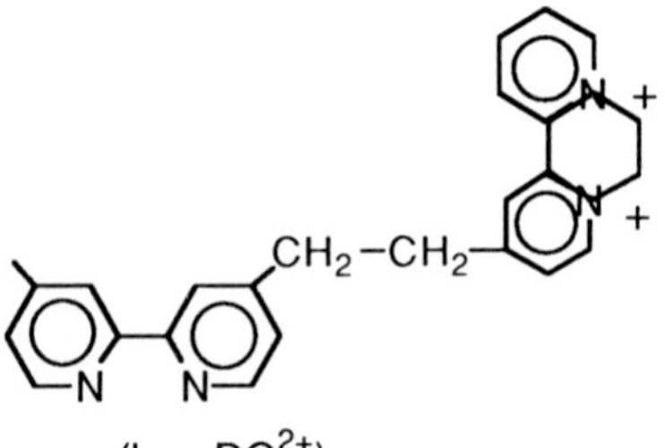

(bpy-DQ^{2+})

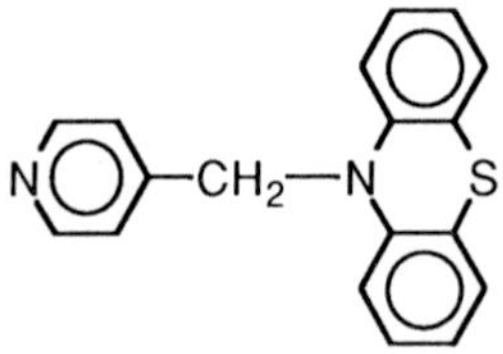

(py-PTZ)

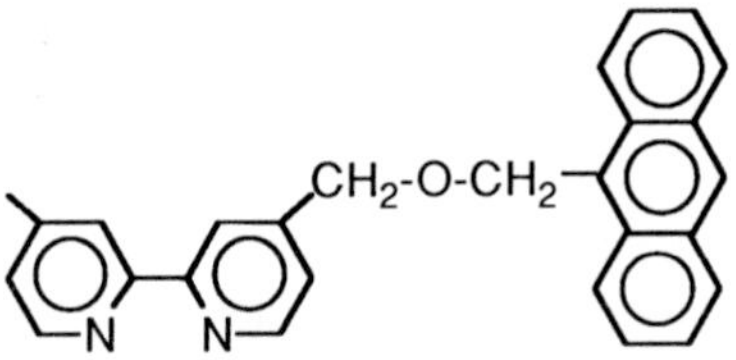

(bpy-An)

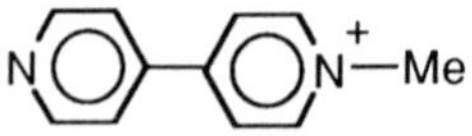

(MQ$^+$)

These ligands provide a useful starting point for the assembly of intramolecular electron or energy transfer components.

Synthetic routes are also available for assembling components into multi-molecular arrays. In our own work, we are exploring a number of approaches to the problem of assembly.[21,22] In one approach we are attempting to fabricate controlled microstructures in thin polymeric films. In another, the goal is to grow complex molecular structures directly on the surfaces of electrodes by using peptide links. We are also exploring the preparation of complex, soluble, molecular assemblies by using two different approaches. One involves linking metal complexes by ligand-bridges. This is an off-shoot of earlier work in mixed-valence chemistry.[23,24] By using well-established synthetic procedures, it has been possible to prepare ligand-bridged complexes such as the one shown below.

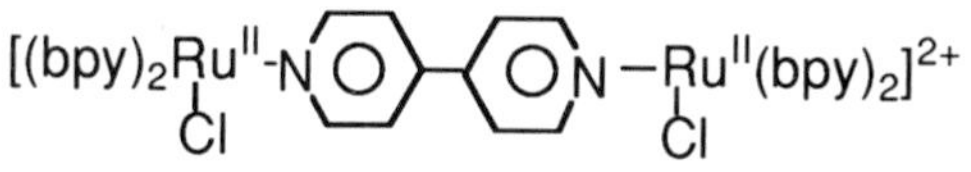

$$[(bpy)_2Ru^{II}\text{-}N\underset{Cl}{\bigcirc}\text{-}\bigcirc N\text{-}\underset{Cl}{Ru^{II}}(bpy)_2]^{2+}$$

Synthetic procedures for the prepartion of oligomers, in which more than two metal complex units are linked, have also been developed. [25] In these complexes, electronic coupling through the bridging ligands provides the electronic basis for electron transfer. The extent of electronic interaction can be controlled by varying the bridging and non-bridging ligands in a systematic way.

A second approach to assembling soluble systems is based on the utilization of modified polymers. There is an extensive background literature in this area as well. [26-28] In our own efforts we have developed the chemistry of modified copolymers of styrene/m, p-chloromethylated styrene through the attachment of metal complex chromophores, and electron or energy transfer acceptors or donors. The synthetic chemistry is based on the nucleophilic displacement of the chloro groups.[28] The structure of a repeating unit containing a Ru-bpy chromophore is illustrated below.

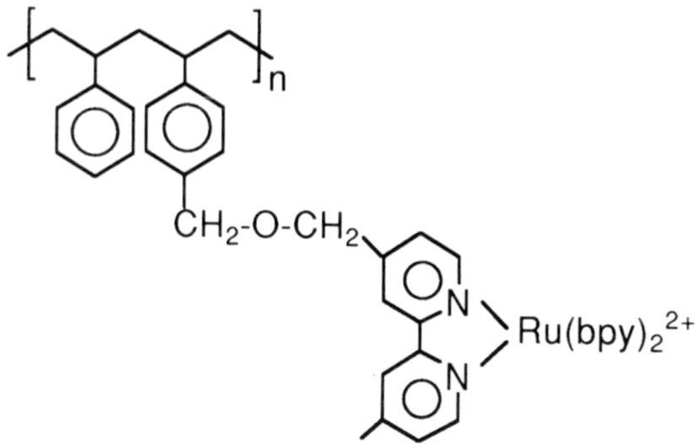

Based on these soluble polymers, a number of interesting materials have been prepared and it has been possible to demonstrate that photochemically-induced, intramolecular, electron or energy transfer processes can occur. In one example, it was possible to demonstrate that long-range energy transfer could occur between metal complex sites through intervening, chemically-linked anthracene groups which were present on the same polymeric backbone. [29]

4.   INTRAMOLECULAR, PHOTOCHEMICAL ELECTRON TRANSFER

With the synthetic chemistry of the coordination complexes and of the ligands in hand, it has been possible to prepare a series of chromophore-quencher complexes. The excited state properties of these complexes have been examined in a systematic way by using laser flash photolysis. An example that has been studied in some detail is illustrated below.

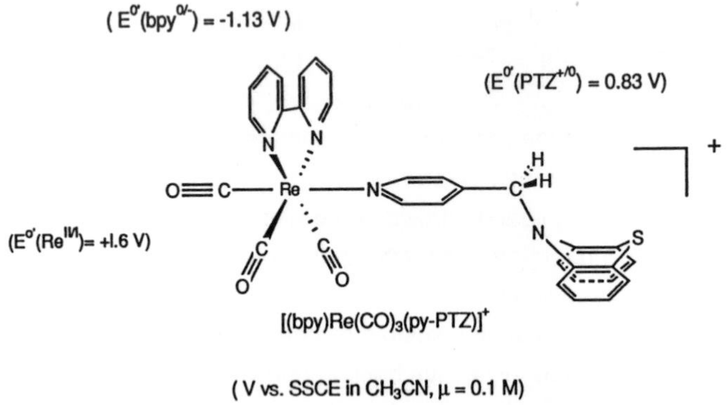

$(E^{\sigma}(bpy^{0/-}) = -1.13 \, V)$

$(E^{\sigma}(PTZ^{+/0}) = 0.83 \, V)$

$(E^{\sigma}(Re^{II/I}) = +1.6 \, V)$

$O \equiv C - Re - N$

[(bpy)Re(CO)$_3$(py-PTZ)]$^+$

(V vs. SSCE in CH$_3$CN, $\mu = 0.1$ M)

Redox potentials for the various, reversible couples that appear in this molecule are also indicated in the structure. They reveal that there is a reversible -PTZ$^{+/0}$ based couple at 0.83 V vs. SSCE in acetonitrile ($\mu = 0.1$ M). In addition, couples for reduction at the bpy ligand and for oxidation of Re^I to Re^{II} appear in the cyclic voltammograms.

In this complex, electronic coupling between the electron transfer donor, -PTZ, and the metal complex chromophore is weak because of the intervening -CH$_2$- link. Following Re^I → bpy excitation, two transient processes are observed. In the first, the characteristic $\pi \to \pi^*$ transition of the bpy$^-$ group in the MLCT excited state is rapidly depleted as an absorbance for -PTZ$^+$ at 510 nm grows in. By time-resolving the two components, it was possible to obtain rate constants for both intramolecular processes that occur following MLCT excitation.[30] These events are summarized in Scheme 1.

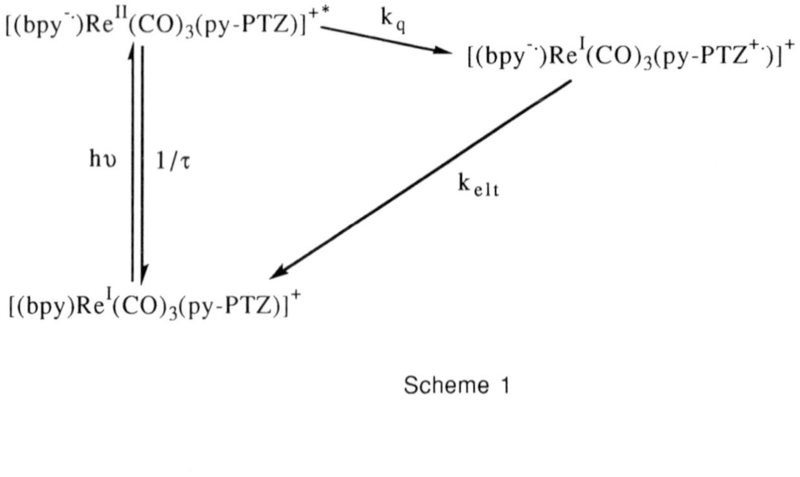

$[(bpy^{\cdot -})Re^{II}(CO)_3(py\text{-}PTZ)]^{+ *}$ $\xrightarrow{k_q}$ $[(bpy^{\cdot -})Re^I(CO)_3(py\text{-}PTZ^{\cdot +})]^+$

$h\upsilon \quad | \quad 1/\tau$

k_{elt}

$[(bpy)Re^I(CO)_3(py\text{-}PTZ)]^+$

Scheme 1

This experiment demonstrated that intramolecular, reductive quenching can occur following MLCT excitation, that the quenching step can be rapid, and that the resulting, redox-separated state, can have a reasonably long lifetime. Our best estimates suggest that the efficiency of electron transfer quenching in this system is very high, well above 50%.

An electrochemical map of a related complex, in which the electron transfer donor -PTZ has been replaced by a pyridinium acceptor, is illustrated in the structure below.

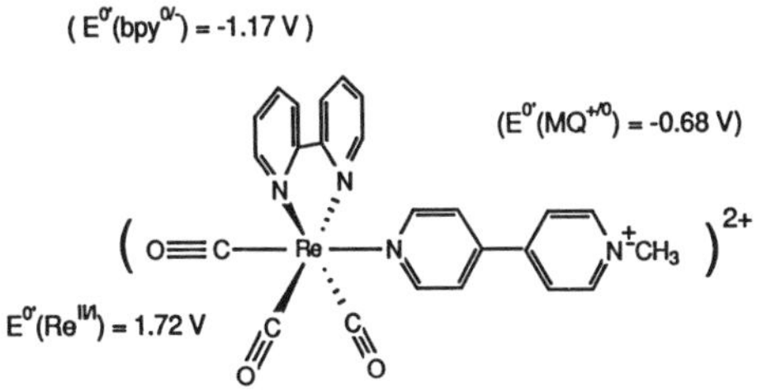

$(E^{\sigma}(bpy^{0/-}) = -1.17\,V)$

$(E^{\sigma}(MQ^{+/0}) = -0.68\,V)$

$E^{\sigma}(Re^{II/I}) = 1.72\,V$

$[(bpy)Re(CO)_3(MQ^+)]^{2+}$

$(V\ vs.\ SSCE\ in\ CH_3CN,\ \mu = 0.1\ M)$

As an electron acceptor, the pyrydinium site is better than bpy by ~0.3 V. Because of strong electronic coupling through the attached pyridyl group, charge transfer bands appear in the absoprtion spectrum arising from both $Re^I \rightarrow bpy$ and $Re^I \rightarrow MQ+$ transitions.

Following $Re^I \rightarrow bpy$ excitation of this complex in polar organic solvents, clear evidence for a transient species is obtained. A characteristic feature that appears in the transient absorbance difference spectrum is a new absorbance at ~610 nm. This feature can be assigned to a $\pi \rightarrow \pi^*$ transition at reduced MQ+. Based on the experimental observations, including the time resolution of the transient absorbance signal, it was possible to construct the diagram shown in Scheme 2.[31]

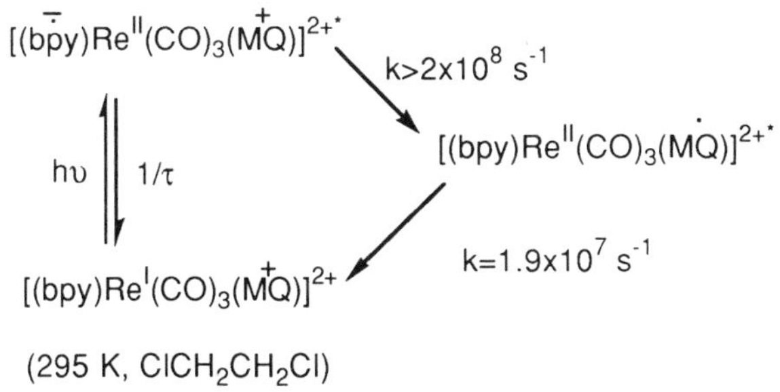

$[(\overline{bpy})Re^{II}(CO)_3(\overset{+}{MQ})]^{2+*}$

$k>2\times10^8\ s^{-1}$

$h\upsilon \quad 1/\tau$

$[(bpy)Re^{II}(CO)_3(\overset{\cdot}{MQ})]^{2+*}$

$k=1.9\times10^7\ s^{-1}$

$[(bpy)Re^{I}(CO)_3(\overset{+}{MQ})]^{2+}$

(295 K, $ClCH_2CH_2Cl$)

Scheme 2

In this scheme, intramolecular electron transfer leads to the lower, MQ-based, MLCT excited state. It decays with a rate constant of 1.9×10^7 s^{-1} in 1,2-dichloroethane at 295 K. Only a lower limit for the preceding bpy → MQ$^+$ electron transfer reaction is available from our measurements. In this case, intramolecular electron transfer is an intraconversion between two different excited states of the system. The redox-separated state is also an excited state because of the strong electronic coupling that exists between Re^{II} and MQ$^.$. The MQ-based excited state is a weak emitter and can be observed directly by emission spectroscopy. Our estimates suggest that intramolecular electron transfer, which in this case is an interconversion between excited states, is highly efficient in this system as well.

It is also possible to combine both the oxidative and reductive features in a single molecule based on a MLCT chromophore as shown by the example below.

Our work on this complex has been a collaborative effort with the research group of Professor C. M. Elliott at Colorado State University. In the complex there are electron transfer donors and acceptors bound to different bipyridyl ligands in a single complex. As inferred by the structure, following Ru^{II} → bpy excitation, a series of rapid, intramolecular electron transfer events occur.[32] In this case there are oxidation state markers in the transient absorbance difference spectrum which point towards the production of separated oxidative equipments on -PTZ and reductive equipments on the bipyridinium. The lifetime of the resulting redox-separated state depends somewhat on the solvent, but is in the range of 100 ns. In this case, there is sufficient flexibility in the -(CH$_2$)$_4^-$ connecting links that back electron transfer may occur following segmental rotational motions which would allow the two sites to come into relatively close contact.

In a somewhat related experiment, it has been possible to show that MLCT excitation can lead to intramolecular redox splitting across a ligand bridge.[33] In the case shown below, the synthetic chemistry was developed for linking two different Re^I groups across 4,4'-bipyridine as the bridging ligand.

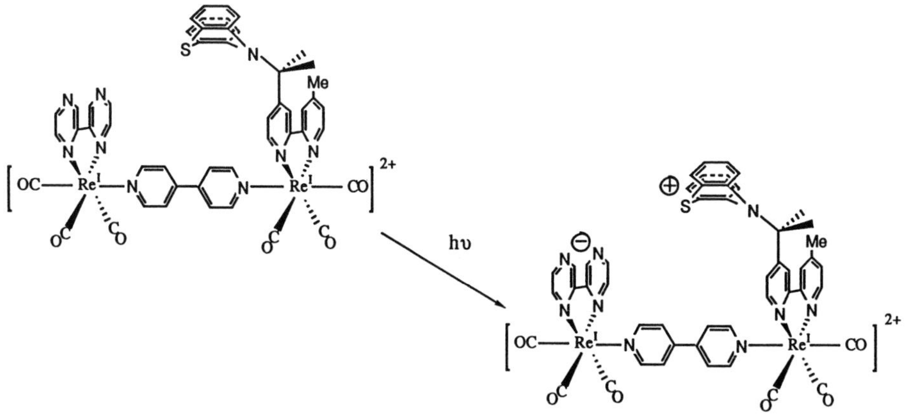

$$[(\overset{+}{PTZ}\text{-bpy})(CO)_3Re^I(4,4'\text{-bpy})Re^I(CO)_3(\overset{-}{bpz})]^{2+} \longrightarrow GS$$

$$k \text{ (DCE, 295} \pm 2 \text{ K)} = 1.33 \times 10^7 \text{ s}^{-1}$$

As suggested in the structure, following MLCT excitation, the photoproduced, oxidative equivalent ends up at -PTZ as -PTZ⁺. The associated, reductive equivalent resides on the bipyrazine ligand. The appearance of characteristic absorbance features in transient absorbance difference spectra allow for the assignment of the oxidation state distribution in the final, redox-separated state . The results of wavelength dependent excitation studies reveal that this state was reached with high efficiency following excitation into Re^I → bpy, Re^I → 4,4'-bpy, or Re^I → bpy-PTZ transitions. Following MLCT excitation, a series of rapid, one-electron transfer events occur which lead to the final state.

In Table 1 are summarized examples of the various types of intramolecular electron transfer processes that have been observed following MLCT excitation of polypyridyl complexes of Ru^{II}, Os^{II}, or Re^I.

TABLE 1. PHOTOCHEMICAL ELECTRON TRANSFER. SUMMARY

- *Reductive quenching*

$$ML\text{-}D \xrightarrow{h\upsilon} M^+L^-\text{-}D \longrightarrow M\text{-}L^-\text{-}D^+$$

- *Oxidative quenching*

$$ML\text{-}A \xrightarrow{h\upsilon} M^+L^-\text{-}A \longrightarrow M^+L\text{-}A^-$$

- *Redox splitting*

$$D\text{-}LML\text{-}A \xrightarrow{h\upsilon} D\text{-}LM^+L^-\text{-}A \longrightarrow D^+\text{-}LML\text{-}A^-$$

- *Redox splitting across a ligand bridge*

$$LM\text{-}ML\text{-}D \xrightarrow{h\upsilon} L^-M^+\text{-}ML\text{-}D \longrightarrow L^-M\text{-}ML\text{-}D^+$$

- *pH-induced oxidative quenching*

$$LML' \xrightarrow{h\upsilon} L^-M^+L' + H^+ \longrightarrow LM^+L'H^-$$

The studies that have been conducted so far demonstrate that, it is possible to prepare molecular fragments based on polypyridyl complexes of Ru^{II}, Os^{II}, or Re^{II} that have the required photochemical, electron transfer properties. Following MLCT excitation of these complexes, separated oxidative and reductive equivalents are produced with kinetic facility and with relatively high efficiencies.

A number of studies on these systems remain to be conducted. One is to turn to shorter time scales in order to chart the sequence of one-electron transfer events that lead to the final redox-separated states. Another is to explore, in a general way, the influence of temperature, and medium effects on both the efficiency of redox separation and on the individual electron transfer steps. This information is required in order to understand in more detail the factors that determine the final separation efficiencies.

In the context of the larger goals stated initially, the results described here are only a beginning. They are encouraging in suggesting that the components that are required to construct complex molecular assemblies are available in the polypyridyl chemistry of Re^I, Ru^{II}, or Os^{II}. They point toward synthetic strategies and approaches that will lead to more complex systems in the future. Although the synthetic challenges are formidable, there is no reason to believe that systems that begin to incorporate many of the features of the working models in Figures 1 or 2 can not be incorporated by utilizing approaches that are currently available.

5. INTRAMOLECULAR, PHOTOCHEMICAL ENERGY TRANSFER

Another problem that we have turned to is how to couple energy transfer into complex, photochemical, redox assemblies. In some applications it might be of value to separate the light absorption and photochemical redox functions by constructing separate molecular units and connecting them by molecular-level, energy transfer conduits. Models of this kind are inspired, in part, by the photochemical apparatus of the reaction center in natural photosynthesis. In that apparatus a separate light harvesting region exists which consists of a closely-packed array of chlorophyll molecules. Following photo-excitation, rapid energy transfer occurs. Ultimately, the excitation energy reaches the reaction center where the photochemical redox events are initiated. The advantage of adding such a region to an already complex device is the enhancement gained in light absorptivity per molecular unit. In an artificial device an array of chromophores must be constructed in which facile, intramolecular energy transfer occurs. In addition, there must be a connecting, energy transfer conduit for transferring the energy to a photoredox assembly.

Progress has been made in constructing molecular building blocks for applications of this kind as well. In one series of experiments, the complex $[Ru(bpy-An)_3]^{2+}$ was prepared. In this complex, an anthryl group is linked chemically to each of the three bipyridyl ligands; the structure of the ligand was illustrated above. It is known by independent measurement that the triplet state of the attached anthryl group lies ~0.3 eV below the Ru^{III}-bpy$\cdot^-$-An, MLCT excited state. Upon $Ru^{II} \rightarrow$ bpy-An excitation of this molecule, the MLCT emission is nearly completely quenched. In transient absorbance difference spectra, a new absorption feature appears within the laser pulse (< 5 ns) which has a maximum absorbance at ~430 nm. This feature persists and decays with a lifetime of 15 μs. The properties of this system following MLCT excitation are consistent with the series of reactions illustrated in Scheme 3.[34]

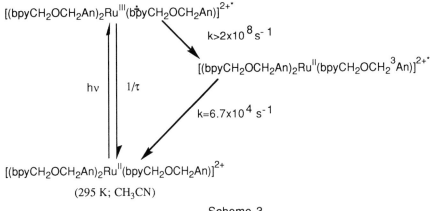

$[(bpyCH_2OCH_2An)_2Ru^{III}(b\overset{\cdot}{p}yCH_2OCH_2An)]^{2+*}$

$k>2x10^8 s^{-1}$

$[(bpyCH_2OCH_2An)_2Ru^{II}(bpyCH_2OCH_2\,^3An)]^{2+*}$

hν   1/τ

$k=6.7x10^4 s^{-1}$

$[(bpyCH_2OCH_2An)_2Ru^{II}(bpyCH_2OCH_2An)]^{2+}$

(295 K; CH$_3$CN)

Scheme 3

Initial MLCT excitation is followed by rapid, intramolecular energy transfer to give the triplet state of the chemically linked anthryl group. The intramolecular energy transfer process appears to be highly efficient.

We have extended our studies on energy transfer to ligand-bridged complexes as well. The study began with the synthesis of a series of complexes of Re^I where the bridging ligand was 4,4'-bipyridine or the dimethyl derivative illustrated below.

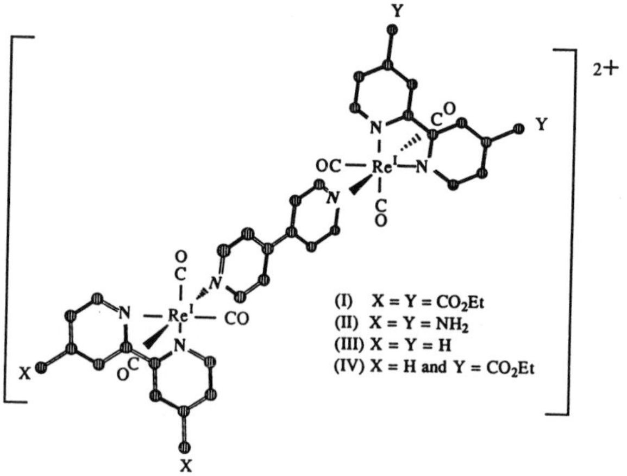

(I) X = Y = CO$_2$Et
(II) X = Y = NH$_2$
(III) X = Y = H
(IV) X = H and Y = CO$_2$Et

$[(4,4'-(X)_2\text{-bpy})(CO)_3Re^I(4,4'\text{-bpy})Re^I(CO)_3(4,4'-(Y)_2\text{-bpy})]^{2+}$

4,4'-bpy   3,3'-(CH$_3$)$_2$-4,4'-bpy

In these complexes, it is possible to develop an asymmetry across the bridge by changing the substituents at the 4,4' positions of the bipyridyl groups at the two separate Re^I sites. For cases where there is a considerable electronic asymmetry, the difference between MLCT excited state energies can be modulated by several tenths of an eV. Electron donating groups such as -NH$_2$ increase the energy of the π^* acceptor orbitals thus increasing the energy of the MLCT excited state. Electronic withdrawing substituents such as -CO$_2$Et stabilize the π^* orbitals and decrease the energy of the excited state.

By a combination of transient absorbance and transient emission measurements it was possible to demonstrate that initial excitation into the high energy, MLCT chromophore was followed by rapid energy transfer across the ligand bridge. An example is shown in eq 7.[35]

$$\xrightarrow{h\upsilon} [(4,4'\text{-}(NH_2)_2\text{-}b\bar{p}y)(CO)_3Re^{II}(4,4'\text{-}bpy)Re^{I}(CO)_3(4,4'\text{-}(CO_2Et)_2\text{-}bpy)]^{2+} \longrightarrow$$

$$[(4,4'\text{-}(NH_2)_2\text{-}bpy)(CO)_3Re^{I}(4,4'\text{-}bpy)Re^{II}(CO)_3(4,4'\text{-}(CO_2Et)_2\text{-}\bar{b}py)]^{2+} \qquad (7)$$

There is a bridging ligand effect in this reaction. When 4,4'-bipyridine is replaced by the methyl derivative, energy transfer does not occur. In the methylated ligand, the pyridine rings can not achieve co-planarity. Based on this and related observations it was concluded that the mechanism for energy transfer might involve initial $(4,4'\text{-}(NH_2)_2\text{-}bpy) \rightarrow 4,4'\text{-}bpy$ electron transfer to give an intermediate, $Re^{II}(4,4'\text{-}bpy^{\cdot})$-based state.

Although the work on energy transfer is still at an early stage, it seems apparent that by using polypyridyl complexes of Ru^{II}, Os^{II}, or Re^{I} it will be possible to manipulate and explore energy transfer processes much as in electron transfer. There are a number of outstanding issues in this area as well. They include acquiring enough information to probe available theories of energy transfer, to prepare complex systems where electron transfer and energy transfer are in competition, and finally, to prepare more complex molecular assemblies where, for example, it may be possible to couple long-range energy transfer to the sensitization of long-range electron transfer.

6. INTRAMOLECULAR, PHOTOCHEMICAL ELECTRON AND ENERGY TRANSFER

Significant progress has been made in exploiting polypyridyl complexes of Ru^{II}, Os^{II}, or Re^{I} in relatively simple molecular assemblies. From the studies that have been conducted, information is emerging concerning the details of the microscopic processes that lead to the directed flow of electron and energy transfer followng MLCT excitation. The fundamental aspects of this problem are intellectually fascinating. By utilizing these systems and the ability to change them synthetically, it is possible to provide an effective working interface between experiment and theory. This will remain an important part of this area in the future. Past a point, the systems described here become "molecular black boxes" for investigating the fundamental details of electron and energy transfer.

It also encouraging that such a variety of processes has been identified and that the synthetic chemistry for preparing even more complex molecular assembles is available. In the future it can be anticipated that it will be possible to utilize systems like these to explore more complex molecular phenomena including directed, long-range energy or electron transfer, the creation of controlled time delays at the molecular level, energy up conversion, or to begin to mimic some of the characteristics of macroscopically based devices, but at the molecular level.

ACKNOWLEDGMENT

Acknowledgements are made to the National Science Foundation under grant no. CHE-8806664 for support of this research.

REFERENCES

1) G. M. Bryant , J. E. Fergusson, and J.K.J Powell, Aust. J. Chem. 24 (l97l) 257;
    G. M. Bryant and J. E. Fergusson, Aust. J. Chem. 24 (l971) 275.

2) a) T. J. Meyer, Pure Appl. Chem. 58 (1986) 1193. b) E. Krausz and J. Ferguson,
    Prog. Inorg. Chem. 37 (1989) 293. c) R. J. Watts, J. Chem. Educ. 60 (1983) 834.
    d) G. A. Crosby, K. A. Highland and K. A. Truesdell, Coord. Chem. Rev. 64 (1985)
    41. e) M. K. DeArmond, K. W. Hanck and D. W. Wertz, Coord. Chem. Rev. 64 (1985)
    65.

3) a) P. G. Bradley, N. Kress, B. A. Hornberger, R. F. Dallinger and W. H. Woodruff,
    J. Am. Chem. Soc. 103 (1981) 7441. b) Y. Komada, S. Yamauchi and N. Hirota,
    J. Phys. Chem. 92 (1988) 65ll.

4) a) A. Juris, V. Balzani, F. Barigelletti, S. Campagna, P. Belser and A. Von Zelewsky,
    Coord. Chem. Rev. 84 (1988) 85. b) K. Kalyanasundaran, Coord. Chem. Rev. 46
    (1982) 159. c) N. Sutin and C. Creutz, Pure Appl. Chem. 52 (1980) 2717. d) N.
    Sutin, J. Photochem. 10 (1979) 19.

5) a) T. J. Meyer, Acc. Chem. Res. 11 (1978) 94. b) T. J. Meyer, Prog. Inorg. Chem. 30
    (1983) 389. c) D. G. Whitten, Acc. Chem. Res. 13 (1980) 83.

6) a) F. Felix, J. Ferguson, H. U. Güdel and A. Ludi, J. Am. Chem. Soc. 102 (1980)
    4096. b) S. Decurtins, F. Felix, J. Ferguson, H. U. Güdel and A. Ludi, J. Am. Chem.
    Soc. 102 (1980) 4102. c) H. Yersin, D. Huber and D. Brauer, J. Phys. Chem. 94
    (1990) 3560 and references therein.

7) a) E. M. Kober and T. J. Meyer, Inorg. Chem. 21 (1982) 3967. b) J. Ferguson and
    F. Herren, Chem. Phys. 76 (1983) 45.

8) a) P. G. Bradley, N. Kress, B. A. Hornberger, R. F. Dallinger and W. H. Woodruff,
    J. Am. Chem. Soc. 103 (1981) 7441. b) O. Poizat and C. Sourisseau, J. Phys.
    Chem. 88 (1984) 3007. c) P. A. Mabrouk and M. S. Wrighton, Inorg. Chem. 25
    (1986) 526.

9) a) Y. Komada, S. Yamauchi and N. Hirota, J. Phys. Chem. 92 (1988) 6511. b) C.
    Creutz, M. Chou, T. L. Netzel, M. Okumura and N. Sutin, J. Am. Chem. Soc. 102
    (1980) 1309. c) S. J. Milder, J. S. Gold and D. S. Kliger, J. Phys. Chem. 90 (1986)
    548. d) T. Hiraga, N. Kitamura, H. B. Kim, S. Tazuke and N. Mori, J. Phys. Chem.
    93 (1989) 2940.

10) a) J. Van Houten and R. J. Watts, J. Am. Chem. Soc. 98 (1976) 4853. b) F.
    Barigelletti, A. Juris, V. Balzani, P. Belser and A. Von Zelewsky, J. Phys. Chem.
    90 (1986) 5190. c) B. Durham, J. V. Caspar, J. K. Nagle and T. J. Meyer, J. Am.
    Chem. Soc. 104 (1982) 4803. d) E. M. Kober, J. L. Marshall, W. J. Dressick, B. P.
    Sullivan, J. V. Caspar and T. J. Meyer, Inorg. Chem. 24 (1985) 2755. e) K. R.
    Barqawi, A. Llobet and T. J. Meyer, J. Am. Chem. Soc. 110 (1988) 7751. f) Z.
    Murtaza, K. R. Barqawi and T. J. Meyer, J. Phys. Chem. in press.

11) a) F. Barigelletti, P. Belser, A. Von Zelewsky, A. Juris and V. Balzani, J. Phys.
    Chem. 89 (1985) 3680. b) F. Barigelleti, A. Juris, V. Balzani, P. Belser and A.
    Von Zelewsky, J. Phys. Chem. 90 (1986) 5190.

12) G. A. Crosby, Acc. Chem. Res. 8 (1975) 231.

13) R. S. Lumpkin, E. M. Kober, L. A. Worl, Z. Murtaza and T. J. Meyer, J. Phys. Chem. 94 (1990) 239.

14) J. V. Caspar, T. P. Westmoreland, G. H. Allen, P. G. Bradley, T. J. Meyer and W. H. Woodruff, J. Am. Chem. Soc. 106 (1984) 3492.

15) a) E. M. Kober, J. V. Caspar, R. S. Lumpkin and T. J. Meyer, J. Phys. Chem. 90 (1986) 3722. b) J. V. Caspar and T. J. Meyer, J. Am. Chem. Soc. 105 (1983) 5583. c) R. S. Lumpkin and T. J. Meyer, J. Phys. Chem. 90 (1986) 5307.

16) a) W. J. Vining, J. V. Caspar and T. J. Meyer, J. Phys. Chem. 89 (1985) 1095. b) J. V. Caspar, E. M. Kober, B. P. Sullivan and T. J. Meyer, J. Am. Chem. Soc. 104 (1982) 630. c) J. V. Caspar and T. J. Meyer, Inorg. Chem. 22 (1983) 2444.

17) a) M. Wrighton and J. Markhaur, J. Phys. Chem. 77 (1973) 3042. b) K. Maudal, T. D. Pearson, W. P. Krug and J. N. Demas, J. Am. Chem. Soc. 105 (1983) 701.

18) a) V. Balzani, F. Bolleta and F. Scandola, J. Am. Chem. Soc. 102 (1980) 2152. b) G. Orlandi, S. Monte, F. Barigelletti and V. Balzani, Chem. Phys. 52 (1980) 313.

19) a) J. C. Luong, Ph.D. Dissertation, Massachusetts Institute of Technology, 1981. b) J. V. Caspar, B. P. Sullivan and T. J. Meyer, Chem. Soc., Chem. Commun. (1984) 403. c) E. M. Kober, J. V. Caspar, B. P. Sullivan and T. J. Meyer, Inorg. Chem. 27 (1988) 4587. d) G. Maecker and F. H. Case, J. Am. Chem. Soc. 80 (1953) 2745. e) R. B. Woodward and E. Wenkert, J. Org. Chem. 48 (1983) 283. f) L. Della Ciana, I. Hamachi and T. J. Meyer, J. Org. Chem. 54 (1989) 1731.

20) a) J. E. Baggott, G. K. Gregory, M. J. Pillling, S. Anderson, K. R. Leddon and J. E. Turp, JCS Faraday Trans. II 99 (1983) 195. b) M. J. Cook, A. P. Lewis, G. S. G. McAuliffe and A. J. Thompson, JCS Perkin Trans II (1984) 1293. c) R. A. Krause, Structure and Bonding 62 (1987) 1.

21) T. J. Meyer, Acc. Chem. Res. 22 (1989) 163.

22) T. J. Meyer, Pure App. Chem. 62 (1990) 1003.

23) C. Creutz, Prog. Inorg. Chem. 30 (1983) 1.

24) a) E. M. Kober, K. A. Goldsby, D. N. S. Narayana and T. J. Meyer, J. Am. Chem. Soc. 105 (1983) 4303. b) T. J. Meyer, Electron Transfer in Mixed-Valence Compounds, in: Mixed-Valence Compounds, ed. D. B. Brown (D. Reidel Publishing Company, Dordrecht, Holland, 1980) pp. 75. c) M. J. Powers and T. J. Meyer, J. Am. Chem. Soc. 102 (1980) 1289.

25) S. A. Adeyemi, E. C. Johnson, F. J. Miller and T. J. Meyer, Inorg. Chem. 12 (1973) 2371.

26) a) Y. Itoh, Y. Morishima and S. Nozakura, Photochem. Photobiol. 39 (1984) 451. b) Y. Itoh, Y. Morishima and S. Nozakura, Photochem. Photobiol. 39 (1984) 603. c) M. Kaneko, A. Yamada, E. Tsuchida and Y. Kurimura, J. Phys. Chem. 88 (1984) 1061. d) M. Kaneko and H. Nakamura, Macromolecules 20 (1987) 2265.

27) a) K. Sumi, M. Furue and S. Nozakura, Photochem. Photobiol. 42 (1985) 485.
b) R. E. Sassoon and J. Rabani, J. Phys. Chem. 89 (1985) 5500.

28) J. N. Younathan, S. F. McClanahan and T. J. Meyer, Macromolecules 22 (1989) 1048. b) S. F. McClanahan, E. Danielson, J. N. Younathan and T. J. Meyer, J. Am. Chem. Soc. 109 (1987) 3297.

29) G. F. Strouse, L. A. Worl, J. N. Younathan and T. J. Meyer, J. Am. Chem. Soc. 111 (1989) 9101.

30) P. Chen, D. Westmoreland, E. Danielson, K. S. Schanze, D. Anthon, P. E. Neveux, Jr. and T. J. Meyer, Inorg. Chem. 26 (1987) 1116. b) P. Chen, R. Duesing, G. Tapolsky and T. J. Meyer, J. Am. Chem. Soc. 111 (1989) 8305.

31) a) T. D. Westmoreland, H. LeBozec, R. W. Murray and T. J. Meyer, J. Am. Chem. Soc. 105 (1983) 5952. b) P. Chen, E. Danielson, H. LeBozec and T.J. Meyer manuscript in preparation.

32) E. Danielson, C. M. Elliot, J. W. Merkert and T. J. Meyer, J. Am. Chem. Soc. 112 (1990) 5378.

33) R. Duesing, G. Tapolsky and T. J. Meyer, J. Am. Chem. Soc. 112 (1990) 5378.

34) S. Boyde, G. F. Strouse, W. E. Jones, Jr. and T. J. Meyer, J. Am. Chem. Soc. 111 (1989) 7448.

35) a) G. Tapolsky, R. Duesing and T. J. Meyer, J. Phys. Chem. 93 (1989) 3885. b) G. Tapolsky, R. Duesing and T. J. Meyer, Inorg. Chem. 29 (1990) 2285.

Photochemical Processes in Organized Molecular Systems
K. Honda (Editor-in-Chief)
© Elsevier Science Publishers B.V., 1991

INTRAMOLECULAR ELECTRON TRANSFER AND ENERGY TRANSFER IN
BIMETALATED COMPOUNDS

T. Ohno, K. Nozaki, and M.-A. Haga*

Chemistry Department, College of General Education,
Osaka Univeristy, Toyonaka, Osaka 560
*Department of Chemistry, Faculty of Education,
Mie Univeristy, Tsu, Mie 514

Intramolecular electron transfers (ET) within bi-chromophoric
compounds have been examined by means of laser photolysis spectroscopy. The chromophores examined are $Ru(II)L_2$ (L = 2,2'-
bipyridine (bpy) or 4,4'-dimethyl-2,2'-bipyridine (dmbpy)) and
$Rh(III)L'_2$ (L' = 1,10-phenanthroline (phen)) which are linked by a
tetradentate ligand, bis-2,2'-(2"-pyridyl)-bibenzimidazole
$(bpbimH_2)$. The $Ru(II)$-site in the MLCT state is a strong electron donor to the $Rh(III)$-site. The intervening ligand itself
consists of two chromophores of 2-(2'-pyridyl)-benzimidazole
(pbimH)$_2$ which acts an electron acceptor in the diprotonated form
$(pbimH_3^{2+})$. Quenching of the excited MLCT state of the $Ru(II)$-
site are ascribed to ET to the $Rh(III)$-site in
$RuL_2(bpbimH_2)RhL'_2^{5+}$ or to $pbimH_3^{2+}$-moiety in the $RuL_2(bpimH-$
$pbimH_3)^{4+}$ on the basis of redox potentials of the sites. Rates
of the $Ru(II)$-to-$Rh(III)$ ET were measured at low temperatures
(165-240 K) by monitorring the decay of excited state absorption
(ESA). Rates of the $Ru(II)$-to-$pbimH_3^{2+}$ ET in $Ru(dmbpy)_2(pbimH-$
$pbimH_3)^{4+}$ were 2.6×10^{10} s^{-1} at room temperature and 1×10^7 s^{-1}
at 165 K. Analysis of the temperature dependence of the rate
affords an activation energy of 0.2 eV for the ET process.
Non-adiabaticity in the $Ru(II)$-to-$Rh(III)$ ET is discussed in
conjunction with the superexchange metal-metal interaction
estimated from the spectroscopic and electrochemical data.

1. INTRODUCTION

It has been proved that the low yield of electron transfer
(ET) products in the bimolecular quenching of a triplet excited
metal compound results from a fast back ET occurring within the
radical pair formed in the collisional quenching.[1-8] The
rates of spin-inverted back ET have been investigated in
terms of Franck-Condon factor and electronic coupling between the
radicals.[1-9]

It is instructive to examine photo-initiated intramolecular ET
followed by a thermal back ET, since electronic coupling between
metal-chromophores in bimetallic compounds can be directly
measured in the mixed valence state. Non-adiabatic ET in bi-
metallic compounds have been studied in order to find a
correlation between the ET rate and the electronic coupling.[10-19]
Metal-metal interaction have been estimated by means of spectroscopy and electrochemistry. Intensities of chromophore-to-

chromophore charge transfer absorption provide the extent of chromophore-chromophore electronic coupling.[14,15]   Chromophore-chromophore coupling is reflected on the redox potentials of metal-chromophores.[16-18]   Weak electronic couplings between chromophores separated by chemical bonds are assumed to decrease with increasing chromophore-chromophore distance by help of quantum chemical calculation.[19-22]

A ruthenium(II) ion in the excited state as one chromophore is capable to undergo an ET to another chromophore because of its low (negative) oxidation potential and high (positive) reduction potential.[23]   A rhodium(III) ion as the other chromophore is linked to the Ru(II) site through an intervening tetradentate ligand, bis-2,2'-(2"-pyridyl)-bibenzimidazole (bpbimH$_2$) as is shown in Figure 1.   The intervening ligand itself consists of two chromophores of 2-(2'-pyridyl)-benzimidazole (pbimH), which is capable to behave as an electron acceptor in a diprotonated

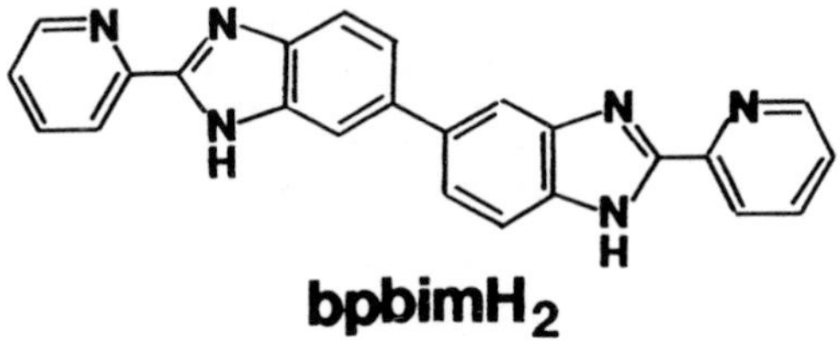

bpbimH$_2$

Figure 1. 2-2'-(2"-Pyridyl)-bibenzimidazole and its abbreviation.

mononuclear compound, RuL$_2$(pbimH-pbimH$_3$)$^{4+}$.   Both photo-initiated ET and subsequent back ET were studied by means of n-sec laser photolysis spectroscopy in a 90-300 K range.   Chromophore-chromophore interaction in binuclear compounds, RuL$_2$(bpbimH$_2$)RuL$_2$$^{4+}$, was estimated by using spectroscopic and electrochemical data, which is discussed in conjunction with the rate of ET between the Ru(II)-site and the Rh(III)-site[24] or pbimH$_3$$^{2+}$-site.[25]

2. EXPERIMENTAL SECTION

2.1. Compounds

Mononuclear ruthenium(II) compounds, RuL$_2$(bpbimH$_2$)$^{2+}$ (L= 2,2'-bipyridine (bpy) and 4,4'-dimethyl-2,2'-bipyridine (dmbpy)) have been prepared as described elsewhere.[26]   Heterobinuclear compounds,   [Ru(dmbpy)$_2$(bpbimH$_2$)Rh(phen)$_2$](ClO$_4$)$_5$ and [Ru(bpy)$_2$(bpbimH$_2$)Rh(bpy)$_2$](ClO$_4$)$_5$, have been synthesized from

$RhL_2(bpbimH_2)Cl_3$ and RuL_2Cl_2.[24]

2.2. Apparatus and Measurements

Phosphorescence of Ru(II) compounds in butyronitrile at 77 K was measured by using a Hitachi spectrofluorometer, MPF-2A. The Q-switched Nd^{3+}-YAG laser (Quantel YG580) was used for excitation of the Ru(II) compound in either acetonitrile, butyronitrile, or a mixture of ethanol and methanol (4:1 v/v). All sample solutions were deaerated by bubbling with nitrogen more than 12 min. Either $HClO_4$ or CF_3COOH of 1 mM was added to suppress deprotonation of the intervening ligand. A transmittance change aquisition system has been described elsewhere.[8] Oxidation potentials of Ru(II) compounds were measured by means of differential-pulse voltammetry in CH_3CN containing 0.1 M tetrabutylammonium perchlorate with a DC pulse polarograph (HECS-312B, Huso, Japan). Reduction potentials of Rh(III) compounds, pbimH and $pbimH_3{}^{2+}$ were obtained from rapid-scan cyclic voltammetry measurements[27] at the scan rate of 10 kV/s in acetonitrile containing 0.5 mol dm^{-3} tetraethylammonium hexafluorophosphate. All potentials are referred to the formal potential of ferrocenium (Fc^+)/ ferrocene (Fc) system which is -0.33 V vs. SCE. The temperature of the sample solutions in 89-300 K region was controlled by using a cryostat, Oxford DN1704, and a controller, Oxford ITC4.

3. RESULTS

3.1. Metal-to-Ligand Charge Transfer State of $RuL_2(bpbimH_2)^{2+}$ [28]

Absorption spectra of $RuL_2(bpbimH_2)RhL'_2{}^{5+}$ are the sum of those of the components, $RuL_2(bpbimH_2)^{2+}$ and $RhL'_2{}^{3+}$.[24] The lowest transition is assigned to a metal-to-ligand charge transfer (MLCT) of Ru→L or Ru→$bpbimH_2$.[26] $RuL_2(bpbimH_2)^{2+}$ exhibited phosphorescence at 77 K whose energy is slightly higher than that at room temperature (see Table I). Laser excitation of $RuL_2(bpbimH_2)^{2+}$ gave rise to a transient absorption spectrum at 300 K. The transient absorption are ascribed to the formation of the excited MLCT state (Figure 2), since the lifetimes (610 ns for L = bpy and 550 ns for L = dmbpy) are in agreement with those of the phosphorescence. An excited state absorption (ESA) difference spectrum of $Ru(dmbpy)_2(bpbimH_2)^{2+}$ in a mixed solvent of ethanol and methanol at 89 K (Figure 3a) was the same as in

CH$_3$CN   at   room   temperature   (Figure   2a).   As   for
Ru(bpy)$_2$(bpbimH$_2$)$^{2+}$, an ESA difference spectrum in the mixed

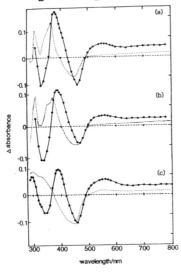

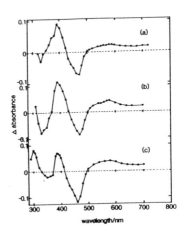

Figure 2. Transient difference absorption spectra of RuL$_2$-(bpbimH$_2$)$^{2+}$ in CH$_3$CN containing HClO$_4$ of 1 mM (solid line) and of RuL$_3$ (dotted line) at 100 ns after the laser excitation.
(a) L = bpy, (b) = dmbpy, and (c) L = phen.

Figure 3. Transient difference absorption spectra of RuL$_2$-(bpbimH$_2$)$^{2+}$ in 1/4 methanol/ethanol containing HClO$_4$ or CF$_3$COOH of 1 mM at 80 K. (a) L = bpy, (b) = dmbpy, and (c) L = phen.

solvent at 89 K (Figure 3b) was not the same as that at room temperature (Figure 2b). The lifetime of excited Ru(dmbpy)$_2$(bpbimH$_2$)$^{2+}$ (1/k$_o$) depending on the temperature was measured in butyronitrile (mp 162 K).

3.2. Rapid Decay of ESA in Ru(Ⅱ)-Rh(Ⅲ) Compounds[24]

Both the phosphorescence and the ESA difference spectrum of RuL$_2$(bpbimH$_2$)$^{2+}$-site were completely quenched in RuL$_2$(bpbimH$_2$)RhL'$_2$$^{5+}$ at room temperature. Bleaching of MLCT band at 460 nm was not evident at all in acetonitrile and butyronitrile. On lowering temperature, the ESA at 410 nm and the bleaching of MLCT at 460 nm appeared and decayed rapidly depending on the temperature. The rapid decay in acetonitrile (mp 230 K) was difficult to be accurately measured. In butyronitrile, the decay of ESA (k$_d$) was measured in 165-240 K region. The recovery of the MLCT band bleached followed the decay of ESA without a time lag.

3.3. Rapid Quenching of ESA of RuL$_2$(bpbimH$_2$)$^{2+}$ by Diprotonation[25]

Addition of $HClO_4$ to the acetonitrile solution of $RuL_2(bpbimH_2)^{2+}$ reduced the intensity of $\pi - \pi *$ band of $bpbimH_2$ in a UV region and then shifted the $\pi - \pi *$ band of $bpbimH_2$ at 340 nm to the longer wavelength. The first stage of absorption change completed in less than 0.1mM of $HClO_4$, which is ascribed to the mono-protonation of $RuL_2(bpbimH_2)^{2+}$. The second stage of absorption change was accompanied by both the enhanced decay rate of the phosphorescence and the reduction in the formation efficiency of ESA. Measurement of the enhanced decay rates with the amount of $HClO_4$ afforded the rate constants of 8.5×10^8 and 8.3×10^8 $dm^3mol^{-1}s^{-1}$ for L = bpy and L = dmbpy, respectively, for the quenching of excited $RuL_2(pbimH-bpbimH_3)^{3+}$ by di-protonation. The formation efficiency of ESA was reduced as the amount of $HClO_4$ added and no formation of ESA was observed with 1 mol dm^{-3} of $HClO_4$. At lower temperature (165 K), the ESA of $Ru(dmbpy)_2(bpbimH_2)^{2+}$ in butyronitrile containing $HClO_4$ of 1 mol dm^{-3} appeared and decayed with the rate of ca. 1×10^7 s^{-1}.

The rapid decay of ESA of $Ru(dmbpy)_2(pbimH-pbimH_3)^{4+}$ at 410 and 560 nm was measured to be 2.6×10^{10} s^{-1} in acetonitrile at room temperature by means of p-sec laser photolysis. The recovery of the MLCT band monitored at 460 nm was not so rapid $(1.9 \times 10^8 \text{ } s^{-1})$.

4. DISCUSSION

4.1. Ergonicities of intramolecular electron transfers

An ET quenching process of the excited metal-chromophore by another metal-chromophore in bi-metallic compounds is followed by the back ET regenerating the original oxidation state. The ergonicity of the initial ET can be estimated from the redox potentials of chromophores and the excitation energy, which are summerized in Table I.[24,25] The reduction potential of bpimH, which is shifted from -2.54 V to -1.02 vs Fc^+/Fc by diprotonation, is utilized for the estimation of ergonicity instead of the redox potential of protonated pbimH moiety of $bpbimH_2$ coordinating to Ru(II). Four ET from the excited Ru(II) site to the Rh(III)-site or the $pbimH_3^{2+}$ moiety are weakly exergonic,

^{3}Ru(II)(dmbpy)$_2$(bpbimH$_2$)Rh(III)(phen)$_2$$^{5+}$
$\rightarrow$ Ru(III)(dmbpy)$_2$(bpbimH$_2$)Rh(II)(phen)$_2$$^{5+}$ $\Delta G°$ = -0.19 eV (1)
^{3}Ru(II)(bpy)$_2$(bpbimH$_2$)Rh(III)(bpy)$_2$$^{5+}$
$\rightarrow$ Ru(III)(bpy)$_2$(bpbimH$_2$)Rh(II)(bpy)$_2$$^{5+}$ $\Delta G°$ = -0.13 eV (2)
^{3}Ru(II)(dmbpy)$_2$(pbimH-pbimH$_3$)$^{4+}$
$\rightarrow$ Ru(III)(dmbpy)$_2$(pbimH-pbimH$_3$)$^{4+}$ $\Delta G°$ = -0.31 eV (3)
^{3}Ru(II)(bpy)$_2$(pbimH-pbimH$_3$)$^{4+}$
$\rightarrow$ Ru(III)(bpy)$_2$(pbimH-pbimH$_3$)$^{4+}$ $\Delta G°$ = -0.26 eV (4).

The back ET from the Rh(II)-site or the pbimH$_3$$^+$ moiety to the Ru(III)-site are strongly exergonic ($\Delta G°$ = -1.7$\sim$-1.9 eV).

Table I   Redox Potentials, Excitation Energies (EE), and pK$_a$

Compounds	$E_{1/2}^{ox}$/mV[a]	$E_{1/2}^{red}$/mV[a]	EE/eV[b]	pK$_a$
Ru(bpy)$_2$(bpbimH$_2$)$^{2+}$	777		2.063	
Ru(bpy)$_2$(bpbimH$_4$)$^{4+}$	783		2.067	1.8
Ru(dmbpy)$_2$(bpbimH$_2$)$^{2+}$	680		2.023	
Ru(dmbpy)$_2$(bpbimH$_4$)$^{4+}$	683		2.013	
Ru(bpy)$_2$(bpbimH$_2$)Rh(bpy)$_2$$^{5+}$	780	-1150	2.058	2.89[c]
Ru(dmbpy)$_2$(bpbimH$_2$)Rh(phen)$_2$$^{5+}$	670	-1165	2.020	2.65[c]
pbimH		-2540		
pbimH$_2$		-1820		
pbimH$_3$		-1020		

a:referred to Fc$^+$/Fc, b:in butyronitile, c: in acetonitrile with buffer (1:1 v/v)[9].

4.2. Electron occupation on bpbimH$_2$ in the MLCT state

The lowest excited states of many ruthenium(II) polypyridine compounds are described as the phosphorescent state of Ru$\rightarrow$ligand charge transfer.[29,30] An excited electron in a localized orbital model occupies an orbital on the ligand which is the most easily reduced.[31-34] Since the first electrochemical reduction of the Ru(II) compounds investigated here is irreversible, the assignment of the electron occupied orbital (ligand) was made by using the ESA spectrum.

Transient absorption spectra following the laser excitation of Ru(bpy)$_2$(bpbimH$_2$)$^{2+}$ and Ru(dmbpy)$_2$(bpbimH$_2$)$^{2+}$ demonstrate the depression of the $\pi - \pi *$ bands of a reduced ligand, which are the strong evidence for the assignment of reduced ligand. A band in 370-420 nm region and an additional band in 500-600 nm region in Figure 2a and 2b are assigned to a $\pi - \pi *$ band of bpbimH$_2^-$ and/or a red-shifted $\pi - \pi *$ of bpbimH$_2$ coordinating to Ru(III). Strong bleaching of $\pi - \pi *$ transition of bpbimH$_2$ at 340 nm in Figure 2a and 2b demonstrates the electron occupation on bpbimH$_2$, too. The bpy$^-$ or dmbpy$^-$-band at 370 nm ($\varepsilon = 2 \times 10^4$ dm^{3}cm^{-1}mol^{-1} [8]), which partly overlapped with the bpbimH$_2^-$ band at 390 nm, is masked by the bleaching of the strong $\pi - \pi *$ band of bpbimH$_2$ ($\varepsilon \sim 4 \times 10^4$ dm^{3}cm^{-1}mol^{-1}).

The assignment of the ligand reduced in the MLCT state was made by analysis of the ESA acquired at 89 K[28] (Figure 3a and 3b). Appearance of the 370 nm-band instead of the bleaching at 340 nm on the excitation of Ru(bpy)$_2$(bpbimH$_2$)$^{2+}$ at 89 K reveals that the excited electron mainly occupies on bpy rather than on bpbimH$_2$. The temperature independent ESA of Ru(dmbpy)$_2$(bpbimH$_2$)$^{2+}$ suggests that the energy level of Ru$\rightarrow$bpbimH$_2$ CT is not higher than that of Ru$\rightarrow$dmbpy CT. This order of the ability of the ligand to accept an electron, bpy > bpbimH$_2$ > dmbpy, is consistent with the decreasing order of the redox potential of Ru(bpy)$_3^{2+}$ (-1.34 V vs SCE[35]) and Ru(dmbpy)$_3^{2+}$ (-1.45 V vs SCE[35]).

4.3. Electron transfer between the Ru(II)-site and the Rh(III)-site

The quenching of Ru(II)-site phosphorescence in Ru(dmbpy)$_2$(bpbimH$_2$)Rh(phen)$_2^{5+}$ is ascribed to ET from the Ru(II)-site to the Rh(III)-site, since the processes are exergonic as described above. An alternative mechanism of the quenching is energy transfer from the Ru(II)-site to the Rh(III)-site. The energy transfer seems endergonic, since the lowest excited state of Rh(phen)$_2$(bpbimH$_2$)$^{3+}$ is inferred to lie near that (2.71 eV[36]) of Rh(phen)$_3^{3+}$.

The decay rate of either the ESA or the phosphorescence was measurable below 230 K. The rate of the ET is evaluated as a difference in the ESA decay rate between the Ru-Rh compound and the reference compound of Ru(II). Temperature dependence of the ET rate gives rise to the rate of 2.4 x 10^8 s^{-1} at 300 K by extrapolation and the activation energy of 0.2 eV[24] (Figure 4). The same analysis of the ET rates obtained for

Ru(bpy)$_2$(bpbimH$_2$)Rh(bpy)$_2$$^{5+}$ affords the   rate of 1.3 x 10^8 s^{-1} at
300 K and the activation energy of 0.22 eV.

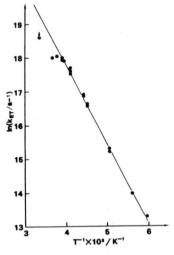

Figure 4.   Temperature   dependence
of electron transfer rate constant
(k$_{ET}$)   of   Ru(II) → Rh(III)   in   the
excited   MLCT   state   of
Ru(dmbpy)$_2$(bpbimH$_2$)Rh(phen)$_2$$^{5+}$.
[HClO$_4$] is 1 mM in butyronitirile.

The   back ET from the Rh(II) to the Ru(III) was too fast to   be
directly   observed.   No time lag was found between the ESA   decay
at   400   nm and the recovery of the MLCT band at   460   nm.   The
possible   slowest rate of the back ET is 1.3 x 10^8 s^{-1} at   300 K
because no ET product   was   detected   by   means   of   n-sec   laser
photolysis.

    4.4.   Electron   transfer   between   the   Ru(II)-site   and   the
pbimH$_3$$^{2+}$-moiety
    The rapid quenching of the excited state of Ru(II)-site by the
diprotonated   pbimH is ascribed to the Ru(II)-to-pbimH$_3$$^{2+}$ ET   by
taking into consideration the ergonicity of eqs. 3 and 4.   Since
the intensity of ESA immediately after the n-sec laser excitation
of   the   diprotonated compounds was reduced with   increasing   the
amount   of   HClO$_4$,   the   ET within the   diprotonated   species   of
RuL$_2$(bpbimH$_2$)$^{2+}$ is considered to occur more rapidly than 10^9 s^{-1}.
The   direct   measurement of the ET   in   RuL$_2$(pbimH-bpbimH$_3$)$^{4+}$ by
means   of p-sec laser photolysis provided the   rate   constant   of
2.6 x 10^{10} s^{-1} at   room temperature.   Cooling   the   sample   in
butyronitrile   to 165 K reduced the rate to ca. 1 x 10^7 s^{-1}.   A
rough   estimation   of the activation energy for   the   ET   process
gives rise to ca. 0.2 eV.

    4.5. Electronic coupling and electron transfer rates
    The   rate (2.3 x 10^8 s^{-1}) of Ru(II)-to-Rh(III) ET is slow   com-
pared   to   that (1.6 x 10^{10} s^{-1}) of Ru(II)-to-pbimH$_3$$^{2+}$ ET.   The

reduction factor, $1/10^2$, can not be accounted for by the difference in the ergonicity (-0.19 eV for the former process and
-0.31 eV for the latter process). The inner-sphere rearrangement
energy involved in Rh(Ⅲ)-Rh(Ⅱ), which has been estimated to be
more than 1 eV[7,37], may make the Ru(Ⅱ)-to-Rh(Ⅲ) ET slow. The
large rearrangement energy could result in a large activation
energy (0.2 eV) for the process compared to that of Ru(Ⅱ)-to-
pbimH$_3{}^{2+}$ ET (ca. 0.2 eV). However, this is not the case for the
Ru(Ⅱ)-to-Rh(Ⅲ) ET.

There is a question left whether the Ru(Ⅱ)-to-Rh(Ⅲ) transfer
is adiabatic or not. The same MLCT phosphorescence of
RuL$_2$(bpbimH$_2$)RhL'$_2{}^{5+}$ as that of the reference Ru(Ⅱ) compound is
indicative of a very weak electronic coupling between the
Ru(Ⅱ)-site and the Rh(Ⅲ)-site. The bi-ruthenium(Ⅱ) compound
of RuL$_2$(bpbimH$_2$)RuL$_2{}^{4+}$, meanwhile, exhibited a small lower-energy
shift of the MLCT phosphorescence (16 meV for L = bpy and 27 meV
for L = dmbpy[28]), which does not mean the existance of considerable Ru(Ⅱ)-Ru(Ⅲ) interaction. The superexchange interaction
between Ru(Ⅱ) and Ru(Ⅲ) is estimated to be very weak (0.1 meV)
from the intensity and transition energy ($\sim$1 eV) of inter-
valence band of Ru(Ⅱ)(bpy)$_2$(bpbimH$_2$)Ru(Ⅲ)(bpy)$_2{}^{5+}$ in a near-
infrared region.[38] The presence of the very weak electronic
coupling between the Ru-sites suggests weak electronic coupling
between the pbimH-moieties in bpbimH$_2$, which is manifested as a
red-shift of the $\pi - \pi *$ transition of pbimH. Consequently, the
chemical bond between the pbimH-moiety and Rh(Ⅲ) ion is
inefficient for the electronic coupling compared with the bond
between Ru(Ⅱ) and pbimH. The very weak electronic coupling
between Rh(Ⅲ) and the pbimH-moiety is consistent with non-commu-
nication between the excited state ligand and the ground state
ligand coordinating to Rh(Ⅲ).[39] It is concluded that the non-
adiabaticity of Ru(Ⅱ)-to-Rh(Ⅲ) ET is responsible for the slow
rate compared with the Ru(Ⅱ)-to-pbimH$_3{}^{2+}$ ET. The common value
of the activation energies for the ET processes (0.2 eV) suggests
that the electronic coupling between the pbimH moieties of
bpbimH$_2$ is dependent on temperature.

ACKNOWLEDGEMENT

We thank Prof. N. Mataga and Mr. T. Asahi of Osaka University
for their help in performing the p-sec laser experiment.

REFERENCES

1) M.A. Hoselton, C-T. Lin, H.A. Schwartz, and N. Sutin, J. Am. Chem. Soc. 100 (1978) 2383.

2) T. Ohno and N.N. Lichtin, J. Am. Chem. Soc. 102 (1980) 4636.

3) T. Ohno and N.N. Lichtin, J. Phys. Chem. 86 (1982) 354.

4) T. Ohno, S. Kato, A. Yamada, and T. Tanno, J. Phys. Chem. 87 (1983) 775.

5) T. Ohno, A. Yoshimura, H. Shioyama, and N. Mataga, J. Phys. Chem. 91 (1987) 4365.

6) T. Ohno, A. Yoshimura, S. Tazuke, Y. Kawanishi, N. Kitamura, J. Phys. Chem. 93 (1989) 3546.

7) T. Ohno, A. Yoshimura, and N. Mataga, J. Phys. Chem. 90 (1986) 3295.

8) T. Ohno, A. Yoshimura, and N. Mataga, J. Phys. Chem. 94 (1990) 4871.

9) G. Winter and U. Steiner, Ber. Bunsenges. Phys. Chem. 84 (1980) 1203.

10) S.S. Isied, A. Vassilian, R.H. Magnuson, and H.A. Schwartz, J. Am. Chem. Soc. 107 (1986) 7432.

11) S.S. Isied, A. Vassilian, J.F. Wishart, C. Creutz, H.A. Schwartz, and N. Sutin, J. Am. Chem. Soc. 110 (1988) 635.

12) C. Creutz, Prog. Inorg. Chem. 30 (1983) 1.

13) T.J. Meyer, Acc. Chem. Res. 11 (1978) 94.

14) G.C. Allen and N.S. Hush, Prog. Inorg. Chem. 8 (1967) 357.

15) N.S. Hush, Prog. Inorg. Chem. 8 (1967) 391; Coord. Chem. Rev. 64 (1985) 735.

16) D.E. Richardson and H. Taube, Inorg. Chem. 20 (1981) 1278.

17) M.-A. Haga, T. Matsumura, and S. Yamabe, Inorg. Chem. 26 (1987) 4148.

18) D.E. Richardson and H. Taube, Coord. Chem. Chem. 60 (1984) 107.

19) A.P. Szecsy and A. Haim, J. Am. Chem. Soc. 103 (1987) 1679.

20) D. Geselowitz, Inorg. Chem. 26 (1987) 4135.

21) Y. Kim and C.M. Lieber, Inorg. Chem. 28 (1989) 3990.

22) P. Siddarth and R.A. Marcus, J. Phys. Chem. 94 (1990) 2985.

23) A. Juris, V. Balzani, F. Barigelletti, S. Campagna, P. Belser

and A. von Zelewsky,   Coord. Chem. Rev. 84 (1988) 85.

24) K. Nozaki, T. Ohno, M.-A. Haga, and T. Ishizaki, unpublished.

25) T. Ohno, K. Nozaki, and M.-A. Haga, unpublished.

26) M.-A. Haga, T. Ano, T. Ishizaki, and K. Kano, unpublished.

27) K. Nozaki, M. Oyama, H. Hatano, and S. Okazaki,
    J. Electroanal. Chem. 270 (1989) 191.

28) T. Ohno, K. Nozaki, N. Ikeda, and M.-A. Haga, unpublished.

29) R.F. Dallinger and W. Woodruff, J. Am. Chem. Soc. 101 (1979)
    4397.

30) M.K. DeArmond and M.L. Myrick, Acc. Chem. Res. 22 (1989)
    364.

31) P.A. Mabrouk and M.S. Wrighton, Inorg. Chem. 25 (1986) 526.

32) S.F. McClanahan, R.F. Dallinger, F.J. Holler, and
    J.R. Kincaid, J. Am. Chem. Soc. 107 (1985) 4853.

33) M.L. Myrick, R.L. Blakley, M.K. DeArmond, and M.L. Authur,
    J. Am. Chem. Soc. 110 (1988) 1325.

34) L.F. Cooley, C.E.L. Headford, C.M. Elliott, and D.F. Kelly,
    J. Am. Chem. Soc. 110 (1988) 6673.

35) Y. Kawanishi, N. Kitamura, Y. Kim, and S. Tazuke, Riken Q.,
    78 (1984) 212.

36) Y. Komada, S. Yamauchi, and N. Hirota, J. Phys. Chem. 90
    (1986) 6425.

37) C. Creutz, A.D. Keller, N. Sutin, A.P. Zipp, J. Am. Chem.
    Soc. 104 (1982) 3678.

38) M. Haga, T. Ano, K. Kano, and S. Yamabe, to be published.

39) G.A. Crosby and W.H. Elfring. Jr., J. Phys. Chem. 80 (1976)
    2206.

Photochemical Processes in Organized Molecular Systems
K. Honda (Editor-in-Chief)
© Elsevier Science Publishers B.V., 1991

SOLVENT EFFECT ON ELECTRON-TRANSFER QUENCHING OF *Ru(bpy)$_3$$^{2+}$

Haeng-Boo KIM, Noboru KITAMURA[+], and Shigeo TAZUKE[++]

Research Laboratory of Resources Utilization, Tokyo Institute of
Technology, 4259 Nagatsuta, Midori-ku, Yokohama 227, Japan

Solvent effect on reductive and oxidative quenching of the excited state
Ru(bpy)$_3$$^{2+}$ (bpy = 2,2'-bipyridine) was studied. The solvent effects on the
activation parameters for electron transfer between the excited state
Ru(bpy)$_3$$^{2+}$ and an electron donor or an acceptor cannot be explained by
variation of the outer-sphere reorganization energy with solvents but are
correlated with the solvent donor number. For reductive quenching of
*Ru(bpy)$_3$$^{2+}$, the solvent interaction with the complex was reflected on the
activation parameters for an electron transfer step irrespective of
aromatic amine while those were dependent on the solvation around both the
complex and electron acceptors for the oxidative quenching as revealed by
solvent effects on the standard entropy and enthalpy change of an electron
transfer step.

1. INTRODUCTION

Photophysics and photochemistry of transition metal complexes were greatly
advanced in the last decade. Beside the basic interest in photophysics and/or
photochemistry, transition metal complexes have been widely used as a photochemical potential catalyst in many research fields such as solar energy conversion, organic synthesis, electrochemistry, biochemistry, and so forth.
Among various transition metal complexes, polydiimine ruthenium(II) complexes
represented by Ru(bpy)$_3$$^{2+}$ (bpy = 2,2'-bipyridine) have been best studied and
applied to many research fields mentioned above.[1,2] Until now, numerous
studies on photochemistry of Ru(II) complexes, particularly electron-transfer
quenching of luminescent Ru(II) complexes, have been reported aimed at understanding intermolecular electron-transfer processes as well as application to
photosensitizers.

To understand photoinduced electron-transfer reactions, the quenching reactions of excited molecules have been studied in several systems.[3] However,
almost all studies are limited to the measurement of quenching rate constants
(k_q) at a fixed temperature, and the discussion was sometimes developed on the
assumption of k_q to be the actual electron-transfer rate constant. Comparison
of the experimental data with available theories is thus confined to the rela-

---

+ Present address: Microphotoconversion Project, Research Development Corporation of Japan, 15 Morimoto-cho, Shimogamo, Sakyo-ku, Kyoto 606, Japan.
++ Deceased on July 11, 1989.

tion between k_q and the free energy change, and the theories represented by Marcus have been scarcely tested on the activation parameters for photoinduced electron-transfer.

Since the first report on the temperature dependent k_q of $Ru(bpy)_3^{2+}$ from this laboratory,[4] we have reported elaborate studies on photoredox quenching of Ru(II) complexes by various organic electron donors (aromatic amines) and acceptors (nitroaromatics, quinones, and so forth), with special reference to the analyses of the activation parameters for the quenching.[4-11] On the basis of Marcus theory, the activation parameters for electron-transfer process can be predicted by both the relevant thermodynamic parameters and the reorganiza-tion energy (particularly solvent dependent outer-sphere reorganization energy). In this paper, we report solvent effects on the activation parameters for the electron-transfer process and compare the results with those predicted by Marcus theory.

2. ACTIVATION PARAMETERS FOR ELECTRON-TRANSFER PROCESS

The redox quenching of Ru(II) complexes is assumed to proceed by the fol-lowing scheme and k_q is given by eq.1 after correction of diffusional effects (k_{12}).[8,12] k_{23} is the actual electron-transfer rate constant between $^*Ru(II)$ and Q. k_{32} and k_{30} are the rate constants of back electron-transfer to the excited state reactants and of the dissociation to free ions and/or back electron-transfer to the ground state, respectively.

$$k_q = K_{12} \, k_{23} \, \frac{k_{30}}{k_{30} + k_{32}} \tag{1}$$

Depending on the relative magnitude of k_{32} to k_{30}, k_q and experimentally ob-served activation parameters from Eyring plots are given as follows.[5,12]

Case I $k_{32} \ll k_{30}$

$$k_q = K_{12}\, k_{23} \qquad\qquad (2a)$$
$$\Delta H^{\ddagger} = \Delta H_{23}^{\ddagger} \qquad\qquad (2b)$$
$$\Delta S^{\ddagger} = \Delta S_{23}^{\ddagger} \qquad\qquad (2c)$$

Case II $k_{32} \gg k_{30}$

$$k_q = K_{12}\, K_{23}\, k_{30} \qquad\qquad (3a)$$
$$\Delta H^{\ddagger} = \Delta H_{23} + \Delta H_{30}^{\ddagger} \qquad\qquad (3b)$$
$$\Delta S^{\ddagger} = \Delta S_{23} + \Delta S_{30}^{\ddagger} \qquad\qquad (3c)$$

In Case I, the observed activation parameters correspond to those for electron-transfer process (k_{23}). Therefore, a normal positive temperature dependence of k_q is observed and the activation parameters for electron-transfer process can be obtained directly from Eyring plots. In Case II, on the other hand, the observed activation parameters involves contributions from other processes, so that the values do not represent those for electron-transfer process. Furthermore, if ΔH_{23} is sufficiently negative, a negative temperature dependence of k_q is observed ($\Delta H^{\ddagger} < 0$).[9] When ΔH_{23} and ΔS_{23} are known (determined by temperature-controlled electrochemistry) and a "bell-shaped" Eyring plot is obtained, the activation parameters for each process can be determined. Therefore, comparison of the parameters with available electron-transfer theory is possible.

We have already reported detail analyses of electron-transfer quenching of *Ru(II) complexes from the view points of i) mechanistic difference between reductive and oxidative quenching,[4,8,9] ii) charge type of reactants,[6,10] iii) size of Ru(II) complexes (i.e., electron-transfer distance),[7] and iv) ionic strength of medium.[12] Based on these studies, a key factor determining quenching mechanism of *Ru(II) was shown to be the relative magnitude of k_{32} to k_{30}. For oxidative quenching of *Ru(bpy)$_3^{2+}$ by electron acceptors (ΔG_{23} > -2 kcal/mol), we showed that anomalous negative temperature dependence of the quenching ($\Delta H^{\ddagger} < 0$) was responsible for the efficient k_{32} process (Case II; $k_{32} \gg k_{30}$) while the quenching by electron donors proceeded with $k_{30} \gg k_{32}$ (Case I).

3. SOLVENT EFFECTS ON THE ACTIVATION PARAMETERS

The activation free energy of an electron-transfer step, $\Delta G^{\ddagger}$, is in general expressed as in eq.4;[14]

$$\Delta G^{\ddagger} = \Delta G_{trans}^{\ddagger} + \Delta G_{in}^{\ddagger} + \Delta G_{out}^{\ddagger} + w_r \qquad (4)$$

$\Delta G_{trans}^{\ddagger}$, $\Delta G_{in}^{\ddagger}$, $\Delta G_{out}^{\ddagger}$, and w_r are the activation free energies associated with the formation of the activation complex from separated reactants (trans), the inner- (in) and outer-sphere (out) reorganizations of reactants, and electrostatic work necessary for bringing reactants to the close contact distance, respectively. In the present case, all the quenchers used are neutral molecules so that w_r is zero and therefore, discussion on w_r can be omitted.

In the first approximation, the contribution of $\Delta G_{in}^{\ddagger}$ to $\Delta G^{\ddagger}$ is negligibly small as compared with that of $\Delta G_{out}^{\ddagger}$. Employing the Marcus quadratic equation (eq.5) to $\Delta G_{out}^{\ddagger}$, the activation free energy, enthalpy, and entropy can be written as in eq.7 on the assumption of $\Delta S_{out}^{\ddagger} = 0$.[14]

$$\Delta G_{ij}^{\ddagger}(\lambda) = \frac{\lambda}{4}\left(1 + \frac{\Delta G_{ij}}{\lambda}\right)^2 = m^2 \lambda \qquad (5)$$

$$\lambda_o = e^2\left(\frac{1}{2r_R} + \frac{1}{2r_Q} - \frac{1}{d}\right)\left(\frac{1}{D_{op}} - \frac{1}{D_s}\right) \qquad (6)$$

where $m = 1/2 + \Delta G_{ij}/2\lambda$.

$$\Delta G^{\ddagger} = -RT \ln\frac{hZ}{k_BT} + \frac{\lambda_o}{4}\left(1 + \frac{\Delta G}{\lambda_o}\right)^2 \qquad (7a)$$

$$\Delta H^{\ddagger} = -\frac{RT}{2} + \frac{\lambda_o}{4}\left(1 + \frac{\Delta G}{\lambda_o}\right)^2 \qquad (7b)$$

$$\Delta S^{\ddagger} = R \ln\frac{hZ}{k_BT} - \frac{R}{2} \qquad (7c)$$

According to eq.7, the slope of a $\Delta G^{\ddagger}$ or $\Delta H^{\ddagger}$ vs λ_o plot should be 1/4 when $|\Delta G| \ll \lambda_o$ while $\Delta S^{\ddagger}$ is independent of λ_o. For a given $Ru(bpy)_3^{2+}$ - quencher pair, λ_o is a function of the optical (D_{op}) and static dielectric constants (D_s) of the medium (eq.6), so that solvent effects on the activation

Table 1. Solvent Effects on Reductive Quenching of $^*Ru(bpy)_3^{2+}$ by Aromatic Amines.

solv.	Q = p-Anisidine				Q = N,N-Dimethylaniline				Q = p-Toluidine			
	k_q	$\Delta H_{23}^{\ddagger}$	$\Delta S_{23}^{\ddagger}$	$\Delta G_{23}^{\ddagger}$	k_q	$\Delta H_{23}^{\ddagger}$	$\Delta S_{23}^{\ddagger}$	$\Delta G_{23}^{\ddagger}$	k_q	$\Delta H_{23}^{\ddagger}$	$\Delta S_{23}^{\ddagger}$	$\Delta G_{23}^{\ddagger}$
ACN	7.8×10^8	2.1	−13.3	6.1	9.9×10^7	2.8	−14.5	7.1	2.0×10^7	3.7	−15.0	8.2
DMF	1.0×10^9	3.6	−7.6	5.9	3.0×10^7	6.0	−6.4	7.9	5.7×10^7	4.8	−9.4	7.6
PC	4.8×10^8	1.9	−14.8	6.3	4.2×10^7	3.8	−13.2	7.7	6.7×10^6	4.5	−14.6	8.9
DMAA	1.2×10^9	3.0	−9.4	5.8	9.3×10^7	4.9	−8.2	7.3	2.6×10^7	5.1	−10.0	8.1

k_q, $\Delta H_{23}^{\ddagger}$, $\Delta S_{23}^{\ddagger}$, and $\Delta G_{23}^{\ddagger}$ are in $M^{-1}s^{-1}$, kcal/mol, eu, and kcal/mol units, respectively.

parameters should be explained by the variation of λ_o with solvents. In the present study, λ_o in the solvents used were calculated to be 19.3, 17.0, 17.6, and 16.8 kcal/mol for acetonitrile (ACN), N,N-dimethylformamide (DMF), propylene carbonate (PC), and N,N-dimethylacetamide (DMAA), respectively. According to the λ_o values, $\Delta G^{\ddagger}$ and $\Delta H^{\ddagger}$ should increase in the order of DMAA < DMF < PC < ACN as expected from eqs.7a and 7b.

3.1. Reductive quenching

k_q and the activation parameters for reductive quenching of $^*Ru(bpy)_3^{2+}$ by N,N-dimethylaniline, p-toluidine, and p-anisidine were determined in four solvents mentioned above. Since the reductive quenching proceed via Case I mechanism, the activation parameters determined for k_q correspond to those for the forward electron-transfer process ($\Delta H_{23}^{\ddagger}$, $\Delta S_{23}^{\ddagger}$, and $\Delta G_{23}^{\ddagger}$, eq.2). The results are summarized in Table 1.

The activation parameters are plotted against λ_o as shown in Figure 1. Both $\Delta G_{23}^{\ddagger}$ and $\Delta H_{23}^{\ddagger}$ do not increase with increasing λ_o and the slopes in Figure 1 ($\Delta G_{23}^{\ddagger}$, $\Delta H_{23}^{\ddagger}$) do not agree with 1/4. In particular, $\Delta H_{23}^{\ddagger}$ decreases with increasing λ_o which is opposite to what expected from eq.7b. It is obvious that solvent dependence of the activation parameters cannot be explained by the variation of λ_o alone (eq.7).

It was shown that the redox and excited properties of Ru(II) complexes depended on donor-acceptor interaction with solvent.[15-17] In particular, the oxidation potential correlates with Gutmann's solvent donor (or acceptor) number, DN (or AN).[17] This implies that the t_{2g} orbital of the complex strongly interacts with solvent molecules. For reductive quenching of $^*Ru(bpy)_3^{2+}$ by aromatic amines, an electron is transferred from a donor aromatic amine to the unfilled metal t_{2g} orbital of $^*Ru(bpy)_3^{2+}$. Therefore, the activation parameters for the electron-transfer process may be susceptible to DN. Figure 2 shows the relationships between the activation parameters and DN for three quenching systems. The correlation coefficients of the plots in Figure 2 are in general better than those in Figure 1, indicating that the solvent effects

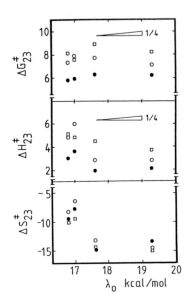

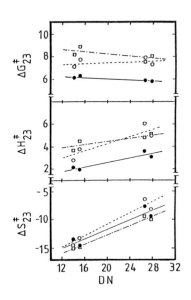

FIGURE 1

λ_o dependence of activation parameters for the quenching by N,N-dimethylaniline (O), p-toluidine (□), and p-anisidine (●).

FIGURE 2

DN dependence of activation parameters for the quenching by N,N-dimethylaniline (O), p-toluidine (□), and p-anisidine (●).

on the activation parameters are explicable by solvent DN.

Figure 2 reveals that both $\Delta H_{23}^{\ddagger}$ and $\Delta S_{23}^{\ddagger}$ increase with increasing DN. Previously, it was shown that $\Delta S_{23}^{\ddagger}$ for reductive quenching of $^*Ru(bpy)_3^{2+}$ in ACN was almost independent of ΔG_{23} (> -5 kcal/mol) at -15 ± 2 eu over 15 aromatic amines.[8] It is important to note that $\Delta S_{23}^{\ddagger}$ in DMF, PC, and DMAA are again almost constant at -6 to -8, -13 to -15, and -8 to -10 eu, respectively, irrespective of aromatic amine. In methanol, analogous results have been also reported by Baggott.[18] These findings indicate that $\Delta S_{23}^{\ddagger}$ **for the reductive quenching is determined by the solvent interaction with $Ru(bpy)_3^{2+}$ with a minor role of the nature of aromatic amine.**

In the reductive quenching, since an electron transferrs from a donor molecule to the t_{2g} orbital of $^*Ru(bpy)_3^{2+}$, the solvent interaction with the t_{2g} orbital is stronger in higher DN solvents. However, when an electron is accepted by the t_{2g} orbital of the excited-state complex, the solvent interaction with the complex will be weakened so that the reduction of $^*Ru(bpy)_3^{2+}$ to $Ru(bpy)_3^+$ by aromatic amine is entropically favorable. The higher the oxida-

Table 2. Solvent Effects on Oxidative Quenching of *Ru(bpy)$_3^{2+}$.

Acceptor solv.	1,4-Naphthoquinone			Duroquinone			m-Dinitrobenzene			Methyl p-nitrobenzoate		
	ACN	DMF	PC	ACN	DMF	PC	ACN	DMF	PC	ACN	DMF	PC
$10^{-9}\,k_q$	13.	3.0	3.2	4.7	2.2	2.9	1.8	1.4		0.53	0.71	0.56
ΔG_{23}	-5.2	-7.5	-5.7	-1.9	-4.0	-2.3	-0.3	-2.4	-0.7	0.6	-1.3	0.1
ΔH_{23}	-16.7	-16.6	-16.7	-11.6	-14.3	-9.7	-6.0	-8.8	-9.1	-6.4	-10.6	-10.6
ΔS_{23}	-38.3	-30.6	-37.0	-33.1	-33.9	-24.6	-19.3	-21.7	-28.4	-23.9	-30.9	-35.7
$\Delta G^{\ddagger}$	4.4	5.2	5.0	5.0	5.6	4.9	5.6	4.7		6.2	5.8	6.0
$\Delta H^{\ddagger}$	1.0	2.5	2.5	-2.2	1.5	-5.2	-2.8	-4.9		-3.4	-6.6	-5.6
$\Delta S^{\ddagger}$	-11.3	-9.2	-8.4	-24.1	-13.1	-33.8	-28.1	-32.1		-32.3	-41.6	-38.7
$\Delta G_{23}^{\ddagger}$	4.4	5.2	5.0	4.4	5.6	5.1	4.4	5.6			5.5	5.8
$\Delta H_{23}^{\ddagger}$	1.0	2.5	2.5	1.7	1.5	3.2	2.6	1.8			2.2	2.4
$\Delta S_{23}^{\ddagger}$	-11.3	-9.2	-8.4	-8.9	-13.1	-6.6	-6.0	-12.7			-11.0	-11.4
$\Delta G_{32}^{\ddagger}$	9.6	12.7	10.7	6.3	9.6	7.4	4.7	7.9			6.9	5.7
$\Delta H_{32}^{\ddagger}$	17.7	19.1	19.2	13.3	15.8	12.9	8.6	10.6			12.8	13.0
$\Delta S_{32}^{\ddagger}$	27.0	21.4	28.6	24.1	20.8	18.0	13.3	9.0			19.9	24.3
$\Delta G_{23}^{\ddagger}(\lambda)$	2.6	1.4		3.9	2.5		4.7	3.2		5.2	3.6	
$\Delta H_{23}^{\ddagger}(\lambda)$	-2.1	-1.3		-1.2	-1.6		1.1	0.2		0.6	-0.9	
$\Delta S_{23}^{\ddagger}(\lambda)$	-16.6	-9.1		-17.8	-13.7		-12.4	-10.0		-15.3	-15.0	
$\Delta G_{32}^{\ddagger}(\lambda)$	7.8	8.5		5.8	6.5		5.0	5.4		4.5	4.9	
$\Delta H_{32}^{\ddagger}(\lambda)$	13.7	14.9		10.1	12.5		7.0	8.9		7.1	9.7	
$\Delta S_{32}^{\ddagger}(\lambda)$	21.7	21.5		15.3	20.2		6.9	11.7		8.6	15.9	

tion number of a molecule, the more favorable the solvation around the molecule in general.[19] Therefore, an increase in the solvent DN brings about the increase in $\Delta S_{23}^{\ddagger}$. On the other hand, the changes in solvation before and after electron transfer will be much larger in higher DN solvents relative to those in lower DN solvents. This will be the primary reason for the increase in $\Delta H_{23}^{\ddagger}$ with increasing DN.

3.2. Oxidative quenching

For oxidative quenching of *Ru(bpy)$_3^{2+}$ by neutral electron acceptors, the activation parameters for k_q ($\Delta G^{\ddagger}$, $\Delta H^{\ddagger}$, and $\Delta S^{\ddagger}$) can be separated into those ($\Delta G_{ij}^{\ddagger}$, $\Delta H_{ij}^{\ddagger}$, and $\Delta S_{ij}^{\ddagger}$) for the forward (ij = 23) and backward electron-transfer processes (ij = 32; see Scheme) only when a "bell-shaped" Eyring plot of k_q is observed.[9] The results in DMF, PC, and ACN are summarized in Table 2. Although the number of the data is limited owing to the reason mentioned above, solvent effects on the activation parameters for both ij = 23 and 32 cannot be explained by the variation of λ_o similar to the reductive quenching. As shown in Figure 3, $\Delta G_{23}^{\ddagger}$ and $\Delta H_{23}^{\ddagger}$ decrease with increasing λ_o in contrasting to what expected from eqs.7a and 7b.

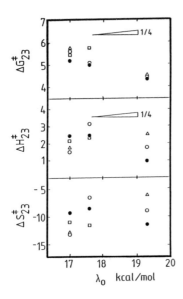

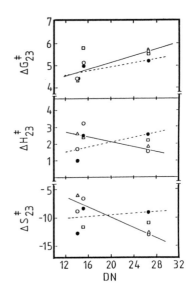

FIGURE 3

λ_o dependence of activation parameters (ij = 23) for the quenching by naphthoquinone (●), duroquinone (○), m-dinitrobenzene (△), and methyl p-nitrobenzoate (□).

FIGURE 4

DN dependence of activation parameters (ij = 23) for the quenching by naphthoquinone (●), duroquinone (○), m-dinitrobenzene (△), and methyl p-nitrobenzoate (□).

Figures 4 and 5 show the DN dependence of the activation parameters for ij = 23 and 32, respectively. $\Delta G_{23}^{\ddagger}$ and $\Delta G_{32}^{\ddagger}$ increase with increasing DN except for $\Delta G_{23}^{\ddagger}$ of methyl p-nitrobenzoate. On the other hand, $\Delta S_{23}^{\ddagger}$ and $\Delta H_{23}^{\ddagger}$ decrease with increasing DN for the quenching by duroquinone (and also by m-dinitrobenzene) whereas those for naphthoquinone (and methyl p-nitrobenzoate) increase with increasing DN. The solvent dependence of $\Delta S_{23}^{\ddagger}$ is quite different from that of the reductive quenching by aromatic amines where $\Delta S_{23}^{\ddagger}$ is almost constant in a given solvent. The DN dependence of $\Delta S_{32}^{\ddagger}$ and $\Delta H_{32}^{\ddagger}$ seems to be reversed to that for ij = 23, exhibiting the decrease or increase (almost constant for $\Delta S_{32}^{\ddagger}$) in $\Delta S_{32}^{\ddagger}$ and $\Delta H_{32}^{\ddagger}$ with increasing DN for naphthoquinone or duroquinone, respectively. For oxidative quenching by quinones or nitroaromatics, $\Delta S_{ij}^{\ddagger}$ and $\Delta H_{ij}^{\ddagger}$ (ij = 23 and 32) are dependent on the solvent properties as well as on the nature of a quencher.

Hupp and Weaver reported the solvent effect on the standard entropy changes (ΔS_{rc}^{o}) of several redox couples such as $Ru(NH_3)_6^{3+/2+}$ and $Ru(en)_3^{3+/2+}$ (en =

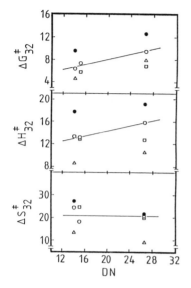

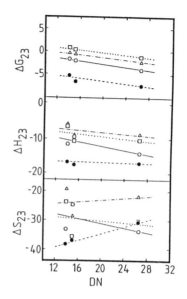

FIGURE 5
DN dependence of activation
parameters (ij = 32) for the
quenching by naphthoquinone
($\bullet$), duroquinone ($\bigcirc$), m-dinitrobenzene ($\triangle$), and methyl p-
nitrobenzoate ($\square$).

FIGURE 6
DN dependence of thermodynamic
parameters (ij = 23) for the
quenching by naphthoquinone
($\bullet$), duroquinone ($\bigcirc$), m-dinitrobenzene ($\triangle$), and methyl p-
nitrobenzoate ($\square$).

ethylenediamine) and they showed that $\Delta S_{rc}^{\,0}$ decreased with increasing the
solvent acceptor number (AN).[20] The correlation of $\Delta S_{rc}^{\,0}$ with AN is attributable to the electron donating abilities of the NH_3 and ethylenediamine
ligands to electron accepting solvents. Their results indicate that the more
the solvation around $Ru(NH_3)_6^{3+}$ or $Ru(en)_3^{3+}$, the more unfavorable $\Delta S_{rc}^{\,0}$. In
the present case, the standard entropy change for electron transfer between
$^*Ru(bpy)_3^{2+}$ and an electron acceptor is expected to decrease with increasing
solvent DN. To confirm the specific solvent interactions with the complex
and/or a quencher, the standard enthalpies (ΔH_{23}) and entropies (ΔS_{23}) of
the forward electron-transfer step were determined by temperature-controlled
cyclic voltammetry[12] and the results are plotted against the solvent DN as
shown in Figure 6.

Figure 6 unequivocally indicates that both positive and negative effects of
DN on ΔS_{23} is observed depending on the nature of an electron acceptor while
both ΔH_{23} and ΔG_{23} decrease with increasing DN. Close inspection of Figures
4 and 6 suggests, furthermore, that the negative and positive slopes of the
ΔS_{23} vs DN plots for the quenching by duroquinone and naphthoquinone, respectively, agrees quite well with the trend of the DN dependence of the relevant

$\Delta S_{23}^{\neq}$. Since ΔH_{23} and ΔG_{23} decrease with increasing solvent DN, solvent effects on ΔS_{23} is the primary origin of those on $\Delta S_{23}^{\neq}$ (and $\Delta S_{32}^{\neq}$). For duroquinone as an example, the decrease in ΔG_{23} with increasing solvent DN brings about the increase and decrease in $\Delta G_{23}^{\neq}$ and $\Delta H_{23}^{\neq}$, respectively. It is clear that the large decrease in $\Delta S_{23}^{\neq}$ with increasing DN is responsible for the increase in $\Delta G_{23}^{\neq}$. For naphthoquinone, however, an increase in DN leads to the increase in $\Delta S_{23}^{\neq}$, $\Delta H_{23}^{\neq}$, and $\Delta G_{23}^{\neq}$. In this case, favorable $\Delta S_{23}^{\neq}$ and unfavorable $\Delta H_{23}^{\neq}$ are compensated with each other, rendering similar DN dependence of $\Delta G_{23}^{\neq}$ with that for duroquinone. The variation of the solvation around a quencher before and after electron transfer is the essential factor governing the oxidative quenching of $Ru(bpy)_3^{2+}$ by quinones and nitroaromatics.

4. ESTIMATION OF ACTIVATION PARAMETERS FROM THERMODYNAMIC PARAMETERS

The activation free energy of an electron-transfer step, $\Delta G^{\neq}$, consists of several terms as introduced in eq.4. Although one of the most important factors determining $\Delta G^{\neq}$ is λ_o, solvent effect on electron transfer can not be explained by solvent dependence of λ_o alone as discussed above. The results suggest that the nature of $Ru(bpy)_3^{2+}$ and/or quencher is worth to be considered (i.e., thermodynamic parameters).

The theoretical expressions of the activation enthalpy and entropy of an electron-transfer step can be obtained by differentiating eq.5 with respect to 1/T and T, respectively.

$$\Delta H_{ij}^{\neq}(\lambda) = \frac{\partial(\Delta G_{ij}^{\neq}(\lambda)/T)}{\partial(1/T)}$$

$$= -m(\Delta H_{ij} - \Delta G_{ij}) + m^2\lambda + (\frac{\Delta G_{ij}}{T} + m\lambda)\frac{m}{Ts}\frac{\partial s}{\partial(1/T)} \qquad (8a)$$

$$\Delta S_{ij}^{\neq}(\lambda) = -\frac{\partial(\Delta G_{ij}^{\neq}(\lambda))}{\partial T}$$

$$= -m[\Delta S_{ij} + (\Delta G_{ij} + m\lambda)\frac{1}{s}\frac{\partial s}{\partial T}] \qquad (8b)$$

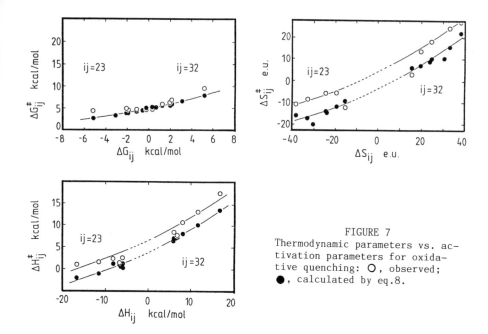

FIGURE 7
Thermodynamic parameters vs. activation parameters for oxidative quenching: O, observed; ●, calculated by eq.8.

where s represents $(1/D_{op} - 1/D_s)$. These equations indicate that the activation parameters for an electron-transfer step can be calculated when ΔG_{ij}, ΔH_{ij}, ΔS_{ij}, and temperature dependence of D_{op} and D_s are known.

The thermodynamic parameters can be obtained by temperature-controlled electrochemistry of the redox couples for $Ru(bpy)_3^{2+}$ and electron acceptors. For reductive quenching by amines, unfortunately, the parameters were not determined owing to the irreversible cyclic-voltammogram. Based on eqs.4 and 8 and the data in Table 2 – 4, $\Delta G_{ij}^{\ddagger}(\lambda)$, $\Delta H_{ij}^{\ddagger}(\lambda)$, and $\Delta S_{ij}^{\ddagger}(\lambda)$, where ij = 23 and 32, were calculated for the oxidative quenching in ACN and DMF (Table 2).

Figure 7 shows the ΔG_{ij}, ΔH_{ij}, and ΔS_{ij} dependence of the observed $\Delta G_{ij}^{\ddagger}$, $\Delta H_{ij}^{\ddagger}$, and $\Delta S_{ij}^{\ddagger}$ (open circles) together with that of $\Delta G_{ij}^{\ddagger}(\lambda)$, $\Delta H_{ij}^{\ddagger}(\lambda)$, and $\Delta S_{ij}^{\ddagger}(\lambda)$ (closed circles), respectively. The most important findings observed in ACN are as follows;

   i) Both $\Delta H_{23}^{\ddagger}$ and $\Delta H_{32}^{\ddagger}$ fall on the same smooth curve.

  ii) $\Delta H_{ij}^{\ddagger}(\lambda)$ almost agree with the observed values within 2 kcal/mol
      in the ΔH_{ij} range between -20 and +20 kcal/mol.

 iii) Analogous results were obtained for $\Delta S_{ij}^{\ddagger}$ as well (errors in 6 eu
      in the ΔS_{ij} range between -40 and +40 eu).

Similarly, the data in DMF were explainable by the Marcus theory.

It is worth noting that the ΔG_{ij} dependence of $\Delta H_{ij}^{\ddagger}$ or $\Delta S_{ij}^{\ddagger}$ is different between ij = 23 and 32. It is apparent that the discussion of the electron-transfer reactions based on ΔG_{ij} alone is not warranted. In the present case, this is readily understandable by the fact that the signs of the standard enthalpy and entropy changes for the backward-electron transfer process (ij = 32) are opposite to those of the forward process (ij = 23). Namely, k_{23} is entropically unfavorable but favored enthalpically while the situation is reversed for k_{32}. The differences in the thermodynamic quantities between k_{23} and k_{32} are attributed to the change in electrostatic solvation entropy (ΔS_{ij}^{es}) and enthalpy (ΔH_{ij}^{es}) before and after electron transfer.[10] The formation of an electrostatically attractive product ion pair $(Ru(bpy)_3^{3+}\cdots Q^-$; ij = 23) is certainly unfavorable in respect to ΔS_{23}^{es} = -8.9 eu and ΔH_{23}^{es} = -0.89 kcal/mol. Taking ΔH_{ij} and ΔS_{ij} into account, the activation parameters for both k_{23} and k_{32} can be explained by the Marcus theory.

5. CONCLUSION

The solvent effects on the activation parameters for electron transfer between $^*Ru(bpy)_3^{2+}$ and an electron donor or an electron acceptor cannot be explained by the variation of the outer-sphere reorganization energy alone. Specific solute-solvent interaction is the reason for this. For reductive quenching of $^*Ru(bpy)_3^{2+}$, the solvent interaction with the metal t_{2g} orbital was reflected on the activation parameters for the electron-transfer step irrespective of aromatic amine. If the entropy change for $Ru(bpy)_3^{2+*/+}$ is correctly estimated, solvent effect on the activation parameters for reductive quenching will be predicted by Marcus theory.

For oxidative quenching, on the other hand, the activation parameters were dependent on the solvation around both $^*Ru(bpy)_3^{2+}$ and an electron acceptor as revealed by solvent effects on the standard entropy and enthalpy changes of the electron-transfer step. The activation parameters were well explained by the standard thermodynamic parameters as well as solvent properties on the basis of the Marcus theory (eqs.4 and 8).

The present study demonstrated the importance of temperature- and solvent-controlled experiments on photoredox quenching reactions. Detailed analysis of an electron-transfer step including molecular level understanding is possible only when temperature- and solvent-controlled experiments are performed. As discussed in the series of publications, the discussion on an electron-transfer process based on the quenching rate constant alone is not guaranteed. The present approaches with combination of temperature-controlled electrochemical measurements are fruitful to develop the efficient photoredox systems as well as to test available electron-transfer theories.

REFERENCES

1) K. Kalyanasundaram, Coord.Chem.Rev. 46 (1983) 159.

2) A. Juris, V. Balzani et al. Coord.Chem.Rev. 84 (1988) 85.

3) Photoinduced Electron Transfer, eds. M.A. Fox and M. Chanon (North-Holland, Amsterdam, 1988)

4) N. Kitamura, S. Okano, and S.Tazuke, Chem.Phys.Lett. 90 (1982) 13.

5) H.-B. Kim, N. Kitamura et al. J.Am.Chem.Soc. 109 (1987) 2506.

6) N. Kitamura, R. Obata et al. J.Phys.Chem. 91 (1987) 2033.

7) N. Kitamura, S. Rajagopal et al. J.Phys.Chem. 91 (1987) 3767.

8) N. Kitamura, H.-B. Kim et al. J.Phys.Chem. 93 (1989) 5750.

9) H.-B. Kim, N. Kitamura et al. J.Phys.Chem. 93 (1989) 5757.

10) N. Kitamura, R. Obata et al. J.Phys.Chem. 93 (1989) 5764.

11) Y. Kawanishi, N. Kitamura et al. Inorg.Chem. 28 (1989) 2968.

12) unpublished results.

13) C.R. Bock, J.A. Connor et al. J.Am.Chem.Soc. 101 (1979) 4815.

14) N. Sutin, in: Tunneling in Biological Systems, eds. B. Chance et al. (Academic Press, New York, 1979) pp.201.

15) P. Belser, A. von Zelewsky et al. Gazz.Chim.Ital. 115 (1985) 723.

16) K. Nakamaru, BullChem.Soc.Jpn. 55 (1982) 1639.

17) N. Kitamura, S. Sato et al. Inorg.Chem. 27 (1988) 651.

18) J.E. Baggott, J.Phys.Chem. 87 (1983) 5223.

19) V. Gutmann, The Donor-Acceptor Approach to Molecular Interactions (Plenum, New York, 1978).

20) J.T. Hupp and M.J. Weaver, J.Phys.Chem. 89 (1985) 2795.

Photochemical Processes in Organized Molecular Systems
K. Honda (Editor-in-Chief)
© Elsevier Science Publishers B.V., 1991

PHOTOSENSITIZED REACTIONS BY [Re(bpy)(CO)₃X]

Chyongjin PAC, Sadamichi KASEDA, Koichiro ISHII,
Shozo YANAGIDA, and Osamu ISHITANI†

Department of Chemical Process Engineering, Faculty of Engineering,
Osaka university, Suita, Osaka 565, Japan

It was found that [*fac*-Re(bpy)(CO)₃X] (bpy = 2,2'-bipyridine; X = Br or Cl) undergoes the following versatile photoreactions in the presence of triethylamine or triethanolamine; (1) the quantitative one-electron photoreduction of 1-benzyl-3-carbamoylpyridinium cation (an NAD^+ model) and pyridine-4-carboxaldehyde, (2) the ethylation of the bpy ligand at C-5 with triethylamine, (3) the formation of CO_2-incorporated Re(I) complexes during the photoreduction of CO_2 in N, N-dimethylformamide, (4) the selective photoevolution of H_2 in tetrahydrofuran, and (5) the efficient photosubstitution of both the Br^- and CO ligands with P(OR)₃ to give a bis-phosphito complex, $[Re(bpy)(CO)_2\{P(OR)_3\}_2]^+Br^-$, in quantitative yield.

1. INTRODUCTION

Photoredox chemistry of transition-metal complexes has received much attention related with visible-light photosensitization of net multi-electron redox reactions in solution, particularly the photoreduction of water and carbon dioxide. Photofixation of CO_2 initiated by electron transfer reminds us of the pioneering works of (the late Professor) Shigeo Tazuke and his coworkers[1] as a milestone in this field of photochemistry. In artifical photosynthetic systems, transition-metal complexes are certainly advantageous as photosensitizer, (1) since a variety of combinations of central metal atoms and ligands can be used to control exitation energies, exited-state lifetimes, and redox potentials of photosensitizers and (2) since single electron transfer often results in ligand substitution with a substrate molecule followed by a chemical change within the coordination sphere which leads to net two-electron redox chemistry.

There have been indeed reported a variety of net two-electron photoreduction reactions[2] of water,[3-10] carbon dioxide,[3-5,11-14] and unsaturated organic molecules[15-17] by one-electron reductants that are photosensitized by transition-metal complexes. In most cases, the photosensitization systems consist of both a

---

† Present adress: National Research Institute for Pollution and Resources, 16-3 Onokawa, Tsukuba, Ibaraki 305, Japan

visible-light photosensitizer (*e.g.* $[Ru(bpy)_3]^{2+}$) acting as a $1e^-$ mediator and a

$$CO_2, H_2O, \;\diagdown C=O \; \xrightarrow[1\,e^-\,donor]{h\nu/[fac\text{-}Re(bpy)(CO)_3X]} \; CO, H_2, \;\diagdown CH\text{-}OH$$

cocatalyst working as a $1e^-/2e^-$ relay (*e.g.* Co(III),[3-5] Ni(II),[6] Pd(I),[7] Ru(II),[12,13] or Rh(III)[15,16] complexes).   Interestingly, however, [*fac*-Re(bpy)(CO)$_3$X] (**1a**, X = Br; **1b**, X = Cl) photosensitizes the selective reduction of CO_2 to CO,[11,14] dihydrogen evolution,[9,10] and the reduction of carbonyl compounds to the corresponding alcohols[17] in the presence of an aliphatic amine without the use of an extra $1e^-/2e^-$ relay.   It is conceivable that electron transfer from an amine to the triplet metal-to-ligand charge transfer state (^{3}MLCT) is responsible for the initiation of the photoreactions.[14]   However, mechanistic pathways following the electron transfer and the chemical behavior of reactive intermediates still remain unknown.   Understanding of mechanistic details of the two-electron redox reactions photosensitized by **1a,b** should provide important information for the construction of optimum photosensitization systems of multi-electron redox reactions using transition-metal complexes.   From this point of view, we have investigated details of the photochemical behavior of **1a,b** in the presence of triethylamine (TEA) or triethanolamine (TEOA).

2. RESULTS AND DISCUSSION

2.1. Photosensitized Reduction of Organic Substrates

It was reported that irradiation of $[Ru(bpy)_3]^{2+}$/Rh(III) complex/TEOA or ascorbate in buffer solution effects the reductions of NAD$^+$ to NADH[15] and of ethylene and acetylene to ethane.[16]   The reduction of 4-benzoyl- and 4-acetyl-pyridine to the corresponding alcohols occurs upon photosensitization by [Re(L)$_2$(CO)$_3$X] (L = the 4-substituted pyridines) in the presence of TEA.[17]   In order to explore the photosensitization capabilities of **1a,b**, we attempted the photosensitized reduction of 1-benzyl-3-carbamoylpyridinium chloride (BNA$^+$Cl$^-$, a typical NAD$^+$ model) and pyridine-4-carboxaldehyde (PA) by TEA.

Irradiation of Ar-purged *N,N*-dimethylformamide (DMF) solution of **1a** (1.0 mM), BNA$^+$ (10 mM), and TEA (1.0 M) at 436 nm quantitatively gave a mixture of the diastereomers of the half-reduced 4,6'-bonded dimers (I and II) and an isomer of the 4,4'-bonded dimers (III) in a 1:2:2 ratio, while BNAH was not formed at all.[18]   The quantum yield for formation of the dimers at 436 nm is 0.13.   Similarly, the photosensitized reaction of PA gave a diastereomeric 1:1 mixture of the pinacol (IV) in quantitative yield without formation of 4-hydroxymethylpyridine.

The luminescence of **1a** was not quenched by BNA⁺ and PA ($K_{sv} \leq 0.2$ M⁻¹) but appreciably by TEA at a rate constant of 8×10^7 M⁻¹ s⁻¹. Therefore, the photo-sensitized reactions can be easily interpreted in terms of electron transfer from TEA to ³MLCT followed by the one-electron reduction of the substrates by **1a⁻·**, as shown in eqs. 1 and 2. The reduction potential of **1a** (ca. -1.32 V vs. SCE)[19,20] is substantially more negative than that of BNA⁺ (ca. -1.0 V)[21] or PA (ca. -1.0 V), indicating that electron transfer from **1a⁻·** to the substrates is certainly efficient enough to predominate over possible ligand substitution via 19e⁻ species **1a⁻·**. It should be noted that the complete lack of two-electron reduction of PA contrasts with the selective alcohol formation from the photosensitized reduction of 4-acylated pyridines by a similar Re(I) complex in the presence of TEA,[17] though the reduction potentials of the pyridines are -0.8 to -1.0 V vs. SCE. Further investigation will be required to explore this difference in the photoredox behavior.

$$\textbf{1a} \xrightarrow{h\nu} \text{³MLCT} \xrightarrow{\text{TEA}} \textbf{1a}^{-}\cdot + \text{TEA}^{+}\cdot \; (\longrightarrow \text{Et}_2\text{N}\overset{\bullet}{\text{C}}\text{HMe} + \text{H}^{+}) \quad (1)$$

$$\textbf{1a}^{-}\cdot + \text{BNA}^{+} \text{ or PA} \longrightarrow \textbf{1a} + \text{BNA}\cdot \text{ or PA}^{-}\cdot \longrightarrow \text{Products} \quad (2)$$

$$\textbf{1a}^{-}\cdot \xrightarrow{\text{very slow}} [\text{Re(bpy)(CO)}_3] + \text{Br}^{-} \text{ or } [\text{Re(bpy)(CO)}_2\text{Br}]^{-} + \text{CO} \quad (3)$$

2.2. Photoethylation of the Bipyridine Ligand with TEA

A photobleaching of **1a,b** occurs upon irradiation of **1a,b** and TEA (or TEOA) in the absence of such an electron-accepting substrate as BNA⁺ or PA, while irradiation of **1a,b** alone results in no photobleaching under similar reaction conditions. We have found that a major pathway of the photobleaching of **1a,b** by TEA is the ethylation of the bipyridine ligand.[22]

Iraradiation of an Ar-purged solution of **1a** (1.0 mM) and TEA (1.0 M) at >400 nm gave the 5-ethyl-2,2'-bipyridine Re(I) complex (**V**) in 10 - 43% yields depending on reaction conditions. Further irradiation resulted in consumption of the **V** formed, accompanied by the formation of 5,5'-diethyl-2,2'-bipyridine complex (VI). It was confirmed that diethylamine is formed while very little evolution of H_2 and CO occurs. The yield of V was improved by the addition of water and Bu_4NBr, reaching a maximum value (43%) in the presence of both water at ca. 1.0 M and Bu_4NBr at 20 - 40 mM. Higher concentration of either resulted in a decrease of the yield. Moreover, the formation of V depends on solvents, favoring acetonitrile, acetone, and tetrahydrofuran but slightly disfavoring DMF. The photoethylation is very inefficient in hexamethyl-phosphoric triamide and did not occur at all in dichloromethane. It was confirmed that the bpy ligand of **1b** is also photoethylated with TEA. Although **1a,b** is comparably photobleached in the presence of TEOA used in place of TEA, no difinite product was detected at all.

$$[Re(bpy)(CO)_3X] + Et_3N \xrightarrow[\text{in THF}]{h\nu} [Re(5\text{-}Et\text{-}bpy)(CO)_3X] + Et_2NH$$
$$\mathbf{V}$$

$$\xrightarrow[\text{TEA in THF}]{h\nu} [Re(5,5'\text{-}Et_2\text{-}bpy)(CO)_3X] + Et_2NH$$
$$\mathbf{VI}$$

The photoethylation should involve a radical coupling between $Et_2\overset{\bullet}{N}CHMe$ and **1a⁻·** formed in eq. 1, probably as the consequence of the localization of an odd electron of **1a⁻·** on the bpy ligand. It was however found that substantial deuterium incorporation in the ethyl group being introduced occurs by the photoreaction in the coexistence of 1.0 M MeOD. The ^1H NMR spectrum of the ethylation product indicates that the ethyl group contains 2.5 D atoms on the methyl part and 1.0 D atom on the methylene part while the bpy ring is not deuterated at all. It was confirmed that no H-D exchange occurs with V in the presence of TEA and MeOD either upon irradiation or in the dark. Probably, there should be involved an intermediate in which the methyl part of the ethyl group being introduced is acidic enough for H-D exchange to be catalyzed by a base (TEA).

2.3. Formation of CO_2-Incorporated Re(I) Complexes

Irradiation of CO_2-saturated DMF solution of **1a,b** and TEA again resulted in a photobleaching of **1a,b** to give several Re(I) products involving V and VI, though the yield of V decreased to one-third of that in an Ar-purged solution. Major products are $[Re(bpy)(CO)_3(O_2CH)]$ (VII),[11,20] $[Re(bpy)(CO)_3(O_2CNEt_2)]$ (VIII), and another CO_2-incorporated, yet unidentified, product abbreviated as $[Re(bpy)(CO)_3Y]$ (IX), in which the latter two are new compounds.[23] These products reveal common spectroscopic features characteristic of the $Re(bpy)(CO)_3$ fragment. The NMR spectra of VIII in $DMSO\text{-}d_6$ show the ^1H resonances of the Et_2N group at δ 0.49 and 2.66 and the ^{13}C resonances of the Et_2NCO_2 ligand at δ 13.80, 40.56, and 160.06 together with those of the bpy and CO ligands.

Treatment of VIII with hydrochloric acid quantitatively gave **1b** and CO_2, whereas methyl *N,N*-diethylcarbamate was formed upon heating VIII in methyl iodide at 70 °C. These data unambiguously support the assigned structure of VIII. The IR spectrum of IX (KBr disc) showed absorptions at 1580 and 1060 cm^{-1} characteristic of C(=O)-O grouping, the former band of which is comparable with that of the C=O group of the diethylcarbamato ligand of VIII at 1590 cm^{-1}. Moreover, IX reveals the ^{13}C resonance at δ 161.07 characteristic of sp^2 carbonyl carbon which is again comparable with that of VIII (δ 160.06), while no extra signal other than those of the bpy, CO, and CO_2-derived ligands appears. Treatment of IX with hydrochloric acid gave **1b** and CO_2 in quantitative yield. Therefore, we tentatively assign IX as the hydroxycarbonyl or bicarbonato Re(I) complex (Y = CO_2H or OCO_2H).

The yield of VIII was 8 - 10% in the photoreaction of **1a** and 1.0 M TEA in CO_2-saturated DMF, but increased to ~ 50% in the coexistance of 1.0 M diethylamine. Presumably, diethylamine might react with CO_2 to give diethylcarbamate anion, which should undergo the ligand exchange following eqs. 1 and 3. The formation of IX was relatively efficient in the photoreaction of **1a** in the presence of TEOA (ca. 50% yield) but was poor with TEA (ca. 5% yield).

$$Et_2NH + CO_2 + Base \; \rightleftharpoons \; Et_2NCO_2^- + Base\text{-}H^+ \qquad (4)$$

$$[Re(bpy)(CO)_3] + Et_2NCO_2^- \xrightarrow{-e^-} [Re(bpy)(CO)_3(O_2NEt_2)] \qquad (5)$$

These products revealed the UV absorption maxima at 360 - 380 nm and very weak emissions at 600 - 620 nm in MeCN at room temperature, which are characteristic of the MLCT transitions. The emission lifetimes at 300 K are shorter than 2 ns in the cases of VIII and IX and 8.9 ns in the case of VII. The short lifetimes of ^{3}MLCT are certainly unfavorable for possible photosensitization reactions by VII - IX. It was indeed confirmed that the CO_2 reduction is not photosensitized by VIII and IX. Therefore, it appears that these Re(I) complexes are dead-end products from **1a,b** in the photosensitized CO_2 reduction.

2.4. Photosensitized Dihydrogen Evolution

Irradiation of **1a,b** in the presence of TEA or TEOA resulted in little H$_2$ evolution in either Ar-purged or CO_2-saturated DMF and MeCN, as reported.[11] By contrast, H$_2$ evolution does occur upon irradiation of **1a,b** and TEA in such ether solvents as tetrahydrofuran (THF), 1,2-dimethoxyethane, 1,4-dioxane, and 2-methyl-THF under Ar and even under CO_2 accompanied by neither the CO_2 reduction to CO nor the formation of VII - IX.[9] During the photosensitized H$_2$ evolution, a photobleaching of **1a,b** again occurred but was 6 - 9 times slower than that in DMF or MeCN. While TEOA is more favorable for the CO_2

Table 1. Effects of Additives on Photosensitized H_2 Evolution [a]

			Additives					
			H_2O (1.0 M)					
	None	Bu$_4$NBr	Temperature/°C			bpy	CO [b]	P(OEt)$_3$
		(10 mM)	0	15	47	(50 mM)		(20 mM)
H_2 ⎡ R(µmol h^{-1})	6.5	≤1	3.8	5.6	7.1	5.6	6.0	<1
⎣ TON	7	<2	8	10	15	9	10	<1
R(-1a/µmol h^{-1})	0.70	>2.5	0.65	0.61	0.55	0.24	0.60	>2.5

a Visible-light irradiation at >400 nm for THF solution of **1a** (1.0 mM), TEA (1.0 M), and an additive. The temperature was usually kept at 15±1 °C unless otherwise stated. b For CO-saturated solution.

photoreduction than TEA, the H_2 evolution favors TEA much more than TEOA. Table 1 summarizes results of the H_2 evolution photosensitized by **1a** in thepresence of 1.0 M TEA under various conditions. Apparent rates of H_2 evolution represented by R(mmol h^{-1}) were obtained from linear portions of plots of H_2 evolved vs. irradiation time. Turn-over numbers abbreviated as TON are defined as molar ratios of H_2 evolved to **1a** consumed at level-off points of time-conversion plots. The photobleaching of **1a** is represented by R(-**1a**/mmol h^{-1}) obtained from linear portions of time-conversion plots.

Notable observations are (1) the enhancement of TON by water added, (2) the increase of both R(mmol h^{-1}) and TON, but slight decrease of R(-**1a**/mmol h^{-1}), with the increase of reaction temparature, (3) little effect of either bpy or CO on the H_2 evolution, and (4) remarkable inhibition of the H_2 evolution and a rapid photobleaching in the presence of Bu$_4$NBr and P(OEt)$_3$. Moreover, it was found that irradiation in undeuterated THF in the presence of 1.0 M D_2O exclusively gave a mixture of D_2 and HD, whereas only H_2 was formed in THF-d$_8$ in the presence of H_2O. This might suggest the participation of a hydride complex after labilization of a ligand following eq. 1. The remarkable effects of Bu$_4$NBr and P(OEt)$_3$ appear to arise from the interception of a reactive intermediate or intermediates, which would shut off the generation of hydride complex but which should open up channels for the degradation of **1a**. However, the liberation of either the bpy or the CO ligand is unlikely to be important in the H_2 evolution, as shown by the lack of formation of free bpy and CO and by little effect of free bpy and CO on the H_2 evolution. The liberation of the Br$^-$ ligand following eq. 1 seems to be a better mechanistic choice. However, it is still unknown why the H_2 evolution favors the ether solvents but not DMF nor MeCN. An attractive speculation is the formation of an exciplex[24] between [3]MLCT of **1a** and TEA in the ether solvents that would lead to pathways for the H_2 evolution unlike the complete electron transfer (eq. 1) occurring in very polar DMF and MeCN, though no exciplex emission was detected at all.

2.5. Photochemical Ligand Substitution with P(OR)₃[25]

The efficient quenching of the photosensitized H_2 evolution by P(OEt)₃ was accompanied by the quantitative formation of [Re(bpy)(CO)₂{P(OEt)₃}₂]⁺Br⁻ (X) along with 0.6 - 0.8 equivalent CO. MeCN and DMF are more favorable solvents for the formation of X than THF and MeOH. The formation of X was linear with irradiation time up to ca. 30% conversions, while we could not detect other significant Re(I) products, *e.g.* monosubstituted Re(I) complexes. These observations clearly indicate that X is the exclusive primary product, but not a secondary one from a monosubstituted complex. The ^{1}H, ^{13}C and ^{31}P NMR spectra demonstrate that X has a structure of C_{2v} symmetry, *i.e.*, the axial trans configuration of the two P(OEt)₃ ligands.

Interestingly, the quantum yields for the formation of X depend on wavelengths of the incident light and also on the solvents used, 0.17 at 366 nm, 0.71 at 405 nm, and 0.32 at 436 nm in degassed MeCN solution of 1.0 mM **1a**, 1.0 M TEA, and 20 mM P(OEt)₃ but ~10⁻³ at 366 nm and 5.3x10⁻² at 405 nm in degassed THF solution. With other P(OR)₃ (R = Me, Bu, and Ph), the photosubstitution again occurred to give [Re(bpy)(CO)₂{P(OR)₃}₂]⁺Br⁻ in 70 - 90% yields.

$$[Re(bpy)(CO)_3Br] + 2P(OEt)_3 \xrightarrow[TEA]{h\nu} [Re(bpy)(CO)_2\{P(OEt)_3\}_2]^+Br^- + CO$$
1a

The photosubstitution should proceed through eq. 1, since no reaction took place in the absence of TEA and since the luminescence of **1a** is not quenched by P(OR)₃. The Br⁻/Cl⁻ ligand substitution of **1a,b** is known to occur via 19-electron spiecies **1a,b**⁻• [11,20] and the substitution of the MeCN ligand of [Re(phen)(CO)₃(NCMe)]⁺ by pyridine and PPh₃ occurs by a chain mechanism initiated by the one-electron reduction of the starting Re(I) complex.[26] By contrast, the present photosubstitution reaction does not afford such a monosubstituted product as [Re(bpy)(CO)₃{P(OR)₃}]⁺ (XI) but directly results in the loss of both the Br⁻ and CO ligands. If the Br⁻ ligand of **1a** is initially substituted with a P(OR)₃ molecule after eq. 1, the axial CO ligand of XI⁻• should be labile enough for the substitution with another P(OR)₃ molecule to predominate over the loss of an electron from XI⁻•. Alternatively, the lack of the formation of XI indicates that **1a,b**⁻• would undergo the initial substitution of a CO ligand with P(OR)₃ followed by the loss of the Br⁻ ligand. This mechanism is of interest related with a suggested mechanism for the photosensitized CO_2

reduction[11] which involves the substitution of a CO ligand of **1b⁻·** with a CO_2 molecule as a key pathway.

However, mechanistic details are still ambigous at the present stage. The wavelength dependence of the photosubstitution reaction can not be attributed to the involvement of different electronic transitions and different excited states. The absorption spectrum of **1a** in DMF, MeCN, or THF is unaffected at all by either or both of TEA and P(OEt)₃. Moreover, the excitation spectra of the **1a** luminescence monitored at different wavelengths are essentially superimposable with the absorption spectra in the respective solvents independently of the presence of P(OEt)₃ and/or TEA. Since the maximum quantum yield for the formation of X is lower than unity, it is unclear whether or not the substitution reaction proceeds through a chain mechanism, *i.e.*, what is the electron acceptor for the loss of an electron from X⁻·.

Finally, it should be noted that P(OEt)₃ inhibits largely the photoethylation of the bpy ligand and almost completely the photosensitized H_2 evolution in THF. Presumably, a common intermediate involved in these different photoreactions seems to be rapidly trapped by P(OEt)₃. On other hand, irradiation of a CO_2-saturated DMF solution in the presence of P(OEt)₃ gave again X in a ~90% yield accompanied by the formation of 1.4 mol-equivalent CO at 100% conversion of **1a** which is twice as efficient as the CO formation in an Ar-purged solution. Presumably, the ligation of CO_2 to a reactive intermediate would compete with that of P(OEt)₃. However, the CO formation in the presence of P(OEt)₃ does not continue to occur but levels off at 100% conversion of **1a** because of the quantitative formation of X, an ineffective complex for the photoreduction of CO_2 by TEA.

ACKNOWLEDGEMENT

This work was supported by Grant-in-Aid for Special Research Project from the Ministry of Education, Science and Culture of Japan (Nos. 61113001, 62113001, and 63104001).

REFERENCES

1) S. Tazuke and N. Kitamura, Nature 275 (1978) 301; N. Kitamura and S.Tazuke, Chem. Lett. (1983) 1109; S. Tazuke, S. Kazama, and N. Kitamura, J. Org.Chem. 51 (1986) 4548.

2) For a review, see; C. Pac and O. Ishitani, Photochem. Photobiol. 48 (1988) 767.

3) J.-M. Lehn and R. Ziessel, Proc. Natl. Acad. Sci., USA 79 (1982) 701; R. Ziessel, J. Hawecker, and J.-M. Lehn, Helv. Chim. Acta 69 (1986) 1065.

4) C. V. Krishnan, B. S. Brunschwig, C. Creutz, and N. Suitin, J. Am. Chem. Soc. 107 (1985) 2005.

5) A. H. A. Tinnemans, T. P. M. Koster, D. H. M. W. Thewissen, and A. Mackor, Recl. Trav. Chim. Pays-Bas 103 (1984) 288.

6) J. L. Grant, K. Gaswami, L. O. Speer, J. W. Otvos, and M. Calvin, J. Chem. Soc., Dalton Trans. (1987) 2105.

7) J. R. Fieser and D. J. Cole-Hamilton, J. Chem. Soc., Dalton Trans. (1984) 809.

8) D. Max Roundhill, J. Am. Chem. Soc.107 (1985) 4354.

9) C. Pac, K. Ishii, and S. Yanagida, Chem. Lett. (1989) 765.

10) M. Tajik and C. Detellier, J. Chem. Soc., Chem. Commun. (1987) 1824.

11) J. Hawecker, J.-M. Lehn, and R. Ziessel, J. Chem. Soc., Chem. Commun. (1983) 536; Helv. Chim. Acta 69 (1986) 1990.

12) J.-M. Lehn and R. Ziessel, J. Organomet. Chem. 382 (1990) 157.

13) H. Ishida, T. Terada, K. Tanaka, and T. Tanaka, Inorg. Chem. 29 (1990) 905.

14) C. Kutal, M. A. Weber, G. Ferraudi, and G. Geiger, Organometallics 4 (1985) 2161; C. Kutal, A.-J. Corbin, and G. Ferraudi, ibid. 6 (1987) 553.

15) R. Wienkamp and E. Steckhan, Angew. Chem., Int. Ed. Engl. 22 (1983) 497; M. Francke and E. Steckhan, ibid. 27 (1988) 265.

16) I. Willner and R. Maidan, J. Chem. Soc., Chem. Commun. (1988) 876.

17) S. M. Fredericks and M. S. Wrighton, J. Am. Chem. Soc. 102 (1980) 6166.

18) O. Ishitani, Y. Wada, and C. Pac, unpublished results.

19) K. Kalyanasundaram, J. Chem. Soc., Faraday Trans. (2) 82 (1986) 2401.

20) J. V. Caspar and T. J. Meyer, J. Phys. Chem. 87 (1983) 952.

21) W. J. Blaedel and R. G. Hass, Anal. Chem. 42 (1965) 918.

22) O. Ishitani, I. Namura, S. Yanagida, and C. Pac, J. Chem. Soc., Chem. Commun. (1987) 1153.

23) O. Ishitani, K. Ishii, S. Kaseda, S. Yanagida, and C. Pac, unpublished results.

24) A. Vogler and H. Kunkley, Inorg. Chim. Acta 45 (1980) L265.

25) C. Pac, S. Kaseda, K. Ishii, and S. Yanagida, in contribution.

26) D. P. Summers, J. C. Luong, and M. S. Wrighton, J. Am. Chem. Soc. 103 (1981) 5238.

Photochemical Processes in Organized Molecular Systems
K. Honda (Editor-in-Chief)
© Elsevier Science Publishers B.V., 1991

MAGNETIC FIELD EFFECTS ON PHOTOINDUCED ELECTRON TRANSFER AND THE
SUCCEEDING PROCESSES IN ORGANIZED MOLECULAR ASSEMBLIES

Taku MATSUO, Hiroshi NAKAMURA, Satoshi USUI,
Hiroaki YONEMURA and Akihiro UEHATA

Department of Organic Synthesis, Faculty of Engineering,
Kyushu University, Hakozaki, Fukuoka 812, Japan

Laser-induced electron transfer in donor-acceptor pairs afforded highly
active radical pairs, whose lifetimes were appreciably extended in the
presence of external magnetic fields. Either porphyrin- or phenothiazine
unit was used as the donor, while the acceptor unit was viologen.
Remarkable magnetic field effects on the radical decay rate were
observed when the porphyrin and viologen units were held in close
distance by either one of the following three methods:(1) covalent
bonding, (2) electrostatic forces, or (3) hydrophobic interactions. In
the case of phenothiazine-viologen linked compounds, photogenerated
radical pairs were observed, when the linked compounds were incorporated
into the cavity of either alpha- or beta cyclodextrin. The magnetic
field effects were useful to elucidate the role of spin interactions in
the radical decay dynamics.

1. INTRODUCTION

The redox potential for photosynthesis in the nature is generated by the
cooperation of photoreaction center and charge-transport systems in thylakoid
membrane of chloroplasts. Detailed studies on photoinduced electron transfer
reactions in donor acceptor pairs in solutions have revealed relevant factors
for charge separation in mobile systems. In the present study, attempts have
been made to construct a thylakoid membrane model system by the use of
appropriately designed donor and acceptor molecules in combination with
various molecular organizates. Laser-induced electron transfer from the donor
(either zinc tetraphenyl-porphinato or phenothiazine unit) to the acceptor
(viologen) afforded triplet radical pairs, and the reverse electron-transfer
processes were examined with and without external magnetic fields (EMF). As a
consequence, the roles of phase boundary in photoinduced electron-transfer
reactions and charge separation steps became clear. Interradical distance and
molecular orientation also play very important roles in photodynamics
involving photoreaction centers. The problem was investigated by the use of
phenothiazine-viologen linked compounds. Complexation of the linked

---

*Contribution No. 935 from the Department of Organic Synthesis, Faculty of
Engineering, Kyushu University.

compounds with cyclodextrins provided efficient means to control steric
factors of the model photoreaction center as described below.

2. KINETIC STUDIES OF PHOTOGENERATED RADICAL PAIRS IN PORPHYRIN VIOLOGEN LINKED COMPOUNDS INCORPORATED INTO MOLECULAR BILAYERS AND REVERSED MICELLES

A series of porphyrin-viologen linked compounds with polymethylene groups
as a spacer were prepared, and the decay process of the laser-generated
radical pair was investigated. The linked compounds (ZPnV) were incorporated
into molecular bilayers of dihexadecylammonium chloride (DHAC) and the decay
rate constant of the radical pair was evaluated[1]. The initial rapid decay
followed the first order kinetics, which could be ascribed to intramolecular
reverse electron-transfer reaction in the donor-acceptor system. The first
order decay rate constant (k_d) was strongly reduced in the presence of EMF as
shown in Figure 1. The k_d-value rapidly decreased with the increase of EMF,
and reached an asymptotic value at EMF above ca 0.2 T. The k_d-value at zero

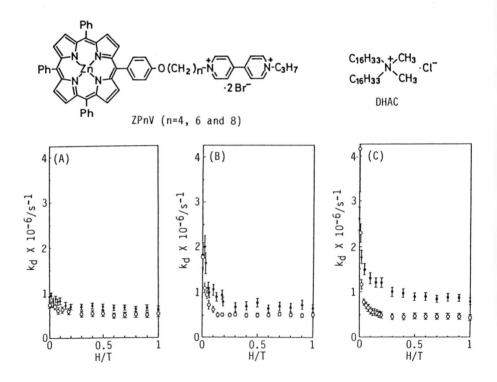

FIGURE 1
The effect of magnetic fields on the rate of k_d-values in the photogenerated
radical pairs of porphyrin-viologen linked compound in aqueous acetonitrile
(⬍) and DHAC molecular bilayers (⬍): (A) ZP4V, (B) ZP6V and (C) ZP8V.

magnetic field increased with the spacer chain length, while the asymptotic value at high EMF was independent of the spacer length.

The above described remarkable, external magnetic field effect (EMFES) on the radical-pair decay rates was further investigated by the use of reversed micelles to elucidate the cause of EMFES[2]. Decay rates of radical pairs obtained with two porphyrin-viologen linked compounds (ZP4V and ZP6V) were examined. The porphyrin moiety of the linked compounds was fixed to the wall of the reversed micelles, while water-soluble viologen moiety was confined to the water pool. General features of EMFES observed with the reversed micellar systems were identical to those in the DHAC molecular bilayers.

Effects of lanthanide ions on the decay rate of the radical pairs provided extremely useful information to elucidate the origin of EMFES. Laser-generated radical pairs were examined in the presence of three different lanthanide ions (La^{3+}, Dy^{3+}, and Gd^{3+}) in the water pool of the reversed micelles. The results are summarized in Figure 2 for ZP4V. In the presence of high EMF (0.5 T), the decay rates considerably increased on the addition of either Dy^{3+} (S= 5/2, and

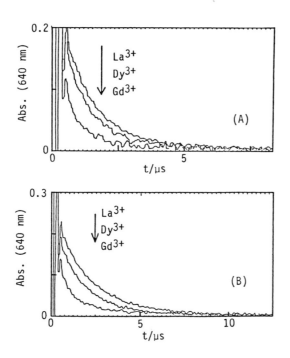

FIGURE 2
Effects of lanthanide ions (0.5 mM) on then decay rates of
radical pairs in the porphyrin-viologen linked compound
(ZP4V): (A) zero magnetic field, and (B) 0.5 T.

J= 15/2) or Gd3+ (S=J= 7/2), while no effect was observed with La3+ (S=J=0).
These data clearly indicate that the radical pairs decay faster in the
presence of lanthanide ions with higher spin multiplicities. Exactly the same
lanthanide effects were observed with the radical pairs for ZP6V at 0.5 T.

The above observations strongly indicate that the EMFES are originated from
Zeeman splitting of triplet sublevels of the radical pairs. The spin flipping
relaxation from the triplet sublevels to the singlet state of the radical
pairs appears to be the rate determining step, which has been well known as
Relaxation Mechanism of EMFES[3].

In the absence of EMF, the lanthanide ions affected the radical decay
process for ZP4V and ZP6V in entirely different manner. As to ZP4V, the decay
rate of the radical pair was affected by the three lanthanide ions to almost
the same extent as observed at high EMF. In the case of the radical pairs
generated from ZP6V, however, the decay rate was hardly affected at 0 T. The
most important difference between ZP4V and ZP6V is the fact that the radical
pair generated from ZP6V decayed much faster than those from ZP4V at 0 T. This
may be related to the difference in the spacer chain length between the
porphyrin and viologen moieties. On going from ZP4V to ZP6V, the spacer chain
length increases by two methylene units, and the electron spin exchange
interaction (ΔE) in the radical pair is expected to decrease accordingly[4]. As
ΔE-value approaches zero, the singlet and the triplet energy levels of the
radical pair become degenerated to each other. In this extreme case, the

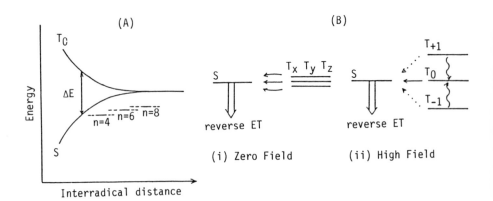

FIGURE 3
Schematic presentation of phenomena relevant to EMFES on the decay rate
of photogenerated geminate radical pairs: (A) variation of the triplet-
singlet energy separation with the interradical distance, and (B) two
extreme modes of radical decay via intersystem crossing at degenerated
S-T levels.

intersystem crossing from the triplet to the singlet is expected to proceed easily via hyperfine coupling- and/or Δg-mechanisms[5], and the rate will not be appreciably affected by paramagnetic additives such as lanthanide ions. The relevant factors for EMFES are schematically shown in Figure 3.

3. GENERATION OF PSEUDO-LINKED RADICAL PAIRS AT THE INTERFACE OF MOLECULAR BILAYERS AND REVERSED MICELLES

Since the above described EMFES are also expected to be observed with other photogenerated radical pairs, laser-induced electron-transfer from amphiphilic derivatives of zinc tetraphenylporphinate (ZPnA) to various viologens was examined by the use of molecular bilayers and reversed micelles.

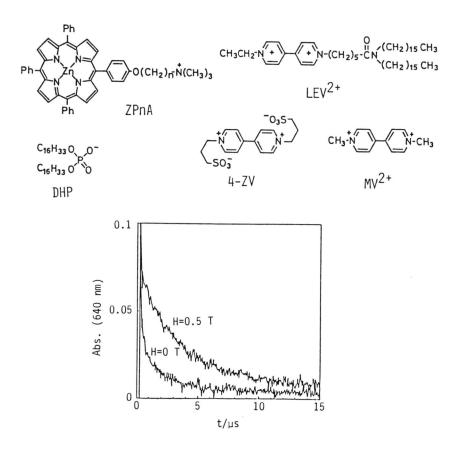

FIGURE 4
Decay profiles of transient absorption due to radical pairs on the laser photolysis of ZP4A-MV^{2+} pair in DHP bilayer system.

An amphiphilic viologen (LEV^{2+}), which was immobilized at the surface of DHAC bilayer together with ZPnA, afforded short-lived radical pairs, and the lifetime was extended by EMF in the same manner as described above[6]. Typical EMFES were also observed with positively charged viologen (MV^{2+}), where the acceptor was electrostatically bound to the negatively charged surface of DHP (Figure 4).

Radical pairs could be generated by laser irradiation of ZP4A in combination with MV^{2+} in DHAC bilayer system. The transient absorption indicated initial, rapid growth of the radical pairs which immediately turned over to slow, bimolecular decay process (Figure 5). No EMFES were observed in this case. The reason may be ascribed to electrostatic repulsion between the photoreduced viologen (MV‡) and the positively charged DHAC surface, which helps MV‡ to escape from the vicinity of the oxidized counterpart of the radical pair (ZP4A‡). In order to be detected by EMFES, the photogenerated radical pairs should be held in a fixed distance at least in a few microseconds after laser pulsing. In other words, EMFES may be taken as an evidence for the presence of pseudo-linked radical pair in the short time domain.

Pseudo-linked radical pairs were also generated by laser irradiation of ZPnA and a zwitterionic viologen (4-ZV) in reversed micellar systems [2]. The zwitterionic viologen was trapped in the water pool as shown in Figure 6. The size of water pools was modified by adjusting ratios between water and surfactant molecules (w=[water]/[surfactant]). The EMFES on the decay rate of

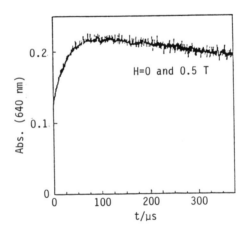

FIGURE 5
Decay profiles of transient absorption due to radical pairs on the laser photolysis of ZP4A-MV^{2+} pair in DHAC bilayer system.

the radical pair were examined with water pool ranging from 1 to 6 nm in radius. In small water pool (w=10), the EMFES indicated that the reduced viologen was tightly bound to the oxidized counterpart (Figure 7). As the size of the water pool was increased, the radical decay rate decreased and the EMFES became less obvious. The viologen molecule (4-ZV) appears to be freely moving around in the large water pools.

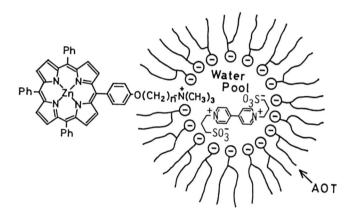

FIGURE 6
Structural formulae for electron donor (ZPnA) and electron acceptor (4-ZV) incorporated into reversed micelles.

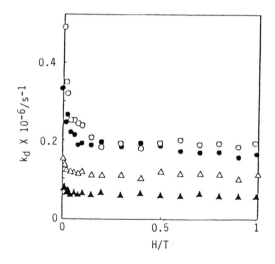

FIGURE 7
External magnetic field effects on the K_d-values for unlinked porphyrin-viologen pairs (ZP6A and 4-ZV) in reversed micelles with various water contents: w=10 (○), w=20 (●), w=30 (△), and w=40 (▲)

4. PHOTOINDUCED ELECTRON-TRANSFER REACTIONS IN PHENOTHIAZINE-VIOLOGEN LINKED COMPOUNDS COMPLEXED WITH CYCLODEXTRINS

Donor-acceptor linked compounds, as represented by porphyrin-viologen pair, are useful model systems for the study of photoreaction center in artificial photosynthesis. The studies have always been plagued with the flexibility of the spacer chain. A novel method to solve the problem was developed by complexing the donor acceptor linked compound with cyclodextrins (abbreviated to CD)[7],[8].

The following phenothiazine-viologen linked compounds were examined to elucidate the effect of spacer chain length. Three CDs were used to examine the effect of pore size on the complexing behavior of the phenothiazine-viologen linked compounds. Photogenerated radical pairs could be observed when either alpha- or beta CD was combined with phenothiazine-viologen pair separated by a relatively long spacer (n =8 or above). Examples of EMFES for PH12V are shown in Figure 8.

In the presence of alpha-CD, the yield of photogenerated radical pairs considerably increased in comparison with the case of beta-CD. No radical was generated with the solution containing PH12V and gamma-CD. Laser excitation of PH4V did not afford radical pair either, in spite of elaborate examination under various conditions (EMF strength and choice of CD).

Intensity of fluorescence emission from phenothiazine moiety also was strongly affected by the choice of the combination between the linked compound and CD. Emission from either PH4V or PH12V was hardly observed in the absence of CD. On the addition of CD, the fluorescence from PH12V was intensified in the following order: alpha-CD > beta-CD > gamma-CD = without CD. In the case of PH4V, the fluorescence intensity did not increase even on the addition of

$$S \overset{}{} N-(CH_2)_n-N^+ \overset{}{} N^+-C_3H_7 \cdot 2Br^-$$

PHnV

$$
\begin{cases}
H \quad CH_2OH \\
HO \quad OH
\end{cases}_n
$$

n = 6, α-CD
n = 7, β-CD
n = 8, γ-CD

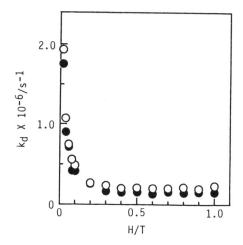

FIGURE 8
External magnetic field effects on the k_d-values for the radical
pairs as evaluated from the absorbance of the photoreduced viologen
units (603 nm) on laser excitation of PH12V: alpha-CD system (●)
and beta-CD system (○).

CD. These data clearly indicate that deactivation of singlet, photoexcited
PH12V is suppressed by complexation of the linked compound with either alpha-
or beta-CD.

The difference in the complexation behavior was also revealed by the
inspection of ^{1}H NMR spectra. When D_2O solution contained PH12V and either
alpha- or beta-CD, distinct signals due to phenothiazine and viologen moieties
in the complexed species were observed apart from the free species (Case A).
The observed spectra are reasonably explained if one assumes that the viologen
moieties are located on the top of the central part of the phenothiazine
moieties in the free species and also that the complex formation enforces
open, flat structure with extended spacer chain in the cavity of CD as in the
following scheme.

SCHEME 1

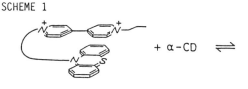

$+ \alpha-CD \rightleftharpoons$

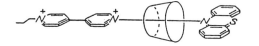

In the case of a solution containing PH12V and gamma-CD, the aromatic proton signals simply shifted, which indicates rapid exchange between the complex and the free species (Case B). Rapid exchange was also indicated in the NMR spectroscopic studies of PH4V-CD systems. These observations are in fair agreement with those of fluorescence studies. The case A complexes are stable enough in NMR time scale, and energy dissipating, direct interaction between phenothiazine and viologen moieties appears to be suppressed in these complexes. As a consequence, the fluorescence intensity increases, and the branching ratio to the triplet manifold leading to the radical pair will also increase. In the rapid exchange limit (Case B), which is usual for ordinary CD complexes, the photoexcited PH4V and PH12V may be easily deactivated by intramolecular interactions between phenothiazine and viologen moieties of the linked compound in the flexible, free species. Relevant intramolecular processes involved in photoinduced electron-transfer and the reverse reactions in the phenothiazine-viologen linked compounds are summarized in Scheme 2.

SCHEME 2

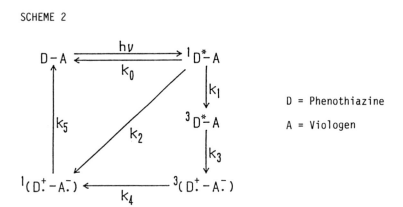

D = Phenothiazine

A = Viologen

5. CONCLUSION

Dynamic behavior of various geminate radical pairs has been elucidated on the basis of EMFES on the reaction kinetics[9]. As to polymethylene-linked donor-acceptor compounds (A=pyrene and D=dimethylaniline), Weller and his associates found remarkable EMFES on the yield of the pyrene triplet state as generated via intramolecular electron-transfer fluorescence quenching followed by intersystem crossing[10,11]. In the present paper, model compounds for the photoreaction center in artificial photosynthesis were constructed by the use of donor-acceptor linked compounds (or pseudo-linked compounds) in various microenvironment. Decay rates of the photogenerated triplet radical pairs were directly detected and enormous EMFES were observed. The EMFES on the reverse electron transfer rate in the photogenerated radical pair elucidated that the spacer between the donor and acceptor moieties could be appropriately adjusted to attain highly efficient charge separation. Both the time domain and the dimension of the space required for the charge separation also became clear by the analysis of EMFES on the reverse electron transfer processes in various microenvironment. The kinetic data are expected to bring a break through in the system design and reaction control of charge separation steps in artificial photosynthesis.

REFERENCES

1) H. Nakamura, A. Uehata, A. Motonaga, T. Ogata and T.Matsuo, Chem. Lett. (1987) 543.

2) A. Uehata, H. Nakamura, S. Usui and T. Matsuo, J. Phys. Chem. 93 (1989) 8197.

3) H. Hayashi and S. Nagakura, Bull. Chem. Soc. Jpn. 57 (1984) 322.

4) Y. Tanimoto, M. Takashima, K. Hasegawa and M. Itoh, Chem. Phys. Lett. 137 (1987) 330

5) N. J. Turro and B. Kraeutler, Acc. Chem. Res. 13 (1980) 369.

6) S. Usui, H. Nakamura, T. Ogata, A. Uehata, A. Motonaga and T. Matsuo, Chem. Lett. (1987) 1779.

7) H. Yonemura, H. Nakamura and T. Matsuo, Chem. Phys. Lett. 155 (1989) 157.

8) H. Yonemura, H. Saito, S. Matsushima, H. Nakamura and T. Matsuo, Tetrahedron Lett. 30 (1989) 3143.

9) U. E. Steiner and T. Ulrich, Chem. Rev. 89 (1989) 51.

10) A. Weller, H. Staerk and R. Treichel, Faraday Disscuss. Chem. Soc. 78 (1984) 271.

11) K. Staerk, W. Kühnle, R. Treichel and A. Weller, Chem. Phys. Lett. 118 (1985) 19.

Photochemical Processes in Organized Molecular Systems
K. Honda (Editor-in-Chief)
© Elsevier Science Publishers B.V., 1991

HIGHLY SPECIFIC ISOMERIZATION OF AROMATIC OLEFIN RADICAL CATIONS PRODUCED BY PHOTOSENSITIZED ELECTRON TRANSFER

Katsumi TOKUMARU, Yasunao KURIYAMA, Tatsuo ARAI,
Igor K. LEDNEV[†], Ryoichi AKABA[‡], and Hirochika SAKURAGI

Department of Chemistry, University of Tsukuba, Tsukuba, Ibaraki 305, Japan

The efficiency of unimolecular cis→trans conversion of the stilbene radical cations is enhanced by introduction of two bromine atoms, two methyl or methoxy groups at the 4- and 4'-positions and by addition of salts like lithium or magnesium perchlorate. On 2,4,6-triphenylpyrylium tetrafluoroborate (TPP[+] BF$_4$[−]) sensitization, only the triplet radical pair is effective for the isomerization though both of the singlet and triplet excited states of the sensitizer are efficiently quenched. The isomerization efficiency increases in a sodium dodecyl sulfate (SDS) micellar solution.

1. INTRODUCTION

Recently, much attention has been paid to photochemical electron transfer-induced isomerization of *cis*-stilbene and its analogues to the corresponding *trans*-stilbenes via their radical cations[1-7], as observed on sensitization with 9,10-dicyanoanthracene (DCA) in acetonitrile[1]. In these reaction systems (Scheme 1), however, the stilbene radical cations can undergo reverse electron transfer with partner radical anions, DCA[−]·, which interferes direct observation of the transient species by laser transient spectroscopy. Addition of biphenyl as a primary electron donor reacting with the electron-accepting sensitizer, DCA, in aerated solution was shown to prevent reverse electron transfer between stilbene radical cations and the acceptor radical anions, since the subsequent electron transfer to the resulting biphenyl radical cations from stilbene gives stilbene radical cations apart from the acceptor radical anions which further transfer electrons to molecular oxygen[2] (Scheme 2). Stilbene radical cations thus generated, however, were reported not to undergo cis→trans conversion in the 100 μs range[3]. This fact tempted one to clarify whether the apparent cis→trans conversion of stilbene radical cations might

---

[†] On leave from the laboratory directed by M. A. Alfimov, Institute of Chemical Physics, USSR Academy of Sciences, Moscow, USSR.
[‡] Department of Chemistry, Gunma College of Technology, Toriba-machi, Maebashi, Gunma 371, Japan.

proceed through a unimolecular thermal process, through further photochemical processes, or by way of bimolecular processes like addition to another *cis*-stilbene molecule followed by elimination of *trans*-stilbene radical cation[4–7].

We now describe that (1) introduction of two bromine atoms, two methyl, or methoxy groups at 4- and 4'-positions of stilbene and addition of salts like lithium and magnesium perchlorate enhance the cis→trans conversion of the radical cations in the μs region[6], and that (2) on 2,4,6-triphenylpyrylium tetrafluoroborate (TPP[+]BF$_4$[−]) sensitization, the triplet radical pair of the reduced sensitizer and the cis radical cation is effective for the isomerization[4], the efficiency of which increases in sodium dodecyl sulfate (SDS) micellar solutions.

2. DIRECT OBSERVATION OF UNIMOLECULAR CIS→TRANS ISOMER-IZATION OF STILBENE RADICAL CATIONS

2.1. Effects of substituents introduced at 4- and 4'-positions of stilbene[6]

DCA (5×10^{-5} mol dm^{-3}) was excited with 406-nm laser in the presence of biphenyl (BP, 0.1 mol dm^{-3}) as a primary electron donor and a cis or trans olefin (1×10^{-3} mol dm^{-3}) in acetonitrile at 23 °C under air. As reported by Lewis et al.[3], *cis*- (c-St) or *trans*-stilbene (*t*-St) gave the transient absorption spectrum due to the stilbene radical cations with a configuration of the starting isomer, and no clear spectral changes indicating cis→trans conversion of the radical cations were observed in the 100 μs time scale of the experiment. The radical cations decayed with second order kinetics through a probable recombination with O$_2$[−·].

trans-4,4'-Dibromostilbene (*t*-DBSt) gave a transient spectrum with an absorption maximum (λ_{max}) at 520 nm, and *c*-DBSt exhibited an absorption with λ_{max} at 540 nm immediately after laser excitation under similar conditions (Figure 1). The short-lived absorption bands at 650–700 nm can be attributed to biphenyl cation radicals[8,9]. The absorption bands observed for *c*-DBSt and *t*-DBSt are nearly identical with those for *c*-St and *t*-St, respectively, in band shape, but their maxima are shifted to longer wavelengths. The bands were efficiently quenched by azulene (E_{ox}=0.95 V/SCE) with rate constants of $\approx 2.4 \times 10^{10}$ dm^3 mol^{-1} s^{-1}. From these observations the 520- and 540-nm bands are reasonably assigned to *t*-DBSt and *c*-DBSt radical cations, respectively. *t*-DBSt radical cations decayed with second order kinetics ($k_2/\varepsilon \approx 2 \times 10^6$ cm s^{-1}, where ε is a molar extinction coefficient of the radical cations), and *c*-DBSt[+·] decayed with bicomponent kinetics of the first (k_1=3×10^5 s^{-1}) and second order ($k_2/\varepsilon \approx 10^8$ cm s^{-1}). The second-order decay of both radical cations can be attributed to recombination with O$_2$[−·].

$$^1DCA^* + c\text{-St} \longrightarrow DCA^{-} + c\text{-St}^{+\cdot}$$

$$c\text{-St}^{+\cdot} \longrightarrow t\text{-St}^{+\cdot}$$

$$t\text{-St}^{+\cdot} + c\text{-St} \longrightarrow t\text{-St} + c\text{-St}^{+\cdot}$$

$$t\text{-St}^{+\cdot} + DCA^{-} \longrightarrow t\text{-St} + DCA$$

Scheme 1

$$^1DCA^* + BP \longrightarrow DCA^{-} + BP^{+\cdot}$$

$$BP^{+\cdot} + c\text{-St} \longrightarrow c\text{-St}^{+\cdot} + BP$$

$$c\text{-St}^{+\cdot} \longrightarrow t\text{-St}^{+\cdot}$$

$$DCA^{-} + O_2 \longrightarrow DCA + O_2^{-}$$

$$c\text{-St}^{+\cdot} + O_2^{-} \longrightarrow c\text{-St} + O_2$$

$$t\text{-St}^{+\cdot} + O_2^{-} \longrightarrow t\text{-St} + O_2$$

Scheme 2

As Figure 1b indicates, the decay of 540-nm band due to c-DBSt$^{+\cdot}$ was accompanied by a rise of 520-nm band due to t-DBSt$^{+\cdot}$. This phenomenon is illustrated in Figure 2 as a decay of the transient absorption at 550 nm and a concurrent rise and subsequent decay at 515 nm. The first-order decay of c-DBSt$^{+\cdot}$ is attributed to its unimolecular isomerization to t-DBSt$^{+\cdot}$. The Arrhenius plot of decay rates of c-DBSt$^{+\cdot}$ at 550 nm at varying temperatures (5–45 °C) gives an activation energy, E_a, of 7.7 ± 3.0 kcal mol^{-1} and a frequency factor, A, of $10^{11\pm2}$ s^{-1} for the cis→trans conversion. The observed frequency factor is in the same order as that for the unimolecular isomerization of the triplet state of 2-(ethenyl-2-d)anthracene, 2-AnthrylCH=CHD, from one form to the other ($A=10^{11.7}$ s^{-1})[10,11].

4,4'-Dimethylstilbene (DMSt) behaves in a similar way. c-DMSt$^{+\cdot}$ ($\lambda_{max}=520$ nm) is converted to t-DMSt$^{+\cdot}$ ($\lambda_{max}=495$ nm) with a rate constant similar to that of DBSt, 3×10^5 s^{-1} at ambient temperature, and with an activation energy of 3.3 ± 0.5 kcal mol^{-1} and a frequency factor of $10^{9\pm2}$ s^{-1}. 4,4'-Dimethoxystilbene (DMOSt) undergoes more rapid conversion (>10^6 s^{-1} at ambient temperature) since c-DMOSt and t-DMOSt exhibits similar spectra even immediately after laser irradiation. This observation is consistent with the results reported on pulse radiolysis by Takamuku et al.[12]; a rate constant of 5.8×10^6 s^{-1} at 200 K and an activation energy of nearly 7 kcal mol^{-1}.

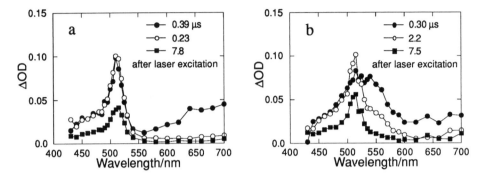

FIGURE 1

Transient absorption spectra observed on 406-nm laser excitation of DCA in the
presence of BP and *t*-DBSt (a) or *c*-DBSt (b) in acetonitrile

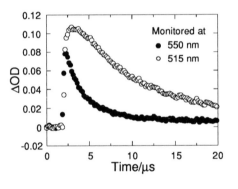

FIGURE 2

Time profiles of the transient absorption spectrum monitored at 515 (o) and 550
nm (•) on 406-nm laser excitation of DCA with BP and *c*-DBSt in acetonitrile

The above results clearly indicate that these substituted stilbenes undergo unimolecular conversion from the cis radical cations to the trans ones. However, the
cis radical cations of 4-bromostilbene (BSt) showed no spectral change.

Combination of electrochemical potentials of DBSt and DMSt (Table 1) and the
observed activation energies together with the estimated ground-state surface
enables us to draw the potential energy surfaces of the radical cations of DBSt and
DMSt as in Figure 3. The energy barrier is too high to be overcome in the conversion from the trans radical cations to the cis ones (11–14 kcal mol[-1]). This makes
the isomerization of stilbene radical cations one-way from cis to trans.

Table 1. Electron transfer-induced isomerization of stilbenes (St),
4-X-C$_6$H$_4$CH=CHC$_6$H$_4$-X'-4', in the presence of biphenyl and
physical properties of their radical cations

| X | X' | $\phi_{c \to t}$ [a) | k_1/s^{-1} [b) | λ_{max}/nm [c) | | $E(St^{+\cdot}/St)/V$ [d) | |
				cis	trans	cis	trans
MeO	MeO	0.56	>10^6		530	1.05	1.03
Me	Me	0.50	3×10^5	520	495	1.44	1.31
H	H	0.04	<10^4	510	475	1.55	1.45
H	Br	0.15	<10^4	540	490	1.62	1.54
Br	Br	0.81	3×10^5	540	520	1.62	1.54

a) Isomerization quantum yields. b) Conversion rate constants. c) Absorption maxima of the radical cations. d) Oxidation potentials of stilbenes vs. SCE.

Table 1 includes isomerization quantum yields ($\phi_{c \to t}$) determined for acetonitrile solutions of DCA (5×10^{-5} mol dm^{-3}), BP (0.1 mol dm^{-3}), and a cis olefin (1×10^{-3} mol dm^{-3}) irradiated with light of wavelengths longer than 400 nm at 23 °C under nitrogen, and shows that introduction of substituents at the 4- and 4'-positions of the phenyl rings of stilbene increases the efficiency of the isomerization irrespective of their electronic property. As to the effect of substituent, Figure 4 plots $\phi_{c \to t}$ and the wave number of absorption maxima ($\tilde{\nu}_{max}$) of the trans radical cations against Hammett's σ-values of the substituents. $\phi_{c \to t}$ and $\tilde{\nu}_{max}$ exhibit a V-shape and an inverted V-shape profiles, respectively; substitution of either electron-donating or electron-accepting group enhances $\phi_{c \to t}$ and reduces $\tilde{\nu}_{max}$.

According to ESR data of *trans*-stilbene radical cations, the hyperfine splitting constant of α-hydrogen for *t*-St (4.53[13], 4.8 G[14]) is larger than those for *t*-DMSt (4.00)[13] and *trans*-di-*t*-butylstilbene (4.50)[13], indicating that introduction of 4,4'-dimethyl or di-*t*-butyl substituents reduces the spin density on the α-hydrogens. The same trend is known for 4-substituted benzyl radicals[15]. Plots of $\phi_{c \to t}$ and $\tilde{\nu}_{max}$ against σ_α values[15] borne out from the hyperfine splitting constants of the benzyl hydrogens in benzyl radicals (Figure 5) illustrate that $\phi_{c \to t}$ tends to increase but $\tilde{\nu}_{max}$ tends to decrease with increasing σ_α. These results clearly show that the decrease of spin density on the ethenyl carbons due to the delocalization effected by substitution of either electron-donating or electron-accepting groups in the radical cations enhances the isomerization efficiency of the stilbene analogues and reduces their absorption energy, $\tilde{\nu}_{max}$.

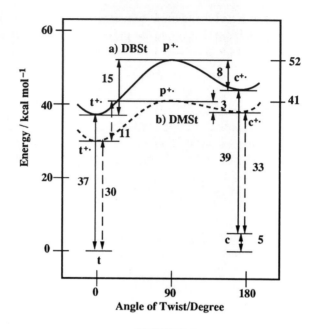

FIGURE 3

Potential energy surfaces of DBSt (a, ——) and DMSt radical cations (b, – – –)

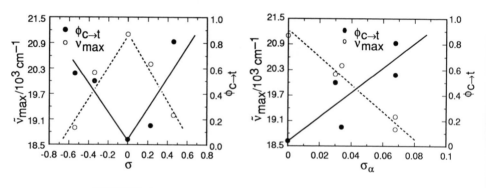

FIGURE 4

Plots of $\tilde{\nu}_{max}$ and $\phi_{c \to t}$ against
Hammett's σ values

FIGURE 5

Plots of $\tilde{\nu}_{max}$ and $\phi_{c \to t}$ against σ_α
values for benzyl radicals

2. 2. Effects of salts added to the solution[16]

The transient spectrum of *cis*-stilbene radical cation (λ_{max}=505 nm) showed no
such spectral change as to indicate formation of *trans*-stilbene radical cation but
decayed with second-order kinetics through recombination with anions (Figure 6a).

However, addition of lithium perchlorate (3×10^{-3} mol dm^{-3}) affected the time profile of spectrum. Figure 6b shows that decrease of the c-St$^{+\cdot}$ absorption intensity is accompanied by appearance of the absorption due to t-St$^{+\cdot}$ (λ_{max}=475 nm) at 6.3 μs after laser excitation. In the presence of the salt, the decay kinetics of c-St$^{+\cdot}$ absorption is comprised of a first-order component due to the isomerization to t-St$^{+\cdot}$ and a second-order component due to the recombination. The addition of salt lowers the second-order rate, which enables the first-order component to be observable. The reduction of the second-order rate was clearly demonstrated in the change of the time profile of t-St$^{+\cdot}$ generated from t-St (Figure 7). The ions of added salt may interact with the radical ions of opposite charge to prevent their recombination.

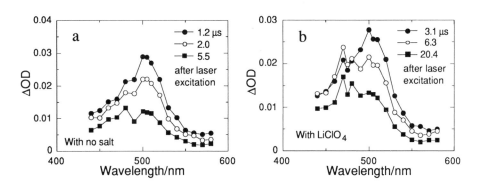

FIGURE 6

Transient absorption spectra observed on 406-nm laser excitation of DCA with BP and c-St in the absence (a) and presence of LiClO$_4$ (b) in acetonitrile

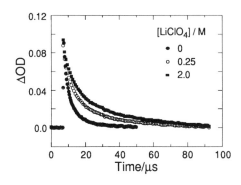

FIGURE 7

Time profiles of t-St radical cations in the presence of LiClO$_4$

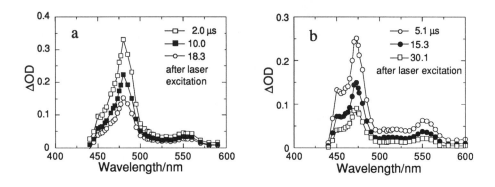

FIGURE 8

Transient absorption spectra observed on 408-nm laser excitation of TPP+ (6.2 ×10^{-5} mol dm^{-3}) in the presence of t-St in concentrations of 4.7×10^{-4} (a) and 4.3×10^{-2} mol dm^{-3} (b) in dichloromethane

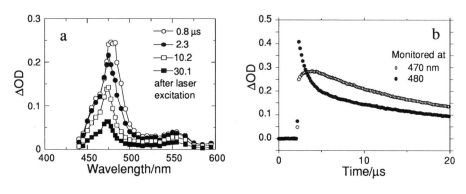

FIGURE 9

Transient absorption spectra observed on 408-nm laser excitation of TPP+ (6.1 ×10^{-5} mol dm^{-3}) in the presence of t-St (1.0×10^{-2} mol dm^{-3}) in CH$_2$Cl$_2$ (a) and the time profiles of transients monitored at 470 and 480 nm (b)

2.3. Formation of dimeric radical cations of $trans$-stilbene[7]

The conversion of c-St$^{+\cdot}$ to t-St$^{+\cdot}$ is slow as a unimolecular process in the absence of added salts. Therefore, in competition with this process, other processes such as addition of c-St$^{+\cdot}$ to c-St followed by rotation and elimination of t-St$^{+\cdot}$ might occur. The following results may support the formation of dimeric radical cations of stilbene.

Excitation of 2,4,6-triphenylpyrylium tetrafluoroborate (TPP$^+$BF$_4^-$) with 408-nm laser pulses in the presence of a low concentration (4.7×10^{-4} mol dm^{-3}) of *t*-St in dichloromethane afforded the transient absorption due to *t*-St$^{+\cdot}$ (strong, λ_{max}=480 nm) together with that of TPP$\cdot$ (weak, λ_{max}=550 nm), as illustrated in Figure 8a. However, in the presence of a higher concentration of *t*-St (4.3×10^{-2} mol dm^{-3}) the above strong band is shifted to shorter wavelengths (λ_{max}=474 nm) with a shoulder (around 445 nm, Figure 8b).

The spectral shift was clearly observed in a *t*-St concentration of 1.0×10^{-4} mol dm^{-3}, as shown in Figure 9. Figure 9b depicts the time profiles of transients monitored at 470 and 480 nm and indicates that the decay at 480 nm is accompanied by a rise at 470 nm. The absorption at 480 nm due to *t*-St$^{+\cdot}$ decayed with two components, a fast component (~10^6 s^{-1}) and a slow component with second-order kinetics, and that at 470 nm consists of a rise (3×10^6 s^{-1}) and a decay component. The latter absorption can be assigned to dimeric species between *t*-St$^{+\cdot}$ and *t*-St. The observed small shift of the absorption from 480 to 474 nm suggests that the dimeric species may be a π-complex between *t*-St$^{+\cdot}$ and *t*-St since a σ-type dimer will exhibit an absorption corresponding to a benzyl radical or a benzyl cation at shorter wavelengths. Measurements of the first-order decay rate at 480 nm in varying concentrations of *t*-St [(0.5–2)×10^{-2} mol dm^{-3}] afforded a rate constant for dimerization as 2×10^7 mol^{-1} dm^3 s^{-1} at ambient temperature.

3. EFFECTS OF TPP$^+$ TRIPLETS AS SENSITIZER AND SDS MICELLES AS REACTION MEDIUM

3.1. Effects of TPP$^+$ triplets on cis→trans isomerization of stilbene[4]

When TPP$^+$BF$_4^-$ is used as sensitizer, both of its singlet and triplet excited states are quenched by *c*-St in acetonitrile as expected from the changes of Gibbs' free energies, –0.92 and –0.42 eV, for electron transfer from *c*-St to singlet and triplet excited TPP$^+$, respectively. However, the apparent quantum yield ($\phi_{c\rightarrow t}$) for cis→trans isomerization tends to increase with decreasing *c*-St concentration. As Figure 10 shows, $\phi_{c\rightarrow t}$ increases linearly with increasing quantum yield (ϕ_q^T) for quenching of the triplet state. ^{1}(TPP$^+$)* is quenched by *c*-St in its presence in high concentrations, whereas *c*-St in lower concentrations quenches efficiently only ^{3}(TPP$^+$)* resulting from ^{1}(TPP$^+$)* escaped from quenching. Actually, laser excitation (406 nm) of 0.1 mol dm^{-3} *c*-St afforded only the absorption due to *c*-St$^{+\cdot}$; however, that of 0.01 mol dm^{-3} *c*-St gave the absorptions due to both *c*-St$^{+\cdot}$ and *t*-St$^{+\cdot}$, as illustrated in Figure 11. These results clearly indicate that only ^{3}(TPP$^+$)* is effective to induce the isomerization of *c*-St$^{+\cdot}$ to *t*-St$^{+\cdot}$ but ^{1}(TPP$^+$)* is not.

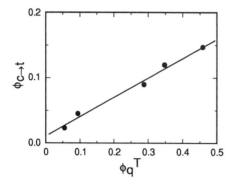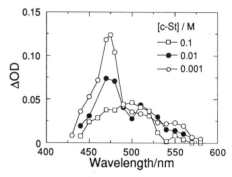

FIGURE 10

Isomerization quantum yields of c-St in varying quantum yields of triplet TPP$^+$ quenching ($\phi_q{}^T$)

FIGURE 11

Transient absorption spectra on TPP$^+$-sensitized excitation of c-St in varying concentrations

3.2. Effects of SDS micelles on cis→trans isomerization of stilbene[17]

The singlet radical pair between c-St$^{+\cdot}$ and TPP$^\cdot$ resulting from singlet quenching will undergo quick reverse electron transfer instead of undergoing isomerization. On the other hand, the triplet radical pair will undergo diffusion or isomerization in competition with the reverse electron transfer.

The effects of micelles on TPP$^+$-sensitized isomerization of c-St to t-St was examined. Sodium dodecyl sulfate (SDS, 0.06 mol dm^{-3}) micellar solutions of TPP$^+$BF$_4{}^-$ (7×10^{-5} mol dm^{-3}) and varying concentrations of c-St [$(0.2$–$4.4)\times10^{-3}$ mol dm^{-3}] in water containing 2% acetonitrile, where the average number of c-St molecules per micelle was 0.2–4.5, were irradiated with 406-nm light from a high-pressure mercury lamp.

Figure 12 plots $\phi_{c\rightarrow t}$ and the fraction of ^{1}(TPP$^+$)*, I/I_0, escaped from quenching by c-St against the number of c-St molecules per micelle (M), p_{cs} (=[c-St]/[M]). This figure indicates that $\phi_{c\rightarrow t}$ increases with increasing p_{cs} to give a considerable value at p_{cs}=1–3 with the maximum value at p_{cs}=2, where ^{1}(TPP$^+$)* is effectively quenched to give only a small amount of ^{3}(TPP$^+$)* (at most 0.2 as ϕ_T, Figure 12).

At a given ratio of quencher to micelle in concentration the quencher molecules are distributed in micelles according to Poisson's distribution[18]; that is, there are micelles containing a definite number of quencher molecules (0, 1, 2, 3, and so forth). The sensitizer (TPP$^+$), in either singlet or triplet excited state, located in an "empty" micelle containing no quencher molecules (c-St) is not quenched at all, and accordingly, ^{1}(TPP$^+$)* emits fluorescence or undergoes intersystem crossing to ^{3}(TPP$^+$)* followed by deactivation. On the other hand, in the micelles containing a number of c-St molecules, ^{1}(TPP$^+$)* will be efficiently quenched before undergoing intersystem crossing. In the micelles containing a small number of c-St molecules, the quenching of ^{1}(TPP$^+$)* by c-St is less effective, and therefore, ^{3}(TPP$^+$)* is afforded in a considerable efficiency.

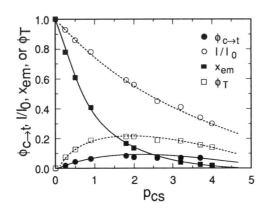

FIGURE 12

Comparison of the cis→trans isomerization quantum yield, $\phi_{c \to t}$, of c-St with the fraction of ^{1}(TPP$^+$)*, I/I_0, escaped from quenching by c-St at various concentration ratios, $p_{cs}=[c\text{-St}]/[M]$

The fraction of "empty" micelle (x_{em}) is also plotted against p_{cs} in Figure 12. The amount of ^{3}(TPP$^+$)* (ϕ_T) to be quenched by c-St was estimated by subtracting x_{em} (=fraction of ^{1}(TPP$^+$)* in empty micelles) from the fraction of ^{1}(TPP$^+$)* (I/I_0) escaped from the quenching, and by multiplying the resulting difference by the estimated intersystem crossing quantum yield (≈ 0.5) of the sensitizer[19]. The result is depicted in Figure 12. It is noticeable that the profile of $\phi_{c \to t}$ against p_{cs} is similar to that of the amount of ^{3}(TPP$^+$)* quenched by c-St. This fact is in good agreement with our previous finding that the isomerization results only from the quenching of ^{3}(TPP$^+$)* but not from the quenching of ^{1}(TPP$^+$)* in homogeneous solutions[4].

The efficiency of isomerization based on ^{3}(TPP$^+$)* quenched by c-St was estimated to be essentially the same value of 0.4 over the examined range of p_{cs} by dividing $\phi_{c \to t}$ by ϕ_T. It is remarkable that this value is nearly four times of that in homogeneous solutions (0.1). These results may reflect again that the micelle prevents diffusion of the triplet radical pair of c-St$^{+\cdot}$ and TPP$\cdot$ and its intersystem crossing to the singlet radical pair followed by reverse electron transfer.

ACKNOWLEDGEMENT

The authors gratefully acknowledge partial support from Ministry of Education, Science and Culture, Japan by Grant-in-Aid for Scientific Research for Special Project Research No. 63104001 (K.T.) and by Grant-in-Aid for Scientific Research No. 02453019 (K.T.), and from the University of Tsukuba by the President's Special Grant for Education and Research, 1990 (K.T.). I.K.L. is grateful for support of his stay at Tsukuba and traffic expense to University of Tsukuba Fund for International Exchange, grants from companies (K.T.), and to the Soviet Academy of Sciences.

REFERENCES

1) F. D. Lewis, J. R. Petisce, J. D. Oxman and M. J. Nepras, J. Am. Chem. Soc. 107 (1985) 203.

2) L. T. Spada and C. S. Foote, J. Am. Chem. Soc. 102 (1980) 391.

3) F. D. Lewis, R. E. Dykstra, I. R. Gould and S. Farid, J. Phys. Chem. 92 (1988) 7042.

4) Y. Kuriyama, T. Arai, H. Sakuragi and K. Tokumaru, Chem. Lett. (1988) 1193.

5) Y. Kuriyama, T. Arai, H. Sakuragi and K. Tokumaru, Chem. Lett. (1989) 251.

6) Y. Kuriyama, T. Arai, H. Sakuragi and K. Tokumaru, Chem. Phys. Lett. 173 (1990) 253.

7) R. Akaba, H. Sakuragi and K. Tokumaru, Chem. Phys. Lett. 174 (1990) 80.

8) T. Shida, Electronic Absorption Spectra of Radical Ions (Elsevier, Amsterdam, 1988).

9) W. H. Hamil, Ionic Processes in γ-Irradiated Organic Solids at $-196°$, in: Radical Ions, eds. E. T. Kaiser and L. Kevan (John Wiley, New York, 1968), p. 405.

10) H. Misawa, T. Karatsu, T. Arai, H. Sakuragi and K. Tokumaru, Chem. Phys. Lett. 146 (1988) 405.

11) T. Arai, T. Karatsu, H. Misawa, Y. Kuriyama, H. Okamoto, T. Hiresaki, H. Furuuchi, H. Zeng, H. Sakuragi and K. Tokumaru, Pure Appl. Chem. 60 (1988) 989.

12) A. Ishida, K. Tabata, S. Toki and S. Takamuku, presented at Annual Meeting of Chemical Society of Japan, Kyoto, April, 1989, and private communication.

13) J. L. Courtneidge, A. G. Davies and P. S. Gregory, J. Chem. Soc., Perkin Trans. 2 (1987) 1527.

14) L. Bonazzola, J-P. Michaut, J. Roncin, H. Misawa, H. Sakuragi and K. Tokumaru, Bull. Chem. Soc. Jpn. 63 (1990) 347.

15) J. M. Dust and D. R. Arnold, J. Am. Chem. Soc. 105 (1983) 1221.

16) Y. Kuriyama, T. Arai, H. Sakuragi and K. Tokumaru, presented at Symposium on Photochemistry, Kyoto, October, 1990. Abstract p. 287.

17) I. K. Lednev, M. V. Alfimov, Y. Kuriyama, T. Arai, H. Sakuragi and K. Tokumaru, unpublished results.

18) M. Tachiya, Chem. Phys. Lett. 33 (1975) 289.

19) S. Tripathi, V. Wintgens, P. Valat, V. Toscano, J. Kossanyi and F. Bos, J. Lumin. 37 (1987) 149.

Photochemical Processes in Organized Molecular Systems
K. Honda (Editor-in-Chief)
© Elsevier Science Publishers B.V., 1991

PHOTOINDUCED SINGLE ELECTRON TRANSFER FRAGMENTATION AND CYCLIZATION
REACTIONS. MEDIUM AND INTERFACIAL EFFECTS*

David G. WHITTEN, Carlos CHESTA,[+] Xiaohong CI, Matthew A. KELLETT and
Vivian W. W. YAM[≠]

Department of Chemistry, University of Rochester, Rochester, N.Y. 14627
U.S.A.

The theme of this paper will be reactions in which excited state single
electron transfer (SET) initiates reaction, to give potentially reactive
oxidized donor (D^+) and reduced acceptor (A^-) fragments which can annihilate
one another, back electron transfer (bet), individually react or react
cooperatively. Depending upon whether the reaction occurs intramolecularly
or intermolecularly as well as on the solvent or medium and the specific path
to generate the pair, D^+, A^-; the important intermediate can be contact
radical ion pairs or exciplexes (CRIP), solvent separated radical ion pairs
(SSRIP) or free ions (FI). We will focus first on a number of specific
reactions in homogeneous solution, with respect to reactivity involving these
different intermediates, and subsequently on how reaction is modified when it
occurs at an interface or in microheterogeneous media.

Intermolecular reactions which will be described are the photochemical
electron-transfer-induced fragmentation of 1,2-aminoalcohols, 1,2-diamines
and pinacols. A number of excited singlet electron acceptors such as the
cyanoaromatics 9,10-dicyanoanthracene (DCA) and 2,6,9,10-tetracyano-
anthracene (TCA) can be quenched by the aforementioned donors in a variety of
media, including nonpolar solvents (benzene), moderately polar organic
solvents (acetonitrile, acetonitrile-water) or microheterogeneous media such
as aqueous micelles or hydrocarbon-continuous reversed micelles. The net
reactivity, although qualitatively similar, in most of these media, shows a
strong variation in quantum efficiency with medium which can, in turn, be
related to the various electron transfer intermediates involved.

Intramolecular reactions have been examined with linked donor-acceptor
molecules including aminoalcohol-nitroaromatics, aminoalcohol-cyanoaromatics
and ketoamides. The two aminoalcohol-acceptor molecules are characterized by
efficient fragmentation accompanied by exciplex fluorescence in a range of
different media. Enhanced reactivity together with reduced fluorescence is
observed in cationic micelles. The ketoamides undergo both cyclization and
fragmentation to give a rich array of products, all ascribable to an electron
transfer mechanism. Reactivity persists when the ketoamide is incorporated
into cyclodextrins but the product distribution is strongly affected by the
medium.

*Research grants: No. DE-FG02-86ER13504 (U.S. Department of Energy) and No.
CHE-8616361 (U.S. National Science Foundation).
[+]Permanent address: Departamento de Quimica y Fisica, Universidad Nacional
de Rio Cuarto, Rio Cuarto 5800, Argentina.
[≠]Permanent address: Department of Chemistry, University of Hong Kong,
Pokfulam Road, Hong Kong.

1. INTRODUCTION

The subject of photoinduced reactions in organized systems - particularly photoinduced electron transfer processes - has been the focus of widespread investigations during the last two decades.[1-11] Many of these studies have been stimulated by the recognition that single electron transfer (SET) quenching of an excited state can generate two highly reactive energy rich products which, if isolated, could be used themselves to store energy or to drive redox processes culminating in the accumulation of kinetically stable but energy storing products. Since the two products - frequently radical ions of opposite charge - can undergo an energy wasting back electron transfer process, often at rates equal to or greater than diffusion controlled, much effort has been spent to design ways of separating them or rapidly utilizing them before this can occur. In several cases it has been shown that the use of an organized molecular assembly or an interface can effectively retard "energy wasting" reverse electron transfer processes or otherwise "compartmentalize" the products generated by excited state electron transfer quenching events.[5-11] The focus of the work described in this contribution is somewhat different in that we will explore situations where net chemical reaction via excited state SET quenching is either dependent on or enhanced by interaction of the two product ions. In these situations overall reactivity is strongly sensitive to the medium and to the extent of solvation of the photogenerated radical ions.

1.1. Photoinduced electron transfer in solution

Photoinduced electron transfer reactions in solution between initially neutral donors (D) and excited acceptors (A^*) can involve at least three distinct sets of intermediates: contact radical ion pairs or exciplexes (CRIP), solvent separated radical ion pairs (SSRIP) or kinetically free ions (FI).[12-18] As far as stability of radical ion pairs is concerned, it has been shown that for a given D,A pair which does not associate in the ground state, but can undergo an energetically favored SET process when one of the molecules (A) is excited, the energies of both CRIP and SSRIP decrease as solvent polarity increases. However, the decrease with increase in ϵ is greater for the SSRIP.[12] Consequently, the CRIP is favored in low polarity ($\epsilon < 7$) solvents while the SSRIP is generally more stable in more polar solvents.[12] Quenching in nonpolar solvents thus generally leads to a CRIP which has little probability of dissociating into free ions ($k_{diss} < 10^6$ s^{-1}) or undergoing solvation to generate a SSRIP.[15] In more polar solvents quenching can directly generate either a CRIP or a SSRIP; solvation of the former to give the SSRIP should be favored and the lifetime of the latter should be determined largely by competition between back electron transfer and diffusion apart of the SSRIP (k ~ 5 x 10^8 s^{-1} for a D‡,A‡ pair in acetonitrile) to form free ions.[15,16] Since the rates of the back electron transfer process within the SSRIP are strongly influenced by energetics,[19,20] the yields of free ions will be strongly system-dependent. For a donor-acceptor combination where yields of free ions are low due to a low yield of cage escape from the CRIP or SSRIP, it is often possible to enhance the formation of free ions via cosensitization, eqs 1-4, where D′ is a "cosen-

$$A^* + D' \longrightarrow A^{\cdot -}, D'^{\cdot +} \qquad (1)$$

$$A^{\cdot -}, D'^{\cdot +} \longrightarrow A^{\cdot -} + D'^{\cdot +} \qquad (2)$$

$$D'^{\cdot +} + D \longrightarrow D' + D^{\cdot +} \qquad (3)$$

$$\text{Net Reaction:} \quad A^* + D \xrightarrow{\ \ (D')\ \ } A^{\cdot -} + D^{\cdot +} \qquad (4)$$

sitizer" and the SSRIP, $A^{\cdot -}$, $D'^{\cdot +}$, has a relatively high cage escape (eq 2) efficiency.[21-24] For several cyanoaromatic sensitizers it has been found that biphenyl is an effective cosensitizer.[21-24] Thus for an intermolecular situation, by varying the solvent environment and by employing both direct and cosensitized electron transfer quenching paths, for a given donor-acceptor combination the three distinct intermediates can be generated and their reactivity investigated as outlined in Scheme I.

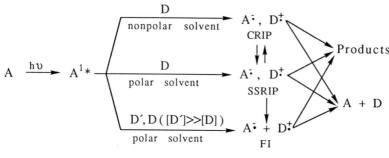

SCHEME I

Different electron transfer intermediates from photolysis of an acceptor, A, and quenching of its excited singlet A^* by SET in solution.

1.2. Photoredox fragmentation

One of the most prominent photoinduced SET reactions is fragmentation of an oxidized donor. This reaction has been observed relatively recently for a wide variety of oxidizable substrates including di- and polyarylethanes,[25-27] organosilanes,[28] organoboranes,[29] aminoalcohols,[30,31] and other substrates.[32-34] A rather high intrinsic reactivity for several of these oxidized donors can be inferred from thermochemical data, which indicate greatly reduced bond dissociation energies for several cation radicals compared with the corresponding

neutral molecules.[35-38] In this paper we compare the reactivity of a series of
donor-acceptor combinations which undergo both fragmentation and, in some cases,
addition reactions which share as a common ingredient the likely involvement of
both reduced acceptor and oxidized donor in the product-determining steps
occurring subsequent to SET quenching of the excited acceptor. The first part
will compare intermolecular reactions in which, depending upon both donor and
acceptor, cooperative reactivity between $D^{\dagger}$ and A^{-} may be crucial to efficient
reaction. In the second part of the paper we will examine linked D-A molecules
in which cooperative reactivity is dominant.

2. INTERMOLECULAR FRAGMENTATION REACTIONS

2.1. Oxidative photofragmentation of 1,2-diheteroatom-substituted ethanes

As outlined above, photoinduced SET reactions can often lead to C-C bond
cleavage reactions in the oxidized donor.[25-34] Depending upon the specific
donor-acceptor combination involved these reactions may lead to net oxidation of
the donor and reduction of the acceptor or to a variety of processes not easily
recognizable as proceeding via one-electron redox initiation including isom-
erizations, additions and substitutions.[39-41] A class of reaction which appears
relatively straightforward in the redox sense is the oxidative fragmentation of
1,2-diheteroatom-substituted ethanes, a reaction shown in eqs 5 and 6 which

$$A \;+\; \underset{:Y-H}{\overset{H-\ddot{X}:}{>C\!\!-\!\!C<}} \;\xrightarrow{h\upsilon}\; AH_2 \;+\; >C\!\!=\!\!\ddot{X}: \tag{5}$$

$$+ \; >C\!\!=\!\!\ddot{Y}:$$

$$A \;+\; \underset{:Y-}{\overset{-\ddot{X}:}{>C\!\!-\!\!C<}} \;+\; 2H_2O \;\xrightarrow{h\upsilon}\; \overset{|}{-\underset{H}{\ddot{X}}:} \quad \overset{|}{\underset{H}{:\ddot{Y}-}} \tag{6}$$

$$>C\!\!=\!\!O \quad O\!\!=\!\!C<$$

$$+ \; AH_2$$

frequently occurs via selective long-wavelength excitation of UV or visible
absorbing acceptors in the presence of donors transparent in that region. This
reaction has been found quite general for X,Y = S,N, or O; in the present paper
we will discuss reactions of aminoalcohols, diamines and pinacols.

2.2. Aminoalcohols

The oxidative cleavage of 1,2-aminoalcohols is observed upon excitation of a
number of acceptors including various metal complexes, cyanoaromatics, quinones
and dyes.[30,31,42,43] Reaction occurs in good chemical yield but with quantum

efficiencies which depend strongly upon the acceptor, substrate and solvent.[30] The overall reaction in the presence of a trace amount of water is shown in eq 7; as indicated in the equation, the reaction probably involves generation

$$A + Ar\text{-}\overset{\overset{\displaystyle OH}{|}}{CH}\text{-}\underset{\underset{\displaystyle NR_2}{|}}{CH}\text{-}Ar' \xrightarrow{h\upsilon} [\, AH\overset{\cdot}{\colon}{}^- + ArCHO + Ar'\text{-}CH{=}\overset{+}{N}R_2\,]$$

$$\Big\downarrow H_2O \qquad\qquad (7)$$

$$AH_2 + ArCHO + Ar'CHO + R_2NH$$

of an iminium ion which is readily hydrolyzed. The oxidation potential of typical 1,2-aminoalcohols is similar to those of corresponding unsubstituted amines and it appears reasonable that SET oxidation generates a cation-radical more or less localized on N. The bond dissociation energy of aminoalcohol cation radicals is expected to be quite low[35,37] and it is reasonable to anticipate a fragmentation according to eq 8 should be facile. In either

$$Ar\text{-}\overset{\overset{\displaystyle OH}{|}}{CH}\text{-}\underset{\underset{\displaystyle \dagger NR_2}{|}}{CH}\text{-}Ar' \longrightarrow Ar\text{-}\overset{\overset{\displaystyle OH}{|}}{CH}\cdot + Ar'\text{-}\underset{\underset{\displaystyle +NR_2}{\|}}{CH}$$

$$\text{or} \qquad (8)$$

$$Ar\text{-}\overset{\overset{\displaystyle +OH}{\|}}{CH} + Ar'\text{-}\underset{\underset{\displaystyle \colon NR_2}{|}}{CH}\cdot$$

instance the neutral radical generated by the cleavage should be a good electron donor and subsequent reduction of an acceptor should culminate in the two-electron redox products shown in eq 7. Mass spectral fragmentation patterns of 1,2-aminoalcohols show that cleavage of these cation radicals is a predominant reaction in the gas phase.[44] The recent fashionable use of "sacrificial" reagents such as triethanolamine as oxidant-scavengers in various schemes to accumulate photogenerated reductants appears due to this potential of amino-alcohols to undergo sequential two-electron oxidation via an electrochemical "ece" process.

1,2-Diamines have been found to undergo analogous reactions, although the scope of the reaction has been much less studied.[45] Pinacols have been found to undergo somewhat similar oxidative cleavage reactions, though in the limited cases studied thus far, the reported products include both simple cleavage as well as addition reactions between the acceptor and pinacols.[32,46,47]

2.3. Comparison of aminoalcohols, diamines and pinacols

In the present study a series of structurally similar aminoalcohols, diamines and pinacols has been investigated with three photoexcited acceptors, thioindigo (TI), 9,10-dicyanoanthracene (DCA) and 2,6,9,10-tetracyanoanthracene (TCA) in nonpolar (benzene) and polar (acetonitrile, acetonitrile-water) solvents under conditions where reaction is initiated only by quenching of acceptor excited singlet states. Application of the equation derived by Weller indicates that for benzene the CRIP should be more stable than the SSRIP by about 0.7 eV; in contrast, for acetonitrile the SSRIP should be more stable than the CRIP by about 0.2 eV. Table I compares the reduction potentials, excited singlet energies and excited state reduction potentials for the three acceptors used. Table II compares the structures, oxidation potentials and free energies for the excited singlet quenching by SET as estimated from the "Weller" equation.[12]

Table I. Reduction Potentials and Excited State Energies of Light-Absorbing Acceptors

Acceptor	$\Delta E^{red}_{1/2}$ [a]	E^{1*}	$\Delta E^{red\ 1*}_{1/2}$
TI	-0.45	2.27	1.82
DCA	-0.90	2.89	1.99
TCA	-0.45	2.90	2.45

[a] Volts vs. SCE

Redox reactions occurring from the different donor-acceptor combinations are generally analogous to those given in eqs 5-7. Thus donors **1-4** are converted to benzaldehyde (and the corresponding amines for **1-3**) while **5** yields benzophenone. TI was found to be reduced upon irradiation with **1** or the diamines to its characteristic (λ_{max} 386 nm benzene) two-electron reduction product TIH$_2$ in both benzene and acetonitrile solution. DCA was also found to photobleach upon irradiation in benzene with the donors concurrent with reduction to its two-electron reduction product, DCAH$_2$. In acetonitrile bleaching was also observed; in several cases (e.g. both diamines) a short-lived (hours) long wavelength absorption attributable to the anion radical DCA$^-$ was observed as the initial photoproduct; on standing in the dark this signal decayed without leading to further decrease or increase in the region where DCA itself absorbs. TCA was observed to photobleach upon irradiation with the various donors in benzene. In contrast, irradiation in acetonitrile or acetonitrile-water (95:5) was found to result in formation of the highly structured visible absorption spectrum characteristic of the anion radical, TCA⁻. The TCA⁻ was found to be stable in the dark indefinitely in the absence of air; aeration of solutions leads to quantitative regeneration of the original absorption of the TCA.[48]

Table II. Oxidation Potentials of Electron Donors and Calculated Free Energies for Excited State Quenching

Donor	$E_p^{ox,a}$	$-\Delta G^{*,b}$					
		TI		DCA		TCA	
		Benzene	CH_3CN	Benzene	CH_3CN	Benzene	CH_3CN
1 HO N(morpholine), diphenylmethyl	0.90	0.56	0.98	0.73	1.15	1.19	1.61
2 H_2N NH_2, diphenyl	0.92	0.54	0.96	0.71	1.13	1.17	1.59
3 $(CH_3)_2N$ $N(CH_3)_2$, diphenyl	0.77	0.69	1.11	0.86	1.28	1.32	1.74
4 HO OH, diphenyl	1.48	-0.02	0.40	0.15	0.56	0.61	1.03
5 HO OH, tetraphenyl	1.90	-0.42	0.02	-0.27	0.15	0.19	0.61

[a]Irreversible oxidation peak potential. Volts vs. SCE.   [b]Free energies for electron transfer quenching of excited acceptor singlet as calculated by equation derived by Weller(ref. 12).

2.3.1. Cosensitized reaction

For both DCA and TCA with donors **1-5** in acetonitrile, it was found that efficient reaction could be obtained under "cosensitized" conditions whereby biphenyl (0.25 M) was the primary quencher of the excited acceptor and the concentration of the "secondary donor" **1-5** was 0.005 M.[21] Under these

D.G. Whitten et al.

conditions net bleaching of DCA was observed (although here again DCA⁻ was observed as a metastable product with **2** and **3**) while TCA was converted cleanly to TCA⁻. Table III compares quantum efficiencies for the various donors with DCA and TCA under the different solvent and quenching (direct vs. cosensitization) conditions. We have previously published data comparing the reactivity of **1** and diamines with the series of acceptors: TI, DCA and TCA and we have measured the effect of changing solvent from benzene to methylene chloride to acetonitrile. These data will be used in developing a picture of the overall reactivity as developed below.

Table III. Quantum Yields of Acceptor Change (Corrected to 100% quenching)[a]

Donor / Acceptor	Benzene[b]	Acetonitrile[c]	Acetonitrile: Water(95:5)[c]	Acetonitrile/ Biphenyl[c]
1 / DCA	0.010	----[d]	0.003	0.46
1 / TCA	0.014	0.035	0.04	0.29
2 / DCA	0.011	0.129	0.77	0.43
2 / TCA	0.10	1.09	0.98	0.38
3 / DCA	0.006	0.068	0.22	0.45
3 / TCA	0.37	0.40	0.20	0.36
4 / DCA	----[d]	----[d]	0.008	0.39
4 / TCA	0.04	0.11	0.40	0.07
5 / DCA	----[d]	----[d]	0.008	0.18
5 / TCA	0.16	0.45	1.20	0.04

[a] Determinations were made using vacuum degassed solutions of the donor(conc. calculated for 50% quenching) and the acceptor(5 X 10^{-5}M), and biphenyl(0.25M). Irradiation was performed using an Hanovia 450W Hg lamp in a merry-go-round. Absorbance change was recorded on an Hewlett Packard 8451A Diode array Spectrophotometer. Quantum yields were determined using K^+ ferrioxalate actinometry.

[b] Quantum yield of Acceptor bleaching.

[c] Quantum yield of Acceptor bleaching for DCA samples, Quantum yield of growth of TCA⁻˙ for TCA samples.

[d] No reaction detected.

The likely reaction mechanism can be summarized very generally in Scheme II

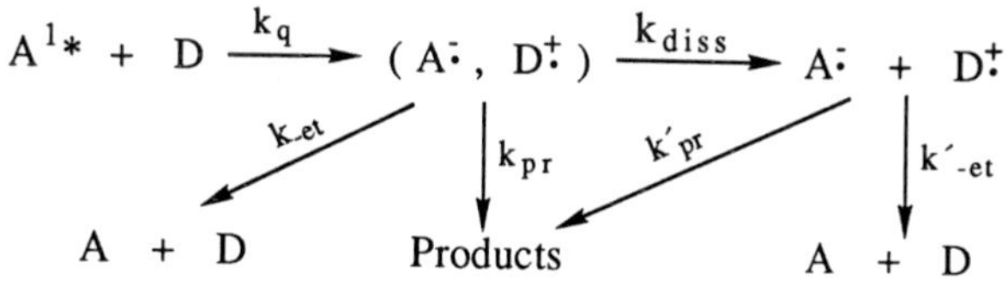

$$A^{1}* + D \xrightarrow{k_q} (A^{\overline{\cdot}}, D^{\overset{+}{\cdot}}) \xrightarrow{k_{diss}} A^{\overline{\cdot}} + D^{\overset{+}{\cdot}}$$

$$\downarrow{k_{-et}} \qquad \downarrow{k_{pr}} \quad \nwarrow{k'_{pr}} \qquad \downarrow{k'_{-et}}$$

$$A + D \qquad Products \qquad A + D$$

SCHEME II

where we consider only the ion pair (without distinction as to CRIP or SSRIP) and free ions as intermediates. We can now examine the possibility first that reaction occurs from **1-5** under conditions where free ions are generated.

Several points emerge from an examination of the cosensitization experiments with biphenyl (BP). Under these conditions the primary reaction is generation of the free ions $D^{\overset{+}{\cdot}}$ (**1-5**) and $A^{\overline{\cdot}}$ as outlined in eqs 1-4. The donor cations, $D^{\overset{+}{\cdot}}$, should either fragment or undergo reverse electron transfer. Although the latter process should occur at diffusion-controlled rates, the relatively low concentrations of ions (under the steady irradiation conditions employed) should make this process relatively slow such that even a relatively slow "unassisted" fragmentation should occur. The limiting value for cage escape of the pair $TCA^{\overline{\cdot}}$, $BP^{\overset{+}{\cdot}}$ in acetonitrile has been shown to be $\phi_{ce} = 0.24$;[21] if we assume each SET quenching event is followed by a dark reduction of TCA by partially oxidized donor,[48] the yield of reduced TCA should be in the range of 0.4 - 0.5 for the BP-cosensitized process. The actual values for donors **1-3** are 0.29, 0.38 and 0.36, respectively, a little lower than the anticipated value while values for the two pinacols are significantly lower. If we examine the situation with the DCA/BP cosensitization in acetonitrile, the limiting yield for separating ions from the pair is 0.75 - 0.83;[21] consequently we would anticipate yields of DCA reduction to $DCAH_2$ should be about equal since one photochemically oxidized donor should furnish two electrons and hence one $DCAH_2$. For the structurally closely related series **1-4** all donors give nearly the same quantum yield $\phi_{-DCA} \cong$ 0.42. Under the conditions used >90% of the quenching of $DCA^{1}*$ is by BP, although a small amount of direct quenching by the "secondary donor" occurs. Although the reasons why the cosensitized quantum efficiencies for **1-3** are slightly lower than the limiting ion yields for TCA/BP and substantially lower for DCA/BP are not yet clear, it is reasonable to conclude from the cosensitization experiments that the free ions, $D^{\overset{+}{\cdot}}$, generated in acetonitrile have a high probability of fragmentation during the lifetime limited by bimolecular ion-recombination. The relatively low quantum efficiencies for both cosensitization reactions with **5** indicate that its corresponding cation radical must fragment much more slowly. The evident discrepancy in cosensitization yields from the hydrobenzoin cation radical $4^{\dagger}$ is less readily explained but it appears reasonable that both pinacol cation radicals may fragment less readily in acetonitrile than those derived from SET oxidation of **1-3**.

2.3.2. Direct electron transfer oxidation of aminoalcohols

Several factors distinguish the reactivity of aminoalcohols such as **1** under the photosensitized electron transfer conditions. Under conditions where electron transfer quenching is complete in benzene the reaction quantum yields show a sharp decrease in the acceptor series TI > DCA > TCA (relative values (1:0.02:0.01)). For a "reactive" acceptor such as TI, the reactivity <u>decreases</u> as solvent polarity increases in the series benzene, methylene chloride, aceto-nitrile. There is furthermore an isotope effect on the alcohol O-H(D) with $\phi^{(H)}/\phi^{(D)}$ in the range 1.3 - 4, depending upon acceptor, solvent and tem-perature. If we consider the reaction in benzene, using the model outlined in Scheme II, several points can be developed. From the reaction exothermicity (1.7-2 eV) it is clear that back electron transfer rates should be near the maximum for either CRIP or SSRIP[49] and thus a rate constant $k_{-et} \sim 10^{10}$ s^{-1} can be estimated for all three acceptors. Since cage escape should be relatively slow in benzene ($k_{diss} \sim 10^6$ s^{-1}), free ion formation should be unimportant and the reaction occurring is probably entirely ascribable to reaction within the CRIP. The strong variation in reaction efficiency with acceptor structure is probably <u>not</u> attributable to variations in k_{-et} since reaction exothermicities are about the same for TI and TCA and comparison of DCA with the former two indicates that its more exothermic back electron transfer should push it into the Marcus-inverted region. If this were the dominant factor a greater re-activity would be expected for DCA than TI which is clearly not the case. The observation of a "kinetic" isotope effect which increases sharply in the series TI < DCA < TCA is in accord with a "cooperative" fragmentation in which the acceptor anion radical $A^{\bar{\cdot}}$ assists the C-C bond cleavage by proton removal (eq 9). Assuming $k_{-et} \sim 10^{10}$ s^{-1} gives estimated values for k_{pr} in the CRIP

$$A^{\bar{\cdot}} \quad H-O \qquad AH\cdot$$
$$\underset{\substack{\vert\\ \dagger NR_2}}{>\!C\!-\!C\!<} \xrightarrow{k_{pr}} >\!C\!=\!O \; + \; \underset{\substack{\vert\\ :NR_2}}{\overset{\cdot}{C}<} \qquad (9)$$

from $1 \times 10^6 - 5 \times 10^8$ s^{-1}; the higher value (for TI) is consistent with a very small activation energy for the cleavage which is probably lower than the bond dissociation energy estimated from thermochemical cycle calculations.[31,37]

The lower efficiencies for reaction of **1** with TI in acetonitrile or aceto-
nitrile:water indicate that reactivity is considerably lower for this donor-
acceptor pair in the SSRIP compared to the CRIP. The lower exergonicity of
the back electron transfer process in acetonitrile for the SSRIP (Table IV)

Table IV. Estimated Back Reaction Exergonicity[a]

Donor	TI		DCA		TCA	
	Benzene	CH_3CN	Benzene	CH_3CN	Benzene	CH_3CN
1	1.71	1.29	2.16	1.74	1.71	1.29
2	1.73	1.31	2.18	1.76	1.73	1.31
3	1.58	1.16	2.03	1.61	1.58	1.16
4	2.29	1.87	2.74	2.32	2.29	1.87
5	2.71	2.29	3.16	2.74	2.71	2.29
BP			2.87		2.40	

[a] Values reported in Volts.

still suggests a value for $k_{-et} \sim 10^{10}$;[49] however the much lower quantum efficiency (3×10^{-5}) suggests the maximum value for $k_{pr} \sim 3 \times 10^5$ s^{-1}, a value more
than 3 orders of magnitude lower than that for reaction in the CRIP. This must
be regarded as a maximum value since cage escape in acetonitrile is estimated to
have a rate constant $k_{diss} \sim 5 \times 10^8$ s^{-1} which could lead to free ions with a
sufficient efficiency (0.05) to account for the observed fragmentation. In any
case it is clear that cooperative reaction such as shown in eq 9, while important within the TI/**1** CRIP, is not very effective in the SSRIP.

2.3.3. Direct electron transfer oxidation of 1,2-diamines

The two 1,2-diamines used in this study show generally similar behavior to
the dimorpholinostyrene investigated earlier. With all three electron acceptors
there is a sharp increase in reactivity as solvent polarity is increased and
little evidence for selective reaction in the CRIP. As suggested earlier, the
fragmentation of the diamines contrasts with that of the aminoalcohol in that
the donor cation radical may be sufficiently reactive due to the presence of a
moderately strong nucleophilic center at the unoxidized N to favor an unassisted
fragmentation (eq 10). Thus reactivity should not depend upon the base strength
of A⁻ nor on whether the electron transfer intermediate is a CRIP or SSRIP.

$$>\!\!C\!-\!C\!\!< \quad \longrightarrow \quad >\!\!C\!=\!\overset{+}{N}R_2 \; + \; \overset{\cdot}{\underset{:NR_2}{C}}\!\!< \qquad (10)$$

The very high quantum efficiencies for reaction of **2** with DCA in acetonitrile: water and with TCA in both acetonitrile and acetonitrile:water indicate that rapid reaction <u>must</u> occur within the SSRIP, since the cage escape yields cannot account for reaction efficiencies of this magnitude. The greater reactivity of **2** compared with **3** can perhaps be attributed in part to the greater exergonicity of back electron transfer with **2** which may place its back electron transfer rates in the "Marcus inverted" region. The very low yields of reaction for **2** and **3** with DCA in benzene may be due to a larger value of k_{-et} in the CRIP compared to the SSRIP.[49] The differences between reactivity of **2** and **3** with DCA and TCA in both benzene and acetonitrile are not readily explained since it would be anticipated that for both the CRIP and SSRIP, the values of k_{-et} estimated purely on the basis of energetics should be larger or equal for TCA compared to DCA. One possibility may be that solvent penetration of the ion-pair is greater for the TCA:diamine radical ion pairs than for those with DCA. The increase in reaction efficiency for both **2** and **3** with DCA between acetonitrile and acetonitrile:water could be due to a solvent or solvent polarity assistance in the fragmentation.

2.3.4. Reactivity of pinacols

The pinacols show trends which are different from those observed with either the aminoalcohol or diamines. Thus hydrobenzoin, which is structurally closely related to **1-3**, shows an increase in reactivity with TCA in the solvent series: benzene, acetonitrile, acetonitrile:water. (The pinacols do not quench DCA or TI fluorescence so no comparison of the reactivity with these acceptors is possible.) A similar increase in reaction efficiency is observed for **5**/TCA and in fact for both pinacols with TCA the direct electron transfer quenching efficiency substantially exceeds that of the BP-cosensitized reaction. That "free" pinacol cation radicals can fragment in acetonitrile seems clear, although the striking differences in reaction between acetonitrile and acetonitrile:water solutions may indicate a greater role of solvent participation in the fragmentation. One source of explanation for the lower apparent fragmentation propensity of **5** in the cosensitized reactions compared with **4**, and perhaps also a source of the increase in the efficiency via direct quenching between acetonitrile and acetonitrile:water, may be the role of charge localization on the fragmentation process. Thus, for the amine donors it is fairly clear that the cation radical is strongly centered on N such that fragmentation mechanistically can be quite favored. For the pinacols the oxidation potentials are much higher, yet the ions may be more delocalized onto the aromatic system and hence less reactive to fragmentation as has been suggested for other cations.[25]

The TCA-mediated fragmentation of the pinacols provide another remarkable example of the range of reactivities of acceptor anion radicals. In these reactions the products are simply those of pinacol 2e-oxidation and TCA 1e-reduction as shown in eq 11. Thus the reaction involves liberation of

$$2 \text{ TCA } + \underset{\underset{\displaystyle OH}{|}}{\overset{\overset{\displaystyle OH}{|}}{>C-C<}} \longrightarrow 2 \text{ TCA}^{\cdot -} + 2 >C=O + 2 H^+ \quad (11)$$

protons and acceptor anion radical and occurs to at least a considerable extent of conversion before any protonation of the TCA$^{\cdot -}$ can be detected.

In summary, for intermolecular reaction we observe that moderate reactivity within the lifetime imposed by the back electron transfer "clock" restricts reactivity of aminoalcohols in photogenerated $D^{\cdot +}, A^{\cdot -}$ pairs to CRIP's in which $A^{\cdot -}$ can induce cooperative reactivity resulting in photofragmentation. In contrast, for diamines the "internal" nucleophilicity of the second nitrogen removes the need for $A^{\cdot -}$ participation and reaction is possible in both CRIP and SSRIP. The trends observed for the pinacols suggest that solvent participation may be more important than cooperative reaction between $A^{\cdot -}$ and $D^{\cdot +}$; however more studies with the pinacols and related compounds are clearly necessary.

Interestingly, the results obtained with the pinacols contrast very strongly with the behavior of aminoketones under SET-quenching of excited singlets of TI and DCA. In these reactions it has been found that cooperative reactivity to produce amides (via a rather complicated sequence)[50,51] occurs cleanly only in nonpolar solvents under conditions where a CRIP is the intermediate. In these cases the predominance of anion-cation or nucleophile-electrophile interactions between "reagents" which can exhibit both ionic and radical reactivity is at least in part attributable to the lack of solvation which might otherwise attenuate ionic reactivity.

3. INTRAMOLECULAR FORMATION OF CRIPS; FRAGMENTATION AND CYCLIZATION REACTIONS

Molecules which contain simultaneously an electron donor and an acceptor often exhibit photophysical and photochemical behavior consistent with efficient SET quenching of "local" excited states of the donor or acceptor. In these cases it is normally to be expected that the type of ion-pair produced is less subject to the medium effects discussed above for radical-ion pairs formed by intermolecular donors and acceptors. Nonetheless, the medium or solvent should produce important effects that, at least in some cases, may be more easy to delineate than for intermolecular systems.

3.1. "Tethered" aminoalcohol-acceptor molecules

We have recently initiated a study of several molecules which contain a light absorbing electron acceptor "tethered" to a potentially fragmentable amino-alcohol donor. We report here some preliminary observations for compounds **6** and **7**. Both of these compounds undergo clean and moderately efficient fragmentation

under direct irradiation of what by absorption spectra appears to be a rel-atively unperturbed acceptor chromophore (Table V) in both nonpolar and polar solvents. Compound **6** shows a very strong fluorescence which shows little solvent dependence of λ_{max} or lifetime. Compound **7** shows a weaker fluorescence and a moderate red shift with increase in solvent polarity. For both **6** and **7** the fragmentation efficiency is moderate in benzene but <u>increases</u> with increase in solvent polarity in the series benzene, CH_2Cl_2, CH_3CN, aqueous micellar cetyltrimethylammonium chloride (CTAC).

Since the amino nitrogen is common to both the acceptor and donor portions of the molecule in **6**, it is not strictly correct to regard this as an independent nonconjugated donor-acceptor molecule. Its fluorescence and absorption are blue-shifted and its fluorescence is strongly reduced in lifetime compared to the "reference" compound 1-amino-4-cyanonaphthalene (Table V). Although the "exciplex" fluorescence observed for **6** is much stronger than any observed for intermolecular cyanoaromatic-aminoalcohol pairs,[52] the short lifetime compared to that of 1-amino-4-cyanonaphthalene is reasonable for a CRIP in which rel-atively exergonic back electron transfer (well into the Marcus-inverted region) is competitive with a fragmentation having a rate constant ca. $10^7 - 10^8$ s^{-1}.

The behavior of **7** is even more interesting since it involves a formally nonconjugated donor and acceptor and the intermolecular photoreactivity of the two reagents is quite different. Nitroaromatics in general are characterized by very weak, short-lived fluorescence and high intersystem crossing efficien-cies.[53,54] Direct irradiation of p-nitrotoluene in the presence of **1** leads to fragmentation of the aminoalcohol in moderate efficiency but to no exciplex fluorescence;[55] presumably reaction may be initiated by triplet quenching which with other acceptors leads to fragmentation of aminoalcohols in very high effi-

Table V. Absorbance, Fluorescence[a], and Lifetimes[b] of Compounds 6 and 7 in Different Solvents

Compound	Solvent	Cyclohexane	Benzene	CH$_2$Cl$_2$	CH$_3$CN	CH$_3$CN+H$_2$O(10%)	SDS	CTAC
1-amino 4-cyano naphthalene	λmax (nm) (absorbance)	334	338	338		340	344	348
	λmax (nm) (emission)	388	402	403		418	418	420
	Lifetime (ns)	τ$_1$=0.93	τ$_1$=2.95	τ$_1$=3.26		τ$_1$=6.62	τ$_1$=6.32	τ$_1$=7.01
6	λmax (nm) (absorbance)	314	316	316		312	316	316
	λmax (nm) (emission)	357	362	361		358	358	359
	Lifetime (ns)	τ$_1$=0.16(86%) τ$_2$=0.77(14%)	τ$_1$=0.16(79%) τ$_2$=1.0(21%)	τ$_1$=0.11(80%) τ$_2$=0.87(20%)		τ$_1$=0.07(67%) τ$_2$=0.94(33%)	τ$_1$=0.18(75%) τ$_2$=0.80(25%)	τ$_1$=0.23(70%) τ$_2$=0.78(30%)
7	λmax (nm) (absorbance)	268	280		278	278	274	280
	λmax (nm) (emission)	382	384		395	416	~420	~420
	Lifetime (ns)	τ$_1$=0.14(74%) τ$_2$=1.5(2%) τ$_3$=25.5(24%)	τ$_1$=0.45(12%) τ$_2$=2.4(27%) τ$_3$=16.2(61%)		τ$_1$=0.05(4%) τ$_2$=1.5(15%) τ$_3$=10.6(81%)	τ$_1$=0.06(6%) τ$_2$=1.5(49%) τ$_3$=4.9(45%)	τ$_1$=0.16(5%) τ$_2$=1.4(24%) τ$_3$=8.4(70%)	τ$_1$=0.03(7%) τ$_2$=0.9(21%) τ$_3$=5.8(73%)

[a] Concentration of substrate was adjusted so that the optical density of chromophore was generally about 0.1-0.3 at the excitation wavelength. Dry gas was bubbled through solution for 20 mins. prior to each measurement. Samples were excited at the maximum absorbance.

[b] Samples were prepared as above. Lifetimes were measured by using picosecond laser Single Photon Counting device with 1ps resolution(mode locked CW Nd: YAG-Dye laser) with excitation at 305nm.

ciency.[56] The observation of exciplex fluorescence seems to indicate a prominent role of the exciplex-CRIP in the intramolecular case. Here again the back
electron transfer process (exergonicity ~ 2 - 2.5 eV, depending upon solvent)
should fall into the Marcus-inverted region with a very roughly estimatable rate
constant, k_{-et} ~ 10^8 s^{-1}; the high reactivity towards fragmentation is consistent with a rate constant in the same range or higher. The increase in reactivity in cationic CTAC micelles compared to anionic micelles, acetonitrile
and nonpolar solvents can be attributed largely to residence of **6** and **7** at very
polar interfacial sites where the cationic surfactant head groups produce an
effectively high local pH which can assist in the deprotonation as shown in eq.
9.

3.2. Ketoamide photoreactions

Ketoamides of general structure **8** have previously been shown to give reaction
on direct photolysis at λ > 300 nm as shown in eq 12 with efficiencies which are

strongly dependent upon the ketoamide structure, and particularly on the groups
attached to the amide nitrogen. In several early investigations[57,58] it was
suggested that the ketoamide reaction was quite analogous to the Norrish Type II
photoelimination of ketones, proceeding via an intramolecular H-atom abstraction, followed by "rearrangement" of the diradical to an intermediate zwitterion
which could subsequently account for most of the products (eq 13). The reaction
is an attractive one from a synthetic point of view in that it affords a good
route to the otherwise difficultly obtainable azetidin-2-one (**9**) and oxazolidin-

$$
\underset{\substack{\text{H} \\ \text{R} \quad \text{C} \quad \text{R}_3 \\ \text{O} }}{\overset{\text{CR}_1\text{R}_2}{\underset{\text{O}}{\text{C}-\text{N}}}} \quad \xrightarrow{h\upsilon} \quad \underset{\substack{\text{R} \quad \text{C} \quad \text{R}_3 \\ \text{O}}}{\overset{\text{HO} \quad \cdot\text{CR}_1\text{R}_2}{\text{C}-\text{N}}} \quad \xrightarrow{\text{SET}} \quad \underset{\substack{\text{R} \quad \text{C} \quad \text{R}_3 \\ \text{O}}}{\overset{\text{HO} \quad \text{CR}_1\text{R}_2}{\text{C}=\overset{+}{\text{N}}}} \qquad (13)
$$

4-one **(10)** derivatives.[59-62] We have recently carried out a number of studies which indicate that the reaction according to eq 12 is clearly a SET process quite analogous to the reactions described above. Moreover we have found that this reaction can be modified to give somewhat greater selectivity in a number of cases by carrying out the photolysis in a controlled microenvironment. The specific compounds we have examined include ketoamides **11** - **14**.

11   **12**   **13**   **14**

11-d   **12-d**   **13-d**

Photolysis products and quantum efficiencies for their formation are given in Table VI. Since the ketoamides are virtually nonfluorescent, their photolysis unquenchable and they yield no transients detectable by ns laser spectroscopy, it is difficult to specify what excited state is the product precursor; however the lack of transients persisting on the ns timescale and relatively high quantum efficiencies for product formation imply that a singlet quenching process producing very short-lived intermediates may be most reasonable. The relatively low reactivity of **11** compared to the other amides studied can be attributed to its existence predominantly in a configuration which, while favorable for excited state quenching, cannot result in facile proton transfer and the subsequent steps leading to product (<u>vide infra</u>). Comparison photolysis of deuterated and undeuterated ketoamides results in very small decreases (< 5%) in the overall quantum efficiencies and to virtually no change in the distribution of products. Relatively little D/H selectivity is observed in transfer of intramolecular H(D) to form the oxazolidin-4-one or azetidin-2-one products. (When the reaction is

Table VI: Disappearance and Photoproduct Quantum Yields
under Different Irradiation Conditions[a]

Amide	Medium[b]	$\Phi_{dis.}$	Φ_9 (Ph–C(OH)–...–N, O)	Φ_{10} (Ph–...–N, O)	Φ_{15} (Ph–C(OH)(H)–N, O)	Φ_{16} (>=O)
11 (Ph–CO–C(O)–N–Ph)	H_2O	0.019	–	0.011	0.002	n.d.[c]
	β-CD	0.005	–	0.002	<0.0001	–
	γ-CD	<0.001	–	<0.001	<0.0001	–
12 (Ph–CO–C(O)–N)	H_2O	0.28	–	0.24	–	–
	β-CD	0.18	–	0.18	–	–
	γ-CD	0.24	–	0.22	–	–
13 (Ph–CO–C(O)–N(Ph)–Ph)	H_2O–H_3COH (1:1), pH=1	0.47	0.09	–	0.32	0.33
	β-CD, pH=1	0.60	0.17	–	0.36	0.38
	H_2O–H_3COH	0.30	0.20	0.07	–	–
	β-CD	0.64	0.56	0.03	–	–
14 (Ph–CO–C(O)–N)	H_2O, pH=1	0.30	0.06	0.04	0.12	n.d.
	β-CD, pH=1	0.24	0.03	0.05	0.07	–
	H_2O	0.27	0.07	0.16	–	–
	β-CD	0.24	0.12	0.13	–	–
	γ-CD	0.35	0.17	0.17	–	–

[a] Determinations were made using argon degassed solutions of the amides
(1×10^{-4} M). Irradiation was performed in a Quanta-Count (PTI) at 300 nm.
The actinometer used was heterocoerdianthrone.

[b] Quantum yields are reproducible at 5% accuracy. Reported values for beta
and gamma cyclodextrins are extrapolated quantum yields to a complexation
fraction of the amide equal to one.

[c] Not detected.

carried out in nondeuterated solvents, the "transferred" D is "washed out,"
while the H/D remaining bound to C is retained. If reaction is carried out with
undeuterated ketoamide in deuterated solvent, deuterium is incorporated into the
product but not in unrecovered starting material.) All of these results are
consistent with a quantum efficiency-limiting electron transfer quenching of the
ketoamide excited singlet to yield consecutively a zwitterion Z_1 (via SET), a
diradical(D) via proton (or D^+) transfer and a second zwitterion Z_2 via a dark
SET process. The failure of deuterated solvents to lead to D incorporation in
recovered starting material indicates that D and Z_2 must not return to the
starting material and hence that the limit in reaction efficiency is probably in
the competition between back electron transfer and the proton transfer to generate D. The lack of D vs. H selectivity or an effect on the quantum efficiency
indicates this $Z_1 \rightarrow$ D conversion must be exergonic and very rapid.

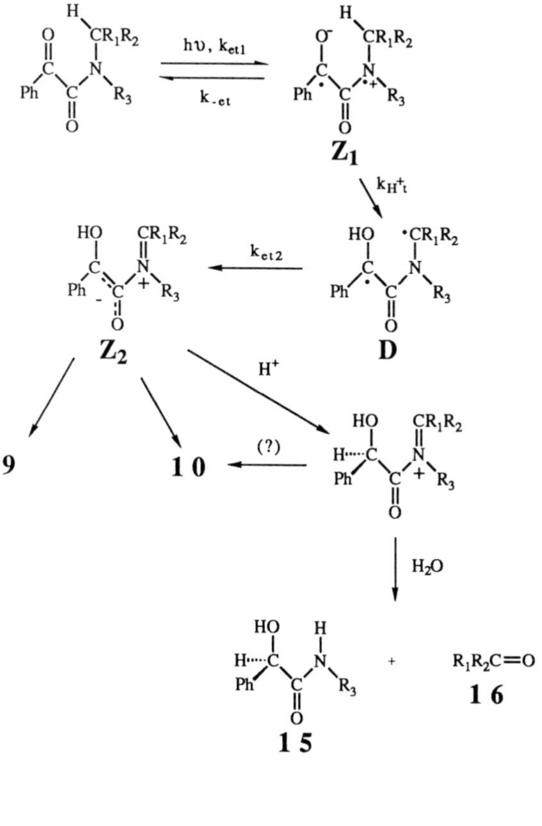

SCHEME III

Although the product distribution varies somewhat with structure and slightly with solvent, ketoamides 11 - 14 give generally good yields of oxazolidin-4-ones and azetidin-2-ones with the former the predominant product for 11, 12 and 14 and the latter the predominant product for 13. Both of these products could be readily formed from Z_2 although the oxazolidin-4-one might also be formed from protonated Z_2. Addition of acid to solutions of 13 and 14 leads to a decrease in the cyclization products upon photolysis with an appearance of the mandel-amide (15) and carbonyl products (16), consistent with an interception of Z_2 in competition with its cyclization.

We have found that all four ketoamides form moderately strong 1:1 inclusion complexes with β and γ cyclodextrins (CD). Not surprisingly complex formation results in a change in the photolysis behavior of 11 - 14; however the changes in photoreactivity are quite specific to the amide investigated. Thus 11 shows a sharp decrease in reactivity with no change in the product formed (predom-inantly 10). No alteration in product distribution is observed upon complexa-tion of 12 with both cyclodextrins; while a moderate increase in the azetidin-2-one quantum efficiency is noticed for ketoamide 14. Most interestingly, 13 shows a sharp increase in reaction efficiency on formation of a complex with β-cyclodextrin and simultaneously an increase in selectivity to form the azetidin-2-one 9. It seems quite clear that formation of a cyclodextrin complex does not lead to any inhibition of the intramolecular excited state electron transfer to give Z_1; the low reactivity of 11 can be ascribed most reasonably to preference for a 11a in the equilibrated (eq 14) cyclodextrin complex. The small effect on reactivity of 12 and 14 could be consistent with a very loose complex structure

$$(14)$$

in which the conformation of the included and free amide remains almost un-changed. However, since cyclic product formation involves bond rotations of the zwitterionic intermediate (Z_2), the rigid cyclodextrin wall plays a key restric-tive role in this process. Although no changes in the photoproduct distribution for 12 is observed upon complexation, photolysis in the presence of β-CD leads to a pronounced increase in the diastereomeric selectivity. Both 9 and 10 exist as mixtures of diastereomers when $R_1 = H$.

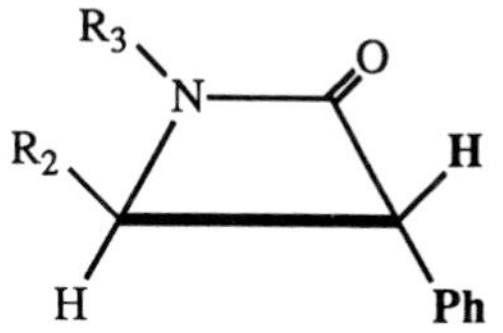

9t

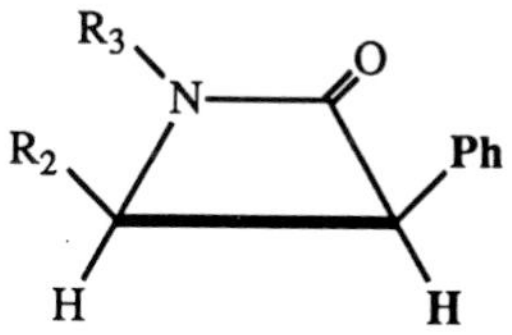

9c

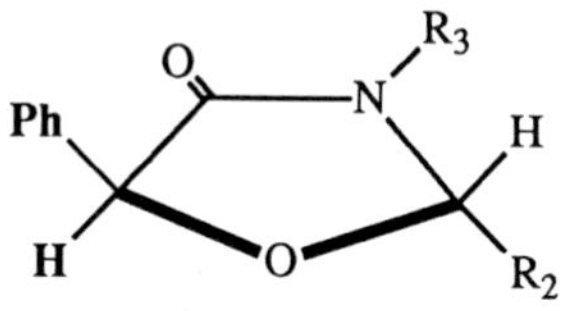

10t

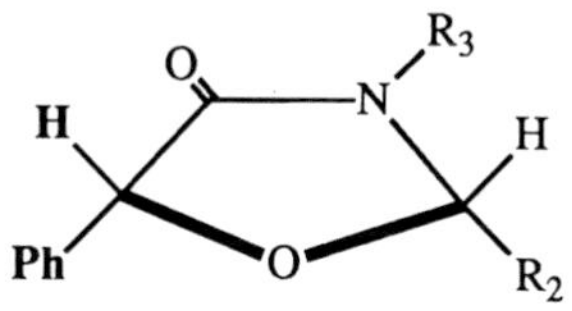

10c

For photolysis of **12** in water, the product distribution is estimated by NMR to be 81% **10t** (in benzene the product **10** is 50% t). However in β-CD the product oxazolidin-4-one is 97% **10t**. While in γ-CD the ratio of **10** is 82t to 18c. Interestingly, while photolysis of **13** in β-CD produces the most marked change in overall quantum efficiency and product selectivity, the diastereomeric selectivity (87% **9t**) is only slightly higher than that obtained in methanol-water and about the same as in benzene. To date we have not detected any indication of chiral selectivity in photoproduct formation for any of the complexes studied.

In summary our preliminary studies of intramolecular donor-acceptor molecules indicate that in a number of cases relatively high yields of net photoproduct formation can be obtained, even for cases where excited singlets are quenched to form what should be very short-lived intramolecular ion-radical pairs. Since in many cases, the precise products formed and their efficiencies depend on much longer-lived intermediates whose fates can be strongly influenced by the environment, it is reasonable to expect that carrying out these reactions in the presence of various hosts or structured interfaces can lead to useful modification of photoreactivity.

ACKNOWLEDGEMENT

We are grateful to the U.S. Department of Energy (studies described in sections 2 and 3.1) (Grant No. DE-FG02-86ER13504) and to the U.S. National Science Foundation (studies described in section 3.2) (Grant No. CHE-8616361) for support of this research. We thank our colleagues William R. Bergmark, Joseph P. Dinnocenzo and Samir Farid for helpful discussions concerning several aspects of these investigations.

REFERENCES

1) M.A. Fox, ed., Organic Phototransformations in Nonhomogeneous Media, ACS Symposium Series, Vol. 278 (ACS, Washington, 1985).

2) M. Anpo and T. Matsuura, eds., Photochemistry on Solid Surfaces (Elsevier, Amsterdam, 1989).

3) M.A. Fox and M. Chanon, eds., Photoinduced Electron Transfer (Elsevier, Amsterdam, 1988).

4) G. McLendon, Acc. Chem. Res. 21 (1988) 160.

5) S. Tazube and N. Kitamura, Pure Appl. Chem. 56 (1984) 1269.

6) A. Heller, Acc. Chem. Res. 23 (1990) 128.

7) N.S. Lewis, Acc. Chem. Res. 23 (1990) 176.

8) M.A. Fox, Pure Appl. Chem. 60 (1988) 7013.

9) T.J. Meyer, Pure Appl. Chem. 58 (1986) 1193.

10) J.K. Thomas, The Chemistry of Excitation at Interfaces, ACS Monograph No. 181 (ACS, Washington, 1984).

11) J.H. Fendler and P. Tundo, Acc. Chem. Res. 17 (1984) 3.

12) A. Weller, Z. Phys. Chem., N. F. 133 (1982) 93.

13) H. Knibbe, D. Rehm and A. Weller, Ber. Bunsenges Phys. Chem. 73 (1969) 839.

14) D. Rehm and A. Weller, Ber. Bunsenges. Phys. Chem. 73 (1969) 834.

15) A. Weller, Pure Appl. Chem. 54 (1982) 1885.

16) I.R. Gould, D. Ege, S.L. Mattes and S. Farid, J. Am. Chem. Soc. 109 (1987) 3794.

17) H. Masuhara and N. Mataga, Acc. Chem. Res. 14 (1981) 312.

18) N. Matuza, Pure Appl. Chem. 56 (1984) 1255.

19) R.A. Marcus, J. Chem. Phys. 43 (1965) 679; R.A. Marcus, Faraday Discuss. Chem. Soc. 74 (1983) 7.

20) I.R. Gould, R. Moody and S. Farid, J. Am. Chem. Soc. 110 (1988) 7242.

21) I.R. Gould, D. Ege, J.E. Moser and S. Farid, J. Am. Chem. Soc. 112 (1990) 4290.

22) S. Mattes and S. Farid, Photochemical electron-transfer reactions of olefins and related compounds, in: Organic Photochemistry, Vol. 6, ed. A. Padwa (Marcel Dekker, New York, 1983) p. 233.

23) A.P. Schaap, S. Siddigin, G. Prasad, E. Palomino and L. Lopez, J. Photochem. 25 (1984) 167.

24) D.R. Arnold and M.S. Snow, Can. J. Chem. 66 (1988) 3012.

25) R. Popielarz and D.R. Arnold, J. Am. Chem. Soc. 112 (1990) 3068.

26) D.R. Arnold and A.J. Maroulis, J. Am. Chem. Soc. 98 (1976) 5931.

27) A. Albini and M. Mella, Tetrahedron 42 (1986) 6219.

28) J.P. Dinnocenzo, S. Farid, J.L. Goodman, I.R. Gould, W.P. Todd and S.L. Mattes, J. Am. Chem. Soc. 111 (1989) 8973.

29) S. Chatterjee, P. Gottschalk, P.D. David and G.B. Schuster, J. Am. Chem. Soc. 110 (1988) 2326.

30) X. Ci, L.Y.C. Lee and D.G. Whitten, J. Am. Chem. Soc. 109 (1987) 2536.

31) R.S. Davidson and S.P. Orton, J. Chem. Soc. Chem. Commun. (1974) 209.

32) L.W. Reichel, G.W. Griffin, A.J. Muller, P.K. Das and S. Ege, Can. J. Chem. 62 (1984) 424.

33) P.G. Gassman and G.T. Carroll, Tetrahedron 42 (1986) 6201.

34) D.F. Eaton, Pure Appl. Chem. 56 (1984) 1191.

35) D.D.M. Wayner, J.J. Dannenberg and D. Griller, Chem. Phys. Lett. 131 (1986) 189

36) D.D.M. Wayner, D.J. McPhee and D. Griller, J. Am. Chem. Soc. 110 (1988) 132.

37) J.P. Dinnocenzo, unpublished results.

38) P. Maslak and S.L. Asel, J. Am. Chem. Soc. 110 (1988) 8260.

39) F.D. Lewis, J.R. Petisce, J.D. Oxman and M.J. Nepras, J. Am. Chem. Soc. 107 (1985) 203.

40) T. Arai, T. Karatsu, H. Misawa, Y. Kuriyama, H. Okamoto, T. Hiresaki, H. Furuuchi, H.L. Zeng, H. Sakuragi and K. Tokumaru, Pure Appl. Chem. 60 (1988) 989.

41) D.R. Arnold and X. Du, J. Am. Chem. Soc. 111 (1989) 7666.

42) X. Ci and D.G. Whitten, J. Am. Chem. Soc. 109 (1987) 7215.

43) X. Ci and D.G. Whitten, J. Am. Chem. Soc. 111 (1989) 3459.

44) X. Ci, unpublished results.

45) M.A. Kellett and D.G. Whitten, J. Am. Chem. Soc. 111 (1989) 2314.

46) H.F. Davis, P.K. Das, L.W. Reichel and G.W. Griffin, J. Am. Chem. Soc. 106 (1984) 6968.

47) A. Albini, Photoinduced electron transfer reactions: aromatics, in: Photoinduced Electron Transfer, Part C, eds. M.A. Fox and M. Chanon (Elsevier, Amsterdam, 1988) p. 88.

48) M.A. Kellett, D.G. Whitten, I.R. Gould and W.R. Bergmark, J. Am. Chem. Soc., in press.

49) I.R. Gould, R.H. Young, R.E. Moody and S. Farid, private communication.

50) W.R. Bergmark and D.G. Whitten, J. Am. Chem. Soc. 112 (1990) 4042.

51) W.R. Bergmark and D.G. Whitten, Molec. Cryst. Liq. Cryst., in press.

52) X. Ci and D.G. Whitten, submitted for publication.

53) A. Cu and A.C. Testa, J. Am. Chem. Soc. 96 (1974) 1963.

54) G.G. Wubbels, W.T. Jordan and N.S. Mills, J. Am. Chem. Soc. 95 (1973) 1281.

55) X. Ci, unpublished results.

56) X. Ci, R.S. daSilva, D. Nicodem and D.G. Whitten, J. Am. Chem. Soc. 111 (1989) 1337.

57) H. Aoyama, T. Hasegawa, N. Watabe, H. Shiraishi and Y. Omote, J. Org. Chem. 43 (1978) 419.

58) H. Aoyama, M. Sakamoto, K. Kuwabara, K. Yoshida and Y. Omote, J. Am. Chem. Soc. 105 (1983) 1958.

59) B. Akermark, N.G. Johansson and B. Sjoberg, Tetrahedron Lett. (1969) 371.

60) N.G. Johansson, B. Akermark and Sjoberg, Acta Chem. Scand., Ser. B, B30 (1976) 383.

61) V. Zehavi, J. Org. Chem. 42 (1977) 2821.

62) K. Shiozaki and T. Hiraoka, Synth. Commun. 9 (1979) 179.

Chapter III:
Photochemistry in Organized
Molecular Systems

Photochemical Processes in Organized Molecular Systems
K. Honda (Editor-in-Chief)
© Elsevier Science Publishers B.V., 1991

LUMINESCENCE SPECTROSCOPY AND POLYMER CHEMISTRY

F.C. De Schryver

Departement of Chemistry, K.U.Leuven, Celestijnenlaan 200 F, B-3001
Heverlee, Belgium

A study of the luminescence properties of an excited species in a polymer
system, either investigating the intensity distribution as a function of
wavelength or measuring at a given wavelength or set of wavelengths the
intensity as a function of time, can provide information on a number of
parameters characteristic of the polymer.
In this contribution the scope of this technique as applied to
macromolecules will be discussed based on a limited set of examples in which
the emitting species is either an intrinsic part of the polymer or is
extrinsic due to a component added to the polymer containing system.
In a first part the dynamics of a polymer chain in solution and the sol-gel
transition are considered. In a second part it is shown that even in the
bulk, polymer properties such as free volume or miscibility can be monitored
if the added or built in probe is adequately designed.

1. INTRODUCTION

The study of the excited state properties of a chromophore not only
provides information on its inherent deactivation pathways but can also report
on the interaction of the chromophore with its surroundings. Luminescence
spectroscopy, being an extremely sensitive technique, is and has been a method
of choice in the probing of the polymer and of its effect on the excited state
properties of an intrinsic or of an added chromophore.

The emissive properties of an excited probe can be studied by measuring the
intensity distribution as a function of the emission wavenumber or wavelength.
The spectral appearance and the change thereof as a function of certain
experimentally variable parameters allow to draw conclusions concerning e.g.
viscosity or polarity of the environment. Alternatively the complimentary
analysis of the time profile of the luminescence intensity at a given
wavelength[1] or more recently the analysis based on a set of decay data at an
ensemble of conditions in which a number of parameters can be linked in global
analysis of single photon timing[2-4] can lead to the determination of all rate
constants of importance. Furthermore related to this analysis, decay
associated spectra as well as emission spectra recorded at different time

intervals after excitation allow the study of the spectral properties of multiple species present as excited states in the system.

In the present contribution no attempt will be made to give an overview of all possibilities neither to review the literature nor to refer to a vast amount of important and interesting papers in this still growing field. For the latter reference can be made to a few recent books[7-9]. In the present contribution a few examples, based on our own research either at Leuven or in collaboration with other groups, will be used to illustrate some of the possible application of luminescence spectroscopy to polymer science.

2. LUMINESCENCE AND POLYMERS IN SOLUTION.

2.1. The polymer backbone is the emitter.

In a first example we consider a rather special situation where the polymer backbone constitutes the absorbing and emitting entity. The spectral properties of the polymer could in such a case be related to conformational aspects of the polymer chain. Such a situation occurs in dialkylpoly[silylenes][10,11].

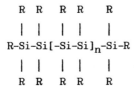

It has been shown that the electronic transition in these polymers is related to the through bond interactions existing along the polymer backbone. Absorption measurements have shown that the maximum of the absorption band as well as the decadic molar extinction coefficient reach a plateau if the chain contains about fifty silicon atoms. Calculations indicate that the all-trans chain conformation has a lower transition energy than the gauche conformation. The experimental absorption and emission spectra of a high molecular weight polymer can be interpreted as originating from an ensemble of all-trans segments with a given length distribution separated by gauche units where fast excitation transfer to the largest segments occurs. These larger all-trans segments would then emit. A recent study in our laboratory[12], in collaboration with R.D. Miller and J. Michl, of the emission maximum and bandwidth as a function of temperature of n-hexyl-n-propylpolysilylene of M_w of a million (Fig.1) reveals that in methylcyclohexane a band narrowing and bathochromic shift of the emission occur upon cooling around 230K. In the same temperature domain the maximum of the excitation spectrum shifts from 325

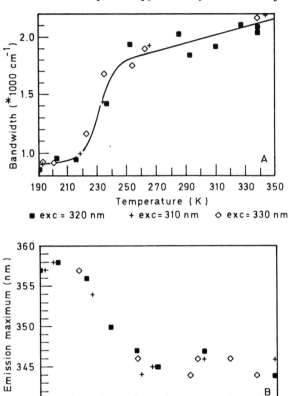

FIGURE 1.
Maximum and bandwith of the emission of poly[hexylpropyl-silylene] at different
temperatures and at different excitation wavelengths in methylcyclohexane.

nm at room temperature to 343 nm at 200 K and the bandwith also narrows upon
cooling.

This suggests that upon lowering the temperature the electronic transition
leading to fluorescence occurs from an ensemble of chains with a more uniform
distribution length and which are in average longer than at higher
temperatures. The sudden change in the spectral properties could then be
related with a structural change of the polymer in solution at 230 K. At 210
K, that is below the transition temperature, the fluorescence decay measured
in the red edge of the emission at 360 nm and excited at 318 nm, where mainly
the short segments absorb, can be globally analyzed as a difference of two
exponentials. The decay times are 226 and 677 picoseconds and the

preexponetials equal respectively - 1.9 and 2.1 suggesting that under those experimental conditions most of the emission arise from excitation transfer from the shorter segments to the longer ones. Since in other dialkylpolysilylenes a concentration dependence of the emissive properties was observed the question of the exact nature of the structural change upon cooling is still a matter of debate.

2.2. A Polymer with a Few Chromophores Attached to its Backbone

Let us first consider a polymer based on a "non absorbing or emitting backbone" that is capped at the end of all chains with a chromophore which can be specifically excited. Over a number of years Winnik[13] has very elegantly shown how end to end cyclization of polymers such as polystyrene, polydimethylsiloxane and poly(bisphenol A-diethylene glycol carbonate) in solution can be probed by studying the fluorescence properties of pyrene end-capped polymers[14]. Combining steady state and decay measurements a detailed description of the reaction kinetics could be obtained. This idea was extended recently to ionomers and more specifically to halato-telechelic polymers[15]. Pyrene labeled quaternized $\alpha,1$-bis(dimethylamino)-polystyrene with a molecular weight M_n of 12000 and a narrow distribution (M_n/M_w=1.1) was dissolved in toluene. The emission spectrum of a 5.8×10^3 gdL^{-1} degassed solution with a chromophore concentration of 5×10^{-6} mol L^{-1} clearly shows a contribution of the locally excited state and the excimer (Fig.2) due to partial intrachain end to end association.

If to this solution increasing amounts of unlabeled polymer of the same size are added, the ratio of the excimer fluorescence intensity over the locally excited state fluorescence intensity decreases. The intensity of the locally excited state increases and that of the excimer decreases indicating that the competition between intrachain association and interchain association of the chain ends tends to shift to the latter. The concentration domain where the relative decrease of the fluorescence intensity ratio is maximal corresponds to that polymer concentration where the onset of an important increase of the viscosity of the solution can be measured.

If a few chromophore molecules whose luminescence properties are sensitive to environmental changes are incorporated as pendant groups along the polymer backbone, one can obtain information on changes in chain conformation for instances as a function of temperature. This approach was used in a study attempting to unravel the mechanism leading to gelation[16] . It is known that upon cooling concentrated solutions of predominantly syndiotactic poly(methylmethacrylate) form transparent gels in xylene and toluene[17]. Spectroscopic techniques had shown that the syndiotactic sequences in this

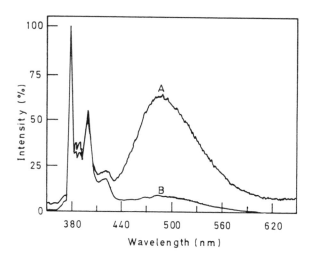

FIGURE 2.
Emission spectrum of polystyrene endcapped by a 1-pyrenylpropyldimethylammo-
nium endgroup in toluene at room temperature and diluted with trimethylammo-
nium endcapped polystyrene of the same molecular weight but without the pyre-
nylgroup. Pyrene concentration 5×10^{-6} molL^{-1}. Polymer concentration (A) 5.8
$\times 10^{-3}$ gdL^{-1}, (B) 2 gdL^{-1}.

polymer form in certain solvents all-trans helixes[18]. Thermoreversible gels
are formed by the aggregation of these all-trans helixes. The inclusion of
solvent in the helix as well as its participation in the formation of
intermolecular associations was suggested[19].

   According to the two step mechanism proposed for the gelation[17], the
polymer first undergoes a conformational change. The polymer chains go from a
random coil conformation to an all-trans conformation. This is confirmed by
IR and nmr measurements. The second step consists of an intramolecular
association leading probably to the formation of physical crosslinks of
crystalline nature. This physical network formation occurs on a much longer
timescale than the second step and can be followed by other techniques as
calorimetry or rheology. This two step mechanism is quite different from the
well known "crystallisation gelation". In this case the random coil diffuses
from the melt or the solution to the crystal surface where it takes a regular
conformation.

   To put this hypothesis to test a syndiotactic copolymer consisting of 1% 1-
pyrenylmethylmethacrylate and of 99% methylmethacrylate was prepared by
Ziegler-Natta polymerization. Incorporation of the label did not influence
the tacticity of the obtained polymer. The photophysical properties in toluene
of the label incorporated in the syndiotactic polymer were compared with those
of an atactic polymethacrylate containing a similar percentage of build in

label and with those of the model 1-pyrenylpivalate. Fluorescence spectra of 0.5% w/w and 5% w/w labeled s-PMMA solutions in toluene show mainly emission of the locally excited state and a minor contribution in the excimer region independent of the concentration suggesting that the excimer is formed between chromophores of the same polymer chain. Based on the molecular weight of the polymer and on the absorption spectrum an average number of four chromophores per polymer chain was obtained. The fluorescence decay at 377 nm could be analyzed as a bi-exponential with a major contribution of the longer lived decay of 215 ns at room temperature. Following a kinetic scheme by Holden et al.[20], the longer lived emission could be attributed to chromophores that can not form an excimer during the excited state lifetime. The shorter lived emission originates from excimer-forming chromofores. Analysing the fluorescence decay at 485 nm confirmed this conclusion. At room temperature a three exponenonential decay is observed with decay times of 215 ns, 30 ns and 128 ns. The ratio of the contribution of the last two components equals -1. This confirms the interpretation of the decays measured at 377 nm. Therefore only the longest component was used to study the gel formation.

In a first series of experiments the fluorescence lifetime of a 5% w/w in toluene is studied as a function of temperature. The longest decay parameter is shorter than the fluorescence lifetime of the pivalate model compound (at 20°C 215 ns instead of 235 ns for the model compound). The logarithm of the inverse of the decay time is plotted as a function of one over the temperature in figure 3.

Upon cooling an increase in this decay time was observed between 50-60°C. In this temperature region the onset of a conformational change was observed in IR[17]. Rheological measurements reveal a much slower formation of the network. This fluorescence behaviour fits exactly in the two step gelation mechanism. The change in decay time could be attributed to the change in conformation. When the polymer chains go from a random coiled state to a more all-trans conformation, the environment of the pyrene chromophores changes. In a random coil the chromophores experience a more polar environment because of the closeness of the ester groups of the polymer. In the extended conformation the less polar solvent surrounds better the probes, consequently their fluorescence lifetime increases. Thomas et al[21]. came to analogous conclusions studying the helix-coil transition of a 1-pyrenylacrylic acid - methacrylic acid copolymer as a function of pH.

This change in fluorescence lifetime is fast and does not vary with time. This suggest that only the change in conformation is responsible for the change in decay time and not the formation of the network. Indeed, if a 0.5% w/w solution, concentration at which no gel is formed is studied, a similar

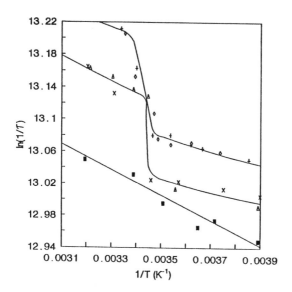

FIGURE 3.
Fluorescence decay times of PMMA as a function of temperature. (■) atactic
PMMA, (Δ) s-PMMA 0,5% heating, (x) s-PMMA 0.5% cooling, (+) s-PMMA 5% heating,
(◊) s-PMMA 5% cooling.

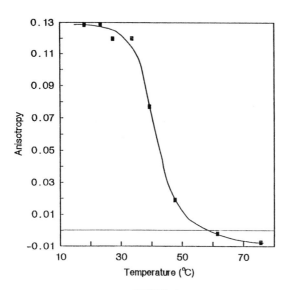

FIGURE 4.
Fluorescence anisotropy of a s-PMMA solution in toluene as a function of temperature.

decrease in decay time is observed. Furthermore a solution of atactic polymethylmethacrylate, that does not undergo such a conformational change, shows a temperature dependence identical to that of the pivalate model.

The absence of any hysteresis between cooling and heating cycles in these data clearly shows that one is dealing with a purely intramolecular phenomenon, not influenced by any intermolecular association. This is to be expected in contrast with the data obtained from IR, calorimetric and rheological observations. The hysteresis observed with these techniques is the consequence of the formation of intermolecular structures, that will disappear only at temperatures above their formation temperature.

The monomer unit with the fluorescence probe represents an irregularity along the chain that can not be incorporated as easily in an intra- or intermolecular structure. Therefore the property change of the probe only reflects the change in molecular mobility, conformation etc. of the regular sequences. A small change in these sequences will immediately result a change in the environment of these pyrenyl substituents and affect their fluorescence behaviour. The absolute values of the fluorescence lifetimes over the whole temperature domain are larger in the more diluted solution and even larger in the atactic polymer. The difference of the decay time between the two polymer concentrations could be related to the difference in overall polarity or to interchain interactions at the higher concentration. The difference in the decay of the random coil of the atactic and the syndio-tactic polymer suggest that the coil of the atactic polymer is much more swollen with solvent.

The second step in the gelation mechanism can also be studied by fluorescence spectroscopy. The anisotropy of the fluorescence of a 5% w/w solution of labeled syndiotactic polymethacrylate can be followed as a function of temperature. A mean value of the anisotropy is calculated over the most intense part of the spectrum and this is plotted in figure 4. At low temperature the solution is rigid and the fluorescence of the probe shows substantial anisotropy. When the gel is heated and melts both the viscosity and the anisotropy of the fluorescence decrease as the network disappears.

Diluted solutions show no anisotropy neither do solutions of labelled a-PMMA, both at high and low temperatures. This shows that the aggregation and not the conformational change leads to the anisotropy.

In a cooling experiment the anisotropy shows a very large hysteresis effect. This extent of the hysteresis is a function of the speed of cooling. This is due to the fact that the values for the anisotropy in this experiment are not equilibrium values.

In Figure 5 the anisotropy is plotted as a function of time at 35°C. One can see that 5 to 6 hours are needed before an equilibrium value is reached.

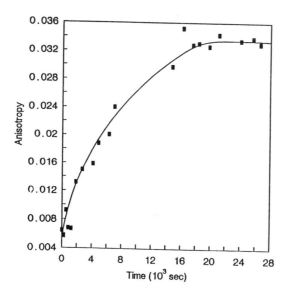

FIGURE 5.

Anisotropy as a function of time at 35°C.

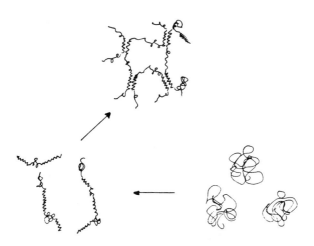

FIGURE 6.

Model for a two step gelation mechanism.

The fluorescence data, together with the other data[17], fully support the two step gelation mechanism as visualised in Fig. 6.

2.3. Each Repetitive Unit in the Chain contains a Chromophore

If the polymer is made up of units (monomers) containing a chromophore that is not part of the main chain a macromolecule is formed with a high local concentration of pendant chromophores.

$$-[-CH_2-CH-CH_2-CH-]_n-$$

| | where R represents a chromophore

R   R

This could result in quite complex photophysics since processes as excitation transfer and complex formation will compete with the monochromophoric events. Furthermore if one considers polymers where in the polymerization process an asymmetric center is introduced in the main chain as in polyvinylaromatic polymers, tacticity will play a role in these photophysical events. In a first approximation we studied the importance of local dynamics on the photophysical behavior using as model compounds 2,4-diarylpentanes [22,23].

$$CH_3-CH-CH_2-CH-CH_3$$

| | 2,4-diarylpentane

R   R

Especially in the study of the fluorescence of polyvinylcarbazole this approach was very fruitful since it allowed the analysis of the complex fluorescence spectrum in terms of two types of excimers related respectively to the meso and the racemic dyad in the polymer. The racemic dyad shows an excimer emission situated at 370 nm while the excimer originating from the meso dyad has a maximum at 420 nm. Furthermore a correlation could be made between the ground state conformation of each configuration and the excited state dynamics of the respective diastereoisomer.

Study of model sytems further showed that complexity in the emissive properties could already arise if a nonsymmetricaly substituted chromophore could reach more than one excimer geometry.

In polyvinylaromatic macromolecules excitation transfer has been the topic of extensive reseach and debate[6]. It is clear that, if excitation transfer occurs , it can and will influence substantially the spectral and even more so the time dependent intensity profile. Excitation transfer of the singlet excited state if faster than or competitive in rate with the other deactivation processes will lead to emission from low energy traps such as preformed excimer sites. In a very elegant contribution C. Frank[24] has developed a model for electronic excitation transfer as a tool in the study of polymer chain statistics .

3. LUMINESCENCE AND POLYMER BULK PROPERTIES

Liquid crystalline side chain polymers show similar deformations as observed for low molecular weight liquid crystals. One of the main deformations known for liquid crystals is the Frederickz transition. Upon applying an electric field the mesogens tend to allign themselves with their director parallel to the field. In this study a polyacrylate copolymer with the following structure was investigated :

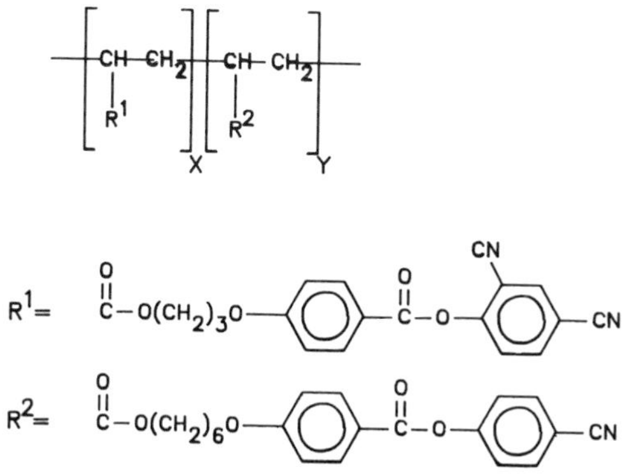

$X = 0.3 \quad Y = 0.7$

This polymer[25] displays a transition from the nematic to the isotropic phase at approximately 90°C. The photophysical active part in this polymer is the dicyano-mesogenic group. In solution upon excitation an intermolecular charge transfer state is formed, as a result the fluorescence spectrum is strongly dependent upon the solvent polarity (the emission maximum shifts from 312 nm in n-hexane to 450 nm in acetonitrile).

At room temperature the fluorescence spectrum of the polyacrylate film shows a fluorescence maximum around 375 nm. Monitoring the fluorescence intensity (1_{EM}) with decreasing temperature, the phase transition is visible as a small discontinuity. Since the specific photophysical properties of the nematic and the isotropic phase are basically the same, only a small change in the emission intensity is observed (Figure 7).

The influence of the electric field on the fluorescence spectrum is shown in figure 8. Upon applying an electric field (AC 1kHz) the intensity decreases and the emission maximum undergoes a small shift to lower energy from 375 nm at 0 V_{eff} to 382 nm at 100 V_{eff}. The electric field has a stabilizing effect

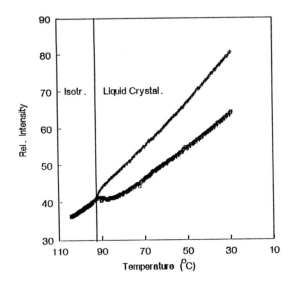

FIGURE 7.
Plot of the fluorescence intensity (l_{EM} = 370 nm) of the liquid crystalline polyacrylate versus the temperature with decreasing temperature : (+) $\vec{E}$ = 0 and (□) AC 1 kHz 100 V_{eff}.

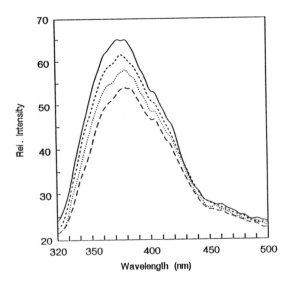

FIGURE 8.
Fluorescence spectrum of the liquid crystalline polyacrylate in the nematic phase (85°C) with varying electric field strengths (AC 1 kHz) : 0 V_{eff} (———); 30 V_{eff} (---); 50 V_{eff} (...); 120 V_{eff} (— — —).

on the ICT state, resulting in a small red shift. The decrease in intensity originates from a decreased number of excited mesogenic groups. The experimental setup is build in the way that at E = 0 the transition dipoles of absorption of the mesogenic groups lie in the plane of polarization of the excitation light. As the electric field is switched on, the mesogenic group tend to allign parallel to the field. In this configuration the transition dipole of absorption is perpendicular to the plane of polarization of the excitation light. The decreased number of excited mesogen results in a decrease in emission intensity. The electric field effect is also visible in figure 7, where the emission intensity at 370 nm is measured with decreasing temperature in the presence of an electric field (AC 1kHz 100 V_{eff}). Above T_{NI} no electric field effect is noticeable. Under T_{NI} a decrease in emission intensity is observed. At room temperature (below Tg) the electric field-induced reorientation is stable and can be stored.

By monitoring the fluorescence intensity as a function of time upon switching on the electric field, the reorientation times (τ_{or}) can be determined. The reorientation time is defined as the time for the fluorescence intensity to decrease to 50% of its initial value on application of the electric field. This was studied as a function of the temperature and the applied electric field. The results are compared with the reorientation times obtained from transmission measurements.

The results of the transmission (o) and fluorescence (D) measurements are summarised in figures 9 and 10. A good correlation between the reorientation times obtained with the two different techniques is achieved. For a fixed voltage and frequency the electro-optic response time decreases markedly with increasing temperature (Figure 9). The reorientation time also decreases with increasing voltage (Figure 10).

This demonstrates that the fluorescence technique can be a useful technique to study liquid crystalline polymers containing mesogenic chromophores. It provides additional information about the local order in the liquid crystalline phase. At the same time it can provide information to characterize reorientations on applying an electric field.

So far emission of probes covalently linked to the polymer has been considered. If well chosen, added probes can also report on interesting properties of the polymer. This will be illustrated by one example while a number of other possible uses can be found in the book by Winnik[6].

In a collaboration with L. Monnerie and L. Bokobza[26] it was shown that the analysis of the emission behavior of small intramolecular excimer forming probes, in particular the meso 2,4-diarylpentanes for which the exact motion to reach the excimer configuration is well understood, is an excellent

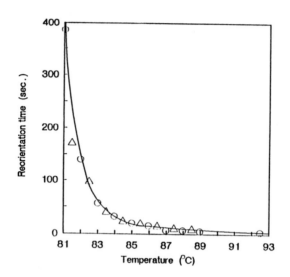

FIGURE 9.
Electro-optic reorientation time for the liquid crystalline polymer as a function of temperature. The applied voltage was 100 V_{eff} at 1 kHz the cell
thickness was $\approx$20 μm.

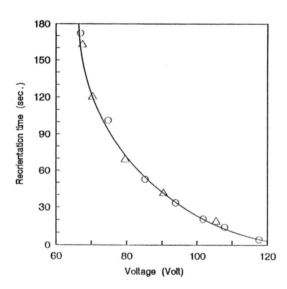

FIGURE 10.
Electro-optic reorientation time for the liquid crystalline polyacrylate as a
function of the applied voltage. (T = 90°C, d $\approx$ 20 μm, f = 1 kHz).

approach to determine chain mobility in a polymer matrix. If the probe motion is only modulated by the mobility of of the matrix, the temperature dependence of its correlation time follows that of the relaxation of the matrix as described by the expression of Williams, Landel and Ferry (WLF)[27]. This proves that the volume swept by the interacting chromophores is equivalent to that determined by viscoelastic measurements.

The original expression of the WLF equation can be transformed in the following way

$$\log\tau_c(T) = \log\tau_c(T_g) - C^g_1 + C^g_1 C^g_2 \, [1/(T-T_2)]$$

In this equation τ_c is the correlation time and equals one over the rate of excimer formation of the bichromophoric probe, (T) refers to the temperature of the measurement, the subscript (T_g) of τ refers to the value of the correlation time at the glass transition temperature, while T_2 is the reference temperature at which the free volume of the systems becomes zero. $C^g_1 \approx 1/f_g$, the coefficient correlates with one over the fractional free volume at the glass transition temperature and $C^g_2 = f_g/\alpha$ where α is the thermal expansion coefficient of the free volume. From the equation it follows that a linear relation should exist between the logarithm of the correlation time at a given temperature and the ratio $1/(T-T_2)$. In Figure 11 a plot is shown of such a correlation for a series of polymers.

As can be seen a linear relation is observed with different slopes for the different polymers. The different slopes correspond to the difference in the product $C^g_1 C^g_2$ which is a characteristic of each polymer. The value of these coefficients can be measured independently by viscoelastic measurements. They are reported in table 1. As can be seen from the table the agreement between the values obtained by both techniques is excellent.

Table 1. Coefficients of the WLF equation

Polymer matrix	T_g K	T_2 K	Predicted $C^g_1 C^g_2$[a] deg	Observed $C^g_1 C^g_2$[b] deg
polybutadieen	164	101	678	604
polypropylene oxide(PPO)	195	174	389	428
poly-iso butylene(PIB)	206	101	1733	1712
poly-isoprene (PI)	211	146	900	914

a) viscoelastic measurements; b) fluorescence measurements

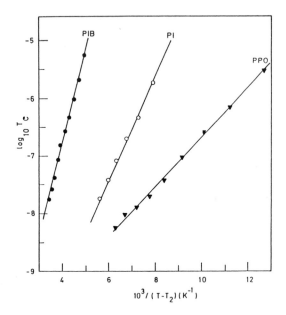

FIGURE 11.

Logarithm of the correlation time for excimer formation of the probe in poly-
isobutylene (PIB), poly-isoprene (PI), polypropylene-oxide (PPO).

These results show that the probe motion is controlled in all the
investigated matrices by the segmental motion involved. However it can be
shown that the intramolecular rotational motion with a given frequency is
achieved for different fractional free volumes of the considered matrices.
This indicates that other factors such as the fluctuation of free volume and
difference in size of the mobile units in each type of chain also play a
role[28].

4. CONCLUSION

This limited set of examples is ample proof that luminescence spectroscopy
is an excellent tool in the study of macromolecules. Either the intrisic
luminescence of the polymer or that of an adequately chosen probe allows the
study of structural aspects and physical properties of polymer in bulk and in
solution.

ACKNOWLEDGEMENTS.

The continous financial support of the N.F.W.O., F.K.F.O and the Ministery
of Scientific Programmation through IUAP3 and Geconcerteerde Onderzoeksactie

are greatfully acknowledged. I am also strongly indebted to all collaborators and colleagues who have been involved in the projects mentioned in this paper.

REFERENCES

1) N. Boens, F.C. De Schryver, Macromol.Chem. Macromol. Symp. 18 (1988) 93-100.

2) N. Boens, L.D. Janssens and F.C. De Schryver, Biophys. Chem. 33 (1989) 77-90.

3) N. Boens, H. Luo , M. Van der Auweraer, S. Reekmans, F.C. De Schryver and A. Malliaris, Chem. Phys.Lett. 146 (1988) 337-342.

4) J.R. Knutson, J.M. Beechem and L.Brand, Chem Phys. Lett., 102 (1983) 501-507.

5) C.E. Hoyle and J.M. Torkelson (ed), In Photophysics of Polymers ACS Symposium Series 358 (1987).

6) M.A. Winnik, ed., In Photophysical and Photochemical Tools in Polymer Science (D. Reidel, Dordrecht, 1986).

7) J.F. Rabek, Mechanism of Photophysical Processes and Photochemical Reactions in Polymers (J. Wiley & Sons, N.Y., 1987).

8) D. Phillips, ed. Polymer Photophysics, (Chapman and Hall, London, 1985).

9) J. Guillet, Polymer Photophysics and Photochemistry (Cambridge University Press, Cambridge, 1985).

10) J. Michl, J.W. Downing, T. Karatsu, A.J. McKinley, G. Poggi, G.M. Walraff, R.D. Miller and R. Sooriyakumaran, Pure Appl.Chem. 60 (1988) 959-972.

11) R.D. Miller, B.L. Farmer, W. Fleming, R. Sooriyakumaran and J. Rabolt, J.Am.Chem.Soc. 104 (1987) 2509-2512.

12) D. Declercq, R.D. Miller, J. Michl and F.C. De Schryver, to be published.

13) M.A. Winnik, Acc.Chem.Res. 18 (1985) 73-79.

14) S. Boileau, F. Mechin, J.M.G. Marthino and M.A. Winnik, Macromolecules 22 (1989) 215-220.

15) M. Granville, R.J. Jerome, P. Teyssie and F.C. De Schryver, Macromolecules 21 (1988) 2894-2896.

16) M. Berghmans, H. Berghmans and F.C. De Schryver, to be published.

17) H. Berghmans, A. Donckers, L. Frenay, W. Stoks, F.C. De Schryver, P. Moldenaers and J. Mewis, Polymer 28 (1987) 97-102.

18) J. Spevcaek, B. Schneider, J. Dybal, J. Stokr, J. Baldrian and Z. Pelzbauer, J. Polym. Sci.,Polym. Phys. Ed. 22 (1984) 617.

19) H. Usuyama, N. Myamoto, Y. Chatani and H. Takodoro, Polymer Communications 14 (1983) 119.

20) D.A. Holden, P.Y-K. Wang and J.E. Guillet, Macromolecules 17 (1980) 2142.

21) C. Deh-Ying and J.K. Thomas , Macromolecules 17 (1984) 2142.

22) F.C. De Schryver, P. Collart, J. Vandendriessche, R. Goedeweeck, A.M. Swinnen and M. Van der Auweraer, Acc. Chem. Res. 20 (1987) 159-166.

23) F.C. De Schryver, P. Collart, R. Goedeweeck, F. Ruttens, F. Lopez Arbelao and M. Van der Auweraer, ACS Symposium Series, 358 (1987) 186-200.

24) C.W. Frank, G.H. Fredrickson and H.C. Andersen, Electronic excitation transport as a tool for the study of polymer chain statistics. In : Photophysical and Photochemical Tools in Polymer Science, ed., M.A. Winnik (D. Reidel, Dordrecht, 1986) pp.485-532.

25) M. Van Damme, Doctoral Thesis, K.U. Leuven (1990).

26) L. Bokobza, C.Pham-Van-Cang, C. Giordano, L. Monnerie, J. Vandendriessche, F.C. De Schryver and E.G. Kontos, Polymer 28 (1987) 1876-1880.

27) M.L. Williams, R.F. Landel, J.D. Ferry, J.Am.Chem.Soc. 77 (1955) 3701-3707.

28) L. Bokobza, Prog. Polym. Sci.15 (1990) 337-360.

Photochemical Processes in Organized Molecular Systems
K. Honda (Editor-in-Chief)
Elsevier Science Publishers B.V., 1991

FLUORESCENCE STUDIES OF HYDROPHOBIC POLYMERS IN WATER: PHASE TRANSITIONS IN AQUEOUS POLY-(N-ISOPROPYLACRYLAMIDE) SOLUTIONS

Françoise M. WINNIK

Xerox Research Centre of Canada, 2660 Speakman Drive, Mississauga Ontario, Canada L5K 2L1

Phase transitions of dye-labeled poly-(N-isopropylacrylamides) (PNIPAM) in dilute solution have been studied by two steady-state fluorescence techniques. These probe a) the relative intensity of pyrene monomer and excimer emissions from a pyrene-labeled polymer (PNIPAM-Py/200) and b) the efficiency of Förster energy transfer between a donor (naphthalene) and an acceptor (pyrene) attached to a doubly-labeled polymer (PNIPAM-Py/366-N/50). The phase transitions were induced either by heating solutions of the polymers in water or by varying the composition of methanol/water polymer solutions. Results are interpreted in terms of a mechanism of phase separation initiated by a gradual shrinking of solvated polymer coils into a collapsed state, followed by aggregation of individual chains into larger particles.

1. INTRODUCTION

Poly-(N-isopropylacrylamide) (PNIPAM) exhibits properties that defy everyday experience. For example, it is well known that the solubility of a substance in a given solvent depends on the temperature: usually a lowering of the temperature results in clustering of the solute and eventually in its precipitation from solution. This is not the case for aqueous PNIPAM solutions: decreasing the solution temperature enhances the solubility of the polymer in water and increasing its temperature induces precipitation. Above a critical temperature (31°C)[1] a clear PNIPAM solution becomes suddenly milky in appearance. The process is reversible: cooling the suspension below 31°C results in instantaneous redissolution of the polymer. Even more puzzling is the behavior of PNIPAM in certain solvent mixtures. This polymer is extremely soluble in pure methanol and in pure water, but it is not soluble in certain mixtures of the two solvents.[2] In fact this effect is so extreme that it has been used as a simple and effective purification technique.[3] The origin of both phenomena can be traced, it seems, to the unusual interactions that occur between PNIPAM and water. It is not uncommon for hydrophobic polymers capable of undergoing strong hydrogen bonding in water to exhibit lower critical solution temperatures (LCST). Hydrogen bonding between the polymer and water lowers the free energy of mixing. It also triggers the formation of a layer of highly organized water molecules around the polymer, resulting in a decrease in the entropy of the system. When the temperature of the solution is raised, bound water is released. The relative values of the thermodynamic functions change: the entropic term becomes dominant, resulting in a positive free energy of mixing. A two-phase system is then favored.

Until recently experimental characterization of LCST phenomena has relied upon classical techniques such as cloud point measurements, calorimetry, and light scattering. It would be interesting to follow the phase separation on a molecular scale, and from the polymer's point of view. This can be achieved if one attaches to the polymer a 'reporter' group which responds in some specific fashion to the changes incurred by the polymer during the transition. This article focuses on the use of fluorescent labels to study PNIPAM as it undergoes phase separation from water. For polymer solutions in organic solvents it is usually safe to assume that the fluorescent label does not perturb the properties of the polymers. In water a hydrophobic dye may modify the behavior of the polymer. Hydrophobic substituents of aqueous polymers tend to associate in water. They create interpolymeric networks or induce the formation of unimolecular polymeric micelles.[4] This effect is particularly strong when the substituents are long alkyl groups, but it occurs also with aromatic groups, albeit to a smaller extent.

One set of experiments described here takes advantage of the association of hydrophobic fluorescent labels. They involve a PNIPAM derivative which possesses a small number of pyrene substituents. In methanol, this polymer exhibits a fluorescence behavior typical of pyrene itself, characterized by an emission due to locally excited pyrenes (monomer emission) and an emission due to pyrene excimers. All spectroscopic evidence points to the fact that these excimers form by a dynamic process in accord with the classical Birks' mechanism.[5] When the same polymer is dissolved in water, the pyrene spectroscopy is different. It exhibits all the features of a system in which some of the pyrene groups exist as isolated species and some are associated in the form of (ground-state) dimers or higher aggregates. The evidence for this conclusion will be presented below. Experiments described here will show that these associated pyrenes are rather labile. They are disrupted during the heat-induced or methanol-induced phase separation of aqueous pyrene-labeled PNIPAM-Py/200 solutions. A more fundamental issue of the LCST phenomena will be addressed by a second set of experiments. These report on changes in the dimensions of a polymer coil as it passes through the critical point. In these experiments one follows the efficiency of energy transfer between donor (naphthalene) and acceptor (pyrene) groups attached to a polymer as the chain dimensions contract or expand.

2. EXPERIMENTAL

2.1. Materials

Water was deionized by a Millipore Milli-Q water purification system. Spectral grade methanol was obtained from Caledon. PNIPAM was prepared by free-radical polymerization of N-isopropylacrylamide in *tert*-butyl alcohol.[6] The labeled polymers were prepared from a copolymer of N-isopropylacrylamide and N-acryloxysuccinimide as described in detail elsewhere.[7] The physical characteristics of the polymers are listed in Table 1.

2.2. Instrumentation

UV-spectra were measured with a Hewlett-Packard 8480A diode array spectrometer. The temperature of the cell was controlled with a Hewlett-Packard 89100A temperature control accessory consisting of a digitally-controlled thermoelectrically heated and cooled cell holder with sample stirring and programmed temperature ramping capability. The temperature of the sample fluid was measured with a teflon-coated probe immersed in the sample fluid. Steady-state fluorescence spectra were recorded on a SPEX Fluorolog 212 spectrometer equipped with a DM3000F data system. The temperature of the water-jacketed cell holder was controlled with a Neslab circulating bath. For measurements at different temperatures samples were allowed to equilibrate for 10 min at a given temperature. The heating rate corresponded to approximately 0.2°C min-1.

2.3. Spectroscopic parameters

Excitation spectra were measured in the ratio mode. The emission spectra were not corrected except for those used in quantum yield determinations. Excitation wavelengths were 290 nm for naphthalene, 330 nm for pyrene in aqueous solutions, and 328 nm for pyrene in methanol solutions. The excimer to monomer ratios (I_E/I_M) for the pyrene emission were calculated by taking the ratio of the emission intensity at 480 nm to the half-sum of the emission intensities at 376 nm and 396 nm. Quantum yields (Φ) were calculated by integration of peak areas of corrected spectra in wavenumber units, using as standards either quinine sulfate in 1 N H_2SO_4 ($\Phi = 0.546$, $\lambda_{exc} = 328$ nm, 25°C)[8] or 2-aminopyridine in 0.1 N H_2SO_4 ($\Phi = 0.60$, $\lambda_{exc} = 290$ nm, 25°C).[9] Beer's law corrections were applied for optical density changes at the excitation wavelength. The fraction (F) of the total light absorbed by naphthalene in the presence of pyrene was calculated using equation 1: where $A_N^{(290)}$ and $A_{Py}^{(290)}$ are the absorbances at 290 nm of naphthalene and pyrene, respectively.[10] It was assumed that this fraction does not vary with solution temperature in the 4 to 40°C range.

$$F = \frac{1 - 10^{-A_N^{(290)}}}{2 - 10^{-A_N^{(290)}} - 10^{-A_{Py}^{(290)}}} \qquad (1)$$

TABLE 1
Physical properties of the polymers

POLYMER	$[\eta]^a$ (cm^3 g^{-1})	M_v^b	[Py] (mol g^{-1})	[N] (mol g^{-1})
PNIPAM	72	920,000	---	---
PNIPAM-N/27	87	1.23×10^6	--	3.3×10^{-4}
PNIPAM-Py/200	103	1.1×10^6	4.2×10^{-5}	
PNIPAM-Py/366-N/50	84	1.2×10^6	2.4×10^{-5}	1.7×10^{-4}

a. From THF solution   b. from $[\eta] = 9.59 \times 10^{-3} M_v^{0.65}$, see ref. 19

TABLE 2
Solutions for spectroscopic measurements

POLYMER (Concentration in g L^{-1})		SOLVENT	[Py] mol L^{-1}	[N] mol L^{-1}
NRET Experiments:				
● *Intrapolymeric:*				
PNIPAM-Py/366-N/50	(0.044)	H$_2$O	1.1×10^{-6}	7.7×10^{-5}
PNIPAM-Py/366-N/50	(0.130)	MeOH	3.2×10^{-6}	2.3×10^{-5}
● *Interpolymeric :*				
PNIPAM-Py/200	(0.024)	H$_2$O	1.1×10^{-6}	3.8×10^{-5}
and PNIPAM-N/27	(0.012)			
Py Excimer-Monomer Experiments:				
PNIPAM-Py/200	(0.110)	H$_2$O	4.2×10^{-6}	----
PNIPAM-Py/200	(0.110)	MeOH	4.2×10^{-6}	----

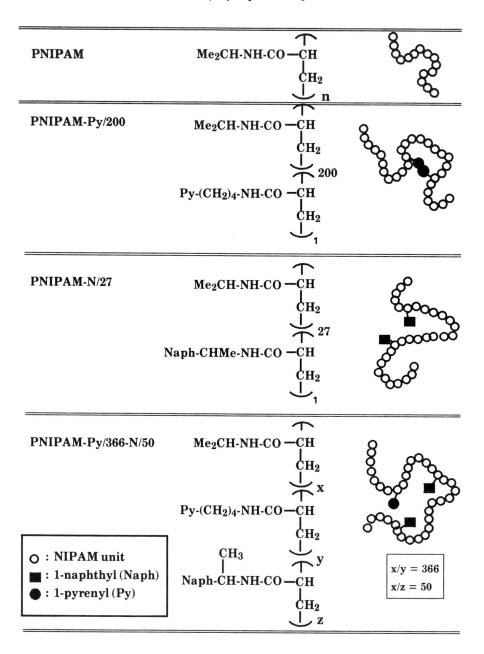

FIGURE 1

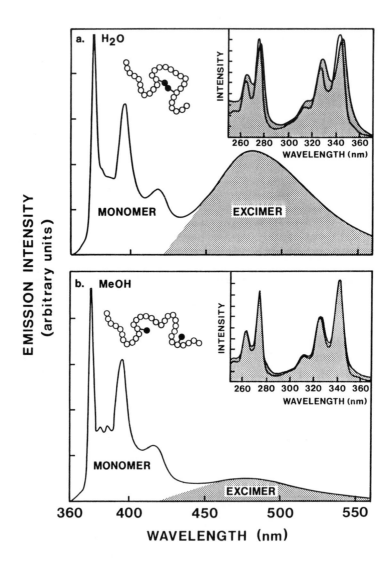

FIGURE 2

Fluorescence spectra of PNIPAM-Py/200 (50 ppm) in water (top) and in methanol (bottom). Insets are the excitation spectra of the solutions monitored for the monomer (λ_{em} = 396 nm) and the excimer (λ_{em} = 480 nm). The excitation spectra are normalized at 345 nm. (Reprinted with permission from ref. 6. Copyright 1990 American Chemical Society).

2.4. Solutions for analysis

Solutions in water were prepared by allowing the polymers to dissolve for 24 hr before they were diluted to a known volume. They were kept at room temperature for at least 2 hr before measurements. In all solutions the concentration of polymer was such that the absorbance at the excitation wavelengths was lower than 0.05. Solution compositions are listed in Table 2. Solutions in methanol were degassed by vigorous bubbling of solvent-saturated argon for 1 min. Solutions in water were not degassed. Cloud points were determined by spectrophotometric detection of changes in turbidity of solutions heated at 0.2°C min-1 in a magnetically stirred cell, as described previously.[11] For measurements below 10°C the cell holder was flushed with nitrogen to prevent moisture condensation.

3. MATERIALS AND SPECTROSCOPY

3.1. Synthesis of the labeled polymers

Attachment of the dyes to the PNIPAM backbone was achieved by reacting a copolymer of N-isopropylacrylamide and N-acryloxysuccinimide with chromophores bearing a short amino-terminated alkyl chain. Specifically, the copolymer was labeled with pyrene (Py) by reaction with 4-(1'-pyrenyl)-butylamine, with naphthalene (N) by reaction with 1-(1'-naphthyl)-ethylamine, and with Py and N by reaction first with 4-(1'-pyrenyl)-butylamine, then with 1-(1'-naphthyl)-ethylamine. In all cases the unreacted N-acryloxysuccinimide groups were converted to isopropyl groups by quenching with excess isopropylamine. Three polymers were prepared: PNIPAM-Py/200, labeled with pyrene only,[5,12] PNIPAM-N/27,[6] labeled with naphthalene only, and PNIPAM-Py/366-N/50,[6] doubly labeled with pyrene and naphthalene (Figure 1).

3.2. Spectroscopy of the labeled polymers in solution

The fluorescence spectra of PNIPAM-Py/200 in water and in methanol are presented in Figure 2. In methanol (bottom) PNIPAM-Py/200 exhibits an emission due to locally excited pyrene chromophores with the (0,0) band located at 376 nm, together with a broad featureless emission centered at 480 nm, which originates from pyrene excimers. Identical excitation spectra (inset) are obtained for both emissions, and their maxima correspond to the UV absorption spectra. Therefore both the monomer and excimer emissions originate from the same species, the excited isolated pyrene chromophores. The dynamic nature of the excimer is confirmed by the time-dependent fluorescence profile of the excimer emission. It shows both a growing-in component and a decaying component. In water (top) the overall fluorescence intensity from PNIPAM-Py/200 is lower, indicating a larger extent of pyrene self-quenching. Also, the contribution from the excimer is much larger than in methanol. The excimer emission in this case originates mostly from preformed pyrene dimers or higher aggregates, as evidenced by the following spectral features. The excitation spectra for the monomer and excimer emissions are different: their general features are similar, but the former is blue-shifted by about 3 nm (Figure 2, top inset). The UV spectrum of the polymer exhibits a hypochromic effect. Furthermore the time-dependent excimer fluorescence profile does not show a growing-in component, at least on the nanosecond time-scale. Pyrene dimers or higher aggregates have been detected in aqueous solutions of other pyrene-labeled polymers.[13] Their existence has been attributed to a gain in free energy of mixing through hydrophobic interactions between the non-polar pyrene groups.[14]

The fluorescence of the naphthalene labeled polymer, PNIPAM-N/27, in methanol exhibits an emission due exclusively to locally excited naphthalenes, with the [0,0] band located at 323 nm. In water the polymer presents a weak structureless emission (350 to 460 nm) attributed to naphthalene excimers, in addition to the structured naphthalene emission (310 to 360 nm).

The polymer PNIPAM-Py/366-N/50 is labeled with a pair of chromophores known to interact as donor (naphthalene) and acceptor (pyrene) in non-radiative energy transfer (NRET).[15] This Förster process originates in dipole-dipole interactions between the excited donor and the acceptor in its ground-state. One important technical requirement for energy transfer experiments is that one should be able to excite the donor in the presence of the acceptor. Here it is possible to choose an excitation wavelength such that most of the light is absorbed by N. At 290 nm 77 % (see Experimental) of the total light absorbed by the solution results in the formation of excited naphthalenes. The residual light serves to excite directly the pyrene chromophores. Their emission will contribute to the total pyrene emission from PNIPAM-Py/366-N/50 excited at 290 nm. The efficiency of energy transfer between the chromophores can be determined either from the increase in pyrene emission or from the quenching of the naphthalene emission in the presence of the acceptor. The second method was used here, with emission from PNIPAM-N/27 taken as a reference. The naphthalene quenching efficiency (E_N) was calculated using Equation 2, where I_D^N and I_S^N are the fluorescence intensities of naphthalene from doubly and singly labeled polymers, with absorbances A_D and A_S at the excitation wavelength (290 nm), respectively.

$$E_N = 1 - \frac{I_D^N A_S}{I_S^N A_D} \qquad (2)$$

The efficiency (E) of energy transfer between two chromophores depends on their separation distance R by a well defined function (Equation 3), where R_o, for rotationally averaged pairs, is the donor/acceptor separation distance for which the energy transfer efficiency is 50 %. The characteristic distance R_o for the N/Py pair of chromophores is 29 Å.[6]

$$E = \frac{R_o^6}{R_o^6 + R^6} \qquad (3)$$

When the donor and acceptor chromophores are attached to a polymer chain, the distribution of donor/acceptor separations is related to the chain dimensions. Factors which swell these dimensions increase the mean separation of the interacting groups. Conversely chain collapse should lead to a pronounced decrease in these distances. As a consequence the efficiency of intramolecular energy transfer should be a sensitive measure of swelling or contraction of chain dimensions, increasing sharply as the chain dimensions decrease.

4. THERMALLY INDUCED TRANSITIONS

4.1. Poly-N-(isopropylacrylamide) (PNIPAM)

Since the early work of Heskins and Guillet on PNIPAM[1] and the seminal paper of Taylor and Cerankowski on LCST phenomena in water,[16] several aspects of the phase separation mechanism of PNIPAM in water have been examined in detail. Results from microcalorimetric measurements have ascertained that the phenomenon involves rupture of hydrogen bonds between the polymer and surrounding water molecules.[3, 17] A transition enthalpy of 1.5 to 5.0 kcal mol[-1] of repeating units is consistent with the loss of about 1 hydrogen bond per NIPAM group.[18] Molecular aspects of the transition have been elucidated by dynamic and static light scattering measurements that probe the changes in the dimensions of isolated polymer chains as a function of the solution temperature.[19] In the case of a 13 ppm solution of a polymer of $M_W \approx 8 \times 10^6$ the hydrodynamic radius was shown to decrease from *ca.* 1000 Å at 30°C to *ca.* 300 Å at 32°C. Therefore the LCST phenomenon for PNIPAM can be viewed as a two stage process, consisting of the collapse of the rather expanded PNIPAM chains into more compact globules, followed by the aggregation of individual collapsed chains.

4.2. Pyrene-labeled polymer (PNIPAM-Py/200)

Cloud point measurements indicate that the pyrene-labeled polymer undergoes precipitation at the same temperature as PNIPAM. Fluorescence measurements performed while heating a PNIPAM-Py/200 solution report the following events. The total quantum yield of emission (Φ_T) is deeply affected: it remains constant (0.51) from 20 to 28°C, then it undergoes a pronounced increase between 29 and 33°C to reach a value of 0.74 (Figure 3). The increase in Φ_T corresponds to a large increase in the quantum efficiency of pyrene monomer emission (Φ_M) and to a large decrease in the excimer quantum yield (Φ_E). These changes in quantum yields are reflected in the changes incurred with temperature by the ratio I_E/I_M. It increases slightly from a value of 0.40 to 0.46 at 29°C. Then it decreases sharply to reach a limiting value of 0.01 at 35°C. The midpoint of the transition (31°C) is slightly lower than the LCST (32.5°C) of the polymer. The excitation spectra are affected as well. The differences between the excitation spectra monitored for the monomer and excimer disappear above the LCST. Both spectra have a maximum at 343.5 nm, which corresponds to the maximum in the UV absorption spectrum of the solution at this temperature. The changes are thermoreversible: for example, upon slow cooling of the solution, the ratio I_E/I_M increases to reach its initial value of 0.40 at 20°C.

These spectral changes reflect the aspects of the LCST phenomena sensed by the labels. Below the LCST the pyrenes exist either isolated or in the form of aggregates. As the temperature of the solution reaches the LCST these aggregates dissociate. The polymer-rich phase which separates from solution provides a hydrophobic environment in which pyrenes are solubilized and isolated from each other. The residual excimers are formed by a dynamic process. It is noteworthy that the I_E/I_M ratio is smaller for aqueous PNIPAM-Py/200 solutions above their LCST (0.01) than for solutions of the same polymer in methanol (0.1). Since in both cases the excimer is formed by a dynamic process, the lower value of the former environment may reflect the higher viscosity experienced by the isolated pyrene groups in the polymer-rich phase.

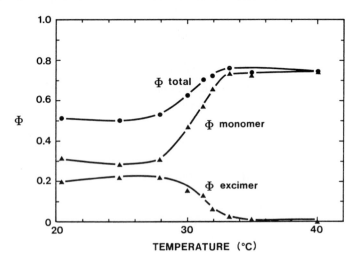

FIGURE 3

Total fluorescence quantum yield (Φ) as a function of temperature for a solution of PNIPAM-Py/200 in water (110 ppm). Also plotted are the contributions of excimer and monomer emissions to the total quantum yield (see text). (Reprinted with permission from ref. 6. Copyright 1990 American Chemical Society).

4.3. Doubly-labeled polymer. (PNIPAM-Py/366-N/50)

This set of experiments focusses on the interactions between the naphthyl and pyrenyl substituents attached to a PNIPAM chain. Here the important parameter is the efficiency of energy transfer between N and Py, determined from the intensity of the emissions of the two labels. As a solution of PNIPAM-Py/366-N/50 is heated from 20 to 31°C, the naphthalene emission decreases slowly with increasing temperature and the pyrene emission intensity increases markedly (Figure 4). Further heating has no effect on the fluorescence intensity of either pyrene or naphthalene. The total pyrene emission is enhanced by a factor of *ca.* 3.3 above the LCST. The simultaneous decrease in total naphthalene emission and increase in total pyrene emission are indicative of an increase in the energy transfer efficiency above the LCST. In order to assess the contribution of energy transfer among chromophores attached to different polymer chains, it is imperative to carry out control experiments with dilute aqueous solutions of the two singly-labeled polymers, PNIPAM-Py/200 and PNIPAM-N/27 (Table 2). No energy transfer occurs in this solution below 31°C. Above the LCST it can be detected, but its efficiency is less than 5 % of the value measured from solutions of the PNIPAM-Py/366-N/50. Therefore it seems reasonable to conclude that measurements on the doubly labeled polymer detect predominantly intramolecular contributions to the energy transfer process.

A detailed analysis of the intramolecular energy transfer data reveals an intriguing feature: the increase in energy transfer efficiency as a function of temperature occurs, not at the LCST itself, but below this temperature and over a broad temperature range. This pattern is seen for example in the plot of the naphthalene quenching efficiency E_N as a function of temperature (Figure 5). Such a gradual increase in this experimental parameter is surprising, as it monitors a property of a solution passing through a critical point. It implies that the average distance between the chromophores is gradually reduced as the temperature increases. This may result either from a continuous decrease in the size of the polymer coil, or from an enhancement in the density fluctuation, giving rise to a higher incidence of naphthalene-pyrene contacts. The precise nature of the conformational changes incurred by the polymer chains cannot be described on the basis of energy transfer experiments alone. The results reported here indicate that conformational reorganization begins well below the LCST. They confirm that a contraction of individual polymer chains precedes interpolymeric aggregation.

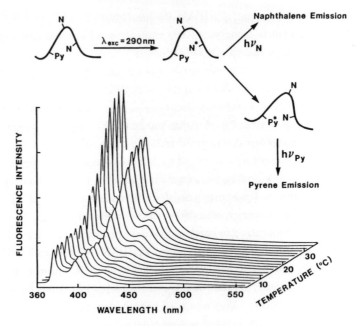

FIGURE 4
Fluorescence spectra of a solution of PNIPAM-Py/366-N/50 in water (44 ppm) at several temperatures between 4 and 40 °C. The 310 to 360 nm range where naphthalene emits is not shown. (Reprinted with permission from ref.7. Copyright 1990 Butterworth Scientific Ltd.).

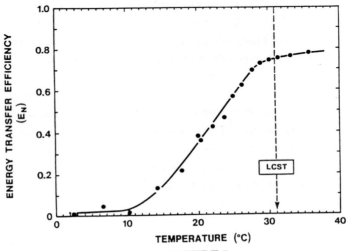

FIGURE 5
Plot of the energy transfer efficiency (E_N) as a function of temperature for an aqueous solution of PNIPAM-PYy/366-N/50 (44 ppm).(Reprinted with permission from ref.7. Copyright 1990 Butterworth Scientific Ltd.).

5. SOLVENT-INDUCED TRANSITIONS

5.1. Poly-(N-isopropylacrylamide)

Addition of methanol to an aqueous PNIPAM solution results, within given temperature and solvent composition ranges, in the precipitation of the polymer. The phase diagram of PNIPAM in water/methanol consists of four solvent composition ranges (Figure 6): (i) for methanol molar fractions (x_M) lower than 0.05, the LCST is hardly affected; (ii) for $0.05 < x_M < 0.35$, the LCST decreases from 31°C to reach a minimum value of -7.5°C; (iii) for $0.35 < x_M < 0.45$, the LCST increases sharply to reach a value of 14.5°C; (iv) for $x_M > 0.45$ (64.4% v/v), it is not possible anymore to detect an LCST. Solutions in this temperature range can be heated to 100°C without showing any turbidity. A subtle interplay between the polymer/water, polymer/methanol, and water/methanol interactions accounts for the complexity of the phase diagram. For solvent compositions of $x_M < 0.05$, the methanol molecules are kept apart from each other by formation of hydration cages around each molecule. Under these circumstances the interactions of the polymer with its environment are essentially undisturbed by the presence of methanol molecules. As the methanol molar fraction increases, there is no longer sufficient water to provide clathrate cavities for all the methanol molecules. These are now free to interact with polymer segments. Changes in LCST observed for $0.05 < x_M < 0.45$ reveal on a macroscopic scale the changes in the relative importance of the free energy of association between methanol-polymer, water-polymer, and methanol-water. In solutions of higher methanol content the methanol-polymer interactions predominate.

5.2. Pyrene-labeled polymer (PNIPAM-Py/200)

At the low level of labeling selected for this study the presence of pyrene substituents does not affect significantly the solubility of PNIPAM in water-methanol mixtures. PNIPAM and PNIPAM-Py/200 solutions exhibit superimposable phase diagrams. Through the fluorescence of the pyrenes, molecular aspects associated with each transition can be observed. One excellent tool to follow the transitions is provided by the ratio I_E/I_M of the pyrene emission, since it takes significantly different values for the polymer in pure water or in pure methanol. An example of the isothermal changes in I_E/I_M as a function of solvent composition is presented in Figure 7 (top). Monitoring the changes in I_E/I_M at 20°C, one can divide the water/methanol composition domain in four sections: (i) for $x_M < 0.07$, I_E/I_M remains constant; (ii) in a very narrow composition range, $0.07 < x_M < 0.13$, centered at $x_M \simeq 0.105$ ($\simeq$ 22 % v/v), I_E/I_M undergoes an abrupt decrease; (iii) for $0.13 < x_M < 0.45$, I_E/I_M increases slightly to reach a value of 0.20; (iv) for $x_M > 0.45$, I_E/I_M levels down to reach its final value (0.10) in pure methanol. Isothermal curves of I_E/I_M *vs* solvent composition exhibit the same overall trends when they are recorded at different temperatures between 6 and 28°C.

**TABLE 3. SOLVENT COMPOSITIONS AT THE PHASE TRANSITION
TEMPERATURE FOR PNIPAM AND PNIPAM-Py/200'**

TEMPERATURE	MeOH CONTENT molar fraction (% v/v)	
(°C)	PNIPAM from cloud point[a]	PNIPAM-Py/200 from I_E/I_M[b]
6	0.23 (40)	0.19 (35)
15	0.16 (31)	0.16 (31)
20	0.13 (27)	0.13 (27)
28	0.06 (14)	0.06 (14)

a. value determined from the phase diagram (Figure 6).
b. see text

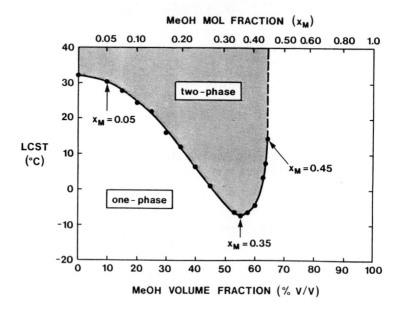

FIGURE 6

Phase diagram of PNIPAM (1 g L^{-1}) as a function of solvent composition in methanol-
water mixtures expressed in MeOH percent volume fraction (bottom x axis) and in
MeOH molar fraction (x_M, top x axis). (Reprinted with permission from ref. 2.
Copyright 1990 American Chemical Society).

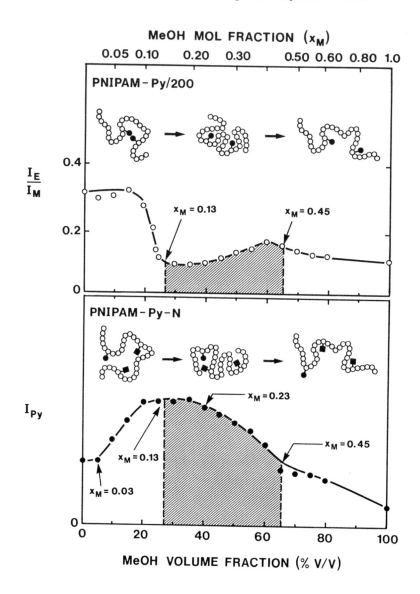

FIGURE 7

Plots of the ratio of excimer to monomer emission intensities (I_E/I_M) for a solution of PNIPAM-Py/200 (110 ppm) (top) and of the total pyrene fluorescence intensity for a solution of PNIPAM-Py/366-N/50 (44 ppm, $\lambda_{exc} = 290$ nm) as a function of solvent composition in methanol-water mixtures.

However there is a temperature-dependance of the solvent composition corresponding to the decrease in I_E/I_M on the methanol-poor side of the diagram. The transition compositions are listed in Table 3. Shown as well are the corresponding compositions at which PNIPAM solutions undergo this phase transition (taken from the phase diagram, Figure 6). There is a remarkable agreement among values determined by the two techniques at each temperature.

The abrupt drop of I_E/I_M on the methanol-poor side of the diagram reflects an increase in pyrene monomer emission at the expense of pyrene excimer emission, in much the same way as in the case of the thermally-induced transition of an aqueous PNIPAM-Py/200 solution. The changes in the excitation spectra are also the same in both situations. Thus the methanol-induced phase transition promotes a disruption of the pyrene aggregates which form in water and in water/methanol mixtures of low methanol content (e.g $x_M < 0.13$, 20°C). When the pyrenes are solubilized, they become isolated from each other in the hydrophobic polymer-rich phase which separates.

5.3. Doubly-labeled polymer (PNIPAM-Py/366-N/50).

The dependence on solvent composition of the extent of energy transfer between N and Py is presented in Figure 7 (bottom) for a PNIPAM-Py/366-N/50 solution at 20°C. The spectroscopic parameter monitored here is the total intensity of the pyrene emission due to excitation of the solution at 290 nm. This parameter is constant for $x_M < 0.03$; it increases gradually and reaches a maximum value at $x_M \simeq 0.13$; it remains constant for $0.13 < x_M < 0.23$, then decreases to reach its lowest value in pure methanol. The total pyrene emission monitored here originates from two sources: 1) directly excited pyrenes and 2) pyrenes excited by energy transfer from excited naphthalenes. Assuming that the first contribution remains small over the entire composition range, the significant increase in pyrene emission at $x_M \simeq 0.03$ reflects an increase in the efficiency of energy transfer between N and Py and, hence, signals a decrease in the dimensions of the polymer chain. The composition range, $0.13 < x_M < 0.23$, for which the pyrene emission is maximum corresponds to the composition domain in which PNIPAM is insoluble, at 20°C, as determined from the phase diagram (Figure 7). Thus the fluorescence of the doubly-labeled polymer provides strong evidence for the occurrence of 1) a collapse of the chain from an expanded coil to a more compact globule at the methanol-poor transition and 2) a swelling of the polymer dimensions during the second transition in the methanol-rich composition. Between the two transitions some reorganization takes place in the aggregated polymer phase, as evidenced by small changes in the energy transfer efficiency.

6. CONCLUSIONS

The investigations of the fluorescence of labeled PNIPAM reported here provide a new insight into the LCST phenomenon. They give strong support to a two-step mechanism. First, polymer chains collapse upon themselves from flexible coils into more compact globules. This collapse is followed by aggregation of individual globules into a polymer-rich phase. In the distance scale of 500 to 1000 Å probed by light scattering experiments, the chain collapse appears to take place abruptly. Fluorescence experiments detect also this contraction of the polymer dimensions. However, in the 30 to 60 Å distance scale probed here, gradual fluctuations in polymer size occur, it seems, well below the transition temperature. This molecular description of the LCST phenomena needs to be clarified further. All the present results concur to highlight a remarkable similarity in the phase transition processes, be they heat-triggered or induced by changes in solvency.

ACKNOWLEDGEMENTS

I would like to thank Mr. J. Venzmer (University of Mainz, Mainz, FRG) for his help in the measurements of cloud points and Professor M. A. Winnik (University of Toronto, Toronto, Canada) for his constructive comments on this manuscripts during the redaction process.

REFERENCES

1.)   M. Heskins and J. E. Guillet, Macrom. Sci., Chem. A2 (1968) 1441.

2.)   F. M. Winnik, H. Ringsdorf, and J.Venzmer, Macromolecules 23 (1990) 2615.

3.)   H. G. Schild and D. A. Tirrell, J. Phys. Chem. 94 (1990) 4352.

4.)   For reviews, see for example: Polymers in Aqueous Media, Advances in
        Chemistry Series 223, ed. J. E. Glass (Am. Chem. Soc., Washington, 1989).

5.)   J. B. Birks, Photophysics of Aromatic Molecules (Wiley-Interscience, London,
        UK, 1970) chapter 7.

6.)   F. M. Winnik, Macromolecules 23 (1990) 233.

7.)   F. M. Winnik, Polymer (1990) in press.

8.)   W. H. Melhuish, J. Phys. Chem. 65 (1961) 229.

9.)   R. Rusakowicz and A. C. Testa, J. Phys. Chem. 72 (1968) 2680.

10.)   G-J. Liu and J. E. Guillet, Macromolecules 23 (1990) 1388.

11.)   F. M. Winnik, Macromolecules, 20 (1987) 2745.

12.)   The digits in the polymer designations refer to the average number of N-
        isopropyl units per chromophore, e.g. PNIPAM-Py/200 has on average 1 pyrene
        per 200 N-isopropyl groups.

13.)   F. M. Winnik, M. A. Winnik, S. Tazuke, and C. Ober, Macromolecules 20 (1987)
        38; H. T. Oyama, W. T. Tang, and C. W. Frank, Macromolecules 20 (1987) 474.

14.)   I. Yamazaki, F. M. Winnik, M. A. Winnik, and S. Tazuke, J. Phys. Chem. 91
        (1987) 4213.

15.)   T. Förster, Discuss. Farad. Soc. 27 (1959) 7; J. R. Lakowicz, Principles of
        Fluorescence Spectroscopy (Plenum, New York, 1982) chapter 10.

16.)   L. D. Taylor and L. D. Cerankowski, J. Polym. Sci., Polym. Chem. Ed. 13 (1975)
        2551.

17.)   S. Fujishige, K. Kubota, and I. Ando, J. Phys. Chem. 93 (1989) 3311; K. Kubota,
        S. Fujishige, and I. Ando, J. Phys. Chem. 94 (1990) 5154.

18.)   J. N. Israelachvili, Intermolecular and Surface Forces (Academic Press, London,
        1985).

19.)   S. Fujishige, Polymer J. 19 (1987) 297.

Photochemical Processes in Organized Molecular Systems
K. Honda (Editor-in-Chief)
© Elsevier Science Publishers B.V., 1991

POLYMER DIFFUSION IN LATEX FILMS

Mitchell A. Winnik, Yongcai Wang, Cheng-Le Zhao*

Department of Chemistry and Erindale College, University of Toronto, Toronto, ON, M5S 1A1, Canada

This paper examines polymer diffusion in films prepared from poly(butyl-methacrylate) [PBMA] latex and annealed for various lengths of time. The films were prepared from equal amounts of phenanthrene- and anthracene-labeled latex. Polymer diffusion was followed by measuring the growth in non-radiative energy transfer by a fluorescence decay method. Diffusion coefficients were calculated from the data and found to be in the range of 10^{-12} to 10^{-16} cm^2 s^{-1}.

1. INTRODUCTION

When aqueous dispersions of soft latex particles are allowed to dry, transparent films are formed.[1] From a conceptual point of view one can think of this process leading first to close-packed spheres, with subsequent water evaporation providing sufficient force via surface and osmotic effects to deform the particles into a tight-packed structure.[2] This process is depicted in Figure 1.

One of the curious features of latex films is that the mechanical properties and barrier properties of the films continue to evolve long after all the water has evaporated. This process of "further coalescence" is an important aspect of all latex coatings and represents an important feature of these coatings that one would like to understand. Many years ago, Voyutskii[3] suggested that this physical aging involved polymer diffusion across the particle boundaries. If this were the case, latex film formation would share similarities with other processes such as sintering of polymer powders, thermal welding of polymer slabs, and healing of cracks, where the development of mechanical properties at the joint is related to the diffusion of polymer molecules across an interface.

Early experiments to test the Voyutskii idea were ambiguous. Transmission electron microscopy [TEM] studies of nascent and aged films sometimes showed the disappearance of particle boundaries,[4] and, in other systems, showed that the particle boundaries persisted for very long periods of time. For example

---

*Current address: BASF Canada, 453 Christina Street, Sarnia, Ontario, Canada, N7T 7Z1

Distler and Kanig[5] studied latex films composed primarily of poly(butyl methacrylate) [PBMA]. These films were aged, sectioned, stained. TEM studies showed that the particle boundaries persisted for months, and the film, when stretched, showed nearly affine particle deformation. There is an obvious temptation to believe that persistence of the particle boundaries in the TEM micrographs implies the absence of interparticle polymer diffusion. Since the Distler and Kanig recipe included small amounts of N-hydroxymethylacrylamide, which can crosslink at the particle surface, one also suspects that the extent of polymer diffusion may be very sensitive to the exact composition and microstructure of the latex.

What one needs, in this situation, is a technique with which to measure directly polymer diffusion. Such experiments require labeled polymers. The first such result was reported by Hahn[6] who mixed deuterated and ordinary PBMA latex and followed polymer diffusion in the film by neutron scattering [SANS]. At almost the same time, we reported the use of direct non-radiative energy transfer [DET] experiments, in conjunction with donor and acceptor dyes to study diffusion in melt-pressed PMMA latex films.[7] This type of experiment allows us to examine many aspects of polymer diffusion in traditional latex films, and also permits us to compare the SANS and DET method for the study of interparticle polymer diffusion.

In this paper we describe the DET method and show that it is a very effective technique for following polymer diffusion across interfaces. To interpret the diffusion measurements quantitatively, one needs to calculate diffusion constants D, and for this purpose, one has to have recourse to a model. We will show, for example, that the data can be fit very effectively to a model based upon Fick's laws of diffusion. These yield D values which depend upon the extent of mixing, but if values at different temperatures are compared at identical extents of mixing, the D values give an excellent fit to the Williams-Landel-Ferry [WLF] expression. On the other hand, one has to consider the possibility that at least the initial stages of the diffusion process are non-Fickian because of the influence of the interface in nascent films on the conformation of nearby polymer chains. Here we see intriguing differences between latex films composed of short chains which are not entangled, and those containing much higher molecular weight polymer.

2. EXPERIMENTAL

2.1. Latex preparation and characterization

Latex samples were prepared by semi-batch emulsion polymerization as previously described.[8] Potassium persulfate served as an initiator, sodium dodecyl sulfate as the surfactant. For the preparation of the low molecular weight samples, dodecanethiol (10 mg/g BMA) was added as a chain transfer agent. To prepare

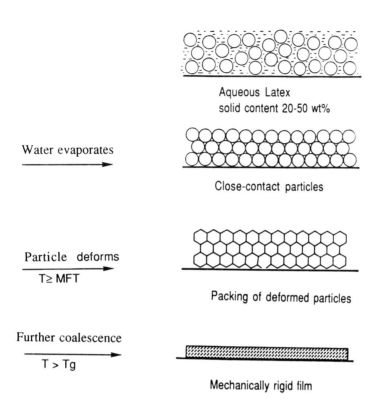

FIGURE 1. A pictorial view of the formation of a film produced by water evaporation from a dispersion of soft latex particles.

Table 1. Size and Composition of Labeled Poly(butyl methacrylate) Latex Particles [a]

Sample	d (nm)[b]	M_w[c]	M_n[c]
Phe-PBMA-1	135	6.3×10^5	2.5×10^5
An-PBMA-1	116	5.5×10^5	1.7×10^5
Phe-PBMA-2	114	8.9×10^4	2.5×10^4
An-PBMA-2	117	6.7×10^4	3.0×10^4

a. Chromophore content in all samples is ca. 1 mol%.
b. Particle diameter.
c. By GPC, based upon narrow molecular weight PMMA standards.

donor labeled particles [Phe-PBMA], 9-phenanthrylmethyl methacrylate (1 mol%/BMA) was added to the monomer feed. For acceptor-labeled particles [An-PBMA], 9-anthryl methacrylate (1 mol%) was added. Pertinent latex characteristics are presented in Table 1. To examine the effects of a coalescing aid, weighed quantities of 2,2,4-trimethyl-1, 23-pentanediol monoisobutyrate (Texanol, Eastman Kodak) were added to measured volumes of the dispersion (37 wt% solids) and stirred gently for 48 hr at 22° to promote equilibration of the system.

2.2. Film preparation and annealing

Films were prepared at ca 26°C by mixing equal volumes of Phe- and An-labeled latex (each 37 wt% solids), and coating each mixture onto a small quartz plate (ca 5mm x 10mm). After drying 1 hr, the first fluorescence decay trace was measured, and the film was placed on a preheated block of aluminum in a microprocessor-controlled oven. The oven was filled with nitrogen, which was circulated over the samples as they annealed. The film samples were removed periodically, cooled quickly to room temperature, placed in quartz test tubes and flushed for a few minutes with argon before the fluorescence decay traces were measured.

2.3. Fluorescence measurements

Fluorescence decay profiles were obtained by the single photon timing technique.[9] Phe was excited at 296 nm and its emission was monitored at 366 nm. The decay traces were fit to models described in the text.

3. MEASURING POLYMER DIFFUSION

Measuring polymer diffusion during latex film formation requires labeled latex particles.

To employ DET measurements one needs a mixture of particles labeled with appropriate fluorescent groups, one with a dye such as phenanthrene which can act as an energy donor, and the other with a dye such as anthracene which can act as an energy acceptor. The DET process can be represented as the reaction

$$\text{Phe}^* + \text{An} \quad \rightarrow \quad \text{Phe} + \text{An}^*$$

where Phe* is the excited phenanthryl groups produced upon excitation with uv light at 290 nm. In our experiments we used the polymerizable methacrylate derivatives 1 and 2 shown above to prepare our labeled particles by emulsion polymerization.

The molecular aspects of polymer diffusion across the particle-particle boundary are shown below in cartoon form. As long as the Phe and An dyes are in separate particles, energy transfer can only occur across the particle interface. One of the interesting features of energy transfer via the dipole-dipole [Förster] mechanism is that it can occur over distances which are large compared to the chromophore dimensions and small compared to the particle diameters.

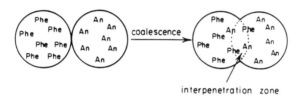

interpenetration zone

3.1. Energy Transfer Kinetics

According to the Förster theory[10,11] of energy transfer by the dipole-coupling mechanism, the rate of energy transfer for isolated donor-acceptor pairs varies as

$$W(r) = \frac{3\kappa^2}{2}\left(\frac{R_o}{r}\right)^6 \frac{1}{\tau_D} \tag{1}$$

where κ^2 is an orientation factor, τ_D is the unquenched donor lifetime, r is the distance between the centers of the transition dipoles, and R_o is a characteristic distance which sets the span of the experiment. The magnitude of R_o, here 26Å for the Phe/An pair, depends exclusively on the spectroscopic properties of the donor-acceptor (D/A) pair. The donor fluorescence decay profile is sensitive to the D/A pair distribution $\Omega(r)$

$$I_D^{ET}(t) = \int_0^\infty \Omega(r) \exp(-W(r)t) \, dr \tag{2}$$

For randomly distributed acceptors in three dimensions and in the absence of donor-donor energy migration, one obtains

$$I_D(t) = B_1 \exp\left(-\frac{t}{\tau_D} - p\left(\frac{t}{\tau_D}\right)^{\frac{1}{2}}\right) \tag{3}$$

$$p = 4\pi^{3/2} N_A R_o^3 [A]/3000 \tag{4}$$

where N_A is Avogadro's number and [A] is the bulk concentration of the acceptor.

In a polymer diffusion experiment, one has the option of introducing an appropriate distribution function $\Omega(r)$ deduced from a diffusion model into equation (2), and fitting the diffusion coefficient as an adjustable parameter. Alternatively, one can make a simplifying assumption about the pair distribution and fit the $I_D(t)$ profile in such a way as to calculate the mass transferred across the interface. The former approach is seemingly more rigorous (and offers the possibility of finding from the experiment the best parameters describing $\Omega(r)$), but has two disadvantages. It makes the entire data analysis model-dependent, and it introduces additional parameters to be fitted in the $I_D(t)$ decay profile. As a consequence, we have taken the second approach and treat the diffusion process as though it generated two populations of donor, a "mixed" population and an as yet "unmixed" population. The details of this model are shown in Figure 2. In this model, the unmixed donor population contributes an exponential term with a prefactor B_2 to the donor decay, so that

$$I_D(t) = B_1 \exp\left(-\frac{t}{\tau_D} - p\left(\frac{t}{\tau_D}\right)^{\frac{1}{2}} \right) + B_2 \exp\left(-\frac{t}{\tau_D} \right) \tag{5}$$

Equation (5), with its three fitting parameters is remarkably effective at fitting both experimental and simulated decays. In mixtures of labeled hard sphere (PMMA) particles, it yields $B_1 = 0$, and for solvent cast films (uniform mixing), it gives $0 \leq B_2 \leq 0.04B_1$. This equation allows us to relate B_1 and B_2 to the amount of polymer transferred across the interface. For equal volumes of D- and A- labeled particles, the apparent volume fraction of mixing $f_m'(t)$ is given by the expression

$$f_m'(t) = \frac{B_1}{B_1 + B_2} \tag{6}$$

which needs to be corrected for the contribution of energy transfer across the interface to the $I_D(t)$ signal,

$$f_m(t) = \frac{f_m'(t) - f_m'(0)}{f_m'(\infty) - f_m'(0)} \tag{7}$$

Note that $f_m'(\infty)$ is essentially unity and, for 100 nm particles, $f_m'(0) \approx 0.15$. A somewhat different approach to analyzing energy transfer measurements of polymer

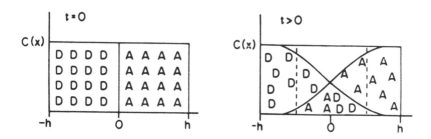

FIGURE 2. Depiction of the initially formed interface (left side) and of the evolution of the label distribution due to polymer diffusion (right side). The dashed vertical lines effectively divide the system into mixed and unmixed regions and simplify the data analysis.

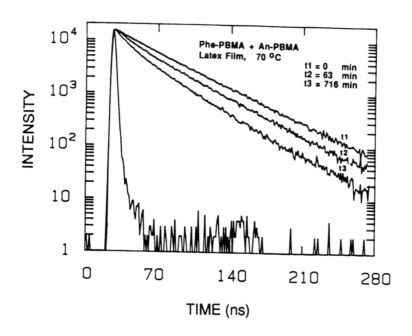

FIGURE 3. Fluorescence decay profiles of PBMA latex films annealed at 70°C for different times.

diffusion across a planar interface was developed several years ago by Tirrell, Adolf, and Prager,[12] following a suggestion by Monnerie. They conceived of the problem in terms of steady-state fluorescence measurements of the intensity I_A^{ET} of acceptor emission due to energy transfer. This is the type of measurement pioneered by Morawetz and his coworkers[13] for application to a wide variety of polymer problems. I_A^{ET} is proportional to the quantum efficiency of energy transfer (ϕ_{ET}) which is in turn proportional to the integrated D/A pair distribution that has evolved due to mixing through polymer diffusion across the interface. If the x axis is normal to the interface, then

$$I_A^{ET} \; \alpha \int_{-\infty}^{\infty} \Omega(x) \, dx \qquad\qquad (8)$$

with $\Omega(x) = C_A(x,t)C_D(x,t)$, the product of the concentration profiles of A- and D-labeled polymers as depicted in Figure 2.

It is worth noting that $B_1/(B_1 + B_2)$ is also a measure of the quantum yield of energy transfer. The energy transfer quantum yield ϕ_{ET} is defined by

$$\phi_{ET} = I_{abs}^{-1} \int_0^{\infty} I_A^{ET}(t) \, dt \qquad\qquad (9)$$

where $I_D^{ET}(t)$ is given by equation (2) and I_{abs} is the photon intensity absorbed by the sample. Quantum yield determinations of non-radiative energy transfer are a viable alternative to our decay profile measurements except for the very delicate problem of correcting for the contribution of radiative energy transfer in the experiment.[14] It was specifically to avoid this problem that we turned to the transient techniques.

3.2. Evidence for Polymer Diffusion

The fluorescence decay profiles from one set of film formation measurements is shown in Figure 3. There are two time scales to consider, the annealing time (hours) and the fluorescence decay time (nanoseconds). In the nascent film (top curve), the log I vs t plot is linear to the eye, and the $f_m'(0)$ value is small. This film sample was heated in an oven at 70°C. It was removed periodically, cooled to room temperature, and its fluorescence decay profile was measured. One sees in Figure 3 that with increased annealing, the fluorescence decay profiles become steeper at early times, indicating an increase in the extent of energy transfer. At long times (200 ns), the decays are essentially parallel. In this time domain it is the fluorescence of Phe from the unmixed domains which makes the largest contribution to the signal. This polymer diffusion rate is very sensitive to temperature, and at 22°C no significant polymer diffusion occurs in these films even on the time scale of months. Removing film samples from the oven is sufficient to cause polymer diffusion to cease, and the fluorescence decay profile provides a snapshot of the extent of mixing in the sample. Figure 3 emphasizes that we use measurements on the time scale of nanoseconds to follow diffusion processes which occur on a time scale of hours to days.

From the fluorescence decay data and equation (7) f_m values can be calculated. In Figure 4 we show examples of the time profile of f_m for films prepared from the lower molecular weight latex samples, Phe-PBMA-2 + An-PBMA-2 described in Table 1. These values increase with time, and for the sample annealed at 90°C, mixing is nearly complete within an hour. At 70°C the rate of mixing is much slower, indicating an important temperature effect on the diffusion process. The lower most curve in Figure 5 shows the time profile of f_m for a film composed of the higher molecular weight latex (cf Table 1) and annealed at 90°C. The slow rate of mixing observed here indicates that molecular weight effects on polymer diffusion during film aging are also very important.

3.3. Coalescing Aids

When latex paints are formulated, small amounts of certain organic solvents are added to improve the film-forming properties of the coating. The coalescing solvent acts as a plasticizer for the latex particle, and lowers the minimum temperature [MFT] at which the dispersion will form a film upon drying. Models for the film formation process suggest that the solvent lowers the elastic modulus of the latex particles.[1] This results in an increased ease of polymer flow; and promotes coalescence during the latter stages of film formation when the particles are in intimate contact.

In order to test the effect of coalescing solvents on the extent of polymer diffusion in latex films, we treated two samples of a PBMA latex dispersion (37 wt% solids) with different amounts of 2,2,4-trimethyl-1, 3-pentanediol monoisobutyrate (TPM, Texanol (Kodak)). This substance was chosen because of its relatively low volatility and its strong tendency to partition into the latex phase. The dispersions were stirred for 48hr at room temperature to allow the additive to equilibrate within the system, and films were cast onto quartz plates. Once dry (1hr), the films were subjected to fluorescence decay measurements to determine $f_m'(0)$ and then were placed in an oven under nitrogen at 95°. For comparison purposes, a film was prepared from the same latex dispersion with no TPM added. The results of these experiments are shown in Figure 5.

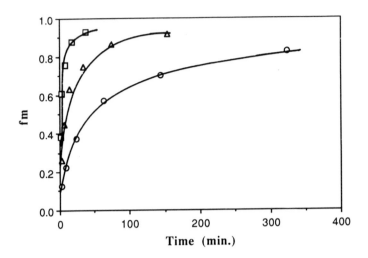

FIGURE 4. Volume fraction of mixing f_m as a function of time for films prepared from latex samples Phe–PBMA–2 + An–PBMA–2, and annealed at (0) 70^0C, (Δ) 80^0C, and ($\square$) 90^0C.

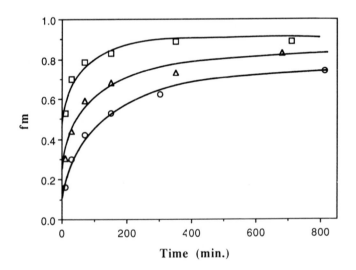

FIGURE 5. Volume fraction of mixing f_m as a function of time for annealed films (90^0C) prepared from latex samples Phe–PBMA–1 + An–PBMA–2 in the presence and absence of TPM: (0) no additive, (Δ) 3wt% TPM, ($\square$) 6wt% TPM.

In these films one observes significant polymer diffusion on the time scale of 30 minutes, which is much more pronounced in the presence of the additive (upper two curves in Figure 4) than in its absence. The amount of TPM added is 1.1 wt% based upon latex solids (middle curve) and 2.2 wt% (upper curve). These values correspond to 3.3 and 6.6 vol% of the additive in the dispersion and 1.2 and 2.4 vol% in the film.

4. DIFFUSION COEFFICIENTS

The parameter of most interest and of greatest theoretical significance is the mean translational diffusion coefficient for the polymer chains in the latex. These diffusion coefficients can be obtained by fitting the f_m data to a diffusion model. One must make some choices about the diffusion model to use. Here we assume that the molecular mixing satisfies Fick's Laws of diffusion. In the nascent film, the particle interfaces are locally planar. As a consequence, one might employ a model for tracer diffusion across a plane. At longer times, the mass diffusion out of a particle will approach spherical symmetry. Here a spherical diffusion model is more appropriate. Fortunately, both models make identical predictions for diffusion over distances up to that of the particle radius as long as one recognizes that in the planar diffusion model, one still has independent diffusion in three dimensions, and that the diffusion coefficient is related to the mean-squared diffusion distance as $\langle l^2 \rangle / 6$.

In the spherical model, the concentration $C(r,t)$ at radius r and time t is given by

$$C(r,t) = \frac{C_0}{2}\left(\text{erf}\left(\frac{R+r}{2\sqrt{Dt}}\right) + \text{erf}\left(\frac{R-r}{2\sqrt{Dt}}\right)\right) - \frac{C_0}{r}\sqrt{\frac{Dt}{\pi}}\left(\exp\left(-\frac{(R-r)^2}{4Dt}\right) - \exp\left(-\frac{(R+r)^2}{4Dt}\right)\right) \tag{10}$$

M_t, the amount of substance which has diffused across the boundary at time t, can be calculated through equation (5) provided D is known:

$$M_t = M_\infty - \int_0^R C(r)4\pi r^2 dr \tag{11}$$

where $M_\infty = (4/3)\pi R^3 C_0$. Experimentally, we determine f_m at different times t, set this value equal to M_t / M_∞, and carry out a numerical integration to find the best D value which satisfies equation (5).

M.A. Winnik, Y. Wang and C.-L. Zhao

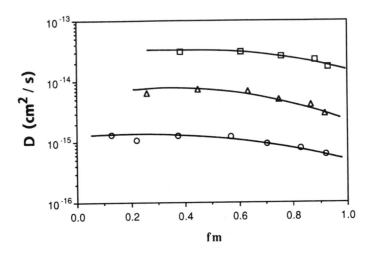

FIGURE 6. D values as a function of f_m for film samples composed
of Phe-PBMA-2 An-PBMA-2 calculated from the data in
Figure 4: (0) 70^0C, (Δ) 80^0C, and ($\square$) 90^0C.

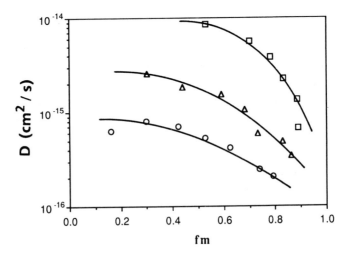

FIGURE 7. D values as a function of f_m calculated from the data in
Figure 5: (0) no additive; (Δ) 3wt% TPM; ($\square$) 6wt% TPM.

In Figure 6 we plot D as a function of f_m for the three samples shown in Figure 4, and in Figure 7 we present a similar plot for the three samples show in Figure 5. One sees that diffusion coefficients in the range of 10^{-14} cm^2 s^{-1} to 10^{-16} cm^2 s^{-1} are readily determined by the DET method. The dynamic range is significantly larger, and D values from 10^{-12} cm^2 s^{-1} to 10^{-19} cm^2 s^{-1} can be determined. These correspond to rms diffusion distances of 100Å/sec to 10Å/day.

A second feature is that the D values decrease with increasing time. The D values obtained from equations (10) and (11) are cumulative D values; the instantaneous values must decrease even further. One explanation for this change in D has its origin in the molecular weight dependence of D, and the molecular weight polydispersity of the latex. In this case, the diffusion of the low molecular weight components dominates the interface healing process at early times. As time evolves, the diffusion of polymers of increasing molecular weight contributes to the measured enhancement in energy transfer. The third feature is the influence of TPM on the polymer diffusion coefficients. At early times, 1.2 vol% TPM leads to a three-fold increase in D, and 2.4 vol% TPM leads to an order of magnitude increase in D.

There are strong suggestions in the literature that the diffusion of entangled chains across an interface will not be Fickian at early times.[15-17] Several factors play a role in affecting this diffusion. First there is the entropy loss suffered by chains adjacent to the interface, since confinement to a half-space restricts the number of available conformations. Second, reptation requires that chain ends diffuse first across the interface, so that the healing of the interface is essentially a sewing process for those chains initially within one radius of gyration of the boundary. These considerations lead to the predictions that in such systems, the mean penetration depth across the interface will vary as $t^{1/4}$ and that the total mass transferred will increase as $t^{3/4}$, instead of $t^{1/2}$ as predicted by Fick's Laws.

In our system, the mass transferred is proportional to f_m. We can explore the applicability of these ideas to polymer diffusion in latex films by examining the time dependence of f_m. For films prepared from low molecular weight latex a plot of f_m vs $t^{1/2}$ is linear (Figure 8) for f_m values up to 0.6. This observation substantiates the use of a simple Fickian diffusion model and suggests that over a substantial time period, the diffusion process is well-described in terms of a single mean diffusion coefficient. The deviations seen in Figure 8 at larger values of f_m are due to molecular weight polydispersity effects contributing to a decrease in D with increasing f_m.

For the higher molecular weight film sample, Figure 9, deviations from a $t^{1/2}$ dependence for f_m occur much earlier, indicating a slowing down of the polymer

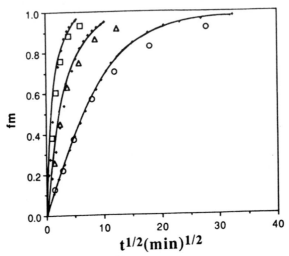

FIGURE 8. A plot of f_m vs $t^{1/2}$ for annealed films composed of
Phe-PBMA-2 + An-PBMA-2, compared to a numerical
evaluation of eqs. (10 and 11) using the values
indicated for D: (O) 70^{0}C, 1.3 x 10^{-15} cm^2 s^{-1}; (Δ)
80^{0}C, 6.6 x 10^{-15} cm^2 s^{-1}; (□) 90^{0}C, 3.05 x 10^{-15}
cm^2 s^{-1}. The open points are experimental data and the
lines refer to the numerical simulation.

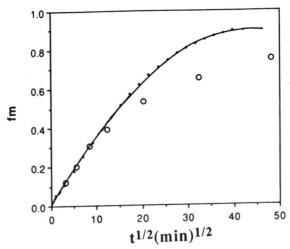

FIGURE 9. A plot of f_m vs $t^{1/2}$ for a film composed of Phe-PBMA-
1 + An- PBMA-1 and annealed at 90^{0}C. The line
indicated by the (+) symbols is calculated from eqs. (10
and 11) using D = 3.3 x 10^{-16} cm^2 s^{-1}.

diffusion with increasing extent of mixing. There are two possible explanations for this phenomenon. First, one expects a larger molecular weight dependence for entangled chains ($D \propto M^2$) than for the lower molecular weight film sample ($D \propto M^1$). In other words, the diffusion is Fickian but D decreases with increasing f_m because of molecular weight effects. Alternatively, confinement effects on the early stages of the diffusion process might operate to make the diffusion we observe non-Fickian. We find, for example, a linear plot of f_m vs $t^{1/4}$ up to $f_m = 0.6$ for the data in Figure 9. Note that this result is different from the $t^{3/4}$ prediction described above, and the $t^{1/4}$ plot may simply act to compress the data on the time axis in Figure 9 without revealing any new fundamental information about the diffusion process. Further experiments are necessary, and in this paper we examine the temperature and diluent effects on diffusion in terms of D values calculated from equations (10) and (11). Since the D values obtained in this way decrease with time, we are careful to compare values for identical magnitudes of f_m.

4.1. Free Volume and Diffusion

Viscoelastic properties of polymers are normally described very effectively by the Williams-Landel-Ferry [WLF] equation.[18] This expression relates temperature effects (above T_g) on the dynamic response of polymers to changes in free volume in the system. For diffusion coefficients, the WLF equation takes the form

$$\log\left(\frac{DT_0}{D_0 T}\right) = \frac{C_1(T - T_0)}{C_2 + T - T_0} \tag{12}$$

Here T_0 is an arbitrarily chosen reference temperature and D_0 is the diffusion coefficient determined at that temperature. In this expression C_1 and C_2 are parameters characteristic of a particular polymer. Their magnitude depends upon the choice of T_0. The WLF model allows one to recalculate the C_1 and C_2 parameters for any other reference temperature, and thus makes it possible to compare experiments carried out in different laboratories, using different techniques, and over somewhat different temperature ranges.

When we fit our diffusion coefficients to equation (12) we get linear plots as long as we are careful to fit D values taken at identical extents of mixing f_m. When we chose $T_0 = 373°K$, we obtain $C_1 = 15.5$, $C_2 = 254°K$, very close to the values of $C_1 = 14.5$, $C_2 = 255°K$ obtained by Child and Ferry[19] from creep compliance measurements on PBMA in bulk.

A corollary of the WLF equation is that Arrhenius plots of dynamic response data will be curved, but will appear linear over a limited range of temperatures.

The apparent activation energy E_a can be described by the expression

$$E_a = 2.302R'C_1C_2T^2/(C_2 + T - T_0)$$
(13)

where R' is the gas constant. The quantity E_a increases with decreasing temperature and at T_g takes the value $2.302\ R'C_{1g}\ T_g^2/C_{2g}\ =\ R'T_g^2\ \alpha_f\ /f_g^2$. Thus E_a is predicted to be independent of chemical structure except as reflected in T_g itself and in minor variations in the fractional free volume f_g and the thermal expansion coefficient α_f in the vicinity of T_g. At T_g, E_a is of the order of 62 kcal/mol if T_g - 200°K and 250 kcal/mol if T_g - 400°K.

In Figure 10 we construct Arrhenius plots of our diffusion data. We have chosen to compare D values associated with equal extents of mixing. These generate three essentially parallel lines and an apparent E_a of 38 kcal/mol. This value can be compared with results of dynamic mechanical measurements of PBMA by the Ferry group.[19,20] When they fit their data to the WLF equation they obtain an apparent activation energy for backbone motion which depends upon temperature and is characterized by a value of 37 kcal/mol at 100°C.

4.2. The Effect of Plasticizers

In order to quantify the effect of coalescing aids on polymer diffusion during aging of latex films, we construct a simple modification of the WLF equation. The introduction of the coalescent will increase the film free volume, and the polymer segments may expected to be more mobile in the presence of the diluent. The effect will be the same as raising the temperature to T', where $T' - T + \Delta T$, and ΔT can be called the temperature shift factor. The WLF equation then becomes

$$\log\left(\frac{D(w_1)T_0}{D_0(T + \Delta T)}\right) = \frac{C_1(T + \Delta T - T_0)}{C_2 + T + \Delta T - T_0}$$
(14)

Here the $D(w_1)$ is the measured diffusion coefficient at temperature T and concentration of coalescing aid w_1. Equation (14) says that the introduction of a small amount of coalescing aid is equivalent to an increase in the temperature for diffusion by ΔT. Combining equation (12) and (14) gives

$$\log\left(\frac{D(w_1)\ T}{D(T + \Delta T)}\right) = \frac{C_1(T + \Delta T - T_0)}{C_2 + T + \Delta T - T_0} - \frac{C_1(T - T_0)}{C_2 + T - T_0}$$
(15)

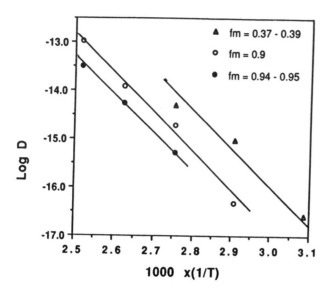

FIGURE 10. Plot of log D vs 1/T for the higher molecular weight
film sample. The slopes correspond to E_a = 38kcal/mol.

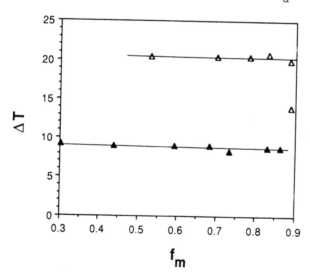

FIGURE 11. Shift factors ΔT calculated from the data in Figure 7
using equation (16), plotted vs f_m. Lower curve,
3wt% TPM; upper curve, 6wt% TPM.

The above equation can be simplified further when the diffusion temperature T is equal to the reference temperature T_o. For this experiment, $T - T_o - 368°K$.

$$\log\left(\frac{D(w_1)\, T}{D\,(T+\Delta T)}\right) = \frac{C_1\,\Delta T}{C_2 + \Delta T} \qquad (16)$$

In Figure 7 we observe that the effect of adding 1.1 wt% and 2.2 wt% TBM to PBMA is to increase the magnitude of D. The model developed above suggests that the three curves shown in Figure 7 can be shifted to generate a single composite curve, and the shift factor necessary can be calculated from equation (9). The shift factors calculated in this way are plotted in Figure 11 as a function of f_m. The fit of the data to equation (16) is excellent. This not only supports the validity of the model, it also indicates that no significant evaporation of TBM from the films occurs during the annealing of these films.

5. SUMMARY

Energy transfer measurements on latex films formed from a mixture of donor- and acceptor-labeled PBMA particles demonstrate the importance of polymer diffusion across the particle boundaries. This diffusion is particularly important during film aging. When the DET data are fitted to a Fickian diffusion model, diffusion coefficients can be calculated, and take values ranging from 10^{-12} to 10^{-19} cm^2/s. These D values exhibit a temperature dependence consistent with WLF behavior and the apparent activation energy (38 kcal/mol) is essentially identical to that reported by Ferry and his coworkers for PBMA as determined from traditional dynamic mechanical measurements. The effect of plasticizers (coalescing aids) on polymer diffusion can also be described in terms of a model based upon the WLF equation. The D values also exhibit a time dependence, which makes it necessary to compare D values obtained for equal extents of mixing.

ACKNOWLEDGEMENTS

The authors would like to thank the Institute of Chemical Sciences and Technology, the NSERC Canada, and the Province of Ontario for their financial support of this research.

REFERENCES

1) T.C. Patton, Paint Flow and Pigment Dispersion (Wiley-Interscience, New York, 2nd Ed., 1979) pp 194-197.

2) S.T. Eckersley and A. Rudin, J. Paint Technol. 62 (1990) 89.

3) S.S. Voyutskii, Autohesion and Adhesion of High Polymers (Wiley-Interscience, New York, 1963).

4) J.W. Vanderhoff, Br. Polym. J. 2 (1970) 161.

5) D. Distler and G. Kanig., Colloid Polym. Sci. 256 (1978) 1052.

6) (a) K. Hahn, G. Ley, H. Schuller and R. Oberthur, Colloid Polym. Sci. 264 (1986) 1092; (b) K. Hahn, G. Ley, and R. Oberthur, Colloid Polym. Sci. 266 (1988) 631.

7) O. Pekcan, M.A. Winnik, and M.D. Croucher, Macromolecules 23 (1990) 2673.

8) C.L. Zhao, Y. Wang, Z. Hruska, and M.A. Winnik, Macromolecules in press (1990).

9) D.V. O'Connor, D. Phillips, Time-Correlated Single Photon Counting (Academic Press, New York, 1984).

10) T. Förster, Discuss Faraday Soc. 27 (1959) 7.

11) See also J. Klafter, M. Drake, Molecular Dynamics in Restricted Geometries (Wiley, New York, 1988).

12) M. Tirrell, D. Adolf, and S. Prager, Springer Lecture Notes Appl. Math. 37 (1984) 1063.

13) (a) H. Morawetz, Science 240 (1988) 172; (b) H. Morawetz in Photophysical and Photochemical Tools in Polymer Science, ed. M. A. Winnik (Dordrecht, D. Reidel, 1986).

14) Y. Wang, J.T. Koberstein, Macromolecules 23 (1990) 3088.

15) P.G. de Gennes, C.R. Acad. Sci. (Paris) 291 (1980) 219.

16) (a) S. Prager, M. Tirrell, J. Chem. Phys. 75 (1981) 5194; (b) H.H. Kausch, M. Tirrell, Annu. Rev. Mater. Sci. 19 (1989) 341.

17) (a) R.P. Wool, K.M. O'Connor, J. Appl. Phys. 52 (1981) 5953; (b) Y.-H. Kim, R. P. Wool, Macromolecules 16 (1983) 1115; (c) R.P. Wool, B.-L. Yuan and O.J. McGarel, Polym. Eng. Sci. 29 (1989) 1340.

18) J.D. Ferry, Viscoelastic Properties of Polymers; 3rd Ed.(Wiley, New York, 1980).

19) (a) W.E. Child, J.D. Ferry, J. Colloid, Sci. 12 (1957) 327; (b) J.D. Ferry, S. Strella, J. Colloid, Sci. 13 (1958) 459.

20) S.J. Chen, J.D. Ferry, Macromolecules 1 (1968) 270.

Photochemical Processes in Organized Molecular Systems
K. Honda (Editor-in-Chief)
© Elsevier Science Publishers B.V., 1991

STEADY STATE INTRINSIC FLUORESCENCE STUDIES OF BLOCK COPOLYMER MICELLES

David A. Ylitalo, Alan S. Yeung, and Curtis W. Frank

Dept of Chemical Engineering
Stanford University
Stanford, CA 94305-5025

We present intrinsic excimer fluorescence data from the polystyrene block of a polystyrene-b-poly(ethylene propylene) block copolymer micelle system to interrogate the state of the micelle core. The excimer bandwidth and bandposition as well as the ratio of the excimer to monomer emission intensities prove to be sensitive measures of the environment around the excimer forming sites. As the bulk concentration is increased, the fluorescence data indicate a decrease in the volume fraction of solvent within the micelle core. As the temperature is increased, a T_g like phenomenon is observed in the micelle core around 60 - 80 °C without an apparent increase in the solvent content in the core. The excimer bandwidth proves to be an especially sensitive measure of the critical micelle temperature, and the thermodynamic parameters associated with the micellization process can be calculated. Finally, a proposal is made to explain the changes in the excimer spectral parameters as the local environment around the excimer forming sites is changed.

1. INTRODUCTION

1.1 Fluorescence

Fluorescence spectroscopy has long been a valuable tool for investigating phenomena on the molecular level. While the most recent trend has been towards more sophisticated nanosecond and picosecond time resolved fluorescence decay data, we present here additional ways of obtaining physical information from carefully done steady state measurements. With prudent use of spectral peak fitting routines we demonstrate the sensitivity of generally overlooked steady state spectral parameters, excimer bandwidth and bandposition, to gain insight into the polymer physics and thermodynamics of a block copolymer micelle system.

In the polymeric systems of most interest to us, two routes are generally available for utilizing fluorescence spectroscopy to elucidate molecular scale structure and morphology. The first method involves introducing a small amount of a fluorescent dye, either free or chemically attached to the polymer chains, into a host polymer system of interest. These experiments have included cyclization dynamics of end tagged polymers[1-3], excitation energy transport along randomly tagged chains[4,5], and anisotropy measurements of solubilized chromophores inside micelles[6]. These experiments have the advantage of being able to choose the chromophore and its density within the system but suffer from the fact that introducing guest molecules perturbs the initial system of study. Indeed, cyclization studies of pyrene end-tagged PEO in water are strongly affected by the influence of hydrophobic forces around the pyrene rather than the physics of the cyclizing chain[3].

The alternative approach which has been of more interest to us in polymer systems is to monitor the intrinsic fluorescence which is associated with many polymers. Ordering phenomena of polyimides during curing[7], changes in conjugation lengths of polyenes with temperature[8], and configurational changes of aromatic vinyl polymers with environment[9] are examples of studies where changes in the intrinsic fluorescence of the polymer reflects some morphological change. Although the number of systems which intrinsically fluoresce is limited, the interpretation of spectral changes can be made without regard for the behavior of a probe.

In our present system of polystyrene-b-poly(ethylene propylene) (PSPEP) block copolymers in linear alkanes, we monitor the intrinsic fluorescence of the phenyl rings on the polystyrene block to elucidate the morphological changes and thermodynamics of microphase separation. It has long been known that the broad structureless fluorescence band centered at about 330 nm in PS is the result of an excimer, a "sandwich type" dimer between an electronically excited phenyl ring and a neighboring ground state phenyl with a separation distance of about 3-4 Å. In PS, the principal process through which the excimers are formed is believed to be through singlet excitation from a lone phenyl ring and subsequent migration of the energy to traps of excimer forming sites (EFS)[10]. The ratio of the excimer to monomer fluorescence intensity ratios, (I_d/I_m) is then a measure of either the population of excimer forming sites or the efficiency of exciton migration to the excimer traps. It is therefore often difficult to interpret I_d/I_m unambiguously in terms of structure or morphology.

To gain more insight into the polymer physics of our system, then, we monitor the excimer bandposition and the bandwidth in addition to the I_d/I_m ratio. We show that the excimer bandwidth, especially, is extremely sensitive to changes in local environment around the EFS, in agreement with earlier work on solvent effects in poly (vinyl naphthalene) excimer formation[11]. We present spectral data, easily obtained with basic fluorescence equipment and judicious use of spectral deconvolution software, that are extremely valuable for observing physical phenomena in the PSPEP system.

1.2 Block Copolymer Micelles.

In our PSPEP block copolymer system, the alkane solvent is selectively poor for the PS block and selectively good for PEP. Over certain ranges of temperatures and concentrations, microphase separation of the insoluble PS block occurs to form spherical micellar structures similar to those formed by amphiphiles in aqueous solution. In contrast to their amphiphilic counterparts, which associate due to a positive entropy of micellization, block copolymer micelles are formed by the decrease in enthalpy associated with the aggregation of the insoluble blocks into a dense nucleus, thus minimizing contact with the solvent[12]. In the case of PSPEP, polystyrene forms the core of the micelle and is surrounded by a shell of expanded PEP chains. A model micelle is depicted in Figure 1.

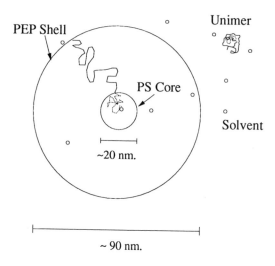

Figure 1. A model PSPEP micelle structure.

PSPEP block copolymer micelles have been observed from scattering and microscopy experiments to have a narrow size distribution with a radius of about 40 nm[13,18], indicative of a type of coordinated aggregation. In such cases it is assumed that a single equilibrium exists between free copolymer chains in solution and micelles with a fixed aggregation number. Systems which follow this equilibrium model are said to undergo a closed association[14]. Characteristic of this model is a critical micelle concentration (cmc) at and above which micelles exist and a critical micelle temperature (cmt) above which the micelles cease to exist and the block copolymers form a homogeneous solution.

The use of excimer fluorescence to study structure and morphology of block copolymer micelles is a powerful method due to its sensitivity to environments on the molecular level. It allows study of various aspects of the micelle core environment which are not directly measurable with the macroscopic techniques, such as scattering, microscopy, and viscometry, which are generally used. In the PSPEP system, the only direct measurement of the state of the micelle core has been the NMR work of Candau and coworkers[15]. In that work they determined that below 50°C the PS aromatic signal had two contributions, one from an effectively rigid component and one from a mobile component. Above 50°C the core of the micelles began to gain mobility, and around 80°C the micelles began to dissociate in octane.

In the PSPEP system in a lubricating oil, Price et al used electron microscopy and light scattering to determine the presence of micelles from -20°C to 150°C with narrow size distributions and spherical structure [13]. They additionally determined the critical micelle temperature for PSPEP in decane using membrane osmometry at elevated temperatures and found it to range from 70 to 100°C for concentrations of to 0.07 to 0.3%wt. Higgins performed small angle neutron scattering experiments at elevated temperatures and indicated that the micelles dissociated somewhere between 100°C and 160°C for concentrated solutions (> 2%)[17]. While these experiments could indicate a dissolution of the micelles, only in the case of the osmometry was a cmt given. With our fluorescence techniques we not only have the ability to simply measure micelle dissolution, but we can also present data on the state of the core of the micelle as a function of temperature.

The effects of bulk concentration on the state of the micelle core and micelle equilibrium was another area that motivated us to study the fluorescence behavior of this system. Higgins et al. showed SANS data which indicated that at high concentrations (10 %wt) an ordered macrolattice of micelles at cubic lattice sites existed[16]. In a more recent study, Higgins and coworkers proposed a model incorporating a hard core and a soft shell to represent interactions between neighboring micelles at higher concentrations[17]. To fit the experimental data, they had to assume that a fraction of the PS blocks had to lie outside of the micelle cores. Candau et al also reported findings from NMR studies of 2.7% PSPEP in octane that a significant portion of the NMR signal from the aromatic rings can be related to the mobility of PS segments[15].

These two studies are consistent with our recent dynamic light scattering, turbidity and viscometry work on the present system[18]. In that work, we detected the presence of a substantial amount of freely dispersed polymer chains at concentrations above a critical concentration of about 0.5% in heptane, an explanation which could account for the neutron scattering results of Higgins and the NMR results of Candau. Fluorescence measurements which could indicate the environment around the EFS reaffirm the proposal that a shift in the micelle-free chain equilibrium exists.

In undertaking our fluorescence studies on the PSPEP system we attempt to describe the changes in micelle core morphology and the micelle-free chain equilibrium discussed above as a function of concentration. Additionally, we are interested in changes in the micelle core as a function of temperature up to the cmt as well as characterizing the cmt using fluorescence techniques. These experiments capture the spirit of this symposium, "Photochemical Processes in Organized Molecular Assemblies", by utilizing the photophysical sensitivity of excimer fluorescence to interrogate the structure of these block copolymer assemblies.

2. EXPERIMENTAL METHODS

2.1 Materials and Sample Preparation.

Poly(styrene-b-ethylene propylene) block copolymer was provided by Shell Development Co. The polymer has a molecular weight of 1.35×10^5 and polydispersity of less than 1.15 as determined by GPC. The styrene content is 37% by weight. Repeated precipitation from tetrahydrofuran into methanol was undertaken to remove trace amounts of styrene and other impurities. Solvents heptane and dodecane were spectroscopic grade and were used as received.

In performing analysis on excimer data one must be certain that no quenchers are present which may retard migration efficiency to excimer traps, especially if these quenchers partition unequally between the micelle core and the dispersed phase. To this end, each sample was degassed by four freeze-pump-thaw cycles at 10^{-5} torr and vacuum sealed in quartz sample tubes. Sealed tubes are important so that the solution can be brought above the critical micelle temperature to completely dissolve and disperse the polymer chains without loss of solvent. We hold each sample at 150°C (40 - 70 °C above cmt.) for six hours and then cool slowly over eight hours to room temperature to allow the micelles to form in an equilibrium structure. Because some measurements are performed well below the T_g of PS it is possible that there are some non-equilibrium effects related to the rate of cooling, so we assume the thermal treatment allows for a consistent thermal history. It should be noted, however, that temperature experiments performed by heating to or cooling from above the cmt show no hysteresis effects when the sample has undergone the above treatment.

2.2 Fluorescence Measurements.

The fluorescence spectra were collected with a spectrofluorimeter that has been previously described [9]. Spectra were recorded with exciting light at 254 nm with bandwidths of 5 nm for both excitation and emission monochromoters. An Optronics Laboratories Standard of Spectral Irradiance Model 220A was used to correct for the spectrofluorimeter response function. Temperature experiments were done with resistance heaters, and at each temperature the sample was allowed to equilibrate for 30 min. The fluorescence spectra were resolved into a monomer peak assumed to have the same shape as the spectrum of 0.2 wt% sec-butyl benzene in heptane and an excimer peak assumed to be Gaussian on an energy scale. A Levenberg-Marquardt non-linear regression routine was used to reconvolve the peaks, and goodness of fit was based on inspection of the weighted residuals and analysis of variance. The program fit the monomer intensity and excimer intensity, bandwidth and bandposition.

3. RESULTS

3.1 Concentration Studies.

Most previous studies which have utilized excimer fluorescence as a tool for investigating polymer structure have made use of I_d/I_m exclusively to monitor configurational or morphological changes. At the same time, excimer bandwidth and bandposition have been neglected except in cases when the changes were substantial, such as the shift seen when going from the solid state to a liquid solution[19]. In most cases the excimer spectral changes resulting from subtle changes in local environment were either overlooked or were too small to be measured with confidence. One goal of this work is to make use of the bandwidth and the bandposition calculated by our peak fitting program in addition to I_d/I_m to monitor changes in environment in the PSPEP system. A representative PSPEP spectrum and the individual monomer and excimer peaks resolved by our program are shown in Figure 2.

Fluorescence spectra were collected for PSPEP in heptane and dodecane at concentrations from 0.001% wt. to 4% wt. at room temperature, as we have reported earlier[21]. As the concentration is increased there exist clear changes in the relative intensities of the excimer band at 330 nm to the monomer band at 285 nm; the ratio of the integrated areas of the resolved peaks are illustrated in Figure 3. It is first interesting to note that we observe no abrupt changes in I_d/I_m in the low

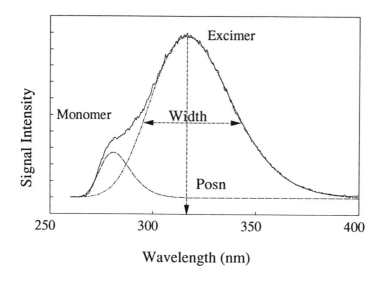

Figure 2. A typical polystyrene fluorescence spectrum with the resolved monomer and excimer peaks.

concentration region around the 0.002% wt. point where light scattering data indicated a cmc. One explanation for this can be made in terms of a photophysical argument based on PS segment density. The excimer emission is strongly affected by the PS segment density because of its effect on the excitation migration efficiency. Thus, if the creation of micelles is not accompanied by a large increase in the local density of phenyl rings, one would not see an abrupt change in I_d/I_m at cmc. However, this would be the case if the freely dispersed chains near the cmc existed in the form of unimolecular micelles with a PS core shielded somewhat by the PEP block, a possibility suggested by Tuzar et al.[20]. For PSPEP in cyclohexane, a theta solvent for PS where no micelles are formed, we found that I_d/I_m = 2.4 - 2.8. In the case of the PSPEP freely dispersed in heptane we would expect a slightly more collapsed PS block so that I_d/I_m may be even higher. Unfortunately we are not able to resolve whether or not we pass through the cmc at 0.002 %wt. or whether the change in local environment is too subtle to measure at room temperature.

From 0.01% to 0.5% of PSPEP, the prominent increase in I_d/I_m indicates an increased migration efficiency associated with an increase in PS segment density within the core or an increase in the population of excimer forming sites (EFS) due to a decrease in the stretching of the core block. At 0.5% in heptane and approximately 0.1% in dodecane I_d/I_m decreases dramatically. This occurs at precisely the same concentration that earlier viscosity, turbidity, and dynamic light scattering studies

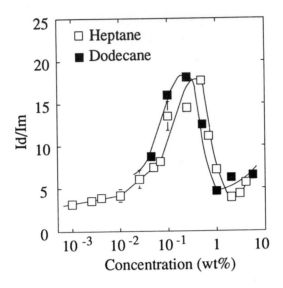

Figure 3. The concentration dependence of Id/Im for PSPEP in heptane and dodecane. From Ref [21].

showed anomalous results[18]. In that work we defined a critical concentration, c_I, above which inter-micellar interactions are significant and the micelle-free chain equilibrium is shifted to a higher number of free chains relative to micelles. This remains consistent with the present data, as I_d/I_m is expected to be lower for chains in the dispersed phase.

The concentration effect on the bandwidth obtained from the resolved excimer peaks is shown in Figure 4. A significant band narrowing is seen from 0.001% to 0.01%, indicating some change in environment around the EFS. Again at 0.5% bulk concentration for heptane and 0.1% for dodecane the bandwidth decreases, reflecting changes which occur at c_I.

Figure 5 shows the excimer bandposition as a function of concentration from our earlier work[21]. From c=0.001% - 0.2% we observe a substantial blue shift in the excimer peak position of approximately 700 cm^{-1}, consistent with what is observed with I_d/I_m and bandwidth. From 0.2% to 0.5% the bandposition is approximately constant, again similar to the bandwidth data. It is apparent that the bandwidth and bandposition reflect similar changes in the environment of the EFS. Finally, at c_I the bandposition shifts to the blue indicating further changes in the environment which occur at c_I.

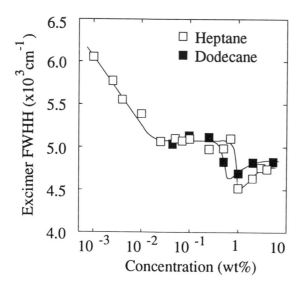

Figure 4. Concentration dependence of excimer bandwidth for PSPEP in heptane and dodecane. From Ref [21].

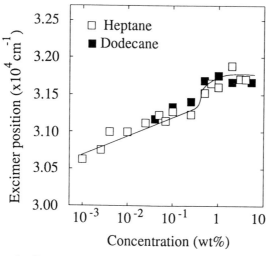

Figure 5. Concentration dependence of excimer position for PSPEP in heptane and dodecane. From Ref [21].

3.2 Temperature Measurements.

Because the boiling point of heptane (98°C) is in the proximity of the cmt, all temperature studies were done in dodecane in sealed quartz tubes. Due to the dramatic change in local environment around the PS block associated with the breakup of the micelles, the bandwidth, bandposition, and I_d/I_m should be sensitive measures of the cmt. Figure 6 shows the temperature dependence of the bandwidth for several concentrations in dodecane[21].

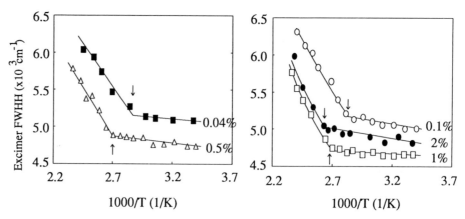

Figure 6. Excimer bandwidth vs 1/T for PSPEP in dodecane. From Ref [21].

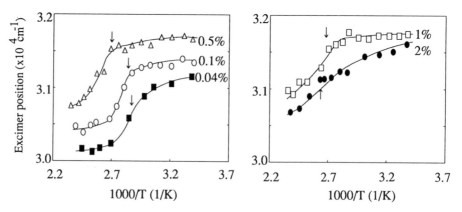

Figure 7. Excimer Position vs 1/T for PSPEP in dodecane. From Ref. [21].

In all cases the bandwidth broadens with increasing temperature and then abruptly changes slope at what we interpret as the cmt. As the micelles dissociate the EFS begin to see an increase in solvent contact which causes the broadening of the bandwidth. This is similar to the increase in bandwidth when the bulk concentration is decreased to near the cmc at room temperature.

The temperature dependence of bandposition is shown in Figure 7[21]. As the temperature is increased, the excimer peak shifts to slightly lower energy, due to either increased solvation within the core or to a decrease in the rigidity within the core which might allow the excimer forming sites to to form lower energy structures. This latter possibility has been suggested to explain shifts in excimer emission energy in rigid matrices where the EFS may not align properly[19]. At the cmt, the peak shifts considerably to lower energies, approaching that of low Mw PS in heptane at similar temperatures (30,700 cm^{-1} at 120 °C for 0.5% 2000 Mw PS).

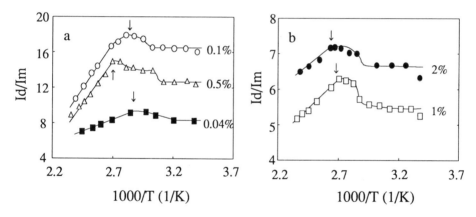

Figure 8. Id/Im ratio vs 1/T for PSPEP in dodecane. From Ref [21].

Figure 8 shows a plot of I_d/I_m vs 1/T for each concentration measured[21]. As the temperature is increased there is a rise in I_d/I_m at 60-80 °C, indicative of a possible rearrangement of PS within the micelle core which increases the population of EFS. At the cmt, depicted in the figure by the arrows, the relative amount of excimer emission decreases due to the decrease in PS segment density. This is in contrast to the cmc region at room temperature where dodecane and heptane are much poorer solvents for PS and the freely dispersed chains may exist as collapsed unimolecular micelles.

The determination of the cmt by excimer fluorescence allows us to follow the formalism proposed by Price [12] for estimating the standard free energy and enthalpy of micellization. For a closed association of micelle formation found in our system, the equilibrium constant between the free chain A_1, and the micelles, A_n, is given by

$$K_n = \frac{[A_n]}{[A_1]^n}$$

such that

$$\Delta G_n^0 = RTln[A_1] - \frac{RT}{n}ln[A_n]$$

where ΔG_n^0 is the standard free energy change per mole of copolymer on formation of an n-mer from unassociated chains.

For block copolymers where n is large ($\sim 10^2$), the second term may be neglected, and $[A_1]$ can be taken as the cmc. Thus,

$$\Delta G_n^0 = RTln(cmc)$$

Since the enthalpy change of micellization, ΔH^0, measured in a direct calorimetric study agrees with that obtained from the van't Hoff equation [12]

$$\Delta H^0 = -RT^2 \frac{d \ln(cmc)}{dT}$$

the enthalpy change of micellization can be estimated from

$$\ln(cmc) = \frac{\Delta H^0}{RT} + C$$

where C is a constant of integration.

A plot of ln cmc against 1/T from the previous paper is given in Figure 9[21]. From the slope of the plot we determine a value for the standard enthalpy of micellization ΔH^0 of -136 kJ per mole of copolymer chains. At 298 K, the standard free energy of micellization ΔG^0 is -52 kJ per mole and the standard entropy contribution -$T\Delta S^0$ is 84.3 kJ per mole. These values are in excellent agreement with those reported by Price [12]. From membrane osmometry they found that ΔH^0 = -130 kJ per mole and ΔG^0 = -42 kJ per mole for a similar PSPEP block copolymer in n-decane.

4. DISCUSSION

4.1 Fluorescence Spectroscopy

While it is apparent that the excimer bandwidth and bandposition as well as the I_d/I_m ratio are sensitive measures of local environment around the EFS, we are forced to speculation on interpreting the results. No theoretical work has been specifically done for the spectral parameters of PS excimers subject to changes in solvent or

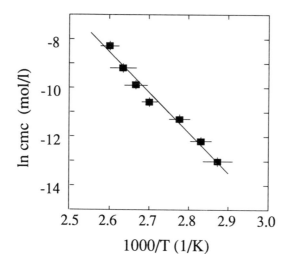

Figure 9. Ln (cmc) vs 1/T for PSPEP in dodecane. From Ref [21].

temperature; in fact, because of their low dipole moments, excimers and their fluorescence emission are generally believed not to be affected by solvation but rather by orientation and alignment affects alone[19]. This was found not to be the case for naphthyl polymers in a homologous series of alkyl benzene solvents where excimer bandwidth and band position were strongly solvent dependent [11]. We therefore put forth a qualitative argument to describe the shift and broadening of the excimer spectra seen in the PSPEP system.

While no theory has been written out for PS, Birks has presented a model to describe both bandwidth and bandposition of pyrene excimers in the crystalline state[23]. To do so, he describes the excimer forming process with an excimer interaction potential V' for two parallel molecules as a function of their separation distance. Figure 10 shows a schematic of such a potential energy diagram. V' represents the attractive excimer interaction potential, and R'+ hv_m is the repulsive van der Waals potential of the excited state molecule and the ground state molecule plus the energy of the 0-0 transition of the isolated molecule. R represents the repulsive intermolecular potential in the ground state. D' is the net excimer energy (V'+ R'). The energy of the excimer in the crystal can then be found by D' - R' at the equilibrium spacing of 3.53 Å for pyrene. Changes in the V' due to improper alignment or solvent effects then directly affect the excimer bandposition.

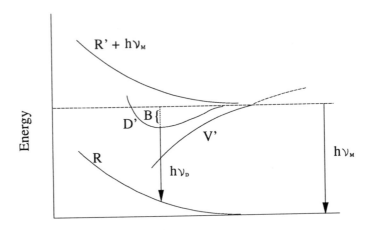

Figure 10.　The Birks excimer potential energy diagram.
Symbols are explained in the text.

To determine V' in a crystal, Birks describes it as

$$V' = V'_\infty - \frac{a}{r^3}$$

Birks and Kazzaz analyzed the excimer bandwidth of the pyrene crystal as a function
of temperature from 4K to 353K to determine the empirical parameters V'_∞ and a[23].
Noting that the potential energy curve was analogous to that of a color center in an
ionic crystal, Birks and Kazzaz used the expression by Williams and Hebb[24] for the
probability of emission as a function of energy P(E). The excimer was treated as a
harmonic oscillator along the separation axis q to give

$$P(E) = \left(\frac{k'}{2\pi kT}\right)^{1/2} \exp\left(\frac{-k'q^2}{2kT}\right)\frac{dq}{dE}$$

where E is the energy of fluorescence from an excimer at displacement q from the
equilibrium spacing and k' is the classical force constant of the oscillator. After
replacing the exponential with a quantum mechanical Slater sum over the
vibrational levels, Birks and Kazzaz could calculate the full width at half maximum
of the emission curve as

$$\Delta P(E) = \Delta P_0(E)\left[\coth\left(\frac{\theta'}{T}\right)\right]^{1/2}$$

with $k\theta'$ being equivalent to the zero point energy of the oscillator with the appropriate frequency. Fitting the experimental bandwidth data to the above gives $\Delta P_0(E)$ and θ', and using the harmonic oscillator model they can then calculate V'_∞ and a for the excimer interaction potential.

While the Birks treatment is an excellent model for the pyrene crystal, it is not easily applied to our system of PS excimer fluorescence in solution. Birks assumes the excimer interaction to be in a vacuum with the surroundings serving the purpose of an isothermal heat bath only. It can not account for solvent effects, migration efficiency, or partial or imperfect overlap of the phenyl rings. Our earlier data on excimer spectral sensitivity of naphthyl polymers in alkyl benzene solvents have indicated that the excimer band position reflects the solvent stabilization of the excimer complex, and that the band width is a measure of solvent shell interactions with the complex [11]. From binding energy measurements in that system we can gain some insight into the potential energy diagram for the excimer. In the case of poly(1-vinyl naphthalene), the binding energy for the excimer complex goes through a minimum as the size of the solvent approaches that of the napthyl ring[25]. At the same time the ground state repulsive interaction was at a maximum for these solvents, and the entropy loss for excimer formation was at a minimum. As the solvent size became similar to the size of naphthalene, the solvent packing around the complex became more ordered, as indicated by the lower ΔS, and the better packing also increased the repulsive potential.

On the potential energy diagram, two possible scenarios can be imagined. If one assumes that the excited state and ground state molecules have identical repulsive potentials, the change in binding energy indicates a change in the slope of the excimer interaction potential V'. Similarly, if the repulsive interactions are affected by the excited state of the molecule, changes in V', R and R' could all occur with solvent, which seems likely due to the lack of a trend of B with band position.

In our PSPEP system, we can qualitatively describe the features seen in our spectral data with simple arguments from the above model. We expect the excimer binding energy to increase in the aliphatic solvent, as evidenced by the increase in B of toluene from 0.17 to 0.26 eV when going from the neat liquid state to hexane solution[22]. Because the excimer interaction energy difference for bringing together a pair of phenyl rings originally surrounded by solvent is greater than that for bringing together rings that are more stabilized by neighboring PS segments, the excimer interaction potential V' must be altered based on the environment around the EFS. In the case of the EFS surrounded by heptane, the interaction potential well is deeper than that for a PS environment, and the excimer emission is red shifted. The broadening of the bandwidth in heptane is the result of the increased slope of the repulsive potential, thereby allowing a broader range of transition energies to be accessible.

4.2 Concentration Dependence

In the low concentration regime of 0.001% to 0.1%, I_d/I_m, excimer bandwidth, and excimer bandposition indicate that the PS segment density in the micelle core is increasing. As the solvent content of the core decreases and phenyl-solvent contacts are replaced by phenyl-phenyl contacts around the EFS, the excimer bandwidth narrows and bandposition blue shifts. Additionally, the migration efficiency is increased as the PS density increases such that I_d/I_m increases. If the assumption is made that the I_d/I_m ratio is affected only by the increased PS segment density, we can utilize the 3-D Energy Migration Model of Gelles[26] to obtain a core volume fraction of PS. As the concentration increases, an estimation of the core PS volume fraction goes from 40% to 90%+ as the concentration is increased from the cmc to 0.5%[21]. Changes in the population of EFS associated with stretching of the PS blocks in the micelle core as well as substantial self absorption effects at higher concentrations could affect the I_d/I_m ratio and may therefore affect the actual values obtained from the migration model. The trend of increased PS volume fraction within the core, however, remains consistent with the bandwidth and bandposition data.

At c_I, however, neither self-absorption nor chain relaxation can explain the dramatic decrease in I_d/I_m or the further changes in excimer bandwidth and bandposition. Rather the argument made previously that the relative amount of free chains is increased provides explanation. The I_d/I_m data reflect an increase in the monomer fluorescence emission from the freely dispersed block copolymers. Interestingly, the blue shift of the excimer peak and band narrowing at c_I goes opposite to what is expected from EFS in a heptane environment. This indicates that excimer emission from the dispersed phase is insignificant, and that the EFS in the micelle core are in a more phenyl-like environment than at lower concentrations. At higher concentrations the bandwidth broadens slightly, possibly due to the increase in excimer contribution from the dispersed phase.

4.3 Temperature Dependence.

The spectral features of PS excimer emission have also proven to be sensitive to changes in core environment as a function of temperature. From room temperature to before the cmt, both excimer bandwidth and bandposition are relatively constant, indicating no significant changes in the solvent content of the core. The increase in I_d/I_m, then, at 60-80 °C reflects an increase in the population of EFS, possibly occurring as the result of increased chain mobility within the core. Candau has looked at a similar PSPEP in octane with NMR and attributed a sharp linewidth transition of the rigid component at 50 °C to increased mobility within the core, possibly by plasticization by solvent[15]. Since we do not observe any bandwidth or bandposition changes at this temperature we propose that this is a T_g type phenomenon in the micelle core. Experiments on quickly quenched, non-equilibrium micelle solutions show a bandwidth transition in this temperature region possibly related to the acquisition of sufficient mobility to rearrange to an equilibrium structure.

At the cmt, the sensitivity of excimer bandwidth to environment manifests itself in a sharp transition in the slope of the bandwidth vs 1/T. Candau has indicated that micelle dissociation occurs over a range from 100 °C to 160 °C for a 2.7% PSPEP solution[15]. The high temperature slope of the excimer bandwidth may then be a measure of the rate at which the micelles dissociate with temperature as the bandwidth approaches that of PS in heptane. The red shift in bandposition at the cmt is expected since a more stable excimer species can be expected with the increase in free volume for bond rotation. Again, the bandposition approaches that of PS in heptane, indicating a breakup of the micelles at high temperatures.

While the cmt is higher for more concentrated solutions as expected, the existence of free chains may complicate the thermodynamic analysis of the cmt. It is unclear whether the free chain-micelle equilibrium is altered such that the free chains are reincorporated into the micelles at some temperature below the cmt. The thermodynamic analysis indicates that the apparent shift in the equilibrium at room temperature does not affect the thermodynamics at higher temperatures. It is also interesting to note that the dissociation of the micelles at the cmt is gradual, in agreement with the results of Candau.

5. CONCLUSIONS

It has been demonstrated that the additional spectral parameters of excimer bandwidth and bandposition are sensitive measures of the local environment around the excimer forming sites of polystyrene. Together with I_d/I_m, these parameters show that the solvent content of the micelle core decreases with increasing bulk concentration and that at a critical concentration the micelle-free chain equilibrium is shifted towards the existence of more free chains. As the temperature of these micelle solutions is increased, the micelle core gains some additional degree of mobility around 60°C, in agreement with previous NMR data presented by Candau. The excimer bandwidth is especially sensitive to the breakup of the micelles and allowed us to measure a cmt with great precision. From those measurements, the thermodynamics of the micelle formation process were calculated and are in agreement with existing literature values. A simple qualitative argument based on the excimer interaction potential diagram of Birks allows us to account for the observed spectral shifts and bandwidth changes in terms of solvent effects on the excimer interaction potential and the pair repulsive interaction potential.

6. ACKNOWLEDGEMENTS

This work was supported by the Polymers Program of the National Science Foundation and in part by Shell Development Company.

7. REFERENCES

1.   Winnik,M.A., Redpath, T., Richards, D.H., *Macromolecules*, **13**,(1980), 328.
2.   Redpath, A.E.C., Winnik, M.A., *J. Am. Chem. Soc.*, **104**, (1982), 5604.
3.   Char, K., Frank, C.W., Gast, A.P., Tang, W.T., *Macromolecules*, **20**, (1987), 1833.
4.   Peterson, K.A., Fayer, M.D., *J. Chem. Phys.*, **85**, (1986), 4702.
5.   Holden, D., Guillet, J.E., *Macromolecules*, **13**, (1980), 289.
6.   Ediger, M.D., Donimgue, R.D., Fayer, M.D., *J. Chem. Phys.*, **80**, (1984), 1246.
7.   Wachsman, E., Frank, C.W., *Polymer*, **29**, (1988), 1191.
8.   Linton, J.R., Frank, C.W., *Macromolecules*, to be submitted.
9.   Fitzgibbon, P.D., Frank, C.W., *Macromolecules*, **14**, (1981), 1650.
10. Semerak, S., Frank, C.W., *Adv. Poly. Sci.*, **54**, (1983), 31.
11. Frank, C.W., *ACS Org. Coat. Plast. Chem. Preprints*, **45**, (1981), 433.
12. Price, C., Kendall, K.D., Stubbersfield, R.B., Wright, B., *Poly. Comm.*, **24**, (1983), 326.
13. Price, C., Kendall, K.D., Stubbersfield, R.B., Wright, B., *Poly. Comm.*, **24**, (1983), 200.
14. Price, C.,"Colloidal Properties of Block Copolymers", *Developments in Block Copolymers* ed. I Goodman, Elsevier, New York, (1982).
15. Candau, F., Heatley, F., Price, C., Stubbersfield, R.B., *Eur. Poly. J.*, **20**, (1984), 685.
16. Higgins, J.S., Blake, S., Tomkins, P.E., Ross-Murphy, S.B., Staples, E., Penfold, J., Dawkins, J.V., *Polymer*, **29**, (1988), 1968.
17. Higgins, J.S., Dawkins, J.V., Maghami, G.G., Shakir, S.A., *Polymer*, **27**, (1986), 931.
18. Yeung, A.S., Frank, C.W., *Polymer*, in press.
19. Martic P.A., Daly, R.C., Williams, J.L.R., Farid, S., *J. Poly. Sci. Poly Lett.*, **17**, (1979), 305.

20. Tuzar, Z., Kratochvil, P., *Adv. Coll. Int. Sci.*, **6**, (1976), 201.

21. Yeung, A.S., Frank, C.W., *Polymer*, in press.

22. Birks, J.B., "Photophysics of Aromatic Molecules", Wiley, New York, (1970).

23. Birks, J.B., Kazzaz, A.A., *Proc. Roy. Soc. A.*, **304**, (1968), 291.

24. Williams, F.E., Hebb, M.H., *Phys. Rev.*, **84**, (1951), 1181.

25. Frank, C.W., unpublished data.

26. Gelles, R., Frank, C.W., *Macromolecules*, **16**, (1983), 1448.

Photochemical Processes in Organized Molecular Systems
K. Honda (Editor-in-Chief)
© Elsevier Science Publishers B.V., 1991

INTERFACIAL CHARACTERISTICS OF DOPED POLYMER FILMS: TOTAL INTERNAL REFLECTION FLUORESCENCE SPECTROSCOPIC STUDY

Akira ITAYA and Hiroshi MASUHARA

Department of Polymer Science and Engineering,
Kyoto Institute of Technology, Matsugasaki, Kyoto 606, Japan

Fluorescence spectra and decay curves of pyrene in poly(methyl methacrylate) films, cast and spin-coated on sapphire substrates, were measured under total internal reflection and normal conditions and compared with each other. From these measurements, it was concluded that micropolarity around pyrene molecules in the vicinity of the sapphire/polymer interface is higher than that of the bulk and that an aggregate state of pyrene is different between the vicinity of the interface and the bulk. Similar phenomena were also observed for an air/polymer interface of the cast film. Influence of casting solvents on these phenomena was also investigated. Interfacial characteristics of molecularly doped polymer films are discussed, and the formation process of these films is considered.

1. INTRODUCTION

In 1980 one of the authors (HM) and the late Prof. S. Tazuke discussed the future development in photochemistry of organized molecular systems in the flight from Denver to Boston. Topics were mainly concerned with photophysics of polymer films, and the studies on carrier generation and its transport, excitation energy relaxation, and electron transfer were reviewed. Since these mechanism are dynamic in nature, it was quite reasonable to consider that photophysical processes should be followed in real time. Another important point we discussed is on homogeneity of the systems. Most of the published works have been made on the assumption that photoactive chromophores are homogeneously distributed in polymer films and that microenvironment around the chromophores is common for all positions in polymer films.

Surface and interface in polymer systems have received a great attention in recent science and technology. Conformation and orientation of polymers, and dopant distribution may change from the bulk to the interface, which results in characteristic physical and chemical properties of their solutions and solids. Powerful tools used for these studies are electron spectroscopy

such as ESCA, SIMS, RBS, *etc.*, and the dry surface can be
elucidated[1]. Optical spectroscopy is also fruitful in the studies
of solid/solid, solid/liquid, and liquid/vapor interfacial
systems. Particularly, attenuated total internal reflection
infrared spectroscopy and total internal reflection (TIR) Raman
spectroscopy have been applied to various kinds of polymer
systems[2,3]. However, all these methods cannot provide direct
information on dynamic process in the nano- and picosecond time
regions.

On the basis of these discussion, HM and Prof. S. Tazuke
considered that fluorescence spectroscopy, which has high time-
resolution, is the most fruitful tool for elucidating
photophysical processes in inhomogeneous polymer films. This was
realized by developing time-resolved TIR fluorescence
spectroscopy[4,5]. When light traveling in a material with higher
refractive index (n_1) is incident upon an interface at an angle
greater than the critical angle, although the light undergoes
total reflection, there is some penetration of the excitation
light into the material with the lower refractive index (n_2).
This evanescent wave can be used to excite selectively or
stimulate polymers which are located in the vicinity of the
interface. The fluorescence spectroscopy under total internal
reflection condition is probably the most sensitive one[6,7], and we
improved its time-resolution up to 10 ps[5]. Now time-resolved TIR
fluorescence spectroscopy is an effective tool for elucidating
molecular and electronic properties of polymer in the surface and
interface.

The TIR fluorescence spectroscopy was applied to poly(methyl
methacrylate) (PMMA) films doped with pyrene[8], which is the most
representative system studied for a long time[9,10]. This molecule
shows unique excited properties such as sensitivity of the
fluorescence vibronic structure to the environment[11] and has the
longest lifetime as the singlet excited state. Furthermore, it is
known that pyrene forms excimer in concentrated solution and that
it forms dimer or/and aggregate in a rigid matrix at high
concentration[10,12-17]. The change of micropolarity around pyrene
molecule and its aggregation near the interface have been
elucidated for the first time by applying time-resolved TIR
fluorescence spectroscopy.

2. EXPERIMENTAL

Sapphire was used as an internal reflection element (n_1 = 1.81 at 310 nm). The refractive index of PMMA films (n_2) is 1.53 at 310 nm, so that the critical angle (θ_c) given by $\sin\theta_c = n_2/n_1$ was calculated to be 57.7°.

Schematic diagram for the measurement is shown in Fig. 1. Fluorescence spectra and decay curves were measured under both optical conditions and compared with each other. An incident angle (θ) was adjusted to 68.0° for PMMA system, and the TIR phenomenon was confirmed by the naked eye.

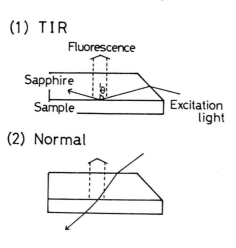

FIGURE 1
Schematic diagram of fluorescence measurement.
(1) Total internal reflection condition (TIR). θ : incident angle.
(2) Normal excitation condition (Normal).

3. RESULTS AND DISCUSSION

3.1. Micropolarity around Pyrene Molecules

Fluorescence spectra of films which were cast from chlorobenzene solution containing PMMA and a known amount of pyrene are shown in Fig. 2. These spectra were normalized at the third vibronic band (0-2) of the monomer fluorescence. The film with low pyrene concentration shows a structured monomer fluorescence spectrum. For high concentration sample, a red-shifted broad excimer band was observed in addition to the monomer one. It has been reported that the excimer fluorescence did not arise from microcrystalline but from pairs of molecule with a parallel configuration[10]. Here we notice the following two

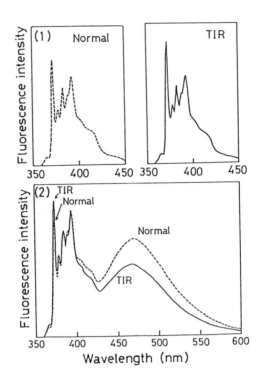

FIGURE 2

Fluorescence spectra of PMMA films doped with pyrene under the TIR and Normal conditions. Concentration: (1) 7.29×10^{-4} and (2) 5.55×10^{-2} mol/MMA unit mol.

points. One is that the fluorescence intensity ratio of the first (I) to the third (III) vibronic bands (I_I/I_{III}) is larger under the TIR condition than under the normal one for both samples. Another is that the intensity ratio of the monomer to the excimer fluorescence under the TIR condition is larger than that under the normal one.

Concerning the former, a dependence of the I_I/I_{III} ratio on the concentration of doped pyrene was investigated. The result indicated that the low intensity of the first band under the normal condition is not due to reabsorption and a superposition of the excimer fluorescence. It has been reported that the third vibronic band, which is strong and vibronically allowed, shows minimal intensity variation with polarity and the first band, which is forbidden vibronic one, shows a significant intensity enhancement in polar solvent. This is so-called Ham effect.

Thus, the intensity ratio of I_I/I_{III} of this molecule has been used as a probe for polarity of surrounding environments[11]. According to this viewpoint, the present result means that micropolarity around pyrene molecules in the vicinity of the sapphire/polymer interface is higher than that of the bulk. We consider that the carbonyl group of PMMA is the origin of the micropolarity and that the degree of its orientation around pyrene molecules is higher near the interface compared with in the bulk, as shown in Fig. 3. This interpretation was supported by the fact that the difference in the I_I/I_{III} ratio between both optical conditions was not observed for polystyrene films doped with pyrene.

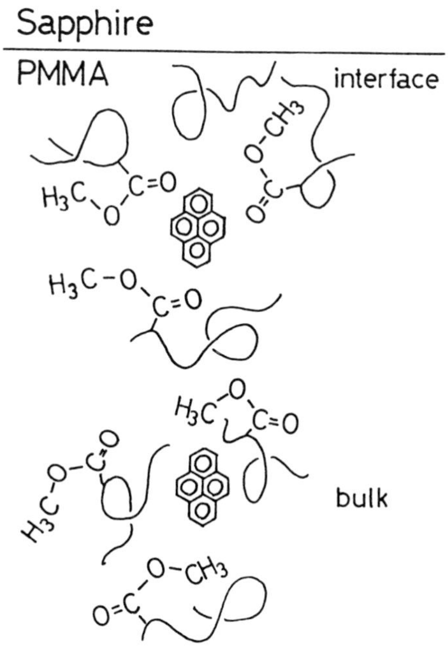

FIGURE 3
Model showing orientation of carbonyl groups of PMMA around pyrene molecules in PMMA films.

3.2. Pyrene Aggregation

As mentioned above, the intensity ratio of the excimer to the monomer fluorescence under the normal condition is larger than that under the TIR one. For the wide concentration range where the excimer fluorescence is observed, the same phenomenon was

observed.   This suggests that the concentration of excimer forming
pairs in the vicinity of the sapphire/polymer interface is lower
than that in the bulk.

The monomer and excimer fluorescence decay curves were measured
by monitoring at 374 and 520 nm, respectively, under the both
optical conditions.   Typical decay curves are shown in Fig. 4.
Most of them obey non-exponential functions, so that the 1/e
lifetime ($\tau_{1/e}$) of the fluorescence is used as a measure and
plotted against the pyrene concentration in Fig. 5.   The 1/e
lifetime under the TIR condition is smaller than that under the
normal one for all concentration.

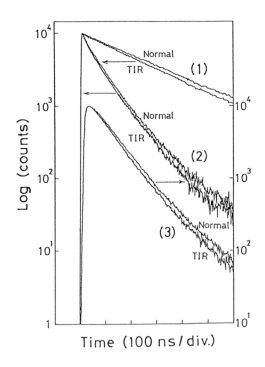

Time (100 ns / div.)

FIGURE 4
Fluorescence decay curves obtained by monitoring at 374 nm
(Concentration: (1) 2.48×10^{-3}, and (2) 5.58×10^{-2} mol/MMA unit mol)
and (3) 520 nm (Concentration: 4.90×10^{-2} mol/MMA unit mol).

Many years ago Avis and Porter reported that an increase in the
pyrene concentration in the same PMMA matrix leads to both an
increase in the excimer fluorescence intensity and a decrease in a
1/e lifetime of both the monomer and excimer non-exponential

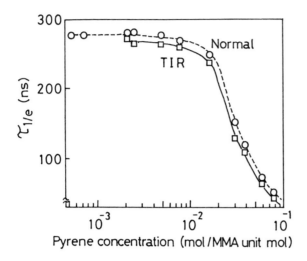

FIGURE 5

Dependence of the 1/e lifetime of the monomer fluorescence on the concentration of doped pyrene. Decay curves of low concentration films under the TIR condition could not be detected because of low fluorescence intensity.

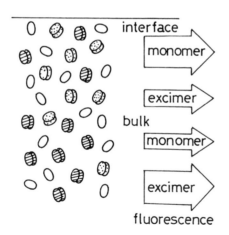

FIGURE 6

Model showing distribution of pyrene molecules in PMMA films.

fluorescence decays and that excimer forming molecules (pairs
whose geometry is close to the excimer configuration) and monomer
molecules undergo relaxation processes almost independently. The
decrease in the 1/e lifetime of both fluorescence was considered
to be attributed to a formation of dimer which is responsible for
quenching of the fluorescence[10]. According to this explanation,
the present results indicate that excited pyrene molecules near
the interface are more efficiently quenched by non-fluorescent
pyrene dimer, compared to that in the bulk. That is,
concentration of pyrene dimers leading quenching of the
fluorescence is higher in the vicinity of the interface than in
the bulk, while the concentration of excimer forming dimer is
lower in the vicinity of the sapphire/polymer interface as
compared with the bulk. This interfacial aggregation is
schematically shown in Fig. 6.

3.3. Air/polymer interface

We have also investigated the fluorescence behavior in the
vicinity of an air/polymer interface. The difference in the
intensity ratio (I_I/I_{III}) of the vibrational band of the monomer
fluorescence was hardly observed under both TIR and normal
conditions. On the other hand, the intensity ratio of the excimer
to the monomer fluorescence for the film under the normal
condition was larger than that under the TIR one. The result on
the 1/e lifetime of the monomer fluorescence at 374 nm was also
the same as that for the polymer/sapphire interface. That is, the
aggregate state of pyrene in the vicinity of the air/polymer
interface is also different from that in the bulk.

3.4. Spin-coated film

Preparation methods of polymer film may be responsible to the
present interfacial characteristics, so that we have investigated
spin-coated films doped with pyrene (film thickness: ca. 2.8 μm).
All of the differences in fluorescence properties observed for the
sapphire/polymer interface of cast films between TIR and normal
optical conditions were also confirmed for spin-coated films. It
is concluded that the micropolarity around pyrene molecules and
their aggregation near the polymer/sapphire interface in the case
of spin-coated films is similar to that in the case of cast films.
This suggests that the formation process of film from solution is
common for both films.

3.5. Influence of casting solvent

Conformation of polymers in solution is considered to be an important factor for these phenomena. By changing casting solvents, we have investigated pyrene-doped PMMA films (3.9×10^{-2} mol/MMA unit mol) in problem. In Fig. 7, the intensity ratio of I_I/I_{III} is plotted against the Grunwald parameter $(\varepsilon-1)/(2\varepsilon+1)$, where ε is solvent dielectric constant. This parameter concerns with an effective electric field in a spherical cavity formed in homogeneous solvent dielectric[18]. Fluorescence spectra of pyrene in the solvents were also measured, and the intensity ratio is plotted for comparison. The change of the ratio for the solution system is reasonable, since the ratio increases with an increase in the parameter. The degree of the change for PMMA films is smaller than that for the solution system. Moreover, the ratio was not changed even if sample films were dried again *in vacuo* for several hours. Therefore, influence of residual solvent in films on the fluorescence spectrum of films is neglected.

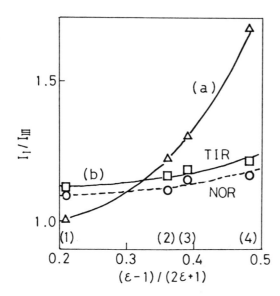

FIGURE 7
Dependence of the fluorescence intensity ratio of the first to the third vibronic bands (I_I/I_{III}) on the Grunwald parameter of casting solvents.(a) in diluted solution and (b) in PMMA films (Concentration: 3.9×10^{-2} mol/MMA unit mol). Solvent: (1) toluene, (2) chloroform, (3) ethyl acetate, and (4) acetonitrile.

For all films, the ratio is larger under the TIR condition than
under the normal one. This indicates that, irrespective of
casting solvents, micropolarity around pyrene molecules in the
vicinity of the sapphire/polymer interface is higher than that of
the bulk. The intensity ratio of PMMA films observed under both
optical conditions increases with an increase in the Grunwald
parameter of casting solvents, of which the fact means that
orientation of carbonyl groups of PMMA around pyrene molecules is
enhanced by the polarity of casting solvents.

The intensity ratio of the excimer to the monomer fluorescence
(I_{470}/I_{384}) and the 1/e lifetime of the monomer fluorescence under
both conditions are plotted against the value of α, which is a
constant in the Mark-Houwink-Sakurada equation, $[\eta]=KM^{\alpha}$, and
depends upon the particular polymer and the solvent used (Fig. 8).
Values of 1.0 and 0.5 of the α mean good and poor solvents for
PMMA, respectively. The ratio of I_{470}/I_{384} and the 1/e lifetime
of the monomer fluorescence under the TIR condition is smaller
than those under the normal one for all films, of which the fact
is the same as PMMA film cast from chlorobenzene. That is,
irrespective of casting solvents, in the vicinity of the
sapphire/polymer interface, the concentration of excimer forming
dimer is lower and the concentration of the non-fluorescent dimer
which is responsible for the quenching is higher as compared with
the bulk.

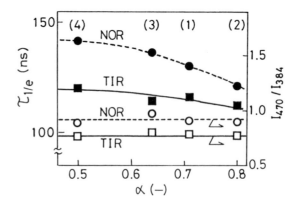

FIGURE 8
Dependence of the fluorescence intensity ratio of the excimer (470
nm) to the third vibronic band of monomer (384 nm) and the 1/e
lifetime of the monomer fluorescence on the α value of casting
solvents. Concentration and casting solvents are the same as
those in FIGURE 7.

The ratio of I_{470}/I_{384} under both conditions is a common value for all the casting solvents. However, the 1/e lifetime decreases with an increase in the α value. The decrement under the normal condition is larger than that under the TIR one. These results suggest that formation of the non-fluorescent dimer which is responsible for quenching is influenced by conformational structure of PMMA chains in solution.

3.6. Thickness of interface layer

We estimate here the depth from which   fluorescence we observed. Under the TIR optical condition, the effective thickness of the interface layer was calculated by applying the followings[2,3,5],

$$I = I_0 \exp(-2\gamma z)$$

$$\gamma = (2\pi n_1/\lambda)(\sin^2\theta - (n_2/n_1)^2)^{0.5}$$

where z is the depth from the interface, I and I_0 are intensities of the excitation light at the interface and at the depth z, respectively, θ is an incident angle of the beam, λ is its wavelength, and n_1 and n_2 are the refractive indices of the sapphire and polymer, respectively. Although these equations hold only for the light with s-polarization, qualitative estimations are possible. The intensity of the evanescent light becomes 1/e of that of the interface at the depth of 0.036 μm under the present experimental condition.

The thickness under the normal condition depends upon absorbance, namely, the concentration of dopant and the excitation wavelength, and the depth where the excitation light intensity is 1/e of the initial one is used as the effective thickness of the layer. At the excitation wavelength of 310 nm, the thickness is 0.476 μm for the highest (8.1×10^{-2} pyrene mol/MMA unit mol) concentration sample investigated. We should point out that the fluorescence observed under the normal condition include the surface contribution which was obtained selectively under the TIR condition. At the present stage, we cannot argue whether the change of micropolarity and aggregation occurs continuously from the sapphire/polymer interface to the bulk or only at the interface. At least, we can say safely that micropolarity and aggregation in PMMA films change in the depth region less than 0.476 μm from the polymer/sapphire interface.

3.7. Interfacial Problem and Film Formation Process

It is reported that there is a nonuniform concentration profile

of polymer solution in the vicinity of a solid/liquid interface.
The presence of the depletion layer of the polymer solution near
nonadsorbing walls was confirmed experimentally by using a steady-
state TIR fluorescence spectroscopy by F. Rondelez et al.[18-21].
The depletion layer thickness decreases with increasing the
concentration of the polymer solution. When the film was cast on
the sapphire plate from the polymer solution as in the present
work, it is considered that the depletion layer of the PMMA
solution is formed in the vicinity of the sapphire/solution
interface, and that its thickness decreases with an evaporation of
the solvent. Local orientation of PMMA around pyrene in this
region should be different from that of the bulk, and aggregation
and its relative geometrical structure are also affected to a
greater extent by this depletion layer.

On the other hand, the interfacial problem for the air/polymer
interface is considered as follows. As soon as the sample
solution containing PMMA and pyrene is dropped on the sapphire
plate, evaporation of the solvent begins. The polymer
concentration at the top of the solution layer reaches immediately
a critical value for which the solution becomes gel, and a crust
is formed. After then, the formation of the film proceeds under
superposition of two phase, solid and fluid. The structure of the
film (crust) formed by the sudden gel formation at the
air/solution interface may be different from that by the slow
evaporation of the solvent in the bulk. We consider that this
difference in the mechanism of film formation is responsible for
the difference in the aggregate state of pyrene between the
air/polymer interface and the bulk. It is worth noting that
micropolarity effect revealed by monomer fluorescence near the
sapphire/polymer interface is appreciable, but not clear for the
air/polymer interface. We consider that the orientation of
carbonyl groups is related to the solid surface properties, while
the dopant aggregation is determined by the polymer itself.

ACKNOWLEDGMENTS

The authors are greatly indebted to the late Prof. Tazuke for
his stimulating discussion in developing time-resolved TIR
fluorescence spectroscopy. We worked together in the project
research on "Development of Fluorescence Microprobe Method for
Analyzing Surface of Polymeric Materials" (1984-1987) supported by

the Grant-in Aid for Developmental Scientific Research from the Japanese Ministry of Education, Science, and Culture (59850146). The present work was partly defrayed by the Grant-in-Aids for Scientific Research on Priority Area for Macromolecular Complexes (63612510) and on Special Project Research for Photochemical Processes (64104007), and the Grant-in-Aid for Scientific Research (63430003) from the Japanese Ministry of Education, Science, and Culture.

REFERENCES

1) For example: D. M. Brewis, Surface Analysis and Pretreatment of Plastics and Metals (Applied Science, London, 1982).

2) N.J. Harrick, Internal Reflection Spectroscopy (Wiley-Interscience, New York, 1967).

3) R. Iwamoto, M. Miya, K. Ohta, and S. Miwa, J. Chem. Phys. 74 (1981) 4780.

4) H. Masuhara, N. Mataga, S. Tazuke, T. Murao, and I. Yamazaki, Chem. Phys. Lett. 100 (1983) 415.

5) H. Masuhara, S. Tazuke, N. Tamai, and I. Yamazaki, J. Phys. Chem., 90 (1986) 5830.

6) N.J. Harrick and G.I. Loed, Modern Fluorescence Spectroscopy 1, ed. E.L. Wehry (Plenum Press, New York-London, 1976) p. 211.

7) B.K. Lok, Y-L. Cheng, and C.R. Robertson, J. Colloid Interface Sci. 91 (1983) 87.

8) A. Itaya, T. Yamada, K. Tokuda, and H. Masuhara, Polymer J. 22 (1990) 697.

9) N. Mataga, H. Obashi, and T. Okada, J. Phys. Chem. 73 (1969) 370.

10) P. Avis and G. Porter, J. Chem. Soc., Faraday Trans. 2, 70 (1974) 1057.

11) K. Kalyanasundaram and J.K. Thomas, J. Am. Chem. Soc. 99 (1977) 2039.

12) J.B. Birks, Photophysics of Aromatic Molecules (Wiley-Interscience, London, 1970, Chapter 7).

13) N. Mataga, Y. Torihashi, and Y. Ota, Chem. Phys. Lett. 1 (1967) 385.

14) G.E. Johnson, Macromolecules 13 (1980) 839.

15) Y. Taniguchi, M. Mitsuya, N. Tamai, I. Yamazaki, and H. Masuhara, Chem. Phys. Lett. 132 (1986) 516.

16) M. Mitsuya, M. Kiguchi, Y. Taniguchi, and H. Masuhara, Thin Solid Films 169 (1989) 323.

17) A. Itaya, T. Kawamura, and H. Masuhara, Thin Solid Films 185 (1990) 307.

18) N. Mataga and T. Kubota, Molecular Interactions and Electronic Spectra (Marcel Dekker, New York, 1970) p.378.

19) C. Allain, D. Ausserre, and F. Rondelez, Phys. Rev. Lett. 49 (1982) 1694.

20) D. Ausserre, H. Hervet, and F. Rondelez, Macromolecules 19 (1986) 85.

21) F. Rondelez, D. Ausserre, and H. Hervet, Ann. Rev. Phys. Chem. 38 (1987) 317.

Photochemical Processes in Organized Molecular Systems
K. Honda (Editor-in-Chief)
© Elsevier Science Publishers B.V., 1991

ENERGY TRANSFER AND MIGRATION IN POLYMER LB FILMS

Masahide YAMAMOTO, Shinzaburo ITO, and Satoru OHMORI

Department of Polymer Chemistry, Faculty of Engineering,
Kyoto University, Yoshida Sakyo-ku, Kyoto 606, Japan

The LB films of poly(vinyl octal) that have chromophores show efficient interlayer energy migration and transfer. This phenomenon can be ascribed to the uniform distribution of the chromophores and the thinness of the LB films. These properties of polymer LB films can be utilized to design new photofunctional molecular assemblies. The interlayer energy transfer manifests structural relaxation of the layered structure of LB films.

1. INTRODUCTION

The Langmuir-Blodgett (LB) technique is a useful method to prepare artificial molecular assemblies. Using this method we can design the spatial arrangement of photofunctional molecules in the dimension of molecular size. After the pioneering work of Kuhn and his co-workers, extensive studies on LB films have been made.[1,2] As for photophysical processes, Yamazaki et al. studied the chromophore distribution in mixed LB films by means of picosecond fluorescence decay analysis.[3,4] In their system, the amphiphilic pyrene moieties were dispersed in the LB film of long-chain fatty acids but aggregates or ground-state dimers of pyrene moieties were easily formed. They concluded that the LB film of long-fatty acids exhibits a nonuniform distribution of pyrene moieties, e.g., a fractal distribution.

As for LB materials, some preformed polymers have been found to form a stable monolayer at the air-water interface and to be transferable to solid substrates.[5-17] One of these polymers is poly(vinyl alkylal) reported by Ogata et al.[7,8] This polymer is prepared by the acetalization of poly(vinyl alcohol) with alkylaldehydes. Various kinds of functional groups can be easily introduced to this polymer through the acetalization with the mixture of alkylaldehyde and chromophoric aldehyde.

Recently we prepared poly(vinyl octal) having photofunctional chromophores and have studied energy transfer and migration in the preformed polymer LB films.[9-12] Through this study, we found that the formation of excimer-forming sites is considerably suppressed in these polymer LB films, and that energy migration and transfer take place efficiently. In this report, energy migration and transfer in the layered LB films are reviewed with special emphasis on several features of polymer LB films, and structural relaxation of the layered LB films is mentioned briefly.

2. SAMPLE PREPARATION AND CHARACTERIZATION OF LB FILMS

The samples were prepared by acetalization of poly(vinyl alcohol) with octanal and chromophoric aldehyde where chromophore is pyrene (Py) for system I; fluorene (F: donor), and anthracene (A: acceptor) for system II; phenanthrene (P: donor) and anthracene (A: acceptor) for system III.

Table 1 shows the composition of the sample polymers for system I (Figure 1). The fractions of octanal unit and pyrene unit were calculated from the carbon content determined by elemental analysis and from UV absorbance. The last column in Table 1 shows the glass transition temperature Tg. The Tg's of these polymers increase with the increase of pyrene concentration. LB films were prepared as follows: The benzene solution of the acetalized polymer (0.01 wt %) was spread on the surface of pure water in a Langmuir trough. The spread polymer was compressed at the rate of 1 cm min^{-1} at 19 $^{\circ}$C. The surface pressure-area isotherm (F-A isotherm) was recorded by using a Wilhelmy type film balance

Table 1. Characterization of Polymers.

sample	X %	Y %	X/(X+Y) %	pyrene chromophore 10^{12}cm^{-2}	Tg $^{\circ}$C
P0	0.0	73.0	0.0	0.0	25
P1	0.14	64.8	0.2	0.6	28
P2	1.35	64.7	2.0	5.8	31
P3	2.91	60.3	4.6	12.7	35
P4	5.58	56.3	9.0	24.3	45
P5	9.07	63.6	12.5	34.7	50

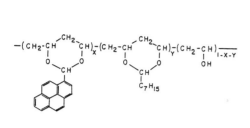

Figure 1
Structural formula of system I.

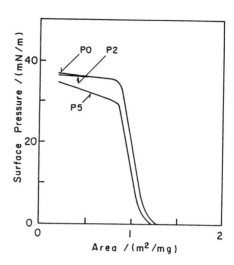

Figure 2
Surface pressure-area isotherms
at 19 °C.

(Shimadzu ST-1). The substrate was a quartz plate (1x4 cm), which was made hydrophobic by dipping it in a solution of trimethylchlorosilane for 20 min. The film was transferred on a quartz plate as a Y-type built-up film by dipping it vertically at the rate of 2 cm min.$^{-1}$ The transfer ratio was of unity. To exclude the effect of the interface of a substrate, four layers of poly(vinyl octal) (P0) were deposited on the substrate beforehand. After the deposition of chromophoric sample polymer, four layers of P0 were again deposited to avoid the effect of the air interface. During the deposition, the surface pressure of the spread film was maintained at a fixed value (20-25 mN m^{-1}).

Figure 2 shows the surface pressure-area isotherms of each polymer. As the surface film was compressed on water, the surface pressure increased sharply at the area of 1 m^2 mg^{-1}. By further compression, the monolayer film began to collapse at the plateau region of the F-A isotherms. With the increase of the concentration of pyrene moieties from P0 to P5, the surface pressure in the collapsed region decreased from 35 to 30 mN m^{-1} and the limiting area slightly decreased from 1.13 to 1.09 m^2 mg^{-1}. When the limiting area of poly(vinyl alcohol) unit is assumed to be 0.12

nm^2 $unit^{-1}$, the limiting area of acetalized unit can be
calculated.   The area is betweeen 0.35 and 0.36 nm^2 $unit^{-1}$
irrespective of the content of pyrene moieties.   This value is
consistent with the one reported by Ogata for poly(vinyl
acetals).[7]

     To verify the deposition of the LB films on the substrate, the
absorbance and the fluorescence intensity of transferred LB films

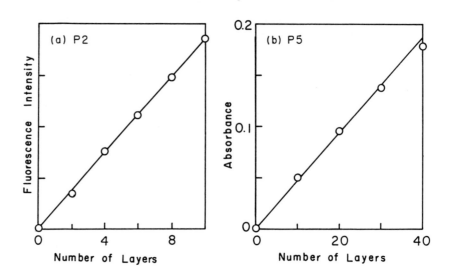

Figure 3
Fluorescence intensity (a) and UV absorbance (344 nm) (b) of
built-up films as a function of layers (on one side).

were measured.   Figure 3 (a) shows the relation between the
fluorescence intensity of P2 and the number of layers on one side.
Here polymer P2, which does not emit excimer emission even in
multilayered films, was chosen.   The fluorescence intensity
increases proportionally with the number of layers.   Figure 3(b)
shows the relation between the absorbance of P5 films at 344 nm
and the number of layers.   This figure also shows a proportional
relation, that is, the surface film can be transferred
satisfactorily at least up to 40 layers on one side.   From the
absorbance of the LB films, the plane density of pyrene moieties
per single layer was calculated under the assumption of the random
orientation of pyrene moieties.   The values are given in Table 1.

3. UNIFORM DISTRIBUTION OF CHROMOPHORES

Figure 4 shows the fluorescence spectra and the excitation spectra of 6-layer films. Structured emission at 375, 386, and 395 nm is the monomer fluorescence. In the samples having pyrene moieties lower than 2 mol%, the shape of the emission spectra was the same as that of P2. The normal excimer emission at 470 nm is observed for the samples of pyrene concentrations above P3. In addition to this emission there is a slight increase of the intensity around 420 nm with the increase of pyrene concentration. This emission is probably due to the partial overlapping excimer whose existence is reported in the LB film of fatty acids.[3] Although the pyrene chromophore easily forms excimers, the yields of these two types of excimer emission are much less in this polymer LB film in comparison with those in the LB film of fatty acids at the same plane density of pyrene moieties in a layer. Figure 4 also shows the excitation spectra of a 20-layer LB film of P5 monitored at 375 nm. The excitation spectrum has the same peak position as the UV absorption spectrum. The excitation spectra are not much different for each polymer having different

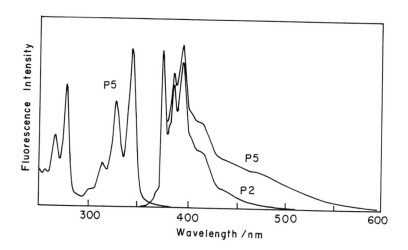

Figure 4
Fluorescence emission spectra of 6-layer (on one side) films and the excitation spectrum of a 20-layer film of P5 monitored at 375 nm.

pyrene contents, and there is no spectral shift even if the
spectra are monitored at 420, 470, or 375 nm. The time-resolved
fluorescence spectroscopy shows that there is no spectral shift
either in the short time region before 1 ns nor in the long time
region after 100 ns: i.e., no dimer band was observed, either in
the excitation spectra, or in the time-resolved fluorescence
spectra. This indicates that there is no specific interaction of
pyrene moieties in the ground state, and that the excimers are
formed dynamically in the excited state. As for the excimer
formation process, the time-resolved spectroscopy also shows that
the fraction of the excimers gradually increases with time in the
1-100 ns region. The result can be explained by the mechanism
that the excimers are formed by the energy migration from the
monomer state to excimer-forming sites.

Energy trap densities in a plane for 2-layer films of each
sample were evaluated by the decay curve analysis, which was based
on the Förster type transfer model.

$$I(t) = I_1 \exp[-t/T_D - 2g(t/T_D)^{1/3}] + I_2 \exp(-t/T_D)$$
$$g = (2/3)n_a \pi R_{DA}^2$$

where T_D is the lifetime of the donor fluorescence without trap
site, R_{DA} (cm) is the Förster radius between donor and acceptor
chromophores, n_a is the number of trap sites per unit area (cm^{-2}).
The lifetime of P2, which has no excimer-forming site, was
substituted for T_D: T_D = 280 ns. For R_{DA}, the Förster radius of
pyrene moieties: R_{DA} = 1.1 nm was used.[18] The decay curve analysis
showed that I_2 should be set to zero to obtain a good fit with the
observed decay curves, that is, there is no isolated monomer state
in this system. The analysis showed that trap sites for P4 and P5
are 3.5×10^{12} and 6.8×10^{12} cm^{-2}, respectively. For P2 and P3, the
fluorescence was not quenched enough to obtain an accurate value.
The results show that trap sites for P5 are formed by half of the
pyrene moieties compared with the chromophore density of pyrene
moieties. This value is one-tenth of the trap density in the
fatty acid LB films and is close to the number expected for a
random distribution of the chromophore.[3,4] However, the value of
the trap site is much higher than that expected from the excimer
emission. The reason is that the energy trap site is not
necessary to emit excimer fluorescence, because in LB films the
molecular motion is suppressed and some fraction of pyrene

moieties cannot take an emissive excimer conformation, then the quantum yield of excimer emission is considerably reduced. The results show that the chromophore distribution in polymer LB films is uniform. Since pyrene moieties are linked to the polymer chain with covalent bonds, pyrene chromophores attain a statistically uniform distribution and excimer emission relative to monomer emission is much less than that for LB films prepared from a fatty acid with pyrene chromophore at the same concentration of the chromophore in a plane.

4. INTERLAYER ENERGY MIGRATION

Figure 5 shows the fluorescence spectra of LB films of P5 for various numbers of transferred layers. The relative intensity of excimer fluorescence against the monomer emission increases with the increase in the number of layers and the ratio approaches that of the spin-coated film. This indicates the existence of interlayer energy migration. In Figure 6 the ratio of fluorescence intensity at 480 nm (excimer emission) to that at 375 nm (monomer emission) was plotted against the number of layers for the samples of various pyrene concentrations. The ratio for P3 is small, and there is no marked change with the increase in the number of layers. On the contrary, the ratio for P5 increases with the increase in the number of layers and there is still an

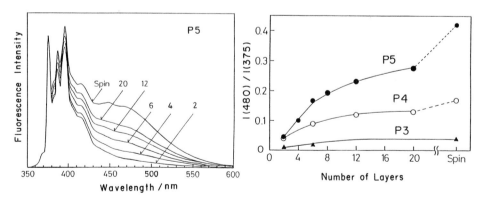

Figure 5
Fluorescence spectra of polymer LB films of P5 and spin-coated films. Numerals are the number of layers.

Figure 6
Ratio of excimer emission at 480 nm to the monomer emission at 375 nm as a function of the number of layers.

increase even in the 20-layer film. In accordance with this
result, the decay curves of monomer fluorescence also depend on
the number of layers. The decay rates increase with an increase
in the number of layers and there is still a difference between
12-layer and 20-layer films.

The Förster radius between pyrene moieties is only 1.1 nm.[18]
and the thickness of the monolayer is also ca. 1.1 nm.[8] The
distance between pyrene moieties in a plane is a few nanometers.
Therefore, for the samples having higher concentrations of pyrene
moieties, interlayer energy migration takes place in addition to
intralayer energy migration where a few steps of intralayer energy
migration may assist the interlayer one.

5. INTERLAYER ENERGY TRANSFER

The sample polymer is the poly(vinyl octal) containing fluorene
(F) or anthracene (A) (system II, Figure 7). The composition of
the polymer, X/(X+Y) is 0.12 for both the F polymer and the A
polymer. Figure 8 shows the energy transfer between a pair of
energy donor polymer (F) and acceptor polymer (A) spaced by
multilayered films of inert poly(vinyl octal). The notation FnA
denotes a layered film of two layers of the F polymer, n layers of
poly(vinyl octal) and two layers of the A layer. The thickness of
each layer is ca. 1.1 nm. The Förster radius between the F and A
units is 2.2 nm.[18] The figure shows that the transfer efficiency
is controlled by the separation distance. In the sample F0A that
has a contact layer of F and A polymers, the emission from the F
unit (300-330 nm) is quenched to about 25 % of the unquenched
fluorescence intensity and the sensitized emission from the A unit
is observed around 390-460 nm. On the other hand, for the samples
having a longer spacer the energy transfer becomes less efficient.
It is noted that weak emission of the A unit for F8A and F10A is
due to direct excitation of the A unit.

This shows that interlayer energy transfer can be controlled by
the distance between a donor layer and an acceptor layer. This
interlayer energy transfer can be extended further to a sequential
energy transfer for multiple chromophoric layers, e.g., triple
chromophoric system, donor - acceptor 1 - acceptor 2, where the
donor is excited.[10] Therefore, the interlayer energy transfer in
polymer LB films is considerably efficient. This is because the

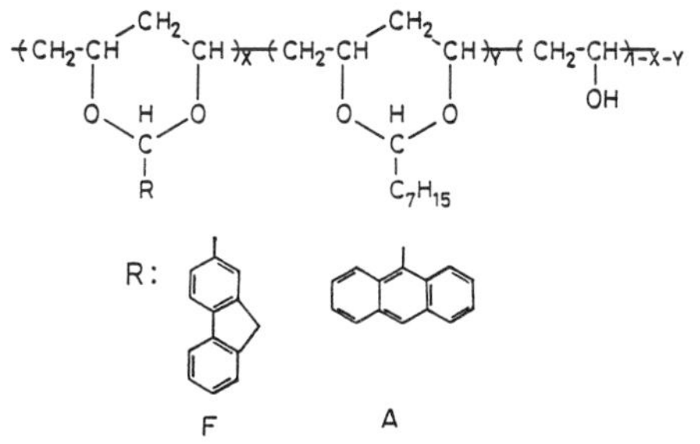

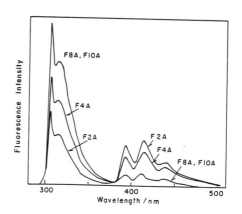

Figure 7
Structural formula of system II.

Figure 8
Interlayer energy transfer for FnA

thickness of polymer LB films is rather thinner than that of the conventional LB films made from long-chain fatty acids.

6. RELAXATION OF LAYERED STRUCTURE STUDIED BY INTERLAYER ENERGY TRANSFER

Interlayer energy transfer can be utilized to evaluate the structure of polymer LB films. In this study poly(vinyl octal) films were sandwiched by phenanthrene (P) layers and anthracene (A) layers: PnA system (n = the number of spacer layers) (system III, Figure 9). The compositions of the polymers, X/(X+Y) are 0.12 for the P polymer and 0.17 for the A polymer, respectively. The Förster radius of the present donor-acceptor pair P-A is 2.12 nm.[18] Figure 10 shows the time course of energy transfer efficiency, the ratio of fluorescence intensity at 438 nm (anthracene) to that at 350 nm (phenanthrene), I_{438}/I_{350}. The time is that during which the LB films were left at room temperature after the LB films were deposited. P4A, P6A, and P8A show a rapid increase of energy transfer efficiency at an early stage after the deposition. After a few days, the increase slowed down. This phenomenon is related to the structural relaxation of polymer LB films. For the films of P4A, P6A, and P8A, the distance between the pair of P and A layers may become closer due to the segmental diffusion of polymer chains and this structural relaxation of the LB films makes the energy transfer efficiency

higher.   On the other hand, POA shows a continuous decrease of
the energy transfer efficiency.   This suggests that the distance
increases with time, because the chromophores exist initially in
adjacent layers.   In the case of P2A, the situation is
intermediate and both effects overlap.   A quantitative study of
the relaxation of the layered structure is now in progress.

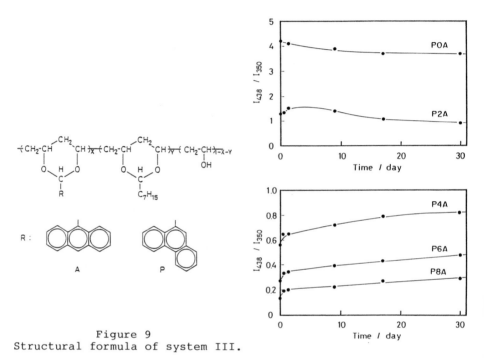

Figure 9
Structural formula of system III.

Figure 10
Time dependence of the interlayer
energy transfer efficiency, the
ratio of fluorescence intensity
of anthracene (438 nm) to that of
phenanthrene (350 nm).

7. CHARACTERISTICS OF POLYMER LB FILMS

From the standpoint of photophysical processes, polymer LB
films have several advantages compared with conventional LB films
of long-chain fatty acids.   One is a uniform distribution of
chromophores in the monolayer, i.e., polymer LB films have a
fairly low concentration of energy trap sites such as excimer-
forming sites.   This is because the polymer films are essentially
amorphous and each chromophore is linked to the polymer chain with

covalent bonds. Therefore the attachment of chromophores to polymer chains is a powerful way for preventing the chromophores from forming ground state dimers or aggregates, and for attaining a uniform distribution of chromophores in a monolayer. The other is thinness of polymer LB films. The monolayer of polymers is stabilized by a hydrophilic group on the main chain and hydrophobic group on the side chain, and has a thickness of ca. 1 nm which is thinner than the conventional LB films made from long-chain fatty acids (Figure 11). The thinness of LB films favors energy transfer and migration between neighbouring layers.

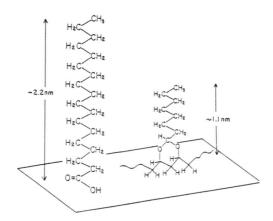

Figure 11
Comparison of poly(vinyl octal)
with stearic acid.

8. CONCLUSION

The results obtained with chromophoric poly(vinyl octal) show that the polymer LB films have favorable characteristics for photophysical and photochemical processes: the uniform distribution of chromophores and thinness of the LB films. These properties of polymer LB films can be utilized to design new photofunctional molecular assemblies. On the other hand, interlayer energy transfer can be applied to characterize the structure of the LB films. This is one of the most sensitive tools to investigate the structural relaxation of layered LB films.

ACKNOWLEDGMENT

This work was partially supported by a Grant-in-Aid for Scientific Research on Priority Areas, New Functionality Materials-Design, Preparation and Control (No.02205071) and a Grant-in-Aid for Scientific Research (No.01550692), from the Ministry of Education, Science and Culture of Japan.

REFERENCES

1) H. Kuhn, D. Möbius and H. Büchner, Spectroscopy of monolayer assemblies, in: Physical Methods of Chemistry, Vol. 1, Part 3B, eds. A. Weissberger and B.W. Rossiter, (Wiley, New York, 1972) pp. 577-702.

2) H. Kuhn, Thin Solid Films 178 (1989) 1.

3) I. Yamazaki, N. Tamai and T. Yamazaki, J. Phys. Chem. 91 (1987) 3572.

4) I. Yamazaki, N. Tamai and T. Yamazaki, Picosecond fluorescence spectroscopy on molecular association in Langmuir-Blodgett films, in: Ultrafast Phenomena V, Vol. 46, eds. G.R. Fleming and A.E. Siegman (Springer-Verlag, Berlin, 1986) p. 444.

5) R.H. Tredgold and S.C. Winter, J. Phys. D 15 (1982) L55.

6) R.H. Tredgold, Thin Solid Films 152 (1987) 223.

7) K. Oguchi, T. Yoden, K. Sanui and N. Ogata, Polym. J. 18 (1986) 887.

8) M. Watanabe, Y. Kosaka, K. Oguchi, K. Sanui and N. Ogata, Macromolecules 21 (1988) 2997.

9) S. Ohmori, S. Ito, M. Yamamoto, Y. Yonezawa and H. Hada, J. Chem. Soc. Chem. Commun. 1989 1293.

10) S. Ito, H. Okubo, S. Ohmori and M. Yamamoto, Thin Solid Films 179 (1989) 445.

11) S. Ohmori, S. Ito and M. Yamamoto, Macromolecules 23 (1990) 4047.

12) S. Ohmori, S. Ito and M. Yamamoto, Macromolecules, 24 (1991) in press.

13) K. Shigehara, M. Hara, H. Nakahama, S. Miyata, Y. Murata and A. Yamada, J. Am. Chem. Soc. 109 (1987) 1237.

14) T. Miyashita, Y. Mizuta and M. Matsuda, Brit. Polym. J. 22 (1990) 327.

15) T. Murakata, T. Miyashita and M. Matsuda, Macromolecules 21

(1988) 2730.

16)   K. Naito, J. Colloid Interface Sci. 131 (1989) 218.

17)   M. Matsumoto, T. Itoh and T. Miyamaoto, Langmuir-Blodgett films of cellulose derivatives, in: Cellulosics Utilization, eds. H. Inagaki and G.O. Phillips (Elsevier, London, 1989) p. 151.

18)   I.B. Berlman, Energy Transfer Parameters of Aromatic Compounds, (Academic Press, New York, 1973).

Photochemical Processes in Organized Molecular Systems
K. Honda (Editor-in-Chief)
© Elsevier Science Publishers B.V., 1991

REGULATION OF LIQUID CRYSTALLINE ALIGNMENT CHANGE BY PHOTOCHROMIC MOLECULAR FILMS

Kunihiro ICHIMURA

Research Institute for Polymers and Textiles, 1-1-4 Higashi, Tsukuba, Ibaraki 305, Japan

A new type of photoresponsive liquid crystalline cells has been developed. The cells were constructed by putting a nematic liquid crystal between two quartz plates covered with azobenzene monolayers, the photoisomerization of which induces reversible alignment change between the homeotropic and parallel modes. Detailed studies revealed that the factors affecting the photoregulation of the alignment involves the molecular structure and occupied area of the azobenzene unit and the coexistence of long alkyl chain. Derivatives of stilbene and α-hydrazono-β-ketoester were found to be also effective for the photoregulation. This suggests intensively that the reversible structural alteration between the rod-like and L-shaped photochromic units plays an essential role in the photoregulation.

1. INTRODUCTION

Reversible structural change of photochromic molecules may bring about various optical properties of materials other than the absorption spectrum when photochromic molecules interact more or less specifically with surrounding molecules, the reorientation or rearrangement of which are induced by the photochromic molecules.[1] The photochromism-based optical responsiveness of organic materials depends intensively upon the ordered structures of matrices as summarized in Table 1. Liquid crystals (LC's) are very attracting reaction media for photochromism in this respect since the molecular reorientation of LC triggered by the structural alteration of photochromic molecules can be amplified because of the unidirectional function of LC molecules. In other words, a combination of photochromics with LC results in the reversible change of optical anisotropy of the system which offers erasable optical memory in which stored information is read out with polarized light. It can be said that the reversible alteration of supramolecular structures caused by photochromic molecules leads to the extension of the concept and the application of photochromism.

Several photosensitive LC systems have been reported so far,[1] as summarized in Table 2. They exhibit the reversible transformation of mesophases between nematic and isotropic and between smectic and nematic phases and the reversible shift of cholesteric pitch band. Reversible mesophase change induced by azobenzene photoisomerization has been extended to polymeric LC's.[9]

Table 1 Dependence of optical responsiveness on ordered structures of matrices

Matrix	Optical property
Amorphous polymer	Absorption
Micelle	Transmission
Inclusion cavity	Emission
Bilayer membrane	Light scattering
Monomolecular layer	Reflection
	Refraction
Stretched film	Birefringence
Liquid crystal(N, Sm)	
Cholesteric liquid crystal	Optical rotatory power
Molecular multilayer	Multi-frequency absorption
Crystal	
Supramolecular system	

Table 2 Mesophase change induced by photochromism

Liquid crystal	Photochromic molecule	Mesophase change	Ref.
Cholesteric	Azobenzene	Alteration of cholesteric pitch	2
Cholesteric	Spiropyran	Alteration of cholesteric pitch	3
Nematic	Stilbenes	N $\longrightarrow$ Iso	4
Cholesteric	Azobenzene	Chol $\rightleftharpoons$ N	5
Smectic	Azobenzene	Sm $\rightleftharpoons$ N	6
Nematic	Chiral azobenzenes	Chol $\rightleftharpoons$ Iso	7
Nematic	Azobenzenes	N $\rightleftharpoons$ Iso	8

During the course of our studies on photochromism in LC media, cholesteric phase of nematic LC's doped with chiral trans-form of azobenzenes (1) having a rod-like shape similar to that of LC was found to result in the elongation of cholesteric pitch and the subsequent destruction of the mesophase upon the trans-cis photoisomerization of the azobenzene (Az).[7] The mesophase change takes place repeatedly and can be interpreted as shown in Figure 1; the bent form of cis-Az molecules leads to the reorientation of surrounding LC molecules.

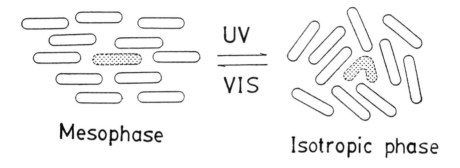

Figure 1   Illustration of reorientation of LC molecules induced by photoisomerized azobenzene molecules

This led us to an idea that certain photoresponse of LC is expected to oc-cur by the photoisomerization of rod-shaped Az (2) attached to substrate sur-face and to the finding that the LC alignment of nematic LC is in fact regu-lated photochemically by Az molecular films.[10] We have proposed to call the substrate surfaces covered with photochromic molecular films "command surfaces" since the structural change of one Az molecule attached to the sur-face results in the reorientation of more than ca. 10^4 LC molecules.

$$CH_3(CH_2)_5-\langle O\rangle-N=N-\langle O\rangle-O\overset{*}{C}H(CH_3)CH_2CH_3 \qquad (1)$$

$$CH_3(CH_2)_5-\langle O\rangle-N=N-\langle O\rangle-O(CH_2)_5CONH(CH_2)_3Si(OC_2H_5)_3 \qquad (2)$$

The reversible change in the LC alignment can be interpreted as illustrated in Figure 2. It is likely that the rod-like Az units in the trans form stand perpendicular to the substrate surface, and the LC molecules are oriented so that their long axis is parallel to that of Az units. This molecular information of the alignment is transferred from molecule to molecule to result in the homeotropic alignment. When the cell is exposed to UV light to form the bent cis Az isomer, the upper part of the Az unit becomes nearly parallel to the surface. The LC molecules surrounding the Az units change their position consequently so that the long axis of the LC molecules also becomes parallel to the substrate surfaces. The information of this alteration of the alignment may be transferred throughout the LC layer to give the parallel alignment.

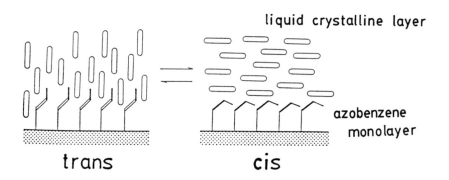

Figure 2 Illustrative representation for the reversible change of LC alignment induced by the photoisomerization of azobenzene units attached to a substrate surface

The principle of our photoresponsive LC systems possessing potential value for photomemory and display devices is thus based on the amplification of the alteration of molecular interaction between the azobenzene units and LC molecules. This paper deals with factors affecting the photoinduced regulation of LC by the command surfaces, focusing on the importance of molecular structures of photochromic monolayers.

2. EXPERIMENTAL
2.1. Preparative methods for command surfaces
Introduction of Az unit on substrate surface has been carried out by the following methods; 1) silylation of silica glass surface with silylating

reagents bearing Az unit,[10] 2) spin coating of Az pendent polymers on substrate plates,[11] 3) transfer of LB monolayers of the Az pendent polymers onto
substrate plates,[12] and 4) coating of substrate plates with PVA films and subsequent surface modification of the PVA films by acylation or acetalization
with Az modifiers.[13] Although the first method is applicable only for silica
glass substrates, various flat and transparent substrates including plastic
films can be employed for the surface modification with Az by the latter three
methods, and they are of practical significance. This paper deals solely with
the photoregulation of LC alignment by command surface prepared by the first
method.

2.2. Cell construction and photoresponse measurement
    A cell was constructed by putting a nematic LC, a mixture of
cyclohexanecarboxylates (3), between two Az modified quartz plates. The cell
thickness was adjusted with use of small glass rods of a diameter of 8 μm.

$$R_1-\bigcirc-COO-\bigcirc-OR_2 \qquad (3)$$

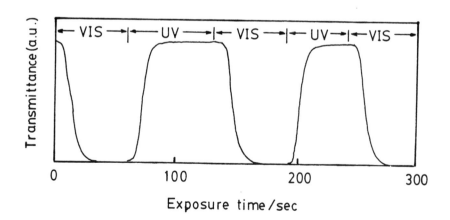

Figure 3 Reversible change of transmittance of He–Ne laser light through an LC
cell constructed with Az modified quartz plates and a crossed polarizer.

The reversible alignment alteration was monitored by following the light intensity of polarized He-Ne laser beam through the cell and a crossed polarizer upon alternate exposure of the cell to uv (365 nm) and visible (ca. 440 nm) light. Typical example is shown in Fig. 3. In the initial state, the cell does not transmit any laser beam, indicating that the rod-shaped LC molecules are arranged in the homeotropic mode, perpendicularly to the glass surfaces when monolayered Az is in the trans form. When the cell is exposed to uv for the isomerization, it becomes bright and allows the laser beam to pass through the cell and the crossed polarizer. This reflects that the LC alignment is changed into a parallel mode where the long axes of the LC molecules are oriented parallel with the surfaces when Az photoisomerizes into the bent cis isomer.

Table 3 Effect of the molecular structure of azobenzene silylating reagents (4) on the alignment regulation of cyclohexanecarboxylate-type nematic liquid crystal

$$R_1-\bigcirc-N=N-\bigcirc-R_2-CONH-R_3-\overset{X}{\underset{Z}{Si}}-Y \quad (4)$$

	R_1	R_2	R_3	X	Y	Z	Photoregulation
Az-1	C_6H_{13}	$O(CH_2)_5$	$(CH_2)_3$	OC_2H_5	OC_2H_5	OC_2H_5	+
Az-2	C_8H_{17}	$O(CH_2)_5$	$(CH_2)_3$	OC_2H_5	OC_2H_5	OC_2H_5	+
Az-3	$\langle H \rangle$	$O(CH_2)_5$	$(CH_2)_3$	OC_2H_5	OC_2H_5	OC_2H_5	+
Az-4	Cl	$O(CH_2)_5$	$(CH_2)_3$	OC_2H_5	OC_2H_5	OC_2H_5	−
Az-5	H	$O(CH_2)_5$	$(CH_2)_3$	OC_2H_5	OC_2H_5	OC_2H_5	−
Az-6	CH_3O	$O(CH_2)_5$	$(CH_2)_3$	OC_2H_5	OC_2H_5	OC_2H_5	−
Az-7	$C_6H_{13}O$	$O(CH_2)_5$	$(CH_2)_3$	OC_2H_5	OC_2H_5	OC_2H_5	−
Az-1	C_6H_{13}	$O(CH_2)_5$	$(CH_2)_3$	OC_2H_5	OC_2H_5	OC_2H_5	+
Az-8	C_6H_{13}	$O(CH_2)_3$	$(CH_2)_3$	OC_2H_5	OC_2H_5	OC_2H_5	+
Az-9	C_6H_{13}	$O(CH_2)_2$	$(CH_2)_3$	OC_2H_5	OC_2H_5	OC_2H_5	+
Az-10	C_6H_{13}	OCH_2	$(CH_2)_3$	OC_2H_5	OC_2H_5	OC_2H_5	+
Az-1	C_6H_{13}	$O(CH_2)_5$	$(CH_2)_3$	OC_2H_5	OC_2H_5	OC_2H_5	+
Az-11	C_6H_{13}	$O(CH_2)_5$	$(CH_2)_3$	CH_3	OC_2H_5	OC_2H_5	−
Az-12	C_6H_{13}	$O(CH_2)_5$	$(CH_2)_4$	CH_3	CH_3	OC_2H_5	−
Az-13	C_6H_{13}	$O(CH_2)_{10}$	$(CH_2)_3$	CH_3	CH_3	OC_2H_5	+

3. FACTORS AFFECTING LIQUID CRYSTALLINE ALIGNMENT REGULATION BY AZOBENZENE UNITS

3.1. Molecular structure of Az unit

A variety of Az derivatives (4) having different head group (R_1), spacer (R_2 and R_3) and silylating unit (SiXYZ) were prepared by the condensation of Az carboxylic acid with aminoalkylsilane and used for surface modification of quartz plates to construct LC cells. The results are compiled in Table 3. LC cells for Az-1, Az-2 and Az-3 were all quite dark between a couple of crossed polarizer, indicating that LC molecules are in the homeotropic alignment. In contrast to this observation, LC cells for Az-4, Az-5, Az-6 and Az-7 were inhomogeneously bright between polarizers, suggesting that the trans form of these Az units does not bring about the homeotropic alignment and that the mode of LC alignment is randomly parallel.

As seen in Table 3, the photoregulation of LC alignment is dependent on the nature of the head group (R_1). When the cells constructed with plates covered with Az modifiers bearing alkyl substituents (Az-1, Az-2 and Az-3) were exposed to uv, they became quite bright. They were reversed into the initial dark states upon subsequent exposure to visible light for the recovery of the trans isomers of Az, and alternate exposure to uv and visible light caused reversible dark/bright change of the cell. On the other hand, no photoresponse was observed for Az derivative without substituent (Az-5) and with chloro (Az-4), methoxy (Az-6) and hexyloxy (Az-7) although the photoisomerization took place in their monolayers. Therefore, the homeotropic alignment due to the effect of the trans Az isomer seems to be one of the necessary conditions for the photoregulation.

The spacer effect is not clear when Az is substituted with hexyl as the head group and has triethoxysilyl unit since the second family in Table 3 (Az-1, Az-8, Az-9 and Az-10) are all active for the photoregulation. The situation becomes quite different when the Az group is attached on a glass surface through monofunctional Si-O bonding. While Az-12 does not result in the photoregulation, Az-13 having a longer spacer makes the cell photoresponsive.

The photoregulation is also influenced by the mode of silylation, as mentioned just above. Mono- and difunctional silylation using Az-12 and Az-11, respectively, failed to afford photoresponsive LC cell. It may be assumed that the mono- or difunctional silylation results in the reduction of the angle formed between the Az molecular axis and the glass surface while trifunctional silylation using Az-1 is favorable for introducing Az units perpendicularly to the surface, but no direct evidence has not yet been obtained.

3.2. Occupied areas of azobenzene unit

The amount of Az attached to a quartz surface was estimated
spectroscopically using the absorption coefficient of Az in solution to give
occupied area per Az. The validity of the estimation was confirmed by
treating the quartz plate with a strongly alkaline methanol solution to detach
the Az unit from the quartz surface and by measuring the absorbance in the
solution. The area occupied by Az unit derived from Az-1 was 36 A^2.

More densely packed monomolecular film was obtained by the Langmuir-
Blodgett (LB) technique employing an amphiphilic Az derivative (C_6H_{13}-   -N=N
-$O(CH_2)_{10}COOH$) which gave occupied area of 28 A^2. The LB film transferred on
a quartz plate was inactive for the photoregulation of LC alignment simply
because no photoisomerization took place from the trans isomer to the cis
isomer which requires a larger occupied area.[10]

The area occupied by Az was controlled by treating quartz plates with a
mixture of Az-1 and ethyltriethoxysilane. LC cells were made from these Az
modified plates and exposed alternately to uv and visible light to observe the
photoresponse. Table 4 tells us that there is a critical area per Az of 120
A^2 for the photoregulation. This critical value is quite in line with that
observed in one of the other techniques for the preparation of command
surfaces. Poly(vinyl alcohol) substituted with Az units has amphiphilic
character and affords a polymeric LB membrane incorporating Az units. The
area per Az is readily controlled in this case by regulating the surface
pressure of the Az pendent polymer at the water-air interface and by the

Table 4 Effect of occupied area of azobenzene unit on the photoresponse

Az-1 in mol%[a) on preparation	O. D.max	Occupied area (A^2)	Photoresponse
100	0.0260	36	+
24	0.0220	43	+
9.6	0.0192	49	+
5.1	0.0132	71	+
3.4	0.0076	120	+
0.64	0.0052	190	−
0.16	0.0020	470	−

a) Mol% in a mixture of Az-1 and ethyltriethoxysilane

subsequent deposition of the membrane onto a quartz plate.[12] This elegant technique has revealed that occupied areas less than 100 A^2 are active for the photoregulation of LC alignment and that no photoresponse was observed when the area is larger than 150 A^2.

3.3. Simultaneous treatment with long alkyl silylating reagents

As already mentioned, homeotropic alignment caused by the trans form of Az seems to be one of the necessary conditions for the photoregulation of LC alignment, and Az-5 having no substituent at Az moiety results in randomly parallel alignment and no photoresponse. Our efforts have been focused on the surface modification condition using Az-5 to realize homeotropic alignment by simultaneous treatment with long alkyl silylating reagents which are known to afford homeotropic alignment.

As summarized in Table 5, whereas C$_8$ modification does not result in the homeotropic alignment, longer alkyl silylation with both C$_{12}$ and C$_{18}$ silylating reagents makes the alignment homeotropic even in the presence of Az-5. However, reversible alignment change was observed only in the case of surface modification with a mixture of Az-5 and C$_{12}$ silylating reagent in 7 : 5 molar ratio. Although the co-treatment with the C$_{18}$ reagent affords the homeotropic alignment and the photoisomerization of Az was observed, no alignment alteration was detected upon exposure to uv.

The distinct dependence of alkyl chain length of the co-modifier on the photoresponse of the LC cells may be interpreted as follows. The chain length of C$_{18}$ is so long that Az units are buried in the C$_{18}$ monomolecular layer and have no chance to interact with LC molecules. On the contrary, the chain

Table 5 Effect of long alkyl silylating reagents on the photoresponse

Molar ratio[a]	Initial alignment	Photoresponse
Az only	parallel	−
Az/C$_8$ = 7/5	parallel	−
Az/C$_{12}$ = 7/5	homeotropic	+
Az/C$_{18}$ = 7/5	homeotropic	−
Az/C$_8$ = 1/9	parallel	−
Az/C$_{12}$ = 1/9	homeotropic	−
Az/C$_{18}$ = 1/9	homeotropic	−

a) Az: Az-5 in Table 3, C$_8$: $C_8H_{17}Si(OC_2H_5)_3$, C$_{12}$: $C_{12}H_{25}Si(OC_2H_5)_3$, C$_{18}$: $C_{18}H_{37}Si(OC_2H_5)_3$

length of C_{12} is roughly equal to that between Si and Az unit of Az-5, as
illustrated in Fig. 4, so that Az units are ready to be exposed to LC
molecules which reorient upon the photoisomerization of Az.

4 EFFECTIVE PHOTOCHROMIC UNIT OTHER THAN AZOBENZENE

4.1. Stilbene

Exploration of photochromic units other than azobenzene is of importance
not only from practical viewpoint, but also from fundamental interest to
elucidate the mechanism of LC alignment alteration by the command surfaces.

Several photochromic units including stilbene (5), cinnamylidene (6),
retinylidene (7), hemithioindigo (8) and nitrospiropyran (9) were introduced

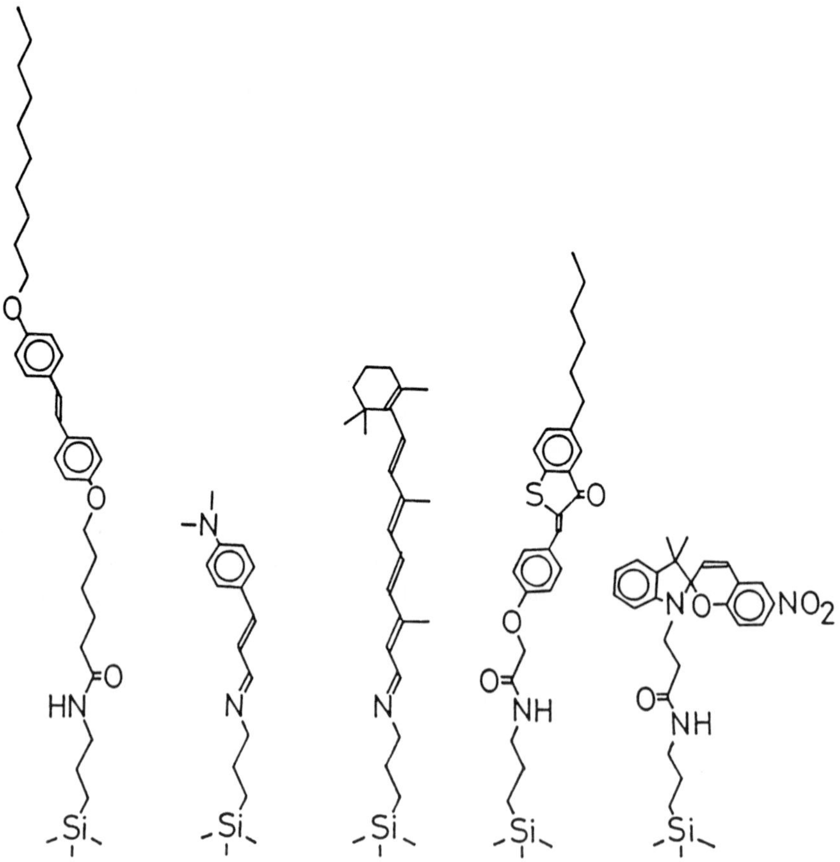

Figure 4   Photochromic units for quartz surface treatment

onto quartz surface by the silylation similar to that of azobenzenes for 5, 8
and 9 or by the condensation of the corresponding aldehyde with
aminopropylated quartz plates for 6 and 7 (Figure 4).[14]

Photoreactivity of the monolayered photochromic units were examined by
exposing the plates to uv. Although hemithioindigo shows excellent fatigue-
resistance upon alternate exposure to UV and visible light in a toluene
solution, as described in our recent report,[15] the uv exposure brought about
monotoneous disappearance of the chromophore which was not recovered upon
exposure to visible light for the reverse reaction, indicating anomalous
irreversible photoreaction. Except the hemithioindigo, the
photoisomerizations were observed for the other photochromic units.

Table 6   Photochemistry and photoregulation of LC alignment of monolayered
photochromic units

Photochromic unit	Alignment	Photoisomerization	Photoregulation
stilbene	homeotropic	+	+
cinnamylidene	parallel	+	−
retinylidene	parallel	+	−
hemithioindigo	homeotropic	−	−
nitrospiropyran	parallel	+	−

Nematic LC cells were made from these surface modified plates and placed
between a couple of crossed polarizer. As summarized in Table 6, stilbene (5)
and hemithioindigo (8) substituted with head alkyl group result in the
homeotropic alignment whereas no homeotropic alignment was observed for 6, 7
and 9. This confirms the essential role of the hydrophobic head group in the
homeotropic alignment.

It was found that the stilbene is active for the photoregulation of LC
alignment. The others demonstrated no photoresponse. Figure 5 shows the
reversible transmittance change of the cell made from quartz plates covered
with the stilbene monolayers upon alternate exposure to 313 nm light and 254
nm light for the trans-cis and cis-trans photoisomerization, respectively
(Figure 5). The transmittance was monitored by polarized He-Ne laser beam,
just in the case of the cell modified with azobenzenes.

The observation that the stilbene monolayer affords an effective command surface implies that the change of molecular shape may play an essential role in the photoregulation of LC alignment since both azobenzene and stilbene exhibit reversible photoisomerization between rod-like and L-shaped molecules.

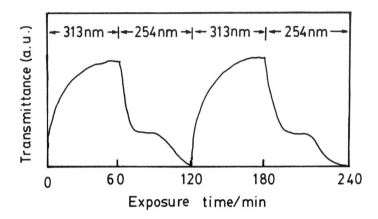

Figure 5   Photoresponse of LC cell made from quartz plates modified with the stilbene unit.

4.2.   α-Hydrazono-β-ketoesters

Photochromism of α-hydrazono-β-ketoester (HKE)[16] involves both geometrical isomerization and switching of hydrogen bonding and attracted our great interest because reversible structural alteration between the rod-like and L-shaped molecules suitable for the command surface can be obtained when R_2 and R_3 are replaced by long alkyl groups, as shown in Figure 6.   We employed hexyl for R_2 whereas triethoxysilyl group was introduced in R_3 residue to attach the photochromic unit to silica surface.   Modification of quartz plates with HKE monolayers was performed by the silylation similar to that for azobenzene monolayers.

LC cells were made in a conventional way and exposed to uv for the isomerization of the chromophore.   As given in Figure 7, the cell of the initial state of HKE is quite dark, indicating the homeotropic alignment. When uv exposure is applied, the cell becomes bright similarly to that of Az modified substrates.

Quite different behavior was observed in the reverse process.   The cell does not become dark even upon prolonged exposure to visible light for the reverse isomerization.   It was confirmed that the reverse isomerization of HKE

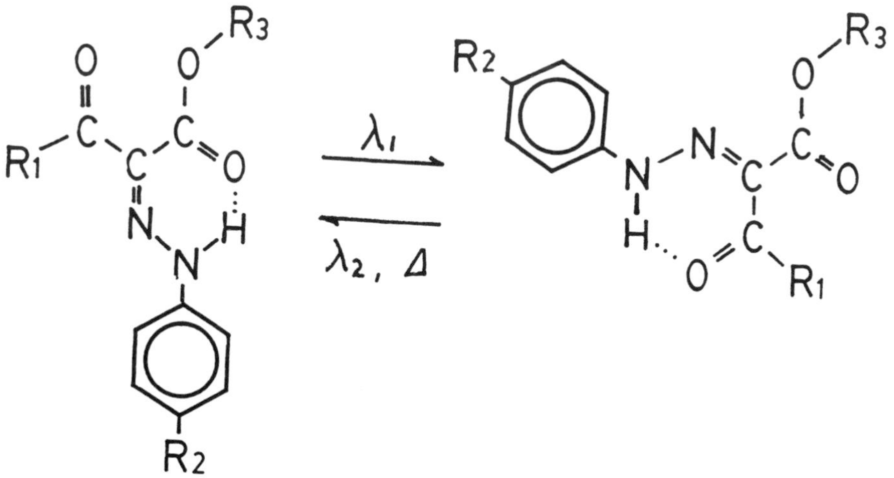

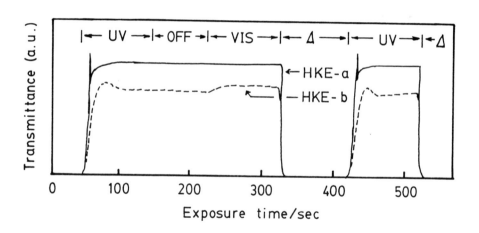

Figure 6 Photochromism of α–hydrazono–β–ketoester

Figure 7 Photoresponse of a cell constructed with quartz plates covered with α
–hydrazono–β–ketoester group

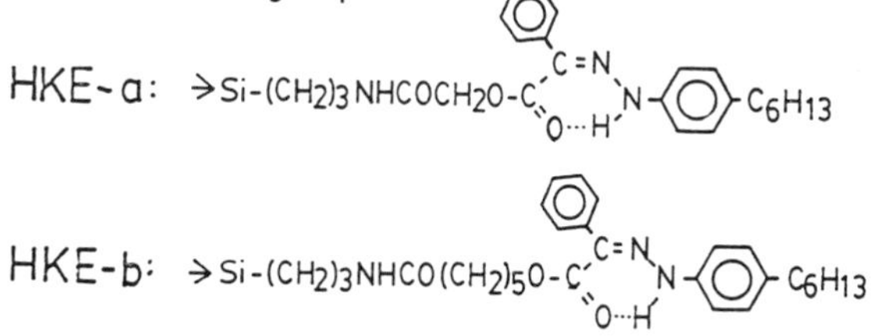

HKE-a: →Si-(CH₂)₃NHCOCH₂O-C ... C=N–N ... –C₆H₁₃

HKE-b: →Si-(CH₂)₃NHCO(CH₂)₅O-C ... C=N–N ... –C₆H₁₃

in the monolayer takes place only thermally in contrast to the solution photochromism which involves the reverse photisomerization upon exposure to visible light. Erasure of photoimage was readily performed by heating the cell. Thus, the combination of uv exposure and heat treatment results in the reversibility of the photoresponse.

5. CONCLUSION

1) Photoisomerization of azobenzene monolayers covalently bonded on silica substrates causes reversible alignment change of nematic LC's of potential value for photomemory and display technologies. This type of photofunctional surfaces have been proposed to refer to "command surfaces".
2) The photoinduced regulation of LC alignment is influenced by various factors including the molecular structure of azobenzene units, occupied area of Az unit, the mode of silylation and simultaneous treatment with alkyl silylating reagents.
3) Quartz surfaces modified with stilbene and α-hydrazono-β-ketoesters have been found to act as command surfaces, implying that the photoregulation is closely related with the reversible change between rod-like and L-shaped photochromic molecules.

ACKNOWLEDGMENT

This work has been carried out in the framework of the project of "photoactive materials" conducted by Agency of Industrial Science and Technology of the Ministry of International Trade and Industry, aiming at establishing fundamental technologies for high density optical memory using photochromism as well as photochemical hole burning. The author would like to devote this paper to the late Prof. S. Tazuke who promoted the project which had started 1985 and gave us fruitful suggestions and encouragement. The author would like to thank all his coworkers for their persistent endeavor.

REFERENCES

1) Photochromism: Molecules and Systems, eds. H. Dürr and H. Bouas-Laurent, (Elsevier, Amsterdam 1990).

2) E. Sackmann, J. Am. Chem. Soc. 93 (1971) 7988.

3) B. Schnuringer and J. Bourdon, J. Chem. Phys. 73 (1976) 795

4) W. E. Haas, K. F. Nelson, J. E. Adams and G. A. Dir, J. Electrochem. Soc. 121 (1974) 1667.

5) S. Sato and H. Ueda, Denshitsushin Gakkaishi J62 (1979) 179.

6) K. Ogura, H. Hirabayashi, A. Uejima and K. Nakamura, Jpn. Appl. Phys. 21

(1982) 969.

7) K. Ichimura, Y. Suzuki and A. Hosoki, in preparation.

8) a) S. Tazuke, S. Kurihara and T. Ikeda, Chem. Lett. (1987) 991; b) T. Ikeda, T. Miyamoto, S. Kurihara, M. Tsukada and S. Tazuke, Mol. Cryst. Liq. Cryst. 182B (1990) 357; c) T. Ikeda, T. Miyamoto, S. Kurihara, M. Tsukada and S. Tazuke, Mol. Cryst. Liq. Cryst. 182B (1990) 373

9) a) T. Ikeda, S. Horiuchi, D. B. Karanjit, S. Kurihara and S. Tazuke, Chem. Lett. (1988) 1679; b) T. Ikeda, S. Horiuchi, D. B. Karanjit, S. Kurihara and S. Tazuke, Macromolecules 23 (1990) 36; c) T. Ikeda, S. Horiuchi, D. B. Karanjit, S. Kurihara and S. Tazuke, Macromolecules 23 (1990) 42.

10) K. Ichimura, Y. Suzuki, T. Seki, Y. Kawanishi and K. Aoki, Langmuir 4 (1988) 1214.

11) K. Ichimura, Y. Suzuki, T. Seki, Y. Kawanishi and k. Aoki, Makromol. Chem. Rapid Commun. 10 (1989) 5.

12) T. Seki, T. Tamaki, Y. Suzuki, Y. Kawanishi, K. Ichimura and K. Aoki, Macromolecules 22 (1989) 3505.

13) K. Ichimura, K. Aoki and T. Seki, in preparation.

14) K. Aoki, K. Ichimura, T. Tamaki, T. Seki and Y. Kawanishi, Kobunshi Ronbunshu 47 (1990) 771.

15) K. Ichimura, T. Seki, T. Tamaki and T. Yamaguchi, Chem. Lett. (1990) 1645.

16) a) P. Courtot, R. Pichon and J. Le Saint, Tetrahedron Lett. (1976) 1177; b) P. Courtot, R. Pichon and J. Le Saint, Tetrahedron Lett. (1976) 1181.

Photochemical Processes in Organized Molecular Systems
K. Honda (Editor-in-Chief)
© Elsevier Science Publishers B.V., 1991

SYNTHETIC DESIGN OF LIQUID CRYSTALS FOR DIRECTIONAL ELECTRON TRANSPORT

Marye Anne FOX* and Horng-Long PAN

Department of Chemistry, University of Texas at Austin, Austin, TX 78712

The construction of a solid state photovoltaic device composed of a thin film of organic semiconductor sandwiched between indium tin oxide electrodes is described. Macroscopic order is achieved in the photoactive thin film by exploiting the inherent tendency of liquid crystals to self-organize: columnar discotic phases with some crystalline dislocations constitute a possible photoinitiated conduction pathway between the illuminated and dark electrodes. A scientific rationale for the unusual photovoltaic effect observed in this symmetrical cell is offered, with particular emphasis on the molecular features necessary for efficient current generation. The efficiency of cells constructed from the zinc and palladium octakis(β-octoxyethyl) porphyrins is compared, and the possibility of analogously employing hexakis(alkoxy)triphenylenes and tetrakis(alkoxymethyl)tetrathiafulvalenes is considered.

1. INTRODUCTION

1.1. Optoelectronic devices

The many practical applications of spatially-controlled photoinduced charge transfer in imaging and current transduction have made the search for the necessary materials a pressing technological issue. The stringent requirements for such components include high extinction coefficients in the desired wavelength region, low processing costs, and easy access to structurally modified materials adaptable to specific applications.

An attractive set of compounds which may satisfy these requirements are organic liquid crystals. Previous investigations have consistently shown that crystalline phases offer obviously superior performance in electronic applications over the same materials disposed in amorphous form. Presumably this effect derives from the lower defect density in well-ordered materials, which in turn allows for improved carrier mobility and enhanced lifetime. Access to large area single crystals of organic or inorganic solids is usually limited, however, and the possibility of employing materials which self-order to reduce the number of structural defects is indeed attractive. Liquid crystals, representing one of the best understood classes of self-ordering materials, spontaneously form partially ordered, but fluid, structures in which many structural defects present in the fluid phases disappear via self-healing. We reasoned, therefore, that large area ordered solids could be attained by freezing an ordered liquid crystal. With discotic phases remaining oriented as in the partially ordered stacked phase, one might expect significant intermolecular orbital overlap which is usually considered a prerequisite for macroscopic conductivity. Such phases would also provide promising vehicles for investigating mechanisms of electronic conduction, energy transfer, and radiationless deactivation of excited states of oriented solids, none of which are well understood at a fundamental level.

1.2. Conductivity in Oriented Organic Solids

Electronic and ionic conduction in single crystals of polycyclic aromatics[1-3] has been known for several years. Usually only small photocurrents are observable, and then only with a bias from the work functions of dissymmetric contacting electrodes.

Much better results are obtained with a discotic liquid crystalline porphyrin[4,5] **1** (M = Zn, R = O(CH$_2$)$_7$CH$_3$) which can be capillary-filled into a solid state photocell via melting

1 (R = OC$_n$H$_{2n+1}$)

to the isotropic liquid phase. An unusual photovoltaic response is observed with such films, whose discotic arrangement is represented schematically in Figure 1, when

FIGURE 1
A schematic diagram of a columnar discotic phase with crystal dislocations[9]

symmetrically described by either an equivalent circuit or by a mathematical model specifying the number and spatial distribution of charge carriers generated in the bulk of the liquid crystalline phase.[6]

Compound **1** was not the first reported liquid crystalline porphyrin exhibiting semiconductive properties. The first member of this family, synthesized by Goodby et al.,[7] was a symmetrical octakis(dodecanol)ester with an liquid crystalline phase stable over a range of only 0.1°C which, in addition, was observed only upon cooling. In 1982, Piechocki et al.[8] synthesized a liquid crystalline phthalocyanine which, when complexed with copper, melted at 53 °C to a columnar discotic mesophase that was stable to its decomposition point at approximately 300°C. However, the viscosity of the columnar phase was so high that it would barely flow, even under pressure, even at temperatures just below its decomposition point. Since this compound possesses no isotropic liquid phase, it could not be capillary-filled into a device.

For this reason, the synthesis of a family of 2,3,7,8,12,13,17,18-octakis(β-octaalkoxyethyl)porphyrin derivatives was undertaken in our research group.[4,5] Most of these porphyrins exhibit a discotic liquid crystalline phase, and all exhibit an isotropic liquid phase. These materials are therefore suitable for testing as photoresponsive components in large-area organic photovoltaic devices. We have also recently prepared two other series of molecules (hexakis(alkoxy)- triphenylenes **2**[9] and tetrakis(alkoxymethyl)tetrathiafulvalenes **3**[9]) possessing, like **1**, an extended planar disc-like core symmetrically substituted with flexible long alkyl chains in the hope that they might similarly exhibit self-ordering and supramolecular conductivity under photochemical excitation.

where R = C_nH_{2n+1}

1.3. Photophysical Properties of the Liquid Crystals

Besides the observed photoeffects in the solid state photovoltaic cell, changes in absorption, fluorescence emission, and fluorescence excitation spectra are also observed in **1** (M = Zn, R = $O(CH_2)_7CH_3$) as the order in the resulting film is increased from amorphous to crystalline.[10,11] These spectral shifts can be rationalized in terms of a molecular exciton model. Photophysical measurements provide evidence for ring-to-ring charge transfer as a function of the organization of neighboring chromophores. With large area thin films of these molecular semiconductors,[4] the efficiency of the photovoltaic cell is thus controlled exclusively by interfacial kinetics at the electrode surface.

Interfacial charge injection in these cells is undoubtedly related to dye sensitization of semiconductor electrode surfaces. For example, lamellar photovoltaic cells containing chlorophyll-a,[12] phthalocyanines,[13] porphyrins,[14] or their metallic complexes as light-absorbing components have been described, and voluminous work[15] has shown that charge injection efficiency is related to the photophysical properties and energy levels of the adsorbed excited state.[16] Three possible mechanisms have been proposed to rationalize dye sensitization: (1) the direct electron transfer from the excited dye to the conduction band of semiconductor; (2) energy transfer from the excited dye to surface states of the contacting electrode, followed by the electron injection to the bulk from the excited surface state; or (3) sequential electron transfer from the excited dye to a surface state and thence to the continuum of the semiconductor. Direct electron transfer from the excited dye to the conduction band of semiconductor surface appears to be most prevalent in systems with appropriately ordered redox levels.[17] Electron injection from either the singlet or triplet excited state has been documented when the energy level of the conduction band edge of the semiconductor lies at a less negative potential than the oxidation potentials of the singlet or triplet excited states, respectively.

Profound changes in the photophysical properties of porphyrins can be induced by altering the identity of the complexed metal. For example, Zn octaethylporphyrin[18] exhibits excited state chemistry dominated by the singlet state (i.e., the triplet state is undetectable at room temperature), whereas intersystem crossing in the palladium analog occurs efficiently, with the triplet excited state being produced with a quantum yield near unity.[19] Furthermore, packing and other intermolecular associations are likely to be sensitive not only to the complexed metal but also to the chain length of the symmetrically appended hydrocarbon chains.

It is unknown at present whether multiplicity or other structural alterations of the component molecules in the discotic liquid crystal will influence the observed efficiency for current generation in solid state photovoltaic devices such as described herein. Specifically, we wished to establish whether the substantial, persistent photovoltaic effect achieved for **1** (M = Zn, R = $O(CH_2)_7CH_3$), i.e., V_{oc} = 0.3 V, i_{sc} = 0.4 mA/cm^2 under white light, 150 mW/cm^2),[9] which presumably derives from a singlet excited state, is maintained in a cell containing **1** (M = Pd, R = $O(CH_2)_7CH_3$), which may involve a triplet excited state. We describe herein a comparison of photovoltaic effects with these liquid crystalline metalloporphyrins, together with a preliminary comparison of the porphyrins **1** with the related disc-shaped molecules **2** and **3**.

2. EXPERIMENTAL SECTION

2.1. Materials

The free base porphyrin and its other metalloporphyrins were prepared and purified as described previously.[5] The syntheses of hexakis(alkoxy)triphenylenes and tetrakis(alkoxymethyl)tetrathiafulvalenes will be described elsewhere.[9] Optical textures were observed as a function of temperature between crossed polarizers in a Leitz polarizing microscope, equipped with a hot stage, at 100 X magnification. Differential scanning calorimetry (DSC) measurements were conducted on either a Perkin-Elmer TG-7 or a Mettler FP 80/84 instrument.

2.2. Preparation of Thin Films and the Photovoltaic Device

Indium-tin oxide (ITO) coated float glass from Delta Technologies was cleaned by a previously reported procedure.[10] The resulting glass was immediately exposed to a dilute (ca. 1 mM) CCl_4 solution of octadecyltrichlorosilane (OTS, Aldrich) in order to modify the surface tension of the glass to obtain more uniform spin-coated thin films for the photophysical measurements. As previously described,[10] the cleaned or silanated glasses were assembled into a cell separated by a 5 μm polymeric spacers which was capillary-filled from an isotropic melt of the photoactive component and allowed to cool slowly through its liquid crystalline phase and ultimately to the solid state. Alternatively, thin films of the molecules of interest could be prepared by spin - coating (as previously described) or by careful evaporation onto ITO or silanated glass at a pressure ca. 10^{-6} torr. The spin coater was a Pine Instruments Co. analytical rotator. The quartz plates were attached to the spinner with double stick tape and one or two drops of coating solution (0.5 M) were applied at the spin rate of 4000 rpm.

2.3. Photocurrent and Spectroscopic Measurements

A 150 watt Xe lamp (Oriel) with a 12 cm water filter was used for most of the experiments involving white light. The intensity of this source at the cell surface was ca. 150 mW / cm^2. For monochromatic illumination, a 250 W Xe lamp equipped with a parabolic reflector and a scanning monochromator were used. The output flux of this system was calibrated with a silicon photodiode detection system (Oriel). Current-voltage measurements were performed with a Princeton Applied Research 175 universal programmer and a Model 173 potentiostat. The signal was recorded on a Houston Instruments 2000 X-Y recorder. An Oriel 7072 detection system was also used to measure photocurrents. Photovoltages were also measured with a Fluka 77 multimeter.

UV-visible spectra of the spun thin films were measured on a Hewlett-Packard

8451A spectrophotometer equipped with a Hewlett-Packard 7470A plotter. An SLM Aminco SPF 500 spectrofluorometer was used to record emission spectra. The hot stage of a Leitz Laborlux D polarizing microscope was mounted in an SLM Aminco SPF 500 spectrofluorometer for the temperature-dependent fluorescence measurements.

3. RESULTS AND DISCUSSION

3.1. Dependence of the Transition Temperature to the Liquid Crystalline Phase on Alkyl Chain Length in 1 (M = Pd)

As has been previously reported for 1 (M = Zn), the family 1 (M = Pd) show the presence of a stable liquid crystalline phase over a workable temperature range, Figure 2. The transition temperatures for the family depend on the length of the alkyl

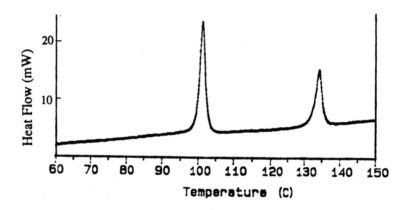

FIGURE 2

Differential scanning calorimetry trace for 1 (M = Pd, R = $O(CH_2)_7CH_3$)

chain as has been previously reported for 1 (M = Zn), Figure 3. Thus, longer chains cause a reduction in the observed phase transition temperatures. Similar effects have been reported for the phase transitions in 2.[21] No evidence for a stable liquid crystalline phase could be observed by differential scanning calorimetry for 3.[9]

As yet, there is no coherent theory available to predict the occurrence of phase transitions or the character of the intermolecular associations encountered within the partially ordered liquid crystalline regimes. Thus, neither the number of the stable phases, the specification of their character, e.g., as nematic, discotic, etc., the temperature range over which these phases occur, the enthalpy changes associated with the transitions of interest, nor the relationship between the phase transition and thermal stability of the component molecule can be predicted by other than purely empirical observation.

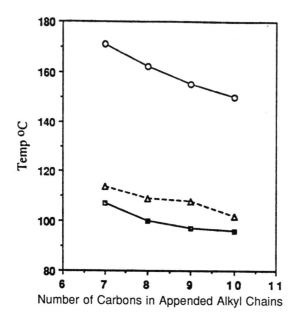

FIGURE 3
Dependence of transition temperatures for the crystal-to-discotic (triangles) and discotic-to-isotropic transitions (squares) for **1** (M = Zn) and the crystal-to-isotropic liquid transition for **1** (free base)

3.2. Photophysical Properties of **1** (M = Zn and Pd), **2**, and **3**

Absorption spectra of **1** (M = Zn, R = $O(CH_2)_7CH_3$) in the solid, liquid crystalline, and isotropic phases, Figure 4, are characterized by small red shifts of the Soret band

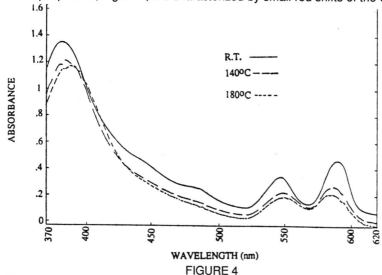

FIGURE 4
Absorption spectra of spun thin films of **1** (M = Zn, R = $O(CH_2)_7CH_3$)

with order.[20] Analogous effects are also observed in its emission spectrum.[9]

Improved spectral resolution is also observed upon heat treatment of the spin-coated films of both the zinc and palladium metallated **1**. Heat-cured films, represented by the dashed line in Figure 5 for **1** (M = Zn, R= $O(CH_2)_7CH_3$) and obtained by heating the as-spun film (solid line) to its isotropic phase before slowly cooling through the liquid crystalline phase to room temperature, show somewhat sharper peaks than films directly spin-coated onto a OTS-pretreated glass, implying the occurrence of self-healing of defects in the as-spun liquid crystal.

Photophysical quenching by electron injection from the adsorbed dye in the vacuum-deposited film is indicated by quenching of fluorescence on native ITO (bottom trace) compared with that observed on OTS-treated glass, Figure 6. The incomplete quenching by ITO (ca. 20 %) presumably derives from the requirement for direct contact of the injecting dye with the semiconductor for efficient interfacial electron transfer. The much weaker fluorescence from the palladium-metallated porphyrin renders the analogous experiment with **1**(M = Pd, R = $O(CH_2)_7CH_3$) impossible.

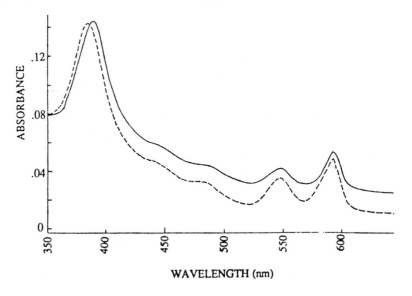

FIGURE 5

Dependence of the absorption spectra of spun thin films of **1** (M = Zn, R= $O(CH_2)_7CH_3$) on pre-annealing

Similar observations have been made with other dyes in contact with poised semiconductors. In particular, J- and K-aggregates are known to be differentially quenched under static conditions which mimic the solid-adsorbate interface encountered in these layers.

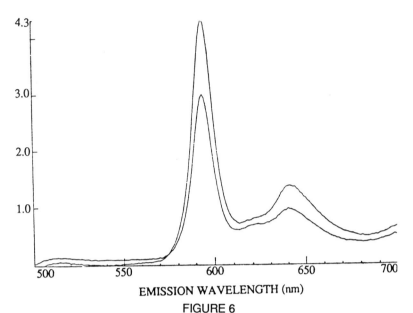

FIGURE 6
Fluorescence intensity of **1** (M = Zn, R = $O(CH_2)_7CH_3$) on ITO (bottom curve) and on glass (top curve).

The absorption spectra of the substituted triphenylenes **2** make them less attractive than **1** as sensitizers for photovoltaic devices employing metal oxide

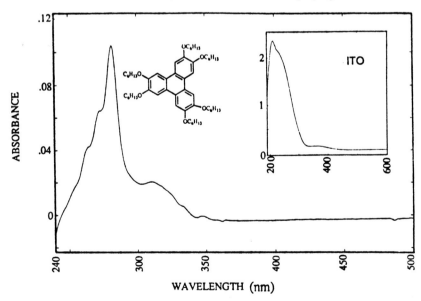

FIGURE 7
Absorption spectrum of **2** (R = C_6H_{13}). Inset: Absorption of ITO glass.

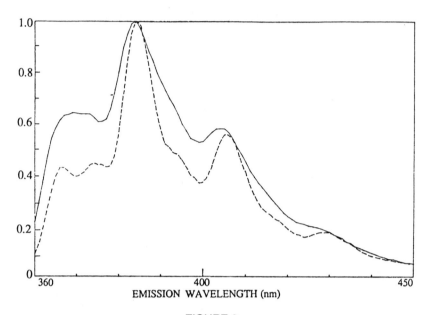

FIGURE 8
Emission spectra of **2** (R = C$_6$H$_{13}$) as a solid thin film (dashed line) and in dilute methylene chloride solution (solid line).

semiconductor transparent electrodes. As seen in Figure 7, these substrates exhibit only trailing absorption at wavelengths longer than 350 nm and their absorption profiles thus overlap strongly with direct band excitation of the semiconductor. Nonetheless, sharpening and similar red shifts in the emission spectra of **2** (R = C$_6$H$_{13}$) as were observed in **1** (M = Zn, R = O(CH$_2$)$_7$CH$_3$), vide supra, are observed, Figure 8.

Lacking a stable liquid crystalline phase, **3** shows no order-related shifts in band positions in the absorption spectrum with temperature. As shown in Figure 9, however, it does develop intense coloration upon charge transfer complexation with tetracyanoquinone (TCNQ). Although such complexes do exhibit interesting materials properties, they cannot be easily ordered via a discotic phase in the manner of **1** and **2**.

The formation of charge transfer complexes is usually accompanied by dramatically increased dark conductivity, when compared with the undoped components of these complexes. It is usually assumed that the degree of conductivity relates to the strength of orbital interaction, particularly in complexes in which segregated or aligned stacks constitute major structural variants of the crystalline phase. The parent of **3**, tetrathiafulvalene, has been extensively investigated in this context, and its charge transfer complex with TCNQ represents a classic example of such alternating stacking. The functionalization of the parent with alkyl chains of variable lengths apparently does not denigrate from its ability to similarly interact with strong electron acceptors, for the observed dark conductivity in the charge transfer complexes of **3**, formed with controlled stoichiometry, very closely mimics that observed in the parent.

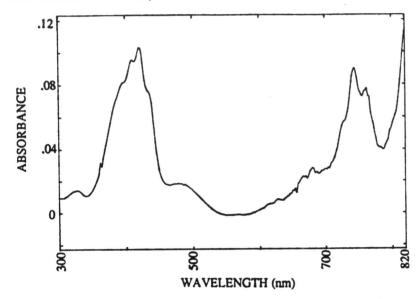

FIGURE 9
Absorption spectrum of a 1:1 complex of **3** ($R = C_6H_{13}$) with TCNQ

3.3. Construction of Solid State Photovoltaic Cells from **1**(M = Zn, R = $O(CH_2)_7CH_3$), **1**(M = Pd, R = $O(CH_2)_7CH_3$), and **2** ($R = C_6H_{13}$)

A large area photovoltaic device in which the photoresponsive liquid crystal is sandwiched between conductive glass plates separated by a polymeric spacer (1 - 5μm) is amenable to easy processing via self-order induced upon slowly cooling through the liquid crystalline phase. When **1**(M = Zn, R = $O(CH_2)_7CH_3$) is loaded into such a device, stable photocurrents are produced under steady state illumination.[10] Photocurrent was observed to increase linearly with light intensity and the observed action spectrum was found to be dependent on the cell thickness, and hence on the spatial penetration of the incident light from the irradiated face.

The observed photovoltaic effects have been interpreted as having arisen from exciton dissociation at the illuminated electrode, which becomes charged negative under illumination because of preferential injection of electrons (rather than holes) at the illuminated electrode-organic interface. The importance of this photoinjection process, which derives from the ordered character of the aligned organic and is controlled entirely by interfacial kinetics, differentiates these cells from less efficient devices containing evaporated films of porphyrins or phthalocyanines.   An approximate mathematical model to describe the requisite photoinjection at the interface and photoconductivity in the bulk of the organic film has been proposed, Figure 10.[10] Therein, light absorption produces an exciton which can diffuse or dissociate to a diffusable charge carrier, either or both of which can migrate to the electrode where preferential electron injection occurs. This process is accompanied by bulk generation and recombination of charge carriers which mediate conductivity through the stacked units of the ordered organic phase and hence electical connection with the dark electrode.

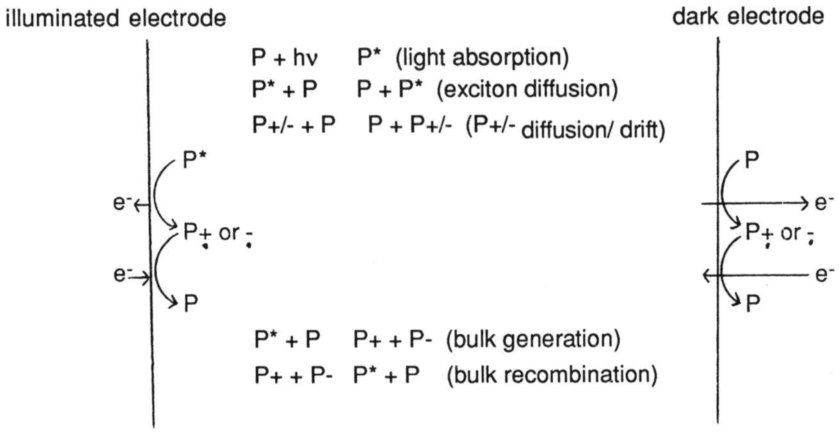

illuminated electrode   dark electrode

$P + h\nu \quad P^*$ (light absorption)
$P^* + P \quad P + P^*$ (exciton diffusion)
$P+/- + P \quad P + P+/-$ (P+/- diffusion/ drift)

$P^* + P \quad P+ + P-$ (bulk generation)
$P+ + P- \quad P^* + P$ (bulk recombination)

FIGURE 10
A proposed model for the photovoltaic effects observed for ordered organic / ITO sandwich devices[10]

Analogous effects, but with a quantum efficiency about 20-fold lower than with its Zn analog, are observed when 1(M = Pd, R = $O(CH_2)_7CH_3$) replaces 1(M = Zn, R = $O(CH_2)_7CH_3$) as the photoactive component of the device. Figure 11 shows a typical current-voltage curve observed in a cell containing 1(M = Pd, R = $O(CH_2)_7CH_3$) upon applying a voltage bias from - 0.5 V to 0.5 V under illumination at three different wavelengths (400, 515 and 550 nm ± 5 nm). The shape of this i - V curve corresponds closely to that previously reported for 1(M = Zn, R = $O(CH_2)_7CH_3$).[10] As before, the action spectrum is a function of cell thickness, with an inverse correlation between the absorption spectrum and the photocurrent wavelength dependence being observed for thick cells. For thinner cells a direct correlation between the observed action spectrum, Figure 12, and the absorption spectrum is observed. As before, substantial open circuit photovoltage and short

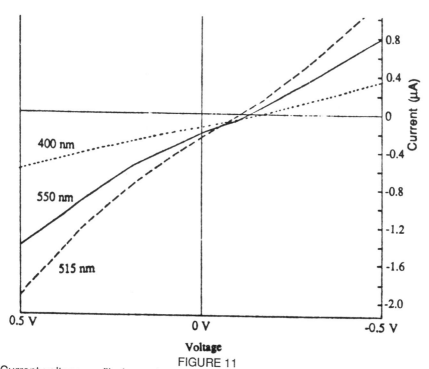

FIGURE 11

Current-voltage profile for a photovoltaic device in which an ordered film of 1(M = Pd, R = O(CH$_2$)$_7$CH$_3$) (ca. 5 μ) is sandwiched between ITO electrodes

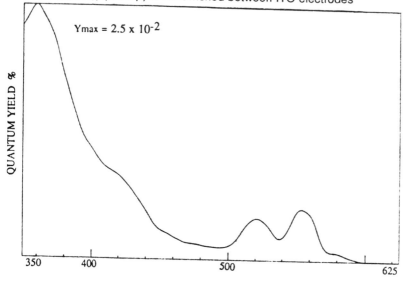

FIGURE 12

Action Spectrum for a thin 1(M = Pd, R = O(CH$_2$)$_7$CH$_3$) · ITO Device

circuit photocurrent is observed in the 1(M = Pd, R = O(CH₂)₇CH₃) cell and steady state illumination produces no significant decrease in the observed photoeffects with illumination time.

The possibility that the observed photocurrents and photovoltages occur as a consequence of the Dember effect is rendered unlikely both from the magnitude of the observed photocurrent densities (which are several orders of magnitude larger than typical Dember effects) and from the directional switching which occurs upon reversal of the face of illumination in these symmetrical cells.

Figure 13 shows the relationship between photovoltage and light intensity observed in this device. The observed photovoltage is almost saturated with an illumination intensity ca. 140 mW / cm² and would therefore be useful for ambient solar exposure.

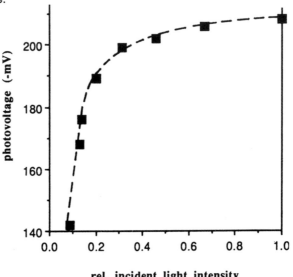

rel. incident light intensity

FIGURE 13

Dependence of open circuit photovoltage in a photovoltaic device in which an ordered film of 1(M = Pd, R = O(CH₂)₇CH₃) (ca. 5 μ) is sandwiched between ITO electrodes on incident light intensity. Full lamp intensity is ca. 450 mW / cm². Attenuation is accomplished with neutral density filters.

Although most treatments of exciton theory have dealt with singlet states as, for example, in numerous studies of the photogeneration of charge carriers in single crystals of aromatic hydrocarbons, these results clearly show that species whose excited state chemistry is likely to have substantial yields of triplet states can also act as excited electron donors at the organic-electrode interface.

When **2** (R = C₆H₁₃) is capillary-filled into an analogous cell, similar (but attenuated) photoeffects are observed, Figure 14. Furthermore, the significant overlap between the absorption spectrum of **2** (R = C$_6$H$_{13}$) and ITO (vide supra) makes difficult the unambiguous assignment of the very small observed photocurrents to either sensitization by the organic layer or trailing absorption by the semiconductor itself. The ordered structure of **2** should permit excitonic migration through stacks, as has been suggested as important for the operation of cells containing **1** (M = Pd or Zn, R = O(CH$_2$)$_7$CH$_3$): hexaalkoxytriphenylenes exhibit columnar mesophases which form coaxial, one dimensional conducting materials.[22-24] The intermolecular distance along the column (0.36 Å)[25,26] permits appreciable orbital overlap between neighboring aromatic rings; the columns are

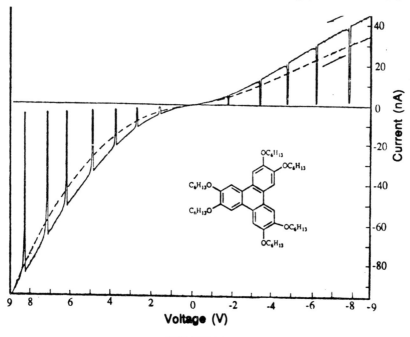

FIGURE 14
Current-voltage profile for a photovoltaic device in which an ordered film of **2** (R = C$_6$H$_{13}$) (ca. 13 μm) is sandwiched between ITO electrodes. The vertical lines indicate manual light chopping of the incident illumination.

hexagonally packed and the distance between two neighboring columns is approximately 20 Å. For non-oriented samples, the coherence length is about 300 Å,[27] and upon doping with bromine and iodine the liquid crystals display semiconducting properties.[28-30] Time-resolved luminescence experiments on **2** (R = C$_6$H$_{13}$) in homogeneous solution, solid, and liquid crystalline phases reveal normal fluorescence, phosphorescence and p-type delayed fluorescence related to energy migration.[31] Clearly, an adequate scientific rationale to permit prediction of molecular features required for effective charge migration in analogous photovoltaic devices is not yet available. On-going work in our laboratories currently seeks to remediate this deficiency.

4. CONCLUSIONS

Liquid crystals with columnar mesophases can be oriented in a symmetrical sandwich photocell to produce significant photovoltaic effects in a novel organic solid state device. Lower efficiencies are observed with liquid crystalline triphenylenes **2** than with liquid crystalline porphyrins **1**, although the signficant spectral overlap of this family with the facing "transparent" metal oxide electrodes compromises this conclusion. The macroscopic order necessary for diffusional migration of charge carriers in these self-ordered arrays can be established spectroscopically from shifts in absorption band positions in the various phases. Order per se, however, is an insufficient criterion for predicting efficiency in such devices. A symmetrically alkylated derivative **3** of tetrathiafulvalene shows no stable liquid crystalline phases and therefore cannot be easily employed as a self-ordering component of such cells.

ACKNOWLEDGEMENT

We are grateful to Dr. Brian Gregg and Professor Allen Bard for fruitful discussions regarding this work. Financial support for this project from the Texas Advanced Research Program is gratefully acknowledged.

REFERENCES

1) F. Gutmann, H. Keyser, and L.E. Lyons, Organic Semiconductors Parts A and B (Krieger, Malabar, Florida, 1981 and 1983).

2) K.C. Kao and W. Hwang, Electrical Transport in Solids (Pergamon, Oxford, 1981).

3) D.P. Craig and S.H. Walmsley, Excitons in Molecular Crystals -Theory and Applications (W. A. Benjamin, New York, 1968).

4) B. A. Gregg, M. A. Fox, and A. J. Bard, Chem. Commun. 15 (1987) 1134.

5) B. A. Gregg, M. A. Fox, and A. J. Bard, J. Am. Chem. Soc. 111 (1989) 3024.

6) B. A. Gregg, M. A. Fox, and A. J. Bard, J. Phys. Chem. 93 (1989) 4227.

7) J.W. Goodby, P.S. Robinson, B.K. Teo, and P.E. Cladis, Mol. Cryst. Liq. Cryst. 56 (1980) 303.

8) C. Piechocki, J. Simon, A. Skoulios, D. Guillon, and P. Weber, J. Am. Chem. Soc. 104 (1982) 5245.

9) M.A. Fox and H.L. Pan, unpublished results.

10) B. A. Gregg, M. A. Fox, and A. J. Bard, J. Phys. Chem. 94 (1990) 1586.

11) B. A. Gregg, M. A. Fox, and A. J. Bard, Mater. Res. Soc. Sympos. Proc. 173 (1990) 199.

12) a) C.W. Tang and A. C. Albrecht, J. Chem. Phys. 62 (1975) 2139.; b) J.P. Dodelet, J. Le Brech, and R.M. Leblanc, Photochemistry and Photobiology 29 (1979) 1135.

13) a) M. Martin, J.J. Andre', and J. Simon, Nouv. J. Chim. 5 (1981) 485; b) J. Noolandi and M. Hong, J. Chem. Phys. 70 (1979) 3230; c) R.O. Loutfy and E.R. Menzel, J. Am. Chem. Soc.102 (1980) 4967; d) R.O. Loutfy and J.H. Sharp, J. Chem. Phys.71 (1979) 1211; e) F.R. Fan and L.R. Faulkner, J. Chem. Phys. 69 (1978) 3334.

14) a) F.J. Kampas and M. Gouterman, J.Phys. Chem. 81 (1977) 690; b) F.J. Kampas, K. Yamashita, and J. Fajer, Nature 284 (1980) 40; c) W.A. Nevin and G.A. Chamberlain, J. Chem. Soc., Faraday Trans. II 85 (1989) 1729; c) ibid, 1747.

15) F. Gerischer and F. Willig, Fortschr. Chem. Forsch. 61 (1976) 33.

16) M.A. Fox, F.J. Nobs, and T. A. Voynick, J. Amer. Chem. Soc. 102 (1980) 4036

17) a) K. Hashimoto, M. Hiramoto, A.B.P. Lever, and T. Sakata, J. Phys. Chem. 92 (1988) 1016; b) K. Hashimoto, M. Hiramoto, and T. Sakata, Chem. Phys. Lett. 148 (1988) 215; c) K. Hashimoto, M. Hiramoto, and T. Sakata, J. Phys. Chem. 92 (1988) 4272.

18) A.T. Gradyushko and M.P. Tsvirko, Opt. Spectrosc. 31 (1971) 548.

19) D. Madge, M.W. Windsor, D. Holten, and M. Gouterman, Chem. Phys. Letts.29 (1974) 183.

20) Although temperature does commonly affect the intensity of absorption and the breadth of a given adsorption band, shifts in band position caused by temperature changes are rare.

21) C. Destrade, M.C. Mondon, and J. Malthete, J. Phys. (Suppl. C3) 40 (1979) 17.

22) O.C. Musgrave and C.J. Webster, J. Chem. Soc. (C) (1971) 1397.

23) C. Destrade, P. Foucher, H. Gasparoux, N.H. Tinh, A.M. Levelut, and J. Malthete, Mol. Cryst. Liq. Cryst. 106 (1984) 121.

24) C. Destrade, N.H. Tinh, H. Gasparoux, J. Malthete, and A.M. Levelut, Mol. Cryst. Liq. Cryst. 71 (1981) 111.

25) A.M. Levelut, J. Phys. (Paris) 40 (1979) L-81.

26) M. Cotrait, P. Marsau, M. Pesquer, and V. Volpilhac, J. Phys. (Paris) 43 (1982) 355.

27) A.M. Levelut, J. Chim. Phys. 80 (1983) 149.

28) L.Y. Chiang, J.P. Stokes, R. Safinya, and A.N. Bloch, Mol. Cryst. Liq. Cryst. 125 (1985) 279.

29) L.Y. Chiang, R. Safinya, N.A. Clark, K.S. Liang, and A.N. Bloch, Chem. Commun., (1985) 695.

30) J. van Keulen, T.W. Warmerdam, R.J.M. Nolte, and W. Drenth, Rec. Trav. Chim. Pays-Bas. 106 (1987) 534.

31) D. Markovitsi, F. Rigaut, and M. Mouallem, Chem. Phys. Lett. 135 (1987) 236.

Photochemical Processes in Organized Molecular Systems
K. Honda (Editor-in-Chief)
© Elsevier Science Publishers B.V., 1991

DIFFUSE REFLECTANCE LASER PHOTOLYSIS OF ADSORBED MOLECULES

Frank WILKINSON and Robert BEER

Department of Chemistry, Loughborough University of Technology, Loughborough, Leicestershire, LE11 3TU, U.K.

Recently we have extended to heterogeneous, opaque and often highly scattering systems the advantages of being able to subject them to flash photolysis investigation by using diffuse reflected light instead of transmitted light as the analysing source on timescales extending from several seconds to picoseconds. Laser induced transient spectra and decay kinetics have been observed from a wide variety of samples including fractions of monolayers of organic molecules adsorbed on catalytic metal oxide surfaces, and included within the hydrophobic man-made zeolite 'Silicalite'. Experimental details are reviewed in brief as are some experimental results concerning a variety of bimolecular reactions which we have studied recently. The potential of the technique to study elementary reactions at interfaces is demonstrated with particular reference to four bimolecular surface reactions : (1) triplet-triplet annihilation in the case of acridine adsorbed on various porous silica surfaces, (2) bimolecular combination of diphenylmethyl radicals on silica gel and included in NaX and 'Silicalite' zeolites, (3) electron - radical cation recombination following multiphoton ionisation of diphenylpolyenes on metal oxide surfaces and (4) triplet energy transfer between eosin and anthracene on silica surfaces.

1. INTRODUCTION

Recently we have developed the technique of diffuse reflectance laser flash photolysis which has the advantage of being able to subject heterogeneous, opaque and often highly scattering samples to flash photolysis investigation by using diffuse reflected light instead of transmitted light as the analysing source on timescales extending from several seconds[1] to picoseconds[2]. Laser induced transient spectra and decay kinetics have been observed from a wide variety of samples including fractions of monolayers of organic molecules adsorbed on catalytic metal oxide surfaces[3,4], and included within zeolites[5], from semiconductor powders[6] and porous electrodes doped and undoped[7,8], from organic and inorganic microcrystals[9,10], and from dyes[11,12] adsorbed on fabrics and chemically bound to polymers. For recent reviews see references 13, 14 and 15.

F. Wilkinson and R. Beer

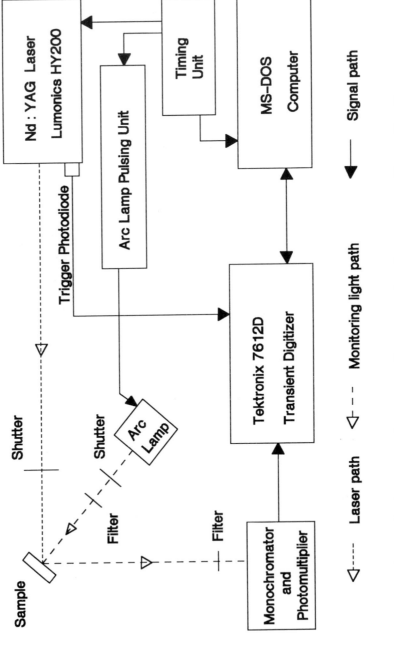

FIGURE 1. SCHEMATIC DIAGRAM SHOWING AN APPARATUS USED IN NANOSECOND DIFFUSE REFLECTANCE LASER FLASH PHOTOLYSIS STUDIES.

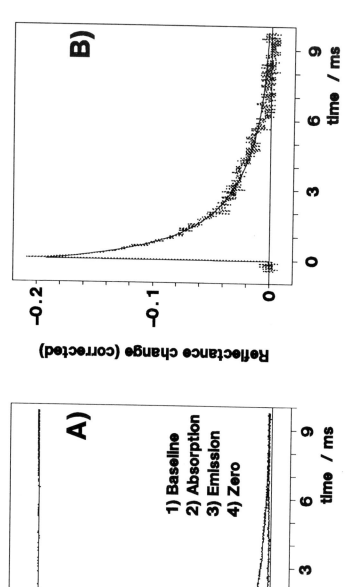

FIGURE 2. DIFFUSE REFLECTANCE LASER FLASH PHOTOLYSIS OF 0.15 μMOL OF ADSORBED EOSIN PER GRAM OF SILICA POWDER, EXCITED AT 532 NM WITH A 8 NS LASER PULSE. A) SEPARATE DIGITIZER TRACES. B) CORRECTED REFLECTANCE CHANGE.

2. EXPERIMENTAL

The equipment used for nanosecond laser flash photolysis in diffuse reflectance mode is identical to that used for studies in transmission mode except for the geometry for collecting the analysing light which is typically as shown in Figure 1. Samples are often held in a powder holder behind a quartz window or in a sealed fluorimeter cell. The observable in diffuse reflectance laser flash photolysis is $\Delta R(t)$ defined as follows:

$$\Delta R(t) = \frac{R_b - R(t)}{R_b} \tag{1}$$

where R_b is the sample reflectance before exposure to the exciting laser pulse and $R(t)$ the reflectance at time t after excitation. ΔR is thus the fractional change in reflectance and $\Delta R \times 100$ is often referred to as the percentage absorption by the transient.

Figure 2A shows the four traces which are recorded at each analysing wavelength necessary to obtain a corrected trace of reflectance change $\Delta R(t)$ as a function of time. The four experimental traces are: 1) Baseline trace i.e. analysing light reflected from sample in the absence of laser excitation. 2) Absorption trace i.e. analysing light reflected from sample before (pretrigger) and after laser excitation. This shows a decrease in reflection due to absorbing transients. 3) Emission trace i.e. any emission caused by laser excitation in the absence of analysing light. 4) Zero trace i.e. the digitizer is fired in absence of either laser excitation or analysing light to establish the zero settings. From these four traces a corrected trace of the reflectance change as a function of time can be obtained (see Figure 2B) at each analysing wavelength. Time resolved spectra are obtained by measuring many such traces at different wavelengths and then plotting the magnitude of the reflectance change as a function of wavelength at a known delay. Full experimental details are given in references 13 - 15.

2.1. Data Analysis of Transient Changes in Diffuse Reflectance

More extensive discussions of the analysis of data are given elsewhere.[16,17] Following the Kubelka-Munk treatment[18] for diffuse reflectance two light fluxes are considered travelling in opposite directions (see Figure 3A) perpendicular to the

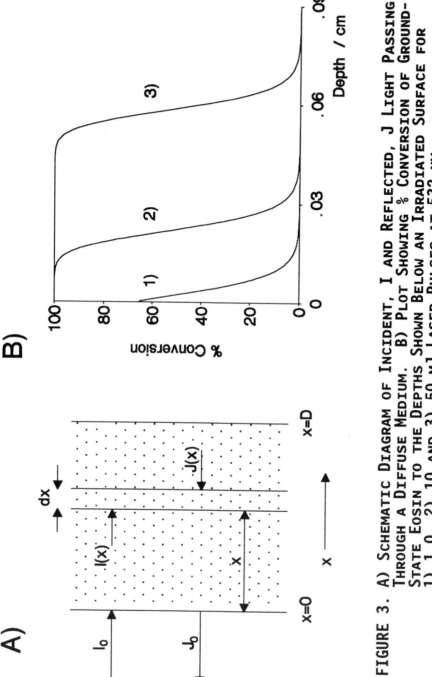

FIGURE 3. A) Schematic Diagram of Incident, I and Reflected, J Light Passing Through a Diffuse Medium. B) Plot Showing % Conversion of Ground-State Eosin to the Depths Shown Below an Irradiated Surface for 1) 1.0, 2) 10 and 3) 50 mJ Laser Pulses at 532 nm.

irradiated surface at x = 0 The attenuation of the incident flux
I depends on the absorption and scattering coefficients K and S
respectively and is given by

$$dI(x) = -I(x) \ (K+S) \ dx + J(x)Sdx \qquad (2)$$

and the generated flux since it passes in the opposite direction
has $\qquad dJ(x) = J(x) \ (K+S) \ dx - I(x)Sdx. \qquad (3)$

The diffuse reflectance R is given by

$$R = \frac{J_o}{I_o} \qquad (4)$$

where I_o and J_o are the incident and reflected fluxes at the
surface. Equations (2) and (3) can be solved for a layer so
thick that any further increase in thickness does not affect R
and provided K and S are independent of x this gives

$$\frac{(1-R)^2}{2R} = \frac{K}{S} = F(R) \qquad (5)$$

F(R), the remission function is linearly dependent on the number
of absorbing chromophores in any sample where S and K are
independent of the penetration depth below the surface.

In diffuse reflectance flash photolysis the initially excited
chromophores are usually homogeneously distributed. However,
photo-excitation produces transient or permanent changes in
absorption, preferentially just below the irradiated surface.
Theoretical treatments[16,17] show that there are two limiting
types of concentration profile produced, namely an exponential
fall off as a function of penetration depth and a homogeneous (or
"plug") profile these are illustrated schematically in Figure 3B
curves 1) and 3) respectively. The latter case is encountered
with large laser fluencies and with low concentrations of ground
state absorbers, where there is total conversion from ground
state to transient to a certain depth below the irradiated
surface. Since a homogeneous concentration of absorbers exists
the Kubelka-Munk theory can be applied [18]. For optically thick
samples at analysing wavelengths where only the transient absorbs
the remission function given by equation (5) is a linear function

of the concentration and can be used for kinetic analysis and for plotting absorption spectra. For the low percentage conversions the concentration of transients decreases exponentially below the irradiated surface. This occurs when there is a high concentration of ground state absorbers and with low laser fluencies. An analytical solution for the change in reflectance expected has been obtained by Lin and Kan [19] and is in the form of a converging series which has been shown[16] to relate ΔR as a linear function of the concentration of transient at values of ΔR less than 0.1.

Between these two limiting cases the change in ΔR with concentration depends on the concentration profile below the irradiated surface. In order to establish if either of the two limiting cases pertain it is necessary to calculate the expected change in transient concentration as a function of distance below the irradiated surface.

2.2 Calculation of Transient Concentration Profiles

To extract kinetic information from $\Delta R(t)$ one must first understand the concentration profile. A detailed description of the modelling procedure, including all relevant equations, has been published[16,17]. A brief outline of the algorithm is given below.

1) The sample is divided into a large number of thin slices i, such that the concentration of the ground-state absorbers stays constant within each individual slice. When the procedure starts (before the sample is subjected to laser excitation) the concentration of ground state absorbers, A in each of the slices is $A_i = A_o$, the initial concentration of ground-state absorbers. The absorption coefficient at the laser wavelength is given by $K^e_{A,i} = 2\epsilon^e_A A_i$ where ϵ^e_A is the extinction coefficient of A at the excitation wavelength. Usually it is assumed that the transient does not absorb significantly at the laser wavelength and that the scattering coefficient S is independent of wavelength.

2) The laser pulse which excites the sample is divided into portions Δt where Δt is generally 1/100 of the total pulse duration.

3) The sample is considered as having being irradiated with a portion Δt of the laser pulse. The concentrations of transient species T_i formed in each slice because of light absorption are determined, and a new set of ground-state concentrations A_i are calculated. In general, after the sample has been exposed to a portion of the laser light, the concentrations A_i are no longer equal, due to the attenuation of the exciting pulse as it passes through the sample. Mass conservation requires that $A_i + T_i = A_o$ in all slices.

4) Using the new values A_i, new absorption coefficients $K^e_{A,i}$ are calculated for each slice.

5) The procedure recycles to step 3), and the next portion of the laser pulse is considered to irradiate the sample.

6) When all of the laser pulse has irradiated the sample, the concentration of ground state absorbers A_i and the concentration of transient species T_i in each slice are known.

Steps 1) through 6) generate the transient concentration profile. To calculate ΔR at an appropriate analysing wavelength the following additional steps are performed.

7) Estimates of the extinction coefficients at the *analysing* wavelength for the transient species, ϵ^a_A are used to calculate the absorption coefficients in each slice due to the ground-state absorbers $K^a_{A,i} = 2\epsilon^a_A A_i$, and transient species, $K^a_{T,i} = 2\epsilon^a_T T_i$.

8) The reflectance of each slice, R_i, can now be calculated using the absorption coefficients $K^a_{A,i}$, $K^a_{T,i}$ and the scattering coefficient S.

9) Finally, the individual reflectances R_i are combined using a recursion formula to give the expected reflectance of the sample.

If the absorption coefficient K is in excess of 10^4 cm^{-1} at the laser excitation wavelength then the penetration depth is only about 1μm and dissipation of laser excitation as heat causes considerable temperature rises (see reference 20). We have not only shown how to predict any temperature rises but have confirmed our calculations by measurements in the case of TiO_2. The possibility of large temperature rises has always to borne in mind and lower laser fluences, lower concentrations or excitation into weaker bands must be employed where necessary to avoid thermal effects. It is important to stress that when the penetration depth of the exciting light is > 0.1 mm, temperature rises in opaque samples using our nanosecond laser system are negligible i.e. < 1°C.

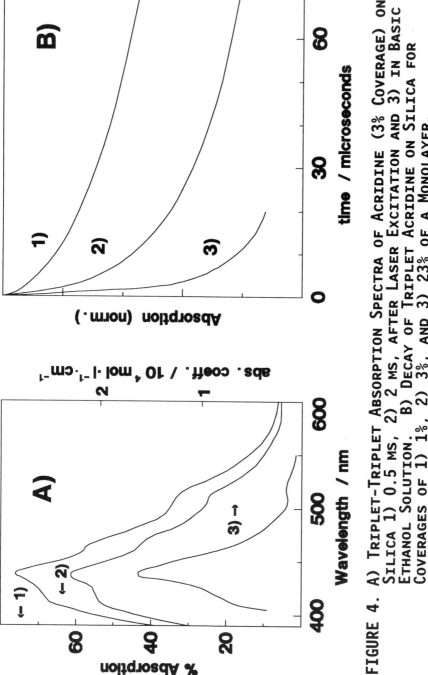

FIGURE 4. A) TRIPLET-TRIPLET ABSORPTION SPECTRA OF ACRIDINE (3% COVERAGE) ON SILICA 1) 0.5 MS, 2) 2 MS, AFTER LASER EXCITATION AND 3) IN BASIC ETHANOL SOLUTION. B) DECAY OF TRIPLET ACRIDINE ON SILICA FOR COVERAGES OF 1) 1%, 2) 3%, AND 3) 23% OF A MONOLAYER.

3. RESULTS AND DISCUSSION

   3.1. Triplet-triplet Annihilation on Surfaces

   The nature and mobility of species present when acridine is
adsorbed from high vacuum and from dry solutions onto thermally
pretreated silica have been investigated by ground and excited
state spectroscopic techniques[4]. The main adsorbed species is
hydrogen bonded acridine which shows strong triplet-triplet
absorption at 493 nm (see Figure 4) which for samples of low
coverage or high pre-treatment temperature has an exponential
decay with a lifetime of 35 ± 2 ms. Such samples do not
show delayed fluorescence. However, for samples of higher
coverage or lower pre-treatment temperature the triplet decay is
faster and non-exponential (see Figure 4) and delayed
fluorescence is observed due to the following reaction

$$A^*(T_1) + A^*(T_1) \longrightarrow A(S_0) + A^*(S_1) \longrightarrow 2A(S_0) + h\nu_F \qquad (6)$$

   The mechanism of elementary bimolecular interactions at
interfaces is by no means fully understood. It can be a
consequence of two-dimensional surface migration (diffusion) and
bimolecular collisions or alternatively a three-dimensional
interaction assuming that the adsorption = desorption equilibrium
inside a pore is not completely shifted to the left. In order to
clarify this problem we have studied a series of silica powders
(particle size ~ 100μm) with very different pore-diameter ranging
from 6 - 100 nm. We have also varied the surface loading on high
surface area silicas between 0.075 mg to 1 mg of acridine per g
of silica.

   The effect of surface coverage on the rate of triplet decay is
shown in Figure 4. At very low loading (< 0.1%) we obtain a
strictly first order decay with a mean lifetime for the triplet
state of 35 ms which is the largest ever reported for acridine at
room temperature. At loadings greater than or equal to 1%
triplet decay is enhanced and the decay becomes strongly non-
exponential. For such non-exponential decays we have evaluated
the first half-life as a function of the volume and surface
concentration assuming statistical adsorption and using the
apparent density of the silica. No correlation is found with
volume concentration however, the decay times correlate well with

the surface concentration. This is consistent with a two-dimensional triplet-triplet annihilation mechanism via surface diffusion and rules out a three-dimensional mechanism via volume diffusion. Detailed examination of the triplet absorption decay curves as well as of decay curves for the delayed fluorescence which is observed at higher fractional coverages yield a two-dimensional bimolecular rate constant for triplet-triplet annihilation of 8×10^{13} dm^2 mol^{-1} s^{-1}.

3.2 Combination of Diphenylmethyl Radicals on Surfaces

Organic photoreactions on zeolite supports has become an area of increasing interest in the last few years[21]. A diffuse reflectance nanosecond laser photolysis study of the ketone, xanthone included within the hydrophobic zeolite Silicalite has yielded some very interesting information relating to the host environment[5]. Silicalite is over 99% SiO_2 and consists of a system of near-circular zig-zag channels, cross linked by elliptical straight channels[22]. The xanthone transient was assigned as the triplet since it has a characteristic maximum at 605 nm. Another observation made was that the decay process extends over a considerable timescale from nano- to micro-seconds. This suggests a variety of lifetimes for this ketone triplet at different surface sites. The growth of this transient on picosecond timescales constituted the first reported example of picosecond diffuse reflectance laser flash photolysis[2].

The α-cleavage of 1,1,3,3-tetraphenylacetone (TPA), reaction (7) has been used as a source of diphenylmethyl radicals in our experiments[23]. The production of the second diphenylmethyl radical via decarbonylation of the initially produced diphenylacetyl radical occurs rapidly[24] ($k = 1.3 \times 10^8$ s^{-1}) and reaction (7) is, thus, an efficient and practically instantaneous source of two diphenylmethyl radicals per ketone molecule.

$$Ph_2CHCOCHPh_2 \xrightarrow{h\nu} Ph_2CHCO + Ph_2CH \longrightarrow 2Ph_2CH + CO \qquad (7)$$

The three solid supports used in these experiments differ considerably in their properties. Silica gel is a porous surface with, in our case, an average pore size of 6 nm and a surface area of 480 m^2/g. Both silica gel and Silicalite contain only silicon and oxygen, although the latter is hydrophobic whereas

F. Wilkinson and R. Beer

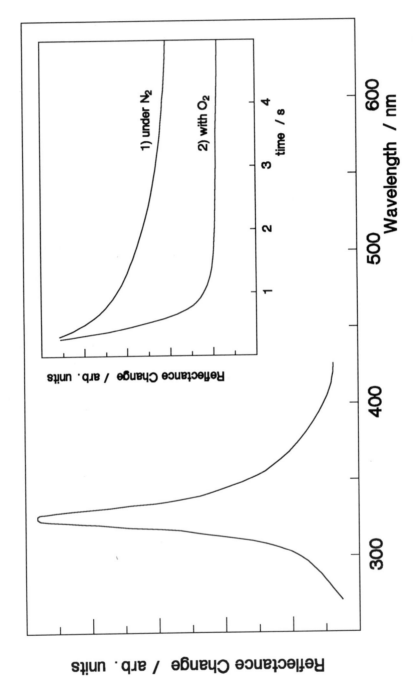

FIGURE 5. TRANSIENT SPECTRUM OF DIPHENYLMETHYL RADICALS PRODUCED BY 266NM EXCITATION OF 3% COVERAGE OF TETRA-PHENYLACETONE ON SILICA. INSET SHOWS TRANSIENT DECAYS MONITORED AT 340 NM.

the former has a large amount of physical adsorbed water on the surface. The structure of the aluminosilicate NaX zeolite is comprised of a three-dimensional network of relatively large cavities or supercages (~1.2 nm) connected by 0.7-0.8 nm pores or channels[24]. The Si/Al ratio of approximately 1.5 results in a large proportion of exchangeable cations (in our case Na) and a strongly hydrophilic zeolite. It should be noted that TPA can easily fit within the pore structure of silica gel and the channel system of Nax. However, the relatively small channels of Silicalite result in adsorption of TPA only on the external surface, although the photoproduced diphenylmethyl radical may migrate into the channels. For all the three supports diphenyl methyl radicals are produced with a characteristic absorption maximum at 335 nm (e.g. see Figure 5) which decays over timescales which vary from hundreds of nanosecond to minutes. The first half-life is much shorter than the second half-life which is shorter than the third half-life etc. These decays can be interpreted[23] in terms of dispersive kinetic analysis[25] to give a distribution of rate constants and a mean value. As can be seen in Figure 5, the radical decay shows oxygen quenching. The efficiency of oxygen quenching increases in going from Silicalite, to NaX, to silica gel, consistent with the greater assessability of oxygen to silica gel pores as opposed to the narrow channels in NaX and very narrow channels in Silicalite.

The above results demonstrate that diphenylmethyl radicals may be readily generated on silica gel, and on the molecular sieves, NaX and Silicalite. The diphenylmethyl radicals show a very wide range of lifetimes on all three supports, as evidenced by the fact that one can obtain lifetimes of anywhere from several microseconds to minutes. Furthermore, for some of the samples there are probably radicals which are decaying on shorter timescales than we can monitor. These results suggest that there are a wide range of possible sites or environments for the radicals. It should be noted that these may reflect largely the distribution of the radicals with respect to each other and the starting ketone rather than particular surface sites on which the radicals are particularly stable. In experiments in which we varied the surface loadings[23], we did not see any evidence for preferential filling of particular sites at low loadings, neither

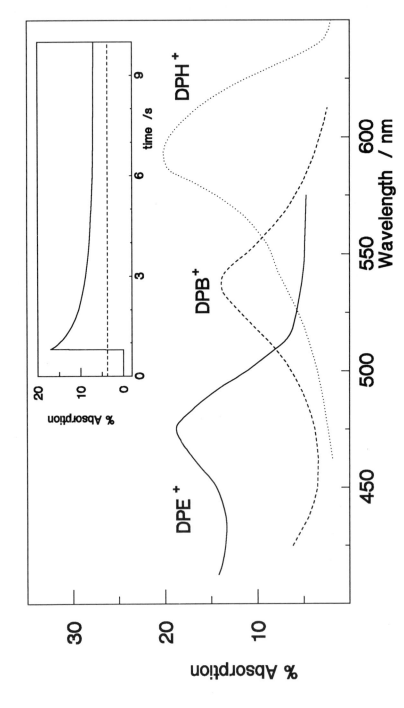

FIGURE 6. TRANSIENT ABSORPTION SPECTRA OF DPE+ ON ALUMINA PRETREATED AT 300ºC, AND DPB+ AND DPH+ ON ALUMINA PRETREATED AT 100ºC. THE SPECTRA WERE MEASURED 1 MS AFTER THE LASER PULSE. INSET SHOWS LONG-LIVED DECAY.

did we see attenuation effects which could be attributed to the dominance of second order components in the decay of the radicals. A lengthening of the transient lifetimes at low loading would have been expected if radical-radical coupling involving diffusion of both partners was determining the transient lifetime. However, product studies[23] indicate that the radical dimerization accounts for most of the products from laser irradiations. It is likely that a large fraction of the radicals decay very rapidly via dimerization and are invisible in our transient experiments.

3.3 Radical Cation-Electron Recombination

Adsorbed radical cations have been detected by the method of diffuse reflectance laser flash photolysis using polycrystalline microporous catalytic metal oxides such as silica and alumina as adsorbents[3]. We have studied several diphenylpolyenes adsorbed on alumina. Typical transient absorption spectra obtained for 1,2-diphenylethene (DPE), 1,4-diphenylbutadiene (DPB) and 1,6-diphenyl-1,3,5-hexatriene (DPH) adsorbed on alumina recorded 1 ms after excitation are shown in Figure 6. These transients are long-lived and decay with a non-exponential decay with the first half-life considerable shorter than the second half-life and so on. In the case of DPH on alumina, we also observed[3] a much shorter lived transient which we have assigned to the adsorbed triplet state of DPH.

There is evidence that the long lived transients in DPE, DPB and DPH are radical cations. Thus for DPE adsorbed on Al_2O_3/SiO_2 the radical cation is known to have its stronger absorption band at 480 nm, while for DPB ad DPH adsorbed on this catalyst the strongest ground-state absorption bands of the radical cations occur at 542 and 602 nm, respectively[26]. In solution a short-lived transient ($\tau < 100$ ns, $\lambda_{max} = 478$ nm) is reported after flashing charge-transfer complexes of DPE in the presence of an electron acceptor such as fumaronitrile,[27]. This transient is assigned as the radical cation. We therefore assigned the long-lived transient in DPE and DPB and of DPH as radical cations. We have proved this assignment by measuring the ESR spectra of the adsorbed radical cations at low temperature[28].

From studies of the dependence of the amounts of radical produced as a function of laser intensity coupled with the

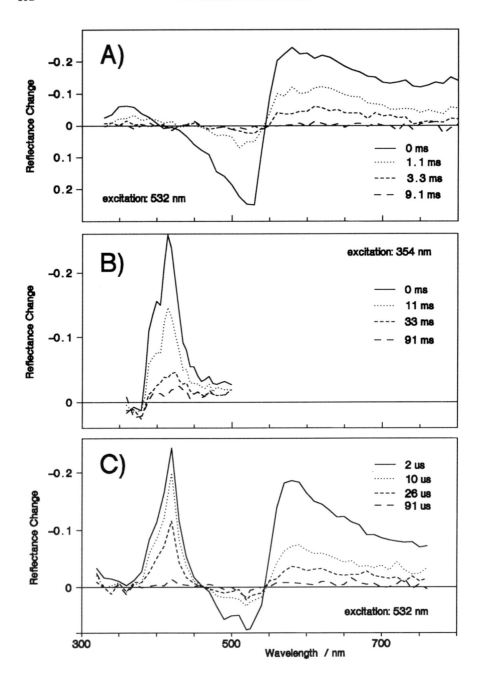

FIGURE 7. TIME RESOLVED TRANSIENT DIFFERENCE DIFFUSE REFLECTANCE
SPECTRA FROM SILICA WITH A) 0.12 µMOL EOSIN, B) 6 µMOL ANTHRACENE
AND C) 0.12 µMOL EOSIN AND 6 µMOL ANTHRACENE PER GRAM OF SILICA.

computer modelling of the transient concentration profile it is possible to show that the production of the radical cation is a multiphoton process,[28] probably involving the consecutive absorption of two photons.

The decay of these transients has been shown to be very dependent on the temperature, on the nature of adsorbent and on its pre-treatment. Repetitive excitation at room temperature of the same sample area at intervals of a few minutes demonstrates that the intensity of absorption and the decay kinetics of the radical cation are reproducible. Thus the radical cations decay almost exclusively to give back the original molecule. Thus the complex decay is due to radical cation - electron recombination. Variation in surface coverage of adsorbates gives only very small differences in the kinetics observed. This suggests that the process of ionization leaves the radical cation and the electron separated by small distances (smaller than the nearest distance between adsorbates) and that geminate pair recombination is the decay mechanism. The decays fit quite well the expression

$$C_0/C = 1 + \alpha t^{\frac{1}{2}} \tag{8}$$

where C_0 and C represent the concentrations of the radical cations formed initially and remaining after time t. The values of α are strongly temperature, adsorbent and adsorbate dependent[28].

3.4 Triplet Energy Transfer on Surfaces

Diffuse reflectance laser flash photolysis has been used by Turro et al[29] and by ourselves[30] to demonstrate triplet energy transfer from triplet benzophenone to naphthalene and some of its derivatives on silica and microcrystalline cellulose via static and dynamic pathways. Recently[31] we have studied energy transfer from triplet eosin to anthracene adsorbed on porous silica and Figure 7 shows some typical results.

Eosin adsorbed on silica exhibits a long lived transient after laser excitation at 532 nm. This we assign to the broad triplet-triplet absorption of eosin with λ_{max} at 600 nm. There is strong laser induced ground state depletion and very good isosbestic points are observed (see Figure 7A). Anthracene adsorbed on silica gives no transient absorption when excited at 532 nm since it does not absorb at this wavelength but excitation at 354 nm yields the characteristic triplet-triplet spectrum of anthracene with λ_{max} = 420 nm (see Figure 7B). Excitation at 532 nm of eosin coadsorbed with anthracene on silica demonstrates that triplet energy transfer occurs since sensitized production of triplet anthracene results (see Figure 7C). In addition, the lifetime of triplet eosin decreases with increasing anthracene concentration due to the process

$$^3Eosin^* + Anthracene \longrightarrow {}^3Anthracene^* + Eosin \qquad (9)$$

On short timescales the growth in sensitized triplet anthracene is detectable. Later on an equilibrium is established between triplet eosin and triplet anthracene and both triplets then decay with the same apparent lifetime. From the quenching of triplet eosin by anthracene and kinetics of the growth of triplet anthracene, it is possible to obtain a two dimensional rate constant of $6 \pm 2 \times 10^{13}$ dm^2 mol^{-1} s^{-1} for triplet energy transfer on a silica surface. This value is close to the value we obtained for triplet-triplet annihilation of acridine on silica (see earlier) and we consider it may well be the two dimensional diffusion controlled rate constant for silica surfaces pretreated at 100°C. There are very few literature values to compare these values with but Turro et al[29] and De Mayo et al[32] report rate constants for quenching of triplet states on silica surfaces which are two orders of magnitude larger than those reported here. This difference could easily be due to the different pretreatment of the adsorbents.

4. CONCLUSIONS

Diffuse reflectance laser flash photolysis has been shown to be a powerful method for studying photoinduced elementary reactions at interfaces and within highly scattering samples. In this work we were able to measure two-dimensional rate constants for certain heterogeneous systems while in other studies dispersive kinetics were observed. It needs to be stressed that this technique can be applied to reactions in all types of heterogeneous systems as well as at gas/solid interfaces as discussed here. The ability to rapidly control the switching on, with a pulsed laser, of every type of elementary reaction at interfaces and in other heterogeneous environments and to follow the resulting reactions is enhancing considerably understanding of such systems.

ACKNOWLEDGEMENTS

It is an honour to dedicate this article to the memory of Professor Shigeo Tazuke and in this way to pay respect to an outstanding scientist and a good friend. This work was supported by SERC, NATO and the EEC. The authors would like to thank Professors D. Oelkrug and J.C. Scaiano and their research groups at Tubingen University, Deutschland and NRC, Ottawa, Canada respectively for productive collaboration.

REFERENCES

1) R.W. Kessler and F. Wilkinson, J. Chem. Soc., Faraday Trans I, 77 (1981) 309.

2) F. Wilkinson, C.J. Willsher, P. Leicester, J.R.M. Barr and M.J.C. Smith, J. Chem. Soc., Chem. Commun., (1986) 1216.

3) F. Wilkinson, C.J. Willsher, D. Oelkrug, G. Krabichler and W. Honnen, J. Phys. Chem., 92 (1988) 589.

4) D. Oelkrug, S. Uhl, C.J. Willsher and F. Wilkinson, J. Phys. Chem. 93 (1989) 4551.

5) F. Wilkinson, C.J. Willsher, M.L. Casal, Linda J. Johnston and J.C. Scaiano, Can. J. Chem. 64 (1986) 539.

6) F. Wilkinson, C.J. Willsher, S. Uhl, W. Honnen and D. Oelkrug, J. Photochem. 33 (1986) 273.

7) J. Pouliquen, D Fichou, P. Valat, J. Kossanyi, F. Wilkinson and C.J. Willsher, J. Photochem. 35 (1986) 381.

8) J. Kossanyi, D. Kouyte, J. Pouliquen, J.C. Ronfard-Haret, P. Valat, D. Oelkrug, U. Mammel, G.P. Kelly and F. Wilkinson. J. Lumin. 46 (1989) 17.

9) F. Wilkinson and C.J. Willsher, Chem. Phys. Letts. 104 (1984) 272.

10) F. Wilkinson and C.J. Willsher, J. Lumin. 33 (1988) 187.

11) F. Wilkinson, C.J. Willsher, J.L. Bourdelande, J. Font and J. Greuges, J. Photochem. 38 (1987) 381.

12) F. Wilkinson and C.J. Willsher, J.Chem. Soc., Chem. Comm. (1985) 142.

13) F. Wilkinson, J. Chem Soc., Faraday Trans II, 82 (1986) 2073.

14) F. Wilkinson and G.P. Kelly, Diffuse Reflectance Flash Photolysis, in: Handbook of Organic Photochemistry, Vol. 1, ed. J.C. Scaiano (CRC Press, Boca Raton, 1990) chap. 12.

15) F. Wilkinson and C.J. Willsher, Diffuse Laser Flash Photolysis of Dyed Fabrics and Polymers, in: Lasers in Polymer Science and Technology: Applications, Vol II, eds, J.P. Foussier and J.F. Rabek (CRC Press, Boca Raton, 1990) chap. 9.

16) R.W. Kessler, G. Krabichler, S. Uhl, D. Oelkrug, W.P. Hagan, J. Hyslop and F. Wilkinson, Optical Acta. 30 (1983) 1099.

17) D. Oelkrug, W. Honnen, F. Wilkinson and C.J. Willsher, J. Chem. Soc., Faraday Trans II, 83 (1987) 2081.

18) P. Kubelka, J. Opt. Soc. Am. 38 (1948) 448.

19) T. Lin and H.K.A. Kan, J. Opt. Soc. Am. 60 (1970) 1252.

20) F. Wilkinson, C.J. Willsher, S. Uhl, W. Honnen and D. Oelkrug, J. Photochem. 33 (1986) 273.

21) H.L. Casal and J.C. Scaiano, Can. J. Chem. 62 (1984) 628.

22) E.M. Flanagen, J.M. Bennett, R.W. Grose, J.P. Cohen, R.L. Patton, R.M. Kirchener and J.V. Smith, Nature 271 (1978) 512.

23) G. Kelly, C.J. Willsher, F. Wilkinson, J.C. Netto-Ferreira, A. Olea, D. Weir, L.J. Johnston and J.C. Scaiano, Can. J. Chem. 68 (1990) 812.

24) I.R. Gould, B.H. Baretz, and N.J. Turro, J. Phys. Chem. 91 (1987) 925.

25) T. Doba, K.U. Ingold, W. Siebrand and T.A. Wildman, Faraday Discuss. Chem. Soc. 78 (1984) 175.

26) G. Kortum and V. Schlichenmaier, Z. Phys. Chem. (Neue Folge) 48 (1966) 27.

27) J.L. Goodman and K.S. Peters, J. Am. Chem. Soc. 107 (1985) 1441.

28) D. Oelkrug, S. Reich, F. Wilkinson, P.A. Leicester, J. Phys. Chem. in print.

29) N.J. Turro, M.B. Zimmt, I.R. Gould and W. Mahler, J. Am. Chem. Soc. 107 (1985) 582.

30) F. Wilkinson and L.P.V. Ferreira, J. Lumin. 49 (1988) 704.

31) F. Wilkinson, R. Beer and P.A. Leicester, in print.

32) P. de Mayo, L.V. Natarajan and W.R. Ware, Chem. Phys. Letts. 107 (1984) 187.

Chapter IV:
Toward Integrated
Molecular Systems

Photochemical Processes in Organized Molecular Systems
K. Honda (Editor-in-Chief)
© Elsevier Science Publishers B.V., 1991

PHOTOCHROMISM OF DIARYLETHENES WITH HETEROCYCLIC RINGS

Masahiro IRIE

Institute of Advanced Material Study, Kyushu University,
Kasuga-Koen 6-1, Kasuga, Fukuoka, Japan

A guiding principle has been proposed for molecular design of
diarylethene type photochromic compounds, both isomers of which
are thermally stable and fatigue resistant. The reaction
mechanism is also discussed.

1. INTRODUCTION

Although photochromic organic compounds have potential for use
in rewritable optical memory media,[1] the compounds still await
the practical applications. The limitation is due to the lack of
suitable compounds which fulfill the requirements for the
reversible recording media. The requirements for the use are as
following.

i)   Archival storage capability(thermal stability).

ii) Low fatigue (can be cycled many times without significant
loss of performance).

iii) High sensitivity at diode laser wavelengths and rapid
response.

iV) Non-destructive read out capability.

Among these requirements, indispensable ones are i) and ii).
In this paper, we propose a new guiding principle for molecular
design of diarylethene type photochromic compounds, both isomers
of which are thermally stable and fatigue resistant. We also
discuss the mechanism of the photochromic reaction.

2. THEORETICAL STUDY[2]

In order to come up with a guiding principle, we carried out
a theoretical study of 1,3,5-hexatriene to cyclohexadiene type
photochromic reactions. Typical compounds of the hexatriene
molecular framework are diarylethene derivatives having benzene
or heterocyclic rings. Semiempirical MNDO calculations were
carried out for the following diarylethene derivatives.

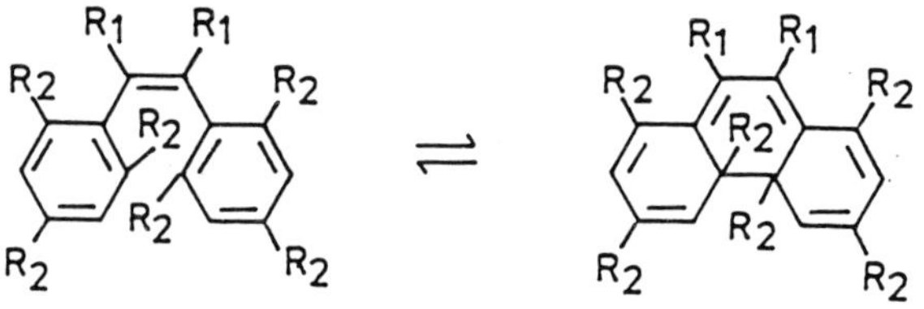

1a, R1=R2=H

2a, R1=R2=CH$_3$

1b, R1=R2=H

2b, R1=R2=CH$_3$

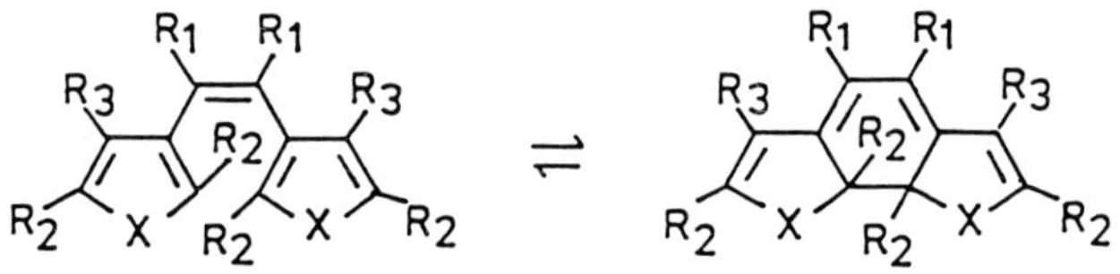

3a, X=O, R1=R2=R3=H

4a, X=S, R1=R2=R3=H

5a, X=S, R1=R2=CH$_3$, R3=H

6a, X=O, R1=R2=CH$_3$, R3=H

7a, X=S, R1=CN, R2=R3=CH$_3$

3b, X=O, R1=R2=R3=H

4b, X=S, R1=R2=R3=H

5b, X=S, R1=R2=CH$_3$, R3=H

6b, X=O, R1=R2=CH$_3$, R3=H

7b, X=S, R1=CN, R2=R3=CH$_3$

Figure 1 and 2 show the state correlation diagrams for the
reactions from 3a to 3b and from 1a to 1b in disrotatory and

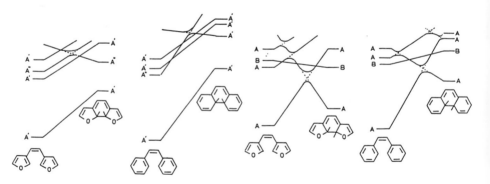

FIGURE 1
The state correlation diagrams
in disrotatory mode for the
reactions from 3a to 3b, and 1a
to 1b.

FIGURE 2
The state correlation diagrams
in conrotatory mode for the
reactions from 3a to 3b, and 1a
to 1b.

conrotatory modes, respectively. According to the state correlation diagrams, orbital symmetry allows the disrotatory cyclization
in the ground state from 3a to 3b and from 1a to 1b. The ground
state energies of the closed-ring forms are, however, much higher
than those of the open-ring forms. The large energy barriers
practically prohibit the cyclization reactions in the ground
states. No such large energy barriers exist in the conrotatory
mode reactions in the excited states from 3a to 3b and 1a to 1b.
This indicates that the cyclization of both 3a and 1a are allowed
in the photochemically excited states.

First, that we should consider is the stability of the photochemically produced closed-ring forms. As seen from Figure 2,
the stability depends on the energy barrier of the cycloreversion
reaction in the ground state. The energy barrier correlates with
the ground state energy difference between the open-ring and the
closed-ring forms, calculated values of which are shown in Table
I. When the energy difference is large, as in the case of 1,2-
diphenylethene, the energy barrier becomes small. On the other
hand, the barrier becomes large when the difference is small, as
shown for 1,2-di(3-furyl)ethene. In this case, the reaction is
hardly expected to occur. The ground state energy difference
controls the thermal stability of the closed-ring forms.

Then, the next question is what kind of molecular property
causes the difference in the ground state energies. We examined
the aromaticity change from the open-ring to the closed-ring
forms. The difference in the energy between the following right
and left side groups was calculated, as shown in Table II.

The aromatic stabilization energy of the aryl groups correlates
well with the ground state energy difference.

From the above consideration, it is concluded that the
thermal stability of both isomers of the diarylethene type photochromic compounds is attained by introducing aryl groups, which
have low aromatic stabilization energy, such as furan and
thiophene rings.

Table I   Relative ground state energy difference between
           the open-ring and the closed-ring forms

compd	disrotatory kcal/mol	conrotatory kcal/mol
1,2-diphenylethene	41.8	27.3
1,2-di(3-pyrrolyl)ethene	32.3	15.5
1,2-di(3-furyl)ethene	27.0	9.2
1,2-di(3-thienyl)ethene	12.1	-3.3

Table II   Aromatic stabilization energy difference

group	energy, kcal/mol
phenyl	27.7
pyrrolyl	13.8
furyl	9.1
thienyl	4.7

3. SYNTHESIS

Based on the guiding principle deduced from the above
theoretical consideration, several symmetric and non-symmetric
diarylethenes having benzene or heterocyclic rings with various
aromatic stabilization energies were synthesized in the hope to
gain access to thermally irreversible photochromic compounds.[3-5]
The general procedure for the preparation of the diarylethene
derivatives is as follows.

8a, X=Y=S, $R\frown R=R\frown R'$ = CH=CH-CH=CH

9a, X=Y=NCH$_3$,
$R\frown R=R\frown R'$ = CH=CH-CH=CH

10a, X=S, Y=NCH$_3$, R=CH$_3$,
$R\frown R'$ = CH=CH-CH=CH

11a, X=Y=S, R=R' =CH$_3$

12a, X=Y=S,
$R\frown R=R\frown R'$ = CH=CH-CH=CH

13a, X=Y=NCH$_3$,
$R\frown R=R\frown R'$ = CH=CH-CH=CH

14a, X=S, Y=NCH$_3$, R=CH$_3$
$R\frown R'$ = CH=CH-CH=CH

Symmetric diarylethene 8a, for example, was prepared by the self coupling reaction of two 2-methyl-3-(cyanomethyl)benzo(b)-thiophene. To 50% NaOH aqueous solution containing tetrabutyl-ammonium bromide was added a mixture of the cyanomethylbenzo(b)-thiophene, benzene, and CCl$_4$, and the solution was stirred for 1 h at 50-60°C. The reaction mixture was poured into water and the product was extracted with benzene. After the solvent was removed, the mixture of the cis and trans isomers were separated by HPLC and purified by recrystallization from a hexane-benzene mixture. The maleic anhydride derivative 12a was obtained by hydrolysis of 8a in a 2-methoxyethanol-water mixture. Other symmetric diarylethenes were also prepared by the similar method.

Figure 3 shows a typical example of the photochromic reactions of symmetric diarylethenes. This exhibits the absorption spectra of 12a(λ max : 417 nm, ϵ : 6.8 x 10^3) in benzene before photo-irradiation and in the photostationary state under irradiation with 405 nm light. Upon irradiation with 405 nm light, the solution became dark red and a new absorption peak appeared at 544 nm (ϵ : 8.7 x 10^3). The new band is ascribable to the closed-ring form of 12a.

Non-symmetric diarylethenes were also prepared by the coupling reactions of two different kinds of cyanomethyl-substituted heterocyclic rings. 10a, for example, was prepared by the coupling reaction of equimolar cyanomethyl-indole and cyanomethyl-thiophene. The reaction yielded a mixture of 7a, 9a, and 10a (4 :

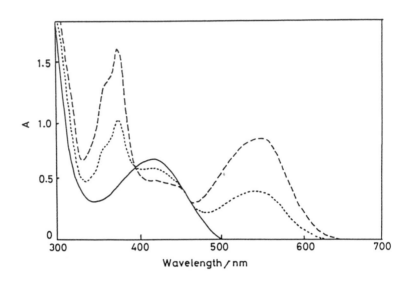

FIGURE 3
Absorption spectra of benzene solutions of 2a(——), the closed-ring form(—--), and photostationary state(·····) under irradiation with 405 nm light.

20 : 1 including the trans forms) . The mixture was separated by HPLC and purified by recrystallization from a hexane-benzene mixture. The non-symmetric dicyano derivative 10a was converted to maleic anhydride derivative 14a by hydrolysis with KOH in a 2-methoxyethanol-water mixture.

14a underwent the cyclization reaction to produce the dihydro-type derivative. Figure 4 shows 14a and the closed-ring form, and the spectrum in the photostationary state under irradiation with 491 nm light in benzene. Upon irradiation with 491 nm light in the presence of air, the solution turned green and a new peak appeared at 595 nm (ε : 5.5 x 10^3). Isosbestic points are observed at 437 and 504 nm. On exposure to visible light(λ >560 nm), the solution became yellow, and the initial absorption of 14a was restored. The non-symmetric diarylethene has photosensitivity at the longest wavelength among the diarylethenes so far synthe-sized. Indole ring is effective to shift the absorption maxima of the closed-ring forms. The cyclization reaction is able to be induced by irradiation with 488 nm Ar ion laser, and the reverse reaction with 633 nm He-Ne laser.

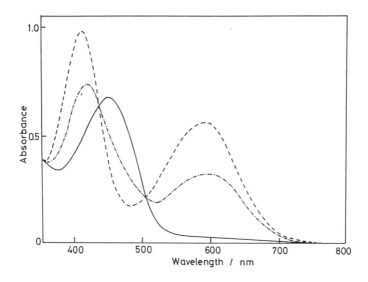

FIGURE 4

Absorption spectra of benzene solutions of 14a(——), the closed-
ring form(-- -), and photostationary state (—·—) under irradiation
with 491 nm light.

4. PROPERTIES

   Next, we examined the validity of the theoretical prediction
for the molecular design of diarylethenes with thermal stability.
Figure 5 shows the thermal stability of photogenerated closed-
ring forms of various diarylethenes at 80°C in toluene solutions.
The stability depends on the aryl groups. When the aryl groups
are furyl or thienyl groups, which have low aromatic stabilization
energy, the closed ring forms are thermally stable. The
absorption intensities of 7b and the closed-ring form of 11a were
found to remain constant for more than 3 months at 80°C. In
addition, the open-ring forms of these diarylethenes did not show
any thermochromic reaction even at 300°C.

   On the other hand, photogenerated closed-ring forms of
diarylethenes with phenyl or indole rings, 2b, the closed-ring
forms of 9a and 13a, are thermally unstable. The photogenerated
yellow color of 2b disappeared with half lifetime of 1.5 min.
at 20°C. The closed-ring form reverts quickly to the open-ring
form. Compounds 9a and 13a also exhibit thermally unstable
photochromic reactions. The closed-ring forms revert slowly

to the open-ring form in the dark.

The different behavior in the thermal stability between diarylethenes with furyl or thienyl groups, and phenyl or indole groups agrees well with the theoretical prediction that the thermal stability depends on the aromatic stabilization energy of the aryl groups. The large aromatic stabilization energy of phenyl and indole rings destabilizes the closed-ring forms.

It is worthwhile to note that the photogenerated closed-ring forms of 10a and 14a were found to be thermally irreversible. The closed-ring forms maintain a constant absorption intensity for more than 12 h at 80°C. The result indicates that the closed-ring forms of non-symmetric diarylethenes are thermally stable when at least one of the heterocyclic rings has low aromatic stabilization energy. In the non-symmetric diarylethenes with an indole ring on one end and a thiophene ring on the other end, the indole ring plays a role to shift the absorption band to longer wavelength, and the thiophene ring to keep the thermal stability.

It is not easy to compare quantitatively the fatigue resistant property, i.e. how many times coloration/decoloration cycles can be repeated without permanent side product formation,

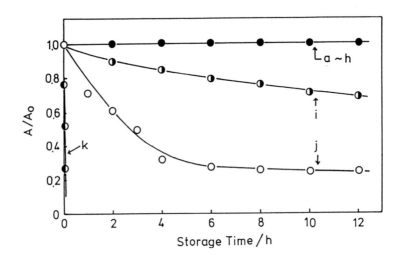

FIGURE 5

Thermal stability of photogenerated closed-ring forms of 5a(a), 6a(b), 7a(c), 8a(d), 10a(e), 11a(f), 12a(g), 14a(h), 9a(i), 13a(j), and 2a(k) at 80°C in toluene.

because the property depends on the environmental conditions,
such as solvent purity or matrix properties. We tried to
measure the fatigue of the compounds so far synthesized as
shown in Table III. The repeatable number indicates the cycle
number when the colored intensity decreases to 80 % of the
first cycle. For the symmetric diarylethenes, the repeatable
number is limited to less than 480 times even in the
absence of oxygen so far the compounds have thiophene rings.
When the thiophene rings are replaced with benzothiophene rings,
the number remarkably increases.

Figure 6 shows an example of the fatigue resistant behavior.
A benzene solution containing 12a was irradiated alternatively

Table III   Fatigue resistant property

compd	repeatable cycle number	
	in the presence of air	in the absence of air
5a	3	–
7a	10	480
11a	70	120
12a	3.7×10^3	1.0×10^4
14a	–	8.7×10^3

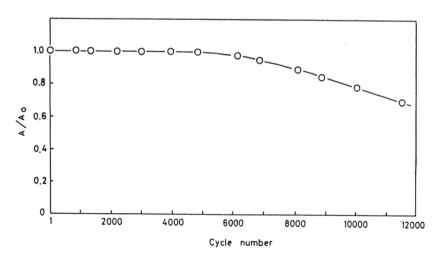

FIGURE 6
Fatigue resistant property of 12a in benzene solution.

with 436 nm and 546 nm lights under deaerated conditions. The
absorption intensity of the closed-ring form remains almost
constant even after 6 x 10^3 cycles. The intensity decreases
to 80 % after 10^4 cycles.

Non-symmetric diarylethenes with an indole ring on one end had
an excellent fatigue resistant property. 14a, for example, kept
the adequate photochromic property even after the cycles of
8.7 x 10^3 times.

5. REACTION MECHANISM

The photochromic reactions of diarylethenes were found to
depend on the medium. For example, the conversion from 11a to
the closed-ring form in the photostationary state under irradiation with 405 nm light decreased with increasing the solvent
polarity. The conversion of 73 % in hexane decreased to 39 % in
THF. This indicates that solvent dependence of the ring-closure
quantum yield is different from that of the ring-opening yield at
the wavelength. Table IV summarizes the solvent dependence of the
quantum yields. The ring-closure quantum yield in THF is one-
third of the value in non-polar n-hexane. In polar acetonitrile,
it is only 0.003. In order to clarify the solvent dependence
spectroscopic measurements were carried out.

Figure 7 shows the absorption and fluorescence spectra of
11a in various solvents. Table V summarizes the maxima. The
fluorescence spectra show remarkable Stokes shifts depending on

Table IV Spectroscopic properties and quantum yields of 11a
          and the closed-ring form(11b)

Solvent	max(nm) 11a	11b	Quantum Yield[a] 11a→11b	11b→11a	11a λ max (nm)[b]	If[c]
Hexane	331	552	0.13	0.16	488	1.0
Benzene	340	564	0.07	0.12	541	0.35
THF	335	560	0.04	0.11	560	0.088
CH$_3$CN	336	563	0.003	0.10	---	<0.01

a)11a→11b:Irradiation with 405nm-light.
   11b→11a:Irradiation with 546nm-light.
b)Fluorescence maxima, excitation wavelength:405nm.
c)Relative fluorescence quantum yield
   (excitation wavelength:405nm)

the solvent polarity. The maximum at 488 nm in n-hexane shifts
to 560 nm in THF. At the same time the intensity decreases.
The fluorescence intensity in acetonitrile is less than 1 %
of the intensity in n-hexane. The result indicates that the
excited state of the open-ring form has a polar structure with
a large dipole moment.

Fluorescence lifetime measurement revealed that the excited
state has two states with different lifetimes, several hundred
picosecond and a few nanosecond. The contribution of the
slowly decaying component increased with increasing the solvent
polarity.

These excited state behavior is similar to that observed in
many TICT(twisted intramolecular charge transfer) molecules.[6]
The TICT molecules have an electron donor group and an electron
acceptor one within a molecule, and in the excited state an
electron donation from the donor to the acceptor brings about a
charge separated polarized state, in which the donor and the
acceptor groups are in perpendicular. Actually 11a consists of
two thiophene rings and an acid anhydride group; the former group
has an electron donating ability and the latter an electron
accepting ability.

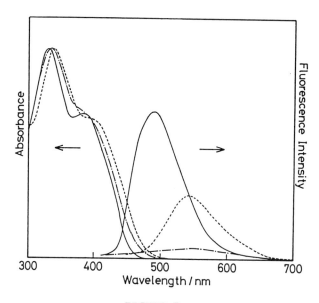

FIGURE 7

Fluorescence spectra of 11a in n-hexane(——), benzene(----), and
THF(—·—).

M. Irie

According to the general consideration of the TICT molecules,
there exist two kinds of conformations in the excited state
of 11a depending on the solvent polarity, as shown in Figure 8
In less polar solvents, a planar conformation is stabler than
twisted polar one. In polar solvents, however, the twisted
polar charge transfer state becomes more stable. Decrease of the
ring-closure quantum yield in polar solvents suggests that the
twisted charge transfer state is not responsible for the photochromic ring-closure reaction. The reaction is considered to
proceed from the less twisted planar conformation in a conrotatory
mode, in which orbital interactions are present between the two
thiophene rings and acid anhydride group. In the twisted conformation, the two chromophores are orbitally decoupled and the
excited state is considered to deactivate to the ground state
slowly by radiative as well as non-radiative transitions
accompanying no reaction.

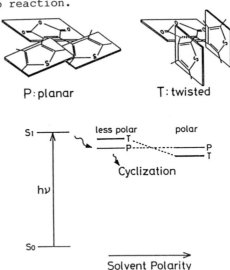

FIGURE 8
Solvent dependence of photochromic reaction of 11a.

REFERENCES

1) A.E.J. Wilson, Phys. Technol. 15 (1985) 232.
2) S. Nakamura, M. Irie, J. Org. Chem. 53 (1988) 2592.
3) M. Irie, M. Mohri, J. Org. Chem. 53 (1988) 803.
4) Y. Nakayama, K. Hayashi, M. Irie, J. Org. Chem. 55 (1990) 2592.
5) K. Uchida, Y. Nakayama, M. Irie, Bull. Chem. Soc. Jpn 63 (1990)
   1311
6) W. Rettig, Angew. Chem. 25 (1986) 971.

Photochemical Processes in Organized Molecular Systems
K. Honda (Editor-in-Chief)
© Elsevier Science Publishers B.V., 1991

PHOTO-INDUCED AND ELECTROCHEMICAL REDOX MOLECULAR SYSTEMS

Takeo SHIMIDZU[§], Tomokazu IYODA[§], Hiroshi SEGAWA[§], and Kenichi HONDA[‡]

[§]Division of Molecular Engineering, Graduate School of Engineering,
 Kyoto University, Sakyo-ku, Kyoto 606, Japan
[‡]Tokyo Institute of Polytechnics,
 Honcho 2-9-5, Nakano-ku, Tokyo 164, Japan

Some aspects of photo-induced and electrochemical redox molecular systems toward molecular devices are described with our recent works, namely, fabrications of conducting polymers, photo-induced electron transfer systems with metalloporphyrin donor and acceptor, and hybridization of conducting polymers and functional dyes.

1. INTRODUCTION

Construction of molecular device has been desired as the breakthrough for the ultimate electronic device in the coming century, since the limits of the integrated circuits made from inorganic materials were pointed out.　From this point of view, considerable efforts have been devoted for making new organic electronics materials in recent years.　The fundamental ideas of the molecular devices are based on the adaptation of the electronic functions within the individual molecules to the electronic elements.　However, the molecular function covers not only an electronic function but also a wide variety of others such as photochemical functions.　More recently, some groups have tried to construct novel molecular systems with the excellent functions which are lacked in the inorganic materials.　In the previous paper, we reported "multi-mode chemical signal transducer" which is a novel molecular system with signal transforming function between photon and electron by the use of the photo-isomerization and the electrochemical redox reaction.[1]　In realizing the conversion of such functional molecules to devices, various kinds of material design can be considered as shown in Figure 1.　As the function of most molecules arises via a deviation or a change in their electronic state or structure, it is considered that a conducting polymer is a suitable matrix for achieving conversion of the functional molecules to the molecular devices.　Therefore we have tried to construct highly ordered conducting polymers as anisotropic electron transfer units.[2]　In this study, we select the photo-induced electron transfer function among various molecular functions and present some methodologies for the material design of photo-induced and electrochemical redox molecular systems.　At the first part, fabrications of the conducting polymers are mentioned, and then homogeneous photo-induced electron

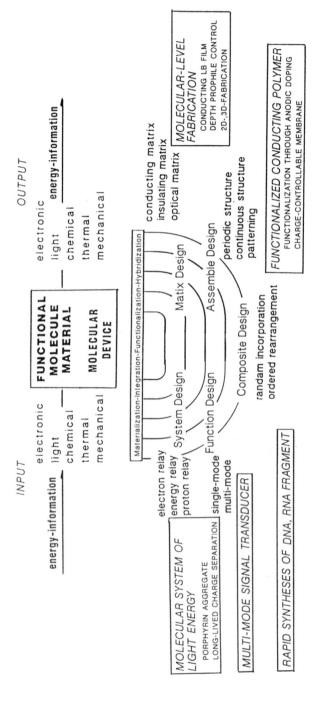

FIGURE 1. Conceptual picture of the production of an ordered molecular device involving functional molecules within a matrix.

transfer system using metalloporphyrin donor and acceptor and heterogeneous photo-induced electron transfer reactions of accordion-type porphyrin cluster are discussed for the design of the effective charge separation units. At the last part, hybridization of conducting polymers and photo-functional dyes are investigated for the combination of photonics and electronics systems.

2. FABRICATION OF CONDUCTING POLYMERS

2.1. Highly anisotropic conducting polymer LB multilayers

For coming electronic materials, current interest has been growing in conducting polymer films with desired structure and in their specific properties. The Langmuir-Blodgett (LB) technique is to be a powerful means for designing the assembly structure of the conducting polymers. We have demonstrated highly anisotropic conducting polypyrrole ultra-thin films by means of LB technique. Here, some preparative methods (Figure 2) and the structural and conducting properties are summarized.[2]

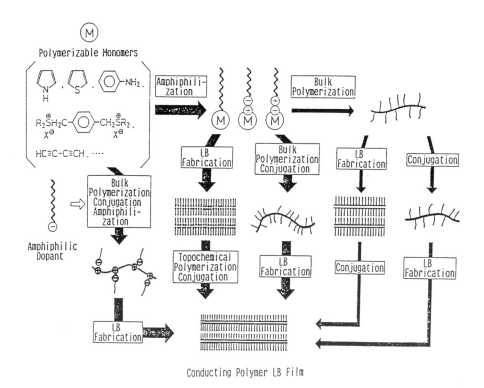

FIGURE 2. Preparative routes of conducting polymer LB films.

The first is electro-polymerization of LB film of amphiphilic monomer for conducting polymer.   New amphiphilic pyrrole derivatives with octadecane provided stable and condensed monolayers on the pure water subphase.   The mixed monolayers were transferred (100-300 layers) onto ITO-deposited polyester or silated ITO-deposited glass substrate as Y-type film by the vertical dipping method.   The pyrrole moiety in the monomeric LB film was electropolymerized very slowly from the liquid surface to the upper direction, when the lower edge of the LB film was immersed CH$_3$CN containing LiClO$_4$ and appropriate voltage was applied.   The doped polypyrrole (PPy) structure in the LB film was confirmed by ATR-IR spectra and visible-near IR absorption spectra.   The X-ray diffraction (XRD) patterns and the TEM cross section of the LB films showed that the layered structure was well preserved even after electropolymerization (Figure 3).   The resulting PPy LB film (200 layers) showed highly anisotropic dc conductivity by ca. 10 orders ($\sigma// = 10^{-1}$ S/cm, $\sigma\perp = 10^{-11}$ S/cm).   The anisotropy was originated from the alternating layered structure of conducting PPy layers and insulating alkyl chain layers.

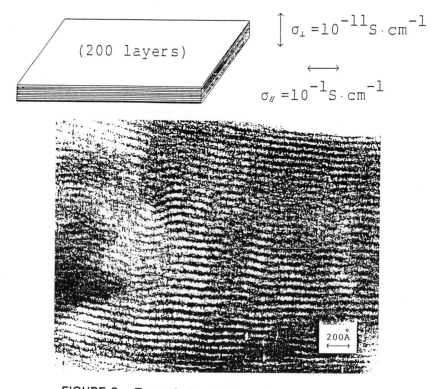

(200 layers)

$$\sigma_\perp = 10^{-11} S \cdot cm^{-1}$$

$$\sigma_{//} = 10^{-1} S \cdot cm^{-1}$$

200Å

FIGURE 3.   Transmission electron micrograph of the electropolymerized conducting polymer LB multi-layer.

The second is to prepare amphiphilic conducting polymer and to transfer it onto a substrate by LB technique.[2] An amphiphilic polyaniline, prepared by chemical polymerization of 2-octadecoxyaniline with $(NH_4)_2S_2O_8$ in THF and 2N HCl mixed solution, was partially oxidized and protonated state. The polymer was soluble in low polar solvents whose dielectric constants ranged from 2 to 12. The peak of the GPC curve of the protonated polyaniline corresponded to eicosamer. The protonation and deprotonation was reversibly occurred by the treatment with HCl and NaOH or pure water, respectively. The polyaniline formed stable monolayer on 0.1N HCl subphase with 45 mN/m of the maximum collapse pressure. The limiting area per an aniline unit estimated from the π-A isotherm was in good agreement with that estimated from the CPK molecular model (23-26 $Å^2$). The horizontal lifting method allowed us to transfer the polymer up to more than 200 layers onto polyester, gold- or ITO-deposited polyester, and silated glass substrates. The XRD pattern and the TEM cross section revealed the ordered layered structure of the film. The d spacing (37 Å) close to the expected thickness of monolayer and the use of hydrophobic substrate illustrated that the LB film would have X-type structure. The dc conductivities of the protonated polyaniline LB film (208 layers) were 10^{-9} S/cm for $\sigma//$ and 10^{-11} S/cm for $\sigma\perp$, respectively. In this material scarcely occurred the insulator-conductor transition by protonation, which is characteristic of ordinary polyaniline. The exposure to iodine vapor led to the increase of $\sigma//$ by ca. 5 orders and $\sigma\perp$ by 1-3 orders, respectively, accompanied with color change of the film from greenish blue to brown. The conductive anisotropy due to the alternating layered structure reached 4-6 orders.

The third is to amphiphilize conducting polymer through anodic doping and then to provide its LB film. To the amphiphilization is applied anodic doping with amphiphilic dopant during electropolymerization, which had been demonstrated in our previous reports.[2] 3,4-Dibutylpyrrole was electropolymerized at 1.0 V vs. SCE in CH_3CN containing perfluorooctanoic acid as supporting electrolyte, so that the amphiphilic polymer composite, soluble in organic solvents whose dielectric constants ranged from 8 to 35, was obtained onto an ITO-deposited glass electrode. The doping ratio per a pyrrole unit was estimated from elementary analyses to be 0.25±0.03. The monolayer of the polymer was spread from the CF_3CH_2OH and benzene mixed solution (2:3) onto the pure water subphase. Transfer of the monolayers onto solid substrates (polyester, ITO- or gold-deposited polyester, and CaF_2 plates) was carried out by the horizontal lifting method at 20mN/m of the surface pressure. The TEM cross section of the LB film showed the uniform thickness of the LB film. The monolayer spacing (16 Å), calculated from the observed thickness divided by the transfer number, is in good agreement with that evaluated by the CPK molecular model (12-16 Å). After exposure to iodine vapor, the parallel conductivity ($\sigma//$) increased by 2-3 orders. The conductive anisotropy of the LB film (50 layers) was 3 orders ($\sigma//$=10^{-6} S/cm, $\sigma\perp$=10^{-9} S/cm).   This anisotropy is considered to be due to the alternating layered structure of conducting PPy layers and insulating perfluoroalkyl and alkyl chain layers.

2.2. Direct photo-patterning of conducting polymers

Fabrication of conducting polymers in 2-dimension and 3-dimension is an important subject in the construction of organic electronics devices.   Although the chemical and electrochemical plymerization has mostly been used so far for the synthesis of the conducting polymers, photochemical polymerization should have great advantages for the 2-D and 3-D fabrication of the conducting polymers.   In this section, a novel photo-sensitized polymerization of pyrrole is presented [3]

Firstly, the photo-sensitized polymerization of pyrrole was investigated in aqueous solution.   A pyrrole saturated aqueous solution containing $Ru(bpy)_3^{2+}$ as a photosensitizer and $Co(NH_3)_5Cl^{2+}$ as a sacrificial oxidant was deoxygenated by bubbling argon and was photo-irradiated by a 500 W Xe lamp.   As the photo-irradiation proceeded, Cl^--doped PPy was produced.   Obtained PPy which was confirmed by i.r. spectrum and elemental analyses had the conductivity above 10^{-4} S/cm at the solid state.   The wavelength dependence of the photo-irradiation on the polymerization rate is arising from the photo-sensitization by $Ru(bpy)_3^{2+}$.   Since the phosphorescence of $Ru(bpy)_3^{2+}$ was strongly quenched by $Co(NH_3)_5Cl^{2+}$, the first step of the photoreaction is considered to be generation of $Ru(bpy)_3^{3+}$ which was oxidized Ru complex through the oxidative quenching process.   The $Ru(bpy)_3^{3+}$ has sufficient oxidizing power for the oxidation of pyrrole and its oligomer.   It was concluded that the phot-sensitized polymerization proceed through the oxidative electron transfer process as below.

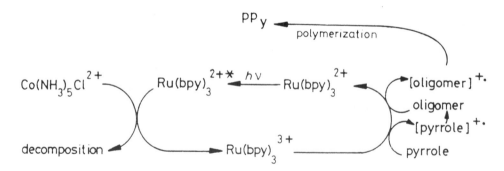

On the basis of the result in the aqueous solution, the photo-sensitized polymerization was performed using an ionic polymer membrane, such as Nafion. The Nafion membrane adsorbing $Ru(bpy)_3^{2+}$ was dipped into a saturated pyrrole solution containing $[Co(NH_3)_5Cl]Cl_2$ and photo-irradiated by a dye laser at 490 nm through a photomask as shown in Figure 4.   The present preliminary study gave a fine PPy pattern with 10 μm line width directly on the Nafion membrane (Figure 4). The present photo-sensitized polymerization is useful for the fabrication of fine conducting polymer patterns on insulating organic materials.

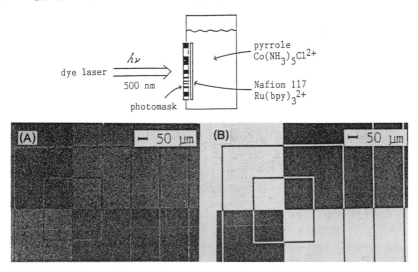

FIGURE 4.   Optical microscope photograph of (A) polypyrrole synthesized on Nafion film by photo-sensitized polymerization and (B) photo-mask used for photo-irradiation (line width: 10 μm).

3   DESIGN OF FUNCTIONAL DYE FOR EFFECTIVE CHARGE SEPARATION

3.1. Metalloporphyrin donor and acceptor

Photo-induced electron transfer reactions between metalloporphyrins play the important roles in the effective charge separation of the primary process in natural photosynthesis.[4]   In homogeneous solutions, the photo-induced electron transfer reactions of the porphyrins with various electron donors or acceptors have been investigated so far,[5] the inter-porphyrin electron transfer reactions were scarcely investigated in spite of its essential importance.   In this study, the photo-induced electron transfer reactions between the metalloporphyrins (Figure 5) were investigated in homogeneous solutions.   In order to accomplish the effective and long-life charge separation, the electron transfer reactions were investigated in aqueous solutions.   Zinc complexes which form triplet excited states in high quantum yield and stable π-radical cations were used as photosensitizers and gold complexes which form stable π-radical anions were used as electron acceptors.[6]

As a typical case, the electron transfer reaction between ZnTSPP and AuTSPP is firstly described in detail.   Zinc and gold complexes of anionic porphyrins tend to form hetero-aggregates with charge transfer interaction in aqueous solution.[7] The fluorescence of the ZnTSPP moiety is strongly quenched according to the hetero-aggregation in static process. The quenching is considered to be the electron transfer quenching. It is suggested that the charge recombination reaction in singlet radical ion pair, generated from the direct excitation of the hetero-aggregate, was considerably fast process.

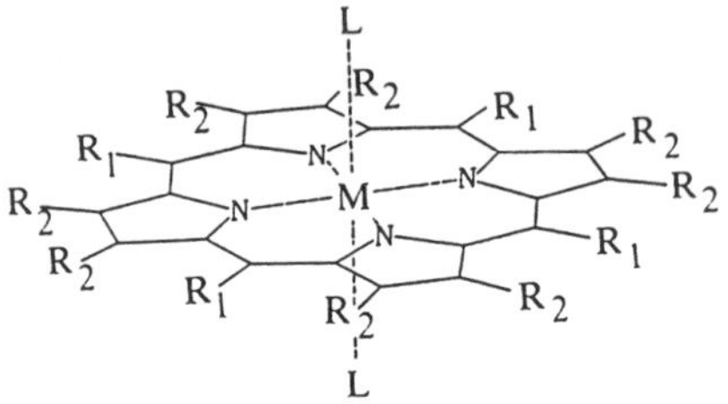

R_1	R_2	abbreviation
$-C_6H_5$	-H	TPP
$-C_6H_4SO_3^-$	-H	TSPP
$-C_6H_4COO^-$	-H	TCPP
$-C_6H_4NCH_3^+$	-H	TMPyP

M = Au Ag Cu Zn Mn Cd Sn Pd Co etc.

FIGURE 5. Structures of metalloporphyrins.

In the highly dilute solution, where the hetero-aggregates were scarcely formed, the singlet excited state was hardly quenched by the AuTSPP in dynamic process and then changed to a triplet excited state with high quantum yield.   The triplet excited state of the ZnTSPP (^{3}ZnTSPP*) was quenched by the AuTSPP in the highly diluted solution, and then long-life component was observed in 600 - 750 nm after the ^{3}ZnTSPP* decay as shown in Figure 6.   Since the absorption feature at 3 ms was resemble to the sum spectrum of π-radical cation of ZnTSPP and π-radical anion of AuTSPP, the long-life component was considered to be charge separation state. The transient absorption decay at 760 nm (decay of the ^{3}ZnTSPP*) and transient absorption rise at 600 nm (generation of the π-radicals of ZnTSPP and AuTSPP) were completely unity in their time constant.   The agreement of the rise and the decay implies that the charge separation was occurred through the quenching of the ^{3}ZnTSPP* by the AuTSPP.   Taking account of the redox potentials of the ZnTSPP ($E_{1/2}$ = 0.87 V for the ring oxidation and -1.16 V for the ring reduction vs. NHE) and the AuTSPP ($E_{1/2}$ = -0.37 V vs. NHE for the ring reduction), the quenching is an oxidative electron transfer process from the ^{3}ZnTSPP* to the AuTSPP.   Surprisingly, the long-life component due to charge recombination process follows first-order kinetics and its life span reached 60 ms.   Furthermore, the charge recombination rate was strongly accelated by the addition of heavy atoms such as Br$^-$ or I$^-$.   The external heavy atom effect implies that the triplet radical ion pair is formed by the oxidative electron transfer process.   The present long-life charge separation was considered to be due to the spin forbidden in the triplet radical ion pair.   The charge separation yield is reached 10 % from the excited ZnTSPP.

The present charge separation process was summarized in Figure 7. The charge recombination rates of the radical ion pairs depended on its electronic spin strongly. In this study, the anomalous long-life charge separations were also accomplished by other conbinations of anionic gold and zinc porphyrins in homogeneous aqueous solutions. For the construction of the devices on the basis of the results in homogeneous solutions, it is suggested that the control of the spin state of the radical ion pair by adequate matrix is important.

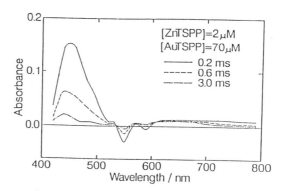

FIGURE 6. Transient difference absorption spectra in the photo-induced electron transfer reaction from ZnTSPP to AuTSPP.

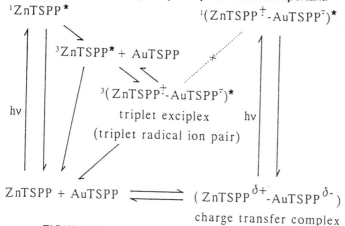

FIGURE 7. Mechanism of the photo-induced electron transfer process and the charge separation process of ZnTSPP and AuTSPP.

3.2. Accordion-type porphyrin cluster

The study on the photo-induced electron transfer reaction using metalloporphyrin assembly are noted in connection with charge separation and electron transfer. As an earlier report,[8] cationic water-soluble porphyrins (TMPyP) and anionic water-soluble porphyrins (TSPP) form hetero-aggregate, namely "accordion-type porphyrin cluster", in aqueous solution (Figure 8). A result of X-ray small angle scattering analysis of the aggregate of ZnTMPyP-PtTSPP suggested the cylindrical structure

with ca. 10 Å of cross
section radius and
several hundreds Å of
length. Assuming
several Å of the intervals
of between the porphyrin

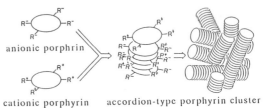

anionic porphrin

cationic porphyrin   accordion-type porphyrin cluster

planes, it was estimated
that the aggregate
consists of over hundreds
of the porphyrins.

aggregation through electrostatic interaction
1:1 composition of cationic and anionic porphyrins
face-to-face-type aggregate
several hundreds of aggregate number
applicable to every free-base and metalloporphyrin

According to increasing
of concentration of the
cationic and anionic
porphyrins, remarkable
hypochromic effects were
observed in Soret-band
and Q-band because of
the π-π interaction. The
face-to-face structure and
the 1:1 composition were
suggested by the
absorption changes due
to aggregation, where the

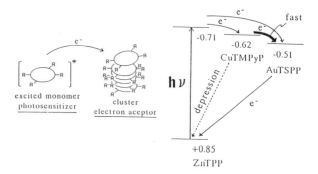

excited monomer
photosensitizer

cluster
electron acceptor

FIGURE 8. Accordion-type porphyrin aggregate.

blue-shifts in Soret-band and red-shifts in Q-band like as the case of other face-to-
face aggregates [9] and the maximum deviation from Beer's law at same
concentrations of each components were observed. The driving force of the hetero-
aggregation is an electrostatic interaction between anionic substituent and cationic
one on the porphyrin periphery.

Photo-induced electron transfer reactions between accordion-type cluster as an
electron acceptor and monomeric porphyrin (MTPP) as a sensitizer were investigated
in DMF. The MTPP in the ground state did not stack with each monomer and cluster
in DMF. In the case of AuTMPyP-AuTSPP cluster and ZnTPP sensitizer, the triplet
excited-state of ZnTPP (ditected by transient absorption spectra at 600 - 800 nm) was
strongly quenched by the cluster, and then a small amount of charge separation into
[AuTMPyP-AuTSPP]$^{\div}$ and [ZnTPP]‡ was observed at 580 - 640 nm. Similar charge
separation through the oxidative electron transfer process were also observed in the
other cluster-sensitizer systems. In the case of the CuTMPyP-AuTSPP cluster and
the CuTMPyP-CuTSPP cluster, short life time components due to electron transfer
quenching process and long life time components due to charge recombination

process were observed. The transient absorption spectra of the charge separation state depended on the central metals of the clusters as shown in Figure 9. The residual absorption spectra at 1 ms after flash excitation were coinsided with sum of the radical cation of ZnTPP and radical anions of energetically low component of the cluster (AuTMPyP 630 nm, AuTSPP 650 nm, CuTMPyP 680 nm). The result suggests the rearrangement of the electron supplied by the triplet excited ZnTPP. The presence of the fast electron transfer process in the cluster was also confirmed by flash photolysis.

The electron transfer rate was depended on the kinds of central metal combination of the cluster, which affects the redox potentials of the porphyrin ring systems. However, the determination of the rate constant was difficult because of the difficulty of determination of the real concentration of the cluster. The stabilization of charge separation state was attributed to the rearrangement of injected electron in the cluster. The present electron transfer reaction in the cluster and between the cluster and monomer provides ideas in the construction of photo-energy conversion assembly system.

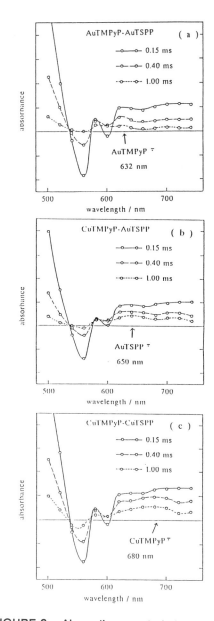

FIGURE 9. Absorption spectral changes due to the quenching of the triplet excited state of ZnTPP (1.3×10^{-6} mol dm^{-3}) by (a) AuTMPyP-AuTSPP cluster, (b) CuTMPyP-AuTSPP cluster, and (c) CuTMPyP-CuTSPP in DMF at room temperature.

4. HYBRIDIZATION OF CONDUCTING POLYMERS AND FUNCTIONAL DYES

Electropolymerization mechanism of pyrrole and thiophene is characterized by anodic doping with supporting electrolyte anion. We demonstrated the preparative method for conducting polymer composite with various anionic functional molecules by electropolymerizing the monomer in the presence of the anionic functional molecule as a supporting electrolyte.[10] Because the conducting polymer film grows with incorporating the anionic functional molecule as a dopant, any anionic molecule, even if large and still more a polymer, was incorporated into the polymer matrix. Especially in case of large dopant its very low mobility in the matrix depresses dopant releasing process even when the composite was treated with electro-reduction (undoping treatment). This behavior is effective for conducting polymer modified electrode with various kinds of dopant. Table 1 shows a list of functional dopant, which was incorporated into PPy matrix, and a classification of their functionalities.

Here, electrochemical photo-sensitization was demonstrated by porphyrin-incorporating PPy composite electrode, among various functional dopants. According to the above procedure, anionic porphyrin (ZnTSPP) was incorporated into PPy with 0.34 of doping ratio and 0.1 S/cm of conductivity. The current efficiency of the incorporation was 7.9×10^{-8} eq. of sulfonate moiety per coulomb. Figure 10 shows the typical photocurrent-potential curve of a ZnTSPP-PPy ITO electrode in water

TABLE 1   Functionalized conducting polymers

Functional dopant/matrix [a]	Charge	Function	Procedure [b]
Anthraquinone 2-sulphonate	Sulphonate	Electrochromism	1, 3
Lu(PTS)$_2$	Sulphonate	Electrochromism	1, 3
Fe(BPS)$_n$(BP)$_{3-n}$	Sulphonate	Electrochromism	1, 3
Ru(BPS)$_n$(BP)$_{3-n}$	Sulphonate	Electrochemiluminescence	1, 3
MTPPS (M = Zn, Pd, etc.)	Sulphonate	Photosensitized electrode	1, 3
MTMPyP	Pyridinum	Photosensitized electrode	2
Rose bengal	Carboxylate	Photosensitized electrode	1, 3
Indigo carmine	Sulphonate	Photosensitized electrode	1, 3
Poly(vinyl sulphate)	Sulphate	Charge-controllable membrane	1, 3
Poly(styrene sulphonate)	Sulphonate	Charge-controllable membrane	1, 3
Nafion[R]	Sulphonate	Charge-controllable membrane	1, 4
Polynucleotide	Phosphate	Nucleic acid sensor	1, 3
Nucleotide	Phosphate	Nucleic acid sensor	1, 3
Porous filter	–	Filtration	4
Phosphotungstate	Heteropoly acid	Electrochromism	1
PtCl$_4^{2-}$	PtCl$_4^{2-}$	Highly dispersed metal	1, 3
AuCl$_4^-$	AuCl$_4^-$	Highly dispersed metal	1, 3

[a] PTS = phthalocyanine tetrasulphonate;   BP = bathophenanthroline;   BPS = BP disulphonate;

[b] Incorporation procedure:

1, electrochemical anodic doping; 2, electrochemical pseudo-cathodic doping;

3, vapour–liquid interface chemical polymerization; 4, bulk chemical polymerization.

containing 3.0mmol/l of $Fe(CN)_6^{3-/4-}$ under >390nm irradiation. At >0.3V appreciable anodic photocurrent (>600 nA/cm²) was observed by photo-ac method using a lock-in amp. and a light chopper (20Hz). The agreement of the action spectrum of the photocurrent with the visible absorption spectrum indicated that the incorporated ZnTSPP played a significant role as a sensitizer for anodic photocurrent generation. The current quantum efficiency at 435 nm exceeded 0.1%. The photocurrent was proportional to incident light

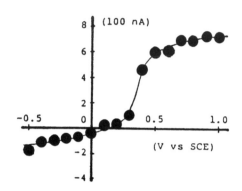

FIGURE 10.   Photocurrent-potential curve of the ZnTSPP-PPy modified ITO electrode in $Fe(CN)_6^{3-/4-}$ aqueous solution under visible light irradiation.

intensity. Other sensitizers, e.g., anionic ruthenium polypyridine complexes, rose bengal, and indigo carmine, were also applicable for this photo-responsive PPy-modified electrode.

On the other hand, we synthesized cationic polypyrrole derivatives in order to hybridize high amount of anionic functional molecules not only through anodic doping process, as the case of our previous report,[10] but also due to electrostatic binding. N-Methyl-3-(N-pyrrylmethyl)pyridinium (MPP) was polymerized chemically with ferric trichloride in water or acetonitrile, so that the polymer (PMPP) was identified by FT-IR. Up to > 10^{-2} mol/l per pyrrole unit the obtained PMPP was soluble into polar solvents with >25.2 of dielectric constsnt.   PMPP has the potential ability to incorporate an anionic functional molecule mainly as counter anion of cationic group and as dopant by anion exchange treatment, different from the case of hybridization through anodic doping on electropolymerization. For example, the addition of polyanion, such as FeBPS, RuBPS to the PMPP aqueous solution resulted in the precipitate of these composites. Since the resulting composite was again dissolved in another mixed solvent such as (conc $HCl:Dioxane:H_2O=2:3:1$). A composite film was obtained by spreading the solution onto the surface of any substrate and then the removing the solvent. The incorporation of the anionic functional molecule was verified by UV-VIS spectra. The composite film was also obtained by electropolymerization of MPP in the presence of polyanion such as FeBPS, RuBPS, and PVSK or hydrophobic salt such as $NaClO_4$, and $NaBF_4$, enough to be insoluble in the electrolytic solvent. The ratio of anion per pyrrole unit, for PMPP obtained by chemical and electrolytic polymerizastion was estimated to be 1.49 and 1.45, which means that PMPP contained ca. 5 times more amount of anion than PPy. These anions were incorporated into the PMPP matrices through anodic doping process and electrostatic binding with cationic group. The PMPP composite disk showed relatively low dc conductivity (10^{-6} S/cm). A remarkable absorbance change in 500-600nm region was observed when the potential of PMPP/FeBPS electrode was swept between 0.5V and 1.3 V in 0.1 mol/l KCl aqueous solution. On the electrode clear color change was observed between red (<0.8V) and almost transparent brown (>1.1V). There was an inflection point of the absorbance change at 540 nm, in agreement with the half wave potential of Fe(BPS)$_3$ incorporated in the PMPP. This evidence indicated that this absorbance change was due to the redox reaction of the incorporated FeBPS. The PMPP / RuBPS electrode in aqueous solution containning Na$_2$SO$_4$ (0.2 mol/l) showed reversible redox wave of Ru(BPS)$_3^{3-/4-}$ at 1.1 V. With 50 mmol/l of K$_2$C$_2$O$_4$ the anodic current increased and

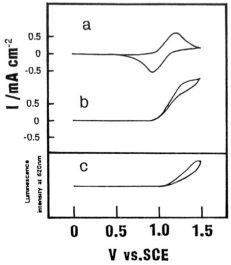

FIGURE 11.   Cyclic voltammograms of the PMPP / Ru(BPS)$_3^{4-}$ on ITO electrode in 0.2 mol/l Na$_2$SO$_4$ aqueous solution (a) and in the presence of 50 m mol/l K$_2$C$_2$O$_4$ and 0.2 mol/l Na$_2$SO$_4$ aqueous solution (b).   The electrogenerated chemiluminescence intensity of (b) at various electrode potential is shown in (c).

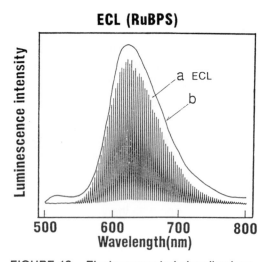

FIGURE 12.   Electrogenerated chemiluminescence of PMPP / Ru(BPS)$_3^{4-}$ on ITO electrode in 0.2 mol/l Na$_2$SO$_4$ and 50 m mol/l K$_2$C$_2$O$_4$ aqueous solution (a).   Luminescence spectra of the PMPP / Ru(BPS)$_3^{4-}$ on ITO electrode (b).

the cathodic one diminished, which indicated that the oxalate was electro-oxidized catalytically via the mediation of $Ru(BPS)_3^{4-}$ in the PMPP (Figure 11).   Orange light, clearly visible to the eye, was emitted from the electrode surface in more than 1.1 V region.   The spectrum of the electrogenerated chemiluminescence (ECL) was agreement with the emission spectra of PMPP/RuBPS electrode under exitation at 450 nm (Figure 12).   The ECL resulted from MLCT excited state of $Ru(BPS)_3^{4-}$ generated through the highly exothermic reaction between the oxdized complex and the CO_2^- with very strong reducing power.   The PMPP had advantages for highly incorporation of anion into the PMPP matrices, solubility into polar solvents, and ion exchanging property.

REFERENCES
1)   (a) T. Iyoda, T. Saika, K. Honda, and T. Shimidzu, Tetrahedron Lett. 30 (1989) 5429.
(b) T. Shimidzu, T. Iyoda, and K. Honda, Pure and Appl. Chem. 60 (1989) 1025.
(c) T. Shimidzu, T. Iyoda, H. Segawa, T. Saika, and K. Honda, MRS Int'l. Mtg. on Adv. Mats. 12 (1989) 261.

2)   (a) T. Iyoda, M. Ando, T. Kaneko, A. Ohtani, T. Shimidzu, and K. Honda, Tetrahedron Lett. 27 (1986) 5633.
(b) T. Iyoda, M. Ando, T. Kaneko, A. Ohtani, T. Shimidzu, and K. Honda, Langmuir 3 (1987) 1169.
(c) T. Shimidzu, T. Iyoda, M. Ando, A. Ohtani, T. Kaneko, and K. Honda, Thin Solid Films 160 (1988) 67.
(d) M. Ando, Y. Watanabe, T. Iyoda, K. Honda, and T. Shimidzu, Thin Solid Films 179 (1989) 225.

3)   H. Segawa, T. Shimidzu, and K. Honda, J. Chem. Soc., Chem. Commun. (1989) 132.

4)   J. Deisenhofer, O. Epp, K. Miki, R. Huber, and H. Michel, Nature 318 (1985) 618.
5)   (a) M. Grätzel et al., "Structure and Bonding. 49", Academic Press, New York (1982).
(b) J.K. Roy, F.A. Carroll, and D.G. Whitten, J. Am. Chem. Soc. 96 (1974) 6349.
(c) E.I. Kapinus, M.M. Aleksankina, V.P. Staryi, V.I. Boghillo and I.I. Dilung, J. Chem.Soc., Faraday Trans. II 81 (1985) 631.   And references their in.

6)   (a) T. Shimidzu, T. Iyoda, H. Segawa, and K. Honda, Nouv. J. Chim. 10 (1986) 213.
(b) T. Shimidzu, H. Segawa, T. Iyoda, and K. Honda, J. Chem. Soc., Faraday Trans. II 83 (1987) 2191.

7)   H. Segawa, H. Nishino, T. Kamikawa, K. Honda, and T. Shimidzu, Chem. Lett. (1989) 1917.

8)   T. Shimidzu and T. Iyoda, Chem. Lett. (1981) 853.

9)   R.F. Pasternack et al., J. Amer. Chem. Soc. 94 (1972) 4511.

10)   (a) T. Iyoda, A. Ohtani, T. Shimidzu, and K. Honda, Synth. Met. 18 (1987) 725.
(b) T. Iyoda, A. Ohtani, T. Shimidzu, and K. Honda, Chem. Lett. (1986) 687.

Photochemical Processes in Organized Molecular Systems
K. Honda (Editor-in-Chief)
© Elsevier Science Publishers B.V., 1991

INFORMATION STORAGE USING PHOTOELECTROCHEMICAL HYBRID SYSTEMS

Z. F. LIU, K. HASHIMOTO and A. FUJISHIMA

Department of Synthetic Chemistry, Faculty of Engineering,
The University of Tokyo, Hongo, Tokyo 113, Japan

Photoelectrochemical behavior of an azobenzene derivative monolayer film has been investigated on a transparent SnO_2 glass electrode. A novel hybrid one-way phenomenon was observed due to the following two experimental results: the first is that the cis form of the azobenzene derivative was reduced to a hydrazobenzene species at substantially more anodic potential than the trans form; the second is that the hydrazobenzene derivative subsequently produced was exclusively oxidized to the energetically stable trans form. On the basis of this hybrid effect, a new type of photon-mode storage technique has been developed, which leads to a photoelectrochemical memory device with an ultra-high storage density and multiple functions being created.

1. INTRODUCTION

High-density information storage is one of the key technologies of modern society. The heat-mode techniques of the current optical memory, in which only the thermal energy of laser light is utilized in recording process and hence the information is usually stored as a physical change of the storage media, are, in principle, subjected to the diffraction limit of light[1]. To achieve higher storage density, a photon-mode technique is indispensable. Since the information is stored as a photon-induced chemical change of the storage media in this case, the frequency domain of light is also available for increasing storage density of information[1]. Although photochromism and photochemical hole burning have widely drawn up people's expectations for the photon-mode technique, a number of intrinsic drawbacks such as thermal instability[2] in the former and extremely low working temperature in the latter[1] have retarded their realization. In this paper, we present a new type of photon-mode storage technique which is established upon a photochemical and electrochemical hybrid phenomenon found in azobenzene system.

As is known, there are two typical reversible reactions, i.e., the photochemical trans/cis isomerization[3] and the electrochemical reduction/oxidation[4], in an azobenzene system. Our basic idea is deliberately organizing these processes to store information photoelectrochemically. This strategy is novel because both photon-process and electron-process are involved in the information recording step, which hence is expected to allow a non-destructive readout of the information stored.

2. EXPERIMENTALS

The azobenzene derivative used in the present study was 4-octyl-4'-(5-carboxy-pentamethylene-oxy)-azobenzene [ABD], the monolayer film of which was deposited onto a transparent SnO_2 glass substrate, the working electrode, in the trans form using the conventional Langmuir-Blodgett method. The surface pressure and subphase temperature for film fabrication were controlled at 25 mN/m and 20 $^{\circ}$C, respectively. Cyclic voltammetry (CV) was used for the electrochemical investigations with a 0.2 M aqueous potassium perchlorate, buffered to pH 7.0 with a Britton-Robinson solution being used as the electrolyte. The reference electrode was Ag/AgCl, and the counter electrode was a Pt wire.   A 500 W xenon lamp and a 14 mW He-Cd laser (325 nm) were used to induce general and localized isomerization of ABD molecules from trans to cis form, respectively. The UV (320-380 nm) and visible (>440 nm) lights of xenon lamp were obtained via glass filters.

3. RESULTS AND DISCUSSION

3.1 Photoelectrochemical hybrid effect in azo system

Figure 1 compares the cyclic voltammetric (CV) behavior of trans- and cis-ABD monolayer films on a SnO_2 glass electrode with the cis-ABD film being obtained after 1-min UV irradiation of the initial trans-ABD film by xenon lamp. In the applied potential range no reduction and oxidation peaks were observed in the trans-ABD film, whereas significantly large reduction and oxidation peaks were obtained in the cis-ABD film. The electrochemical activity found in the latter case has been ascribed to the low-energy reduction of cis-ABD to its

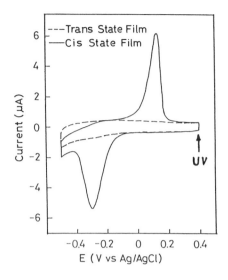

Figure 1 Cyclic voltammograms of trans- and cis-ABD monolayer films on a SnO₂ glass electrode. The cis-ABD was obtained after 1-min UV irradiation of the initial trans-ABD. The sweep rate was 20 mV/sec.

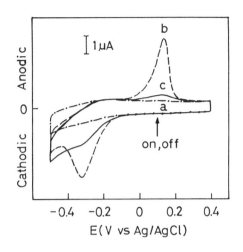

Figure 2 Cyclic voltammograms of an ABD monolayer film-modified SnO₂ glass electrode taken (a) in the dark, (b) after 1-min UV irradiation at 0.1 V, (c) after 1-min UV irradiation followed by 1-min visible light irradiation at 0.1 V.

corresponding hydrazobenzene derivative (hydra-ABD) followed by re-oxidation of the hydra-ABD, on the basis of a series of investigations[5,6]. In other words, cis-ABD molecules are easier to be electrochemically reduced than the trans-ABD. A strong evidence for the cis-ABD reduction mechanism is shown in Figure 2, in which after 1-min UV irradiation(320-380 nm), the ABD monolayer film was then

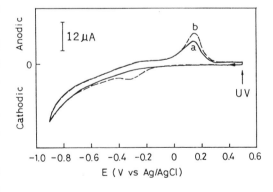

Figure 3 Cyclic voltammograms of an ABD monolayer film-modified SnO₂ glass electrode taken (a) in the dark, (b) after 1-min UV irradiation at 0.5 V.

exposed to   visible irradiation (>440 nm) for another one min.
Since the UV-generated cis-ABD was reconverted to the trans-ABD
with the visible irradiation, almost no reduction and oxidation
peaks were observed in this case (curve c), as compared with the
one that did not experience the visible illumination (curve b).
A similar, but slow reduction of the trans-ABD molecules also
occurred at more negative potential, as evidenced by the broad
reduction peak and the following oxidation peak shown in Figure 3.
The difference in reduction potential $(E_{1/2})$ of the trans- and
cis-ABD molecules was estimated to be more than 400 mV in a
neutral solution.

The hydra-ABD, on the other hand, was found to be exclusively
oxidized to the energetically stable trans-ABD because no
electrochemical activity was observed after its oxidation in the
electro-inactive potential range of trans-ABD.

A similar photoelectrochemical behavior was also obtained for
some other long-chain fatty acid derivatives of azobenzene whose
chromophores are differently located in their alkyl chains.
Additionally, it is noted that the present electrochemical result
on azobenzenes shows some difference from Laviron's one in which
no discernible electrochemical difference between trans and cis
isomers of the naked azobenzene was observed in a similar aqueous
solution[4]. This may be attributed to a structural or orientation
difference of the surface-bound azobenzenes.

Figure 4 summarizes the
photochemical and electrochemical behaviors of the
present system. Evidently, a
3-state and one-way cyclic
process has been formed via
the   sequent isomerization,
reduction and oxidation
reactions, leading to the
photoelectrochemical hybrid
phenomenon what we called. It
is this newly-discovered
hybrid effect strongly implies
the possibility for a
photoelectrochemical information recording.

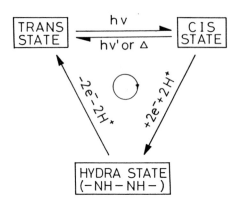

Figure 4   The clockwise hybrid
process obtained in azobenzene
system.

3.2 Information storage using the hybrid phenomenon

3.2.1. Recording scheme

The thermally stable hydra-ABD is used as the storage state for information instead of the thermally unstable cis-ABD. The stability of the storage state molecules has been preliminarily investigated under several conditions. Although these hydra-ABD molecules were slowly oxidized to the trans-ABD at an oxygen ambience, no change was observed for these storage state molecules at deaerated or negatively(-0.3 V)-biased conditions.

The photoelectrochemical hybrid feature of the present system gives rise to two possible means of high-density storage: either high-resolution optical writing or high-resolution electrochemical writing[7] as schematically shown in Figure 5.

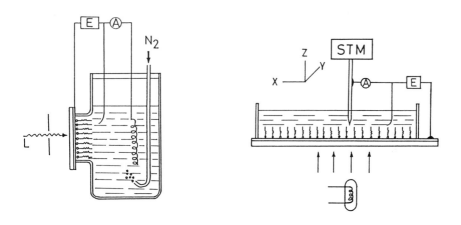

Figure 5   Schematic representations of the optical writing (left) and the electrochemical writing (right).

For the optical writing, a high-density storage is obtained using a fine-beam laser to localize trans to cis isomerization, followed by switching the potential of the entire film to a region where cis-ABD is reduced but not the trans one. Figure 6 shows the experimental results when a He-Cd laser (beam diameter $\sim$ 1 mm) was used to irradiate 6 different areas, with CVs being obtained. The number of electrochemically reacting ABD molecules was calculated by integrating the anodic current-potential curves, which showed a linear increase proportional to the increase in the number of irradiated areas (Figure 7). The small change in the average number of reacting molecules per area is attributed to the thermal cis to trans isomerization which occurred in the previously irradiated areas. A separate experiment was performed as a correction measure to determine the stability of the cis-ABD film in the dark (Figure 7), and it is obvious that only the areas subjected to illumination show an electrochemical response.

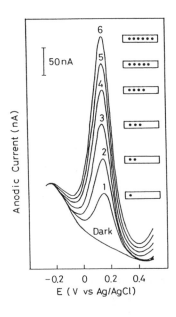

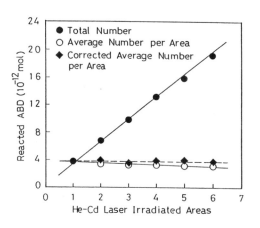

Figure 7 Linear relationship between the reacting ABD and the laser-irradiated areas. (●) total number of reacting ABD molecules; (○) average number per irradiated area; (◆) corrected average number per irradiated area.

Figure 6 Anodic peaks of the cyclic voltammograms at different laser irradiation areas. (●) represents schematically the unit area which was irradiated by one beam spot for 10 sec.

Consequently, the storage density merely depends on irradiation area, and if such operation was carried to the diffraction limit of laser light, one would be able to store 10^8 bits/cm^2 in a two-dimensional configuration. Moreover, if several azobenzene species which have similar hybrid behavior but substantially different spectral responses were located in the storage film, i.e., utilizing the frequency domain of light, the storage density can be further magnified by use of a tunable laser beam.

A high-density electrochemical writing is, on the other hand, obtained by controlling the reduction area of cis-ABD film after uniform UV irradiation of the entire film. This becomes possible, using a sharp glass-coated probe as the counter electrode, with the distance between the probe and the storage film-modified electrode being controlled. The scanning tunneling microscopic technique[8] is available for positioning this probe, and a 10^{12} bits/cm^2 of storage density is principally possible in a two-dimensional configuration without utilizing the frequency domain of light[9].

3.2.2. Readout of information

Using the present photoelectrochemical hybrid system also offers a realistic approach for non-destructive readout of the information stored. Figure 8 shows a comparison of the absorption spectra between trans-ABD, the initial state, and hydra-ABD, the recording state. The spectral change observed is mainly attributed to the   -N=N- to -NH-NH- conversion of the ABD chromophore. On the basis of this large spectral difference, the information

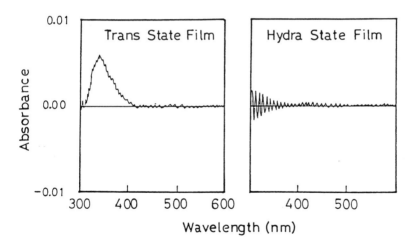

Figure 8   Absorption spectra of trans- and hydra-ABD in the monolayer film. The hydra-ABD was obtained by UV-irradiating the trans-ABD for 5 min with the supporting electrode being biased at -0.2 V vs. Ag/AgCl.

stored in both writing modes can be read out by monitoring optical changes. Since two reaction steps, i.e., the photochemical isomerization and electrochemical reduction, are necessary for formation of the hydra-ABD state, i.e., the storage state, this optical readout method will not destroy the stored information.

The idea that intentionally introduces a hybrid process to

achieve non-destructive readout of the stored information has an
important implication, in view of the intrinsic readout problem
for the current photon-mode techniques based on a one-photon
process. Because no any threshold value of illumination is
available in comparison to the heat-mode recording, optical
readout based on spectral changes is impossible for the current
ones. Additionally, other readout approaches using light
scattering or electric charge may also be possible in the present
hybrid system though the readout based on electric charge will
destroy the information stored.

3.2.3. Electrochemical Erasion

The information stored can be wiped by anodizing the entire
film, whereupon the hydra-ABD molecules are exclusively oxidized
to the original trans-ABD state, indicating that the present
storage system is rewritable. The reversible durability has been
studied by repeating the isomerization $\longrightarrow$ reduction $\longrightarrow$ oxidation
operation. No discernible change in the reaction nature was
observed during several hundred cycles of repetition, though a
little decrease in reduction/oxidation peak currents occurred due
to dissolution of the film molecules to the aqueous electrolyte.
The dissolution problem is expected to be overcome by introducing
a solid electrolyte instead using the present wet system.

3.2.4. Multi-function Memory

The 3-state feature of the present cyclic process may allow a
multi-function memory to be realized since the metastable cis-ABD
film can be used for short-term storage, and the hydra-ABD film
for long-term storage.

3.3 General aspects of the hybrid phenomenon

The idea we proposed for information storage may not be
necessarily limited in the present specific azo molecule, and in
fact, even not the azo system as well. The "-C=N-" and "-C=C-"
systems are also attractive in view of the similar photochemical
and electrochemical behaviors. As a general rule, one may perform
a deliberate structural design to obtain a suitable molecular
system having the expected hybrid effect, or more concretely,
having a substantially different electrochemical behavior between
the stearic isomers of a double-bonded organic molecule.
Clearly, molecular or matrix design will also play an important
role for satisfying the requisition such as long-wavelength
response property of a realistic storage system[10].

4. CONCLUSION

We have proposed a completely new type of photon-mode information storage technique on the basis of the novel photoelectrochemical hybrid phenomenon found in azobenzene system. This technique allows a photoelectrochemical memory device with an ultra-high recording density and multiple functions to be realized. The idea we presented in this paper has supplied a new approach for electrochemistry onto the information storage system, and will open several new avenues of memory studies.

ACKNOWLEDGMENT

The authors gratefully acknowledge Dr. S. Suzuki, Dr. K. Hyodo and Dr. K. Yamamoto of Mitsubishi Paper Mills Co., Ltd. for their helpful discussions.

REFERENCES

1) A. R. Gutierrez, J. Friedrich, D. Haarer and H. Wolfrum, IBM J. Res. Develop. 26(2) (1982) 198.

2) E. Ando, J. Miyazaki and K. Morimoto, Thin Solid Films 133 (1985) 21.

3) See, for example, J. Griffiths, Chem. Soc. Rev. 1 (1972) 481.

4) See, for example, E. Laviron and Y. Mugnier, J. Electroanal. Chem. 111 (1980) 337.

5) Z. F. Liu, Thesis, The University of Tokyo (1990).

6) Z. F. Liu, B. H. Loo, K. Hashimoto and A. Fujishima, J. Electroanal. Chem. 297 (1991) 133.

7) Z. F. Liu, K. Hashimoto and A. Fujishima, Nature 347 (1990) 658.

8) G. Binnig and H. Rohrer, Rev. Mod. Phys. 59 (1987) 615.

9) J. Kwak and A. J. Bard, Anal. Chem. 61 (1989) 1794.

10) S. Tamura, N. Asai and J. Seto, Bull. Chem. Soc. Jpn. 62 (1989) 358.

Photochemical Processes in Organized Molecular Systems
K. Honda (Editor-in-Chief)
© Elsevier Science Publishers B.V., 1991

ELECTROLUMINESCENCE IN ORGANIC THIN FILMS

Tetsuo TSUTSUI, Chihaya ADACHI, and Shogo SAITO

Department of Materials Science and Technology,
Graduate School of Engineering Sciences,
Kyushu University, Kasuga, Fukuoka 816 Japan

Mechanism of electroluminescence in organic dye films,
mechanisms of carrier injection, carrier transport, carrier
recombination, creation of molecular excitons, movement of
molecular excitons, and emission from molecular excitons, is
described. Using high performance electroluminescent
devices, three attempts towards novel fields of photophysics
and photochemistry of organic solids were performed. First,
confinement of molecular excitons within a molecular-size
area was reported. Second, emission from triplet excitons
produced by electric excitation was observed. Third, the
presence of quantum optical size effect of radiation field
in organic thin films was demonstrated.

1. INTRODUCTION

Growing interest has been shown on multilayer-type thin-film
electroluminescent (EL) devices, which exhibit very high
luminance and can be driven with low dc voltage. One expects
that the application of organic thin-film EL devices for large-
area, flat-panel, and full-color displays is promising. We have
reported that a variety of fluorescent dyes could be used for
emitter materials[1-5]. We have also showed that utilization of
appropriate device structures was crucially important for getting
high luminance efficiencies in the EL devices[6,7].

In the development of high performance EL devices, we have
established the method to control the movements of charged
carriers and molecular excitons in organic solids[8,9]. Now we
know how to realize the confinement of charged carriers and
molecular excitons within designed portions in thin film devices.
This means that a novel method to get controlled emission from
organic solids by means of electric excitations in stead of
conventional photo-excitation is at our hands. This new method
gives us a promising tool for fundamental studies in photophysics
and photochemistry on organic solids.

In this paper, we will firstly describe the mechanism of

electroluminescence in organic multilayer EL devices; charge injection, charge transport, charge recombination and creation of molecular excitons, and emission from molecular excitons. Then, we will show our recent approaches to extend the EL phenomena for diversity of research fields in photochemistry and photophysics.

2. ELECTROLUMINESCENT MECHANISM IN ORGANIC MULTILAYER DEVICES

The working mechanism of injection type EL devices is very simple. One can think about a simple cell structure in which an organic dye film is sandwiched between two metal injection electrodes. Holes and electrons are injected from the injection electrodes into the organic layer. Injected holes and electrons move towards counter electrodes by the aid of applied voltage, and they meet and recombine producing molecular excited states, in other words, molecular singlet excitons. Produced molecular singlet excitons also migrate in the organic layer to some extent, and finally fall to their ground states with emission of light. One can observe this emission through a transparent electrode.

In another expression familiar to photochemistry, EL can be ascribed to one of the ways to produce excited states of fluorescent dyes. Thus, EL process can be divided into two parts; one is production of excited states of molecules in which efficiency of production of molecular excitons is very important, and another one is emission from excited states, in which quantum efficiency of fluorescence plays an important role.

Several investigators have shown that bright EL emission could be observed in EL cells with a simple cell structure. However, both the efficiencies of luminance and the stability of cells were far from a satisfaction in the viewpoint of application for display devices[10-13]. Thus many researchers have attempted to improve both the efficiency of production of molecular excitons and the efficiency of emission from produced molecular excitons. Tang and VanSlyke proposed to use multilayer thin films in place of a single thin film, and they succeeded in improving drastically EL cell performances[14].

Based on their idea, we have fabricated multilayer EL devices using variety of organic dyes, and examined their emission properties. Ultimately, we proposed three cell structures shown in Fig. 1. One should use one of the three cell structures

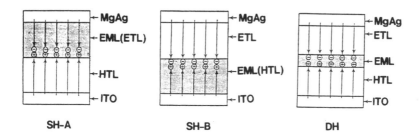

HTL : Hole Transport Layer   ITO : Anode

ETL : Electron Transport Layer   MgAg : Cathode

EML : Emission Layer

FIGURE 1

Three typical cell structures in organic multilayer EL devices.

referring to electronic properties of fluorescent dyes which one wants to use as an emitter layer. In the single hetero-A (SH-A) structure cell, a hole transport layer is deposited by vacuum vapor deposition on a transparent Indium-Tin-oxide (ITO) electrode. On the organic hole transport layer, an emitter dye layer is deposited, and finally a top metal electrode (MgAg alloy) is formed. Holes are injected from an ITO electrode into a hole transport layer, and again injected to an emitter layer. Electrons are directly injected from a metal electrode into an emitter layer and travel towards the boundary of two layers. Holes and electrons meet at around the boundary region within an emitter layer and carrier recombination occurs at this boundary region.

On the other hand, in the single-hetero-B (SH-B) type cells, holes are directly injected into an emitter dye layer which exhibits hole transporting tendency, and electrons are injected from a top metal electrode into an electron transport layer and injected again into an emitter layer. Carrier recombination and emission occur at the boundary region within an emitter layer. Necessity of using different cell structures originates from the difference in electronic properties of emitter dyes; if an emitter dye exhibits electron transporting tendency, the SH-A structure should be used, and the SH-B structure must be used when an emitter has hole transporting tendency.

A double hetero (DH) structure, in which a very thin emitter layer is sandwiched between hole and electron transport layers, is very easy to understand. Electrons and holes are injected from electron and hole transport layers into a thin emitter layer, and recombination of holes and electrons, and emission occurs within this thin layer. The DH structure will be discussed in detail in next section again.

Figures 2 and 3, which show the relations between luminance and current density in EL cells, clearly demonstrate the significance of the three multilayer cell structures. Because luminance proportionally increases with the increase of current density, one can evaluate relative luminance efficiencies from the luminance–current density plots.

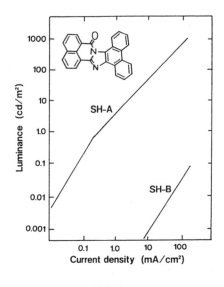 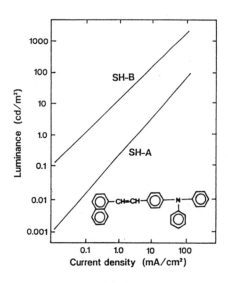

<table>
<tr><td>FIGURE 2</td><td>FIGURE 3</td></tr>
<tr><td>Luminance–current density relations in SH–A and SH–B type cells with a phthalo-perinone derivative as an emitter</td><td>Luminance–current density relations in SH–A and SH–B type cells with a naphtho-styryl derivative as an emitter</td></tr>
</table>

In Fig. 2, we compared the luminous efficiencies of the SH–A and SH–B type cells using the same phthaloperinone derivative dye as an emitter. The EL efficiency of the SH–A type cell was about four orders of magnitude higher than that of the SH–B type cell. This can be interpreted from the viewpoint of electronic properties of phthaloperinone emitter; the phthaloperinone dye has electron transporting tendency. A perfectly reversed example is given in Fig. 3. When another emitter dye, a naphthostyryl derivative (NSD) was used, the situation was perfectly reversed; the SH–B type cell showed about two orders of magnitude higher efficiency than that of the SH–A type cell. These two examples clearly demonstrate the importance of proper selection of cell structures.

In this paper, we will mention very briefly about hole transport, electron transport and emitter materials. Typical dyes for hole transport and electron transport materials which we use are a triphenylamine derivative (TAD) and an oxadiazole derivative (PBD), respectively.

As emitter dyes, one finds various fluorescent dyes in variety of research fields; fluorescent dyes for plastics and fibers, fluorescent pigments, optical brightening agents, organic scintillators, laser dyes, and dyes for fluorescent analyses. Among abundant fluorescent materials, one has to find out proper dyes for emitters. The requisites for emitter dyes are summarized as follows:

(1) Dyes must possess intense photoluminescence in their solid states. One should note that reported high quantum efficiencies of photoluminescence do not necessarily ensure the presence of intense luminescence in solid states.

(2) High quality of vacuum deposited films with the thickness less than 1000 Å can be formed.

(3) Dyes can be purified through sublimation, because purity of dyes is crucially important for emitter materials.

(4) Dyes should possess appropriate semiconducting properties.

In Fig. 4, we show several examples of emitter dyes with various emission colors, which we actually fabricated into EL devices. It should be emphasized that we have very wide selection of emitter materials. This represents one of the most fascinating advantages in utilizing organic materials for electronic devices.

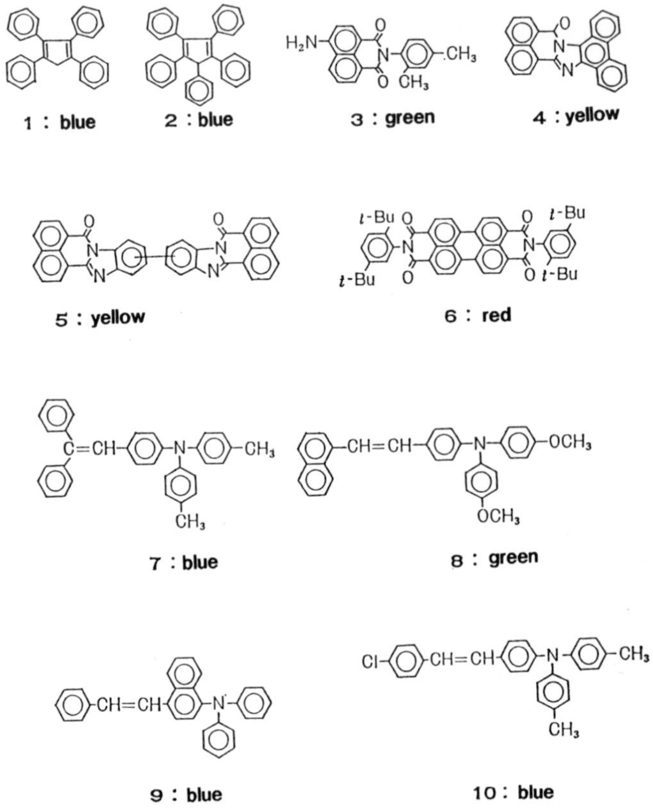

1 : blue   2 : blue   3 : green   4 : yellow

5 : yellow   6 : red

7 : blue   8 : green

9 : blue   10 : blue

FIGURE 4
Fluorescent dyes for emitter layers

3. CONFINEMENT OF CHARGED CARRIERS AND MOLECULAR EXCITONS WITHIN MOLECULAR-SIZE AREA

Figure 5 schematically shows the working mechanism of DH type cells. Holes and electrons are injected from amorphous hole and electron transport layers into a thin emitter layer which consists of a few emitter dye molecules. Holes and electrons located on emitter dye molecules move within emitter layers, and they recombine producing molecular singlet excitons. The produced excitons also migrate to some extent, but excitons do not penetrate into carrier transport layers. Practically, the thickness of an emitter layer can be very thin. We can even think about a monomolecular emitter layer.

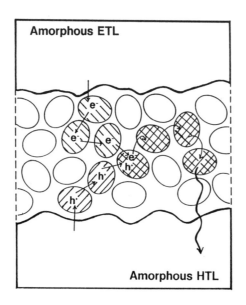

FIGURE 5
Schematic representation of the processes
of charge carrier injection, charge carrier
transport, charge carrier recombination,
exciton migration, and emission in the DH
type cells

Figure 6 shows the relation between luminance at fixed applied
current and the thickness of the emitter dye layer. The case of
the SH–B cell was compared with that of the DH cell. When we
used the SH–B cell structure, the luminance started to drop, when
the thickness of the emitter layer was reduced to less than 200
Å. In the case of the DH cell, in contrast, the luminance
continued to be high, even when the thickness of the emitter
layer was reduced to less than 200 Å.   We observed the case of
the emitter thickness of 50 Å, which was the limit of our
experimental accuracy.   We expect that the DH cells with the
emitter layer thinner than 50 Å will also work well.   The
thickness of 50 Å correspond to the stack of several emitter dye
molecules. Thus, we can even say that we are really dealing with
molecular size electronic devices at least in the direction
perpendicular to the film surface.   The confinement of holes and
electrons and confinement of molecular excitons are perfect in
these molecular size devices[9].

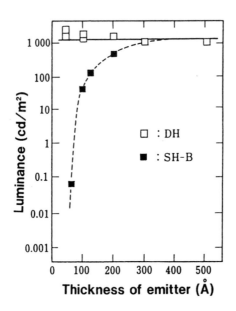

FIGURE 6
Dependence of luminance at the drive
current density of 100 mA/cm^2 upon
the thickness of emitter layer in SH–B
and DH type EL cells

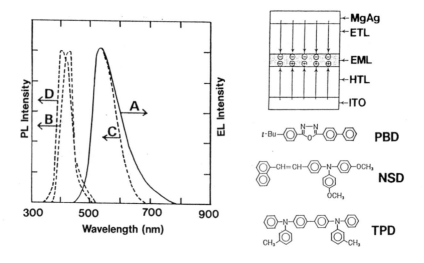

FIGURE 7
EL spectrum of the DH type cell (A), photoluminescene
spectra of, hole transport layer (B), emission layer (C),
and electron transport layer (D)

The attainment of the confinement of molecular excitons within a thin emitter layer can be explained based on photoluminescence spectra of component layers (Fig. 7). The peak of emission spectrum of the emitter layer locates at around 540 nm. This wavelength is more than 100 nm longer than those of hole transport and electron transport layers (at around 400 nm). No energy transfer from the excited states of the emitter layer to the hole or electron transport layer is expected to occur. In other words, molecular excitons on the emitter dye layer can never escape into the hole transport or electron transport layer, via either direct energy transfer or indirect energy migration. It is noteworthy that we now can produce molecular excitons within a molecular size region, and also manage to confine them within very narrow molecular size area.

4. EMISSION VIA TRIPLET EXCITONS

Our second trial is the use of triplet states in EL devices. If one wants to produce triplet excited states via photo-excitation within molecular assemblies, one usually uses inter-system crossing from singlet excited states to triplet excited states. The direct production of triplet excited states by use of photo-excitation is extremely inefficient. In the case of electric excitation, however, the production of triplet excitons is very easy; no spin selectivity is expected in the production of excited states via carrier recombination. Thus almost the same amount of triplet excitons as singlet excitons are produced in EL cells, although almost all the triplet excitons disappear through non-radiative decay pass ways. If we properly design emitter molecules, the emission from triplet excitons may be observed. In other word, we expect we can make use of not only fluorescence but also phosphorescence in organic solids.

Using coumarin dyes, which exhibit phosphorescence life time of several tens of ms, we fabricated EL devices. The EL devices showed slow luminance decays after cessation of applied field, when they were driven with rectangular pulses at liquid nitrogen temperature.   The decay life times were almost the same as those from phosphorescence measurements.   The use of triplet excitons in EL devices, we believe, contribute for further expanding of the research field of organic EL devices.

5. CONTROL OF EMISSION USING MICRO-CAVITY EFFECT

One usually defines natural decay life times of excited molecules and believes that the life times are independent of environmental conditions.   The theory of cavity quantum electrodynamics claims that this is only true when one deals with irreversible emission in free space[15].   If an excited particle is placed within a small spherical mirror, the size of which is close to the wavelength of light, the emission from a particle can not be irreversible any more; one has to consider the interaction of emitted light with radiation field around the particle.   Acceleration or retardation of spontaneous emission is expected to be observed.

For simplicity, we consider the simplest case; excited molecules are placed in front of a metallic mirror with the spacing d.   Excited states of molecules can be produced by photo-excitation or electric excitation (EL).   Electric excitation (EL-mode excitation), however, is much more sophisticated than photo-excitation is, if one premise the application of this phenomenon for some kind of emitting devices.

More than 20 years ago, Kuhn and his coworkers examined this quantum optical phenomenon[16-18].   They observed the emission from europium (Eu) cations excited by photo-irradiation.   They used Langmuir-Blodgett technique for adjusting nm-scale separation between the emission center and a metallic mirror.   The relation between the decay life time of the excited Eu cations and the spacing between emission centers and a mirror was investigated, and large fluctuation of the life time was observed.   They also gave semi-quantitative interpretation based on the classical quantum mechanics.

We performed the experiments of electric excitation case using multilayer EL devices.   First, we used the SH-B structure cell, and we changed the spacing between emission centers and a metal electrode as a mirror by changing the thickness of the electron transport layer.   As is shown in Fig. 8, we found very large change of luminous intensity, when the spacing was varied.

Emission intensity took a maximum at around the spacing of 400 Å.
We assume this observation is related to the effect of the
radiation field. In this case, direct observation of decay life
time was impossible due to very short decay time of fluorescent
dyes used.

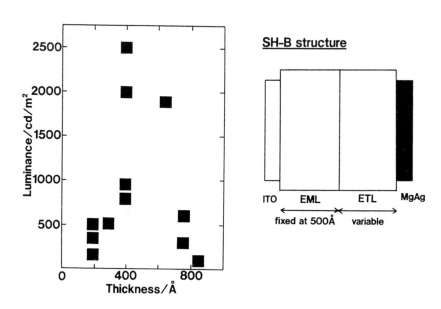

FIGURE 8
Dependence of luminance at the current density of
100 mA/cm2 upon the thickness of the electron
transport layer, which gives the distance between
MgAg electrode as a metallic mirror and the emission
region at the boundary of ETL and EML

By using tribium cations as an emission center, we can observe
emission life time even in the case of electric excitation. We
used tribium-acetylacetonate complex as an emitter and fabricated
DH type cells. The spacing between a mirror and the emission
center was changed by the adjustment of the thickness of the
electron transport layer. Figure 9 shows the relation between the
life time of tribium ions produced with electric excitation and
the spacing between a metallic mirror and emission centers. The
observed life time was around one hundred μs and took a maximum
at around 300 Å. This is very clear evidence for the existence
of the radiation field effect. We believe the datum shown here

is the first one that demonstrate the change of life time of
emission in the case of electric excitation. Although our
experiments are in very preliminary stage at present, we believe
the use of organic EL devices for the study of this fascinating
quantum optical phenomena is promising.

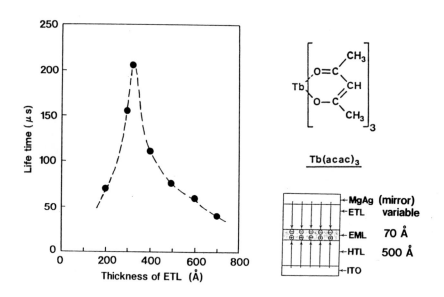

FIGURE 9
Emission life time of the excited state of tribium
cation produced with electric excitation plotted
against the distance between a metallic mirror and
the emission center

5. CONCLUSION

The success in the fabrication of high-efficiency thin film EL
devices with multilayer structures have provided us a new
powerful tool for investigating electronic and optical properties
of organic solids. One can produce molecular singlet and triplet
excitons at specified locations within organic thin films and
observe dynamic behaviors of those molecular excitons. The study
on organic EL devices has been classified into application-
oriented research. The progress of such research development has
actually brought about a promising tool for the basic research
fields in photophysics and photochemistry.

We believe that the progress in basic research on photophysics and photochemistry on organic solids, on the other hands, will surely give significant contribution in the fields of molecular electronics and molecular photonics in near future.

REFERENCES

1) C. Adachi, S. Tokito, T. Tsutsui and S. Saito, Jpn. J. Appl. Phys. 27 (1988) L269.

2) C. Adachi, S. Tokito, T. Tsutsui and S. Saito, Jpn. J. Appl. Phys. 27 (1988) L713.

3) C. Adachi, T. Tsutsui and S. Saito, Electroluminescence in vacuum-deposited organic thin films, in: Springer Proc. in Phys. 38, Eds., S. Shionoya and M. Kobayashi (Springer Verlag, Berlin, 1989) pp.358-361.

4) C. Adachi, T. Tsutsui and S. Saito, Appl Phys. Lett. 56 (1990) 799.

5) T. Tsutsui and S. Saito, Organic thin films for electro-luminescence displays, in: Polymers for Microelectronics, eds. Y. Tabata, et al. (Kodansha, Tokyo, 1990) pp.591-600.

6) C. Adachi, T. Tsutsui and S. Saito, Appl. Phys. Lett. 55 (1989) 1489.

7) C. Adachi, T. Tsutsui and S. Saito, Acta Polytech. Scand., No.170 (1990) 215.

8) C. Adachi, T. Tsutsui and S. Saito, Proc. 9th Int. Display Conf. (October, 1989, Kyoto, Japan) pp.708-711.

9) C. Adachi, T. Tsutsui and S. Saito, Appl. Phys. Lett. 57 (1990) 531.

10) P. S. Vincett, W. A. Barlow, R. A. Hann and G. G. Roberts, Thin Solid Films, 94 (1982) 171.

11) P. H. Partridge, Polymer, 24 (1983) 748.

12) S. Hayashi, T. T. Wang, S. Matsuoka, S. Saito, Mol. Cryst. Liq. Cryst. 135 (1986) 355.

13) S. Hayashi, E. Etoh and S. Saito, Jpn. J. Appl. Phys. 25 (1986) L773.

14) C. W. Tang and S. A. VanSlyke, Appl. Phys. Lett. 51 (1987) 913.

15) S. Haroche and D. Kleppner, Physics Today, Jan. 1989, pp.24-30.

16) K. H. Drexhage, Interaction of light with monomolecular layers, in: Prog in Optics, ed E.W. Wolf (Northholland, New York, 1974) Vol.12, pp.165–229.

17) H. Kuhn, J. Chem. Phys., 53 (1970) 101.

18) R. R. Chance, A. H. Miller, A. Prock and R. Silbey, J. Chem. Phys. 63 (1975) 1589.

Photochemical Processes in Organized Molecular Systems
K. Honda (Editor-in-Chief)
© Elsevier Science Publishers B.V., 1991

ELECTRON-PHONON INTERACTION IN PHOTOCHEMICAL HOLE
BURNING IN PORPHYRIN-POLYMER SYSTEMS

Kazuyuki HORIE

Faculty of Engineering, University of Tokyo,
7-3-1 Hongo, Bunkyo-ku, Tokyo 113, Japan

The phonon frequency, E_s, reflecting the low energy excitation modes of matrix polymers and the temperature dependence of Debye-Waller factor, DW(T), were estimated from photochemical hole burning (PHB) measurements, and are related to the efficiency of hole formation for tetraphenylporphin in various polymer matrices at 4-100 K. The importance of the suppression of laser-induced hole filling for realizing the hole formation above the liquid nitrogen temperature is also emphasized.

1. INTRODUCTION

Photochemical hole burning (PHB) is a phenomenon in which very narrow and stable photochemical holes are burnt at very low temperatures into the absorption bands of guest molecules molecularly dispersed in an amorphous solid by narrow-band excitation with a laser beam[1,2]. Proton tautomerization of free-base porphyrins and phthalocyanines, and hydrogen bond rearrangement of quinizarin are typical photochemical reactions which provide PHB spectra.

The phenomenon of PHB has recently attracted considerable interest not only as a means for frequency-domain high density optical storage, but also as a tool for high-resolution solid state spectroscopy at low temperatures. By using the site selectivity of PHB phenomenon, the energy level and the homogeneous line width of the excited state of guest molecule can be measured as a hole profile in the inhomogeneously-broadened absorption spectrum. The interaction between guest molecule and host matrix plays an important role in PHB. In addition to the spectrum called the zero-phonon line, originating from the guest dye molecule itself, a phonon side band appears reflecting the guest-host interaction and the structure of matrix polymers. The dephasing phenomenon of the excited state due to electron-phonon interaction has been widely studied from the line width measurements of zero-phonon hole spectra[2,3]. We recently showed[4,5] that the electron-phonon interaction in PHB can be studied also by measuring the phonon frequency estimated from the energy difference between a zero-phonon hole and a pseudo-phonon-side hole, E_s, and the Debye-Waller factor, DW(T), corresponding to the fraction of zero-phonon line in the integrated area of zero-phonon line and phonon-side band. In the present paper, the phonon frequency, E_s, and the temperature dependence of Debye-Waller factor, DW(T), obtained from PHB measurements are discussed for various polymers, and are related to the efficiency of hole formation at 4-100 K. The importance of laser-induced hole filling is also emphasized.

2. LOW ENERGY EXCITATION MODES IN AMORPHOUS POLYMERS

The most attractive aspect of PHB from a spectroscopic point of view is that PHB gives information about the homogeneous absorption spectrum of a molecule in the solid state as a hole formed in the usual broad inhomogeneous absorption band. A photochemical hole burnt by laser irradiation consists of three parts: a zero-phonon hole, a phonon side hole, and a pseudo-phonon side hole (cf. Figure 1). The sharp zero-phonon hole and the side hole on the high energy side are due to the excitation and hole burning of guest molecules which have a zero-phonon line at the irradiation wavelength. A pseudo-phonon side hole is formed on lower energy side of the hole band, and the energy difference between the zero-phonon hole and the pseudo-phonon side hole, E_s, is the same as the energy difference between zero-phonon hole and the phonon side hole. The pseudo-phonon side hole is caused by the excitation of photoreactive molecules which have a phonon side band at the irradiation wavelength. The mechanism of pseudo-phonon side hole formation is illustrated in Figure 1. Since coupling between a guest molecule and the amorphous polymer matrix is usually weak, the side hole on the higher energy side is small. But a pseudo-phonon-side hole can be formed to a large extent, and is easily detected, owing to the large number of guest molecules which have an absorption of phonon side bands at the irradiation wavelength.

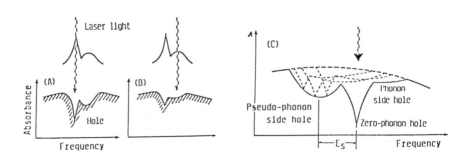

FIGURE 1
Mechanism of the formation of pseudo-phonon side hole. (A) Excitation at zero-phonon line, (B) Excitation at phonon-side band, (C)Overall hole profile.

Low temperature properties of amorphous materials have attracted interest for many years. Many properties of amorphous materials (for example heat capacity at low temperatures) are different from those of crystals, owing to the large density of states for low energy excitation modes in comparison with crystals. A large density of states for low energy excitation modes leads to excess heat capacity at cryogenic temperatures, and the thermal properties differ from those expected by the Debye theory. Low energy excitation modes in amorphous polymers were investigated by heat capacity measurements[6], and by neutron inelastic scattering measurements[7,8]. The extent of electron-phonon interaction with the low energy excitation modes of amorphous polymers can also be evaluated from PHB experiments. It was noticed recently, by comparing the PHB hole profiles of quinizarin and free-base tetraphenylporphin(TPP) in PMMA and some other matrices, that the energy difference between zero-phonon hole and pseudo-phonon-side hole, E_s, does not depend on the nature of the guest molecules, but is specific to the host matrices[4]. The peak energies for low energy excitation modes were also reported recently by using Fourier-transform photon-echo spectroscopy[9].

Figure 2 shows typical hole profiles of TPP in a phenoxy resin (PhR) at various temperatures up to 80 K. A sharp zero-phonon hole is formed at low temperatures in the absorption band of TPP, and the pseudo-phonon side hole is observed at 4-50K on the low energy side. The amount of pseudo-phonon side hole increased with irradiation time, but the value of E_s did not change during the irradiation. The E_s value did not change irrespective of the position of the hole in the absorption band and the burning temperature.

The values of E_s for various systems are summarized in Table 1, together with the energy of low energy excitation modes estimated from heat capacity measurements, E_c, and that measured by neutron inelastic scattering measurements, E_l. The E_s values are constant for any polymers

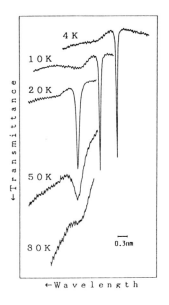

FIGURE 2

Hole profiles formed at various temperatures in the TPP/phenoxy resin system. The holes were formed by 0.75 mW/cm^2 dye laser irradiation for 1 min (4-50 K) or 3 min (80 K).

irrespective of the nature of guest molecules. The E_S values for PMMA and PS agree with the low energy excitation mode, $E_C{}^6$, determined by the heat capacity measurements. The origin of this low energy excitation mode for PMMA is considered to be the rotation of the ester-methyl group of PMMA[6]. The E_s values for polystyrene and for epoxy resin (EpR) almost agree with the peak of the density of states at low energy determined by neutron inelastic scattering measurements[7,8]. Thus one concludes that the energy difference between the zero-phonon hole and the pseudo-phonon-side hole, E_s, is a characteristic parameter of the matrix polymer, and reflects the phonon frequency corresponding to the low energy excitation mode of the amorphous polymer.

TABLE 1

Energies of the low energy excitation modes determined by PHB, E_s, heat capacity, E_c, and inelastic neutron scattering, E_l, measurements.

Chromophore/Amorphous polymer matrix	E_{s-1} (cm^{-1})	E_{c-1} (cm^{-1})	E_{l-1} (cm^{-1})	Ref.
TPPS/PVA	23.5			4
DAQ/PVA	23.0			4
TPP/PhR	15.1			4,11
TPP/EpR(ethylenediamine)	14.5		15	8,12
TPP/EpR(hexamethylenediamine)	17.1			12
TPP/PMMA	13.1	12		4,6
DAQ/PMMA	13.6			4
TPP/PI	13.3			4,12
TPP/LCP	12.6			4
TPP/PET(undrawn)	10.9			5,14
TPP/PET(5-times drawn)	11.8			5,14
TPP/PS	10.1	9.1	12	4,6,7
TPP/PSi	9.8			15

TPP:free-base tetraphenylporphin, TPPS:sulfonated TPP, DAQ:quinizarin, PVA:poly(vinyl alcohol), PhR:phenoxy resin, EpR:epoxy resin, PMMA:poly(methyl methacrylate), PI:polyimide, LCP:liquid crystalline polymer, PET:poly(ethylene terephthalate), PS:polystyrene, PSi:poly(phenylmethylsilane).

Hole formation at 80K in PHB was observed only for sulfonated TPP (TPPS)/poly(vinyl alcohol) (PVA)[10], TPP/PhR[11], and TPP/EpR[12] systems, and was not observed in other systems. The E_s value for TPPS/PVA system is extremely large (23cm^{-1}) and those for TPP/PhR and TPP/EpR are comparatively large. Thus the capability of hole formation at 80K would be related to the high E_s value of the matrix polymers, leading to the slow rate of structural relaxation and hole filling probably caused by the presence of hydrogen bonding in PVA and PhR

and by the presence of hydrogen bonding and crosslinks in EpR. The
coincidence   of E_s values   for   TPP/polystyrene(PS)   and
TPP/polyphenylmethylsilane(PSi) suggests that the low frequency vibrational
mode of the phenyl group would be the origin of low energy excitation modes for
these polymers. The increase in the orderliness in poly(ethylene
terephthalate) (PET) film by drawing results in an increase in its E_s value.

3. DEBYE-WALLER FACTOR

The Debye-Waller factor, DW(T), given by

$$DW(T)=S_0(T)/(S_0(T)+S_p(T))   (1)$$

where $S_0(T)$ and $S_p(T)$ are the integrated intensities for the zero-phonon line
and the phonon side band, respectively, indicates the coupling strength
between a guest and a matrix; and its temperature dependence is affected by the
low energy excitation modes of a matrix
coupled with the electronic trasition
of the guest molecules. As mentioned in
the previous section, relatively high
temperature PHB, i.e., hole formation
and observation above liquid nitrogen
temperature, was reported for TPP in a
phenoxy resin[11,16],TPP in an epoxy
resin[12], and TPPS in poly(vinyl
alcohol)[10] matrix systems. One of the
requisites for high temperature hole
formation is supposed to be a large
Debye-Waller factor and its small
temperature dependence. So the
information on the Debye-Waller factor
is important also for high temperature
PHB.

Figure 3 shows a hole profile burnt
for TPP in the phenoxy resin at 30 K
and its profile after cooling down to 4
K. The hole depth, $\Delta A/A_0$, where ΔA is
the difference in absorbance produced
by hole formation and A_0 is the
absorbance before irradiation, grew
large after cooling down to 4 K, but
the hole width, $\Delta \omega_h$, did not change.

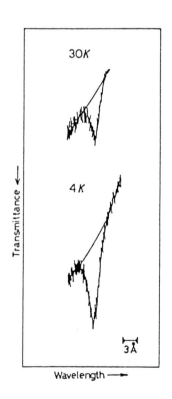

FIGURE 3
A hole profile burnt in TPP/PhR at
30 K with 3.6 mW/cm^2 laser power and
its profile after cooling to 4.2 K.

Since the structural relaxation in the system does not occur during cooling down to 4 K, the hole depth growth after cooling is caused by the change of the Debye-Waller factor. If the oscillator strength for the sum of integrated intensities of a zero-phonon line and a phonon side band is assumed to be constant, $S_0(T)+S_p(T)$ should be constant irrespective of temperature, T, and therefore, the temperature dependence of DW(T) can be determined by the measurement of the temperature dependence of $S_0(T)$. When the hole area burnt at higher temperatures and its change after cooling down to 4 K were measured, the ratio $S_0(T)/S_0(4)$ roughly corresponds to DW(T) due to the weak coupling between molecule and amorphous polymer matrix, as suggested by photon echo technique (DW(4) $\cong$ 0.9 for TPP/PMMA)[9]. Figure 4 shows the temperature dependence of the Debye-Waller factor thus determined from the PHB measurements. Results in Figure 3 agree well with the results measured by photon echo technique for TPP in PMMA and for carbonated TPP in poly(vinyl alcohol)[9].

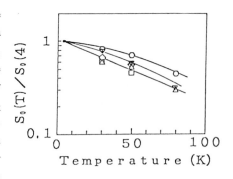

FIGURE 4
Temperature dependence of Debye-Waller factor calculated by $S_0(T)/S_0(4)$ for TPPS/PVA(◯), TPP/EpR(▽),TPP/PhR(△),TPP/LCP(◇), TPP/PET(☐) and TPP/PS(◯).

The temperature dependence of the Debye-Waller factor is expressed by[17],

$$DW(T)=\exp[-C\int\coth(\hbar\omega/2k_BT)P(\omega)g(\omega)d\omega] \quad (2)$$

where C is a constant, k_B is Boltzmann's constant, $(1/2)\coth(\hbar\omega/2k_BT)$ is the average number of phonons in the lattice mode $\hbar\omega$ at temperature T, $P(\omega)$ is Stoke's shift, and $g(\omega)$ is the density of states which have an energy $\hbar\omega$. When the dominant lattice mode $\hbar\omega$ contributing to $g(\omega)$ has large energy, DW(T) shows small temperature dependence resulting from this relationship. A system which has a large E_s value is expected to show a small temperature dependence of DW(T). The poly(vinyl alcohol) matrix systems have larger E_s value than the other matrices and show a smaller temperature dependence of DW(T). This fact is consistent with the theory. In TPP/EpR systems, the E_s value grew large and the temperature dependence of Debye-Waller factor became smaller when the hardener was changed from ethylenediamine to hexamethylenediamine. It suggests that the structure of the cross-linked polymer affects the low energy excitation mode, and improves thermal property of the system.

4. HIGH TEMPERATURE PHB AND LASER-INDUCED HOLE FILLING

The efficiency of hole formation and the temperature dependence of hole profiles burnt at 4 K are important aspects of the PHB phenomenon.   Typical profiles of the holes burnt at 4-80 K for TPP in phenoxy resin (PhR) are shown in Figure 2.   The quantum efficiency for hole formation, Φ , can be calculated from the initial slope of the change in hole depth during laser irradiation by using the following equation[11,18]

$$\Phi = [d(A/A_0)/dt]_{t=0} A_0 \Big/ [10^3 I_0 (1-10^{-A_0}) \epsilon R] \qquad (3)$$

where A is time-varying absorbance, A_0 is the absorbance before irradiation, $[d(A/A_0)/dt]_{t=0}$ is the initial slope of the irradiation time dependence of hole depth, I_0 is incident laser intensity given in einstein/cm^2sec, ϵ is the molar extinction coefficient for inhomogeneous line profile at the hole burning wavelength and temperature, and $R=C_0/C_{0h}= \Delta\omega_i / \Delta\omega_h$ is the reciprocal initial fraction of photoreactive molecules within a homogeneous line width, $\Delta\omega_h$, at the laser frequency.

Figures 5 and 6 show the temperature dependence of hole formation efficiency, Φ , calculated by using eq(3) for TPP in various polymers and for TPPS/PVA, respectively.   In Figure 5 and 6, the temperature dependences of the Debye-Waller factor estimated from the PHB measurements as mentioned above are

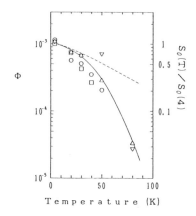

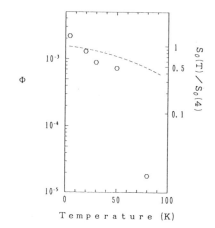

FIGURE 5
Temperature dependence of hole formation efficiency, Φ , for the TPP/PhR($\triangle$), TPP/EpR(∇), TPP/ PMMA($\square$), and TPP/PS($\bigcirc$) at various temperatures. Dashed line in the figure corresponds to the DW(T).

FIGURE 6
Temperature dependence of Φ for the TPPS/PVA. Dashed line in the figure corresponds to the DW(T).

also shown by dashed lines. The temperature dependence of Φ almost agrees with
that of the Debye-Waller factor up to around 30-50 K for each system, but
deviates considerably at 80 K. It indicates that the temperature dependence of
hole formation efficiency is determined not only by the temperature dependence
of the Debye-Waller factor, but also by the temperature dependence of the
structural relaxation of the matrix polymers.

Now, various factors related to the structural relaxation, i.e., thermal or
laser-induced spectral diffusion leading to irreversible broadening of hole
width without change in hole area and thermal or laser-induced hole filling
leading to irreversible change in hole area, should be examined as to their
importance affecting the high-temperature PHB.

Figure 7 shows the annealing temperature dependence of hole area during the
cycle annealing experiments. In the cycle annealing experiments, a hole burnt
at 4.2 K by 0.75 mW/cm² laser irradiation for 1-5 min was annealed at 30, 50,
and 80 K for 30 min and measured at 4.2 K successively. As can be seen from
Figure 7, the hole areas scarcely changed up to 80 K-annealing and almost above
80 percent of initial area was conserved for the all systems. It means that
the thermally activated backward reaction hardly occurs and that spectral
diffusion is the dominant process in this temperature region. This fact is
consistent with the absence of intramolecular proton exchange in free-base TPP

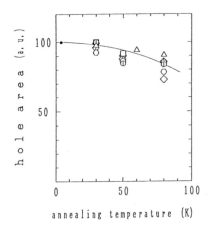

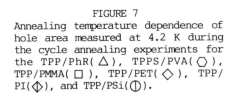

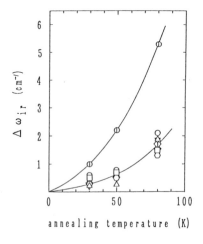

FIGURE 7
Annealing temperature dependence of
hole area measured at 4.2 K during
the cycle annealing experiments for
the TPP/PhR(△), TPPS/PVA(○),
TPP/PMMA(□), TPP/PET(◇), TPP/
PI(◈), and TPP/PSi(⬒).

FIGURE 8
Annealing temperature dependence of
irreversible change of hole width,
$\Delta\omega_{ir}$, during the cycle annealing
experiments for various systems.
Symbols in the figure are the same
as those in Fig. 7.

at least up to 80 K, consistent with results elucidated by NMR spectroscopy[19].

Figure 8 shows the annealing temperature dependence of irreversible change of the hole width, $\Delta\omega_{ir}$, during the cycle annealing experiments. Almost all the systems except for the TPP/PSi showed the same annealing temperature dependence of $\Delta\omega_{ir}$ irrespective of the capability of hole formation at 80 K as mentioned in the previous sections. The extents of the spectral diffusion for these systems are pointed out to be almost the same in this temperature region. It suggests that the spectral diffusion process hardly affects the capability of hole formation at 80 K.

Next, laser irradiation was performed during the cycle annealing experiments for obtaining information on the laser-induced hole filling. Figure 9 shows the annealing temperature dependence of hole area with and without the laser irradiation at elevated temperatures for the TPPS/PVA, TPP/EpR. and TPP/PI. The irradiation condition was 0.75 mW/cm^2 for 1 or 10 min at 30, 50, and 80 K and the energy separation between the frequency of an initial hole burnt at 4.2 K and that of laser irradiations at elevated temperatures was 50-100 cm^{-1}. The irradiations were carried out on the lower energy side of the initial hole except for the case of TPPS/PVA at 80 K (●). No change was observed with the irradiation up to 50 K for TPPS/PVA and

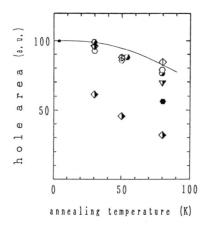

FIGURE 9
Annealing temperature dependence of hole area during cycle annealing experiments for TPPS/PVA(◑,●,○), TPP/EpR(▼), and TPP/PI(◆,◈) with the irradiation at elevated temperatures at low energy side (◑,▼,◆) and at high energy side (●) and without irradiation (○,◈).

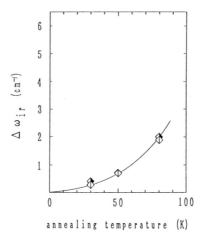

FIGURE 10
Irreversible change of hole width, $\Delta\omega_{ir}$, measured at 4.2 K for the TPP/PI during the cycle annealing experiments with the irradiation (◆) at elevated temperatures and without irradiation (◈).

TPP/EpR, but the hole area decreased for the 10-min-irradiation at 80 K at higher side for the TPPS/PVA. Almost the same result was obtained also for the TPP/PhR for the case of low-energy-side laser irradiation. In the case of TPP/PI, hole area decreased considerably with the irradiation at each temperature even for the low-energy-side irradiation. Figure 10 shows the irreversible change of hole width, $\Delta\omega_{ir}$, for the TPP/PI during the cycle annealing experiments with and without the irradiation. The $\Delta\omega_{ir}$ for both cases have no difference, showing the absence of the change in hole width during the laser-induced filling. The photo-induced structural relaxation leading to hole filling makes the system approach thermal equilibrium and therefore it does not change the width nor the frequency of an initial hole. This laser-induced hole filling might originate from the deactivation process of chromophores which lead to the structural relaxation of spatially neighboring but energetically separating sites in the system.

The systems which have the capability of hole formation at 80 K (TPPS/PVA and TPP/EpR) showed a small decrease of hole area with the irradiation at elevated temperatures compared with the systems which do not have the capability of hole formation at 80 K (TPP/PI). Furthermore, Figures 7 and 9 for TPPS/PVA show the deviation at 80 K from the expected value for hole-formation efficiency based on the Debye-Waller factor and that for hole area in the case of high-energy-side irradiation, respectively. These facts suggest the influence of the laser-induced hole filling at 80 K on the efficiency of hole formation at 80 K[20].

Thus, one of the most important factors affecting the capability of hole formation at 80 K is the rate of laser-induced hole filling at 80 K. The systems which have the capability of hole formation at 80 K have slight laser-induced hole filling at 80 K for the case of low-energy-side irradiation. The laser-induced hole filling process does not change the hole width, it diminishes the hole area alone, and leads to a decrease in hole formation efficiency. Thus, the suppression of the rate of the laser-induced hole filling at 80 K is essential for hole formation at 80 K, and its small rate at 80 K is realized in the TPP/PhR, TPP/EpR, and TPPS/PVA. Hole formation at 80 K is observed mainly in hydrogen bonding matrix systems. So the small rate of the laser-induced hole filling at 80 K would be due to the existence of hydrogen bonding in the matrix polymers.

Finally, various factors affecting the high temperature PHB are summarized in Table 2. The changes in $\Delta\omega_h$ and DW(T), thermally-activated spectral diffusion, and laser-induced hole filling are mainly observed in PHB experiments up to 80 K.

TABLE 2
Factors affecting high-temperature PHB

Factors	Phenomena
Homogeneous width, $\Delta\omega_h$	reversible broadening of hole width
Debye-Waller factor, DW(T)	reversible change in hole area
thermally-activated spectral diffusion	irreversible broadening of hole width after temperature cycle (no change in hole area)
laser-induced spectral diffusion	irreversible broadening of hole width caused by laser irradiation (no change in hole area)
thermally-activated hole filling	irreversible change in hole area after temperature cycle due to structural relaxation and/or backward reaction
laser-induced hole filling	irreversible broadening of hole area caused by laser irradiation due to structural relaxation (no change in hole width) deviation of hole formation efficiency from the temperature dependence of DW(T) at 80 K

REFERENCES

1) J. Friedrich and D. Haarer, Angew. Chem. Int. Ed. Eng., 23, 113 (1984).

2) W.E. Moerner ed., Persistent Spectral Hole-Burning; Science and Applications, Springer, Berlin, (1988).

3) R.M. Macfarlane and R.M. Shelby, J. Luminescence, 36, 179 (1987).

4) A. Furusawa, K. Horie and I. Mita, Chem. Phys. Lett., 161, 227 (1989).

5) A. Furusawa, K. Horie and I. Mita, Jap. J. Appl. Phys., 28 Suppl. 28-3, 19 (1989).

6) W. Reese, J. Macromol. Sci. Chem., A3, 1257 (1969).

7) T. Kanaya, K. Kaji, S. Ikeda and K. Inoue, Chem. Phys. Lett., 150, 334 (1988).

8) H.M. Rosenberg, Phys. Rev. Lett., 54, 704 (1985).

9) S. Saikan, A. Imaoka, Y. Kanematsu, K. Sakoda, K. Kominami, M. Iwamoto, Phys. Rev. B, 41, 3185 (1990).

10) K. Sakoda, K. Kominami and M. Iwamoto, Jap. J. Appl. Phys., 27, L1304

(1988).

11) A. Furusawa, K. Horie, K. Kuroki and I. Mita, J. Appl. Phys. 66, 6041 (1989).

12) A. Furusawa, K. Horie, T. Suzuki, S. Machida, and I. Mita, Appl.   Phys. Lett., 57 141 (1990).

14) K. Horie, K. Kuroki, I. Mita, H. Ono, S. Okumura and A. Furusawa, Polymer, in press.

15) A. Furusawa, K. Horie, and I. Mita, J. Mol. Electr., in press.

16) K. Horie, T. Mori, T. Naito and I. Mita, Appl. Phys. Lett., 55, 935 (1988).

17) K.K. Rebane and L.A. Rebane, in Ref. 2, p.23.

18) W.E. Moerner, M. Gehrtz, and A.L. Huston, J. Phys. Chem., 88, 6459 (1984).

19) H.H. Limbach, J. Henning, R. Kendrick, and C.S. Yannoni, J. Am. Chem. Soc., 106, 4059 (1984).

20) A. Furusawa and K. Horie, J. Chem. Phys., in press.

Photochemical Processes in Organized Molecular Systems
K. Honda (Editor-in-Chief)
© Elsevier Science Publishers B.V., 1991

PHOTO-ELECTRIC CONVERSION WITH LANGMUIR-BLODGETT FILMS

Masamichi FUJIHIRA

Department of Biomolecular Engineering, Tokyo Institute of Technology,
4259 Nagatsuta, Midori-ku, Yokohama 227, Japan

In the photosynthesis, solar energies harvested by antenna pigments are
funneled to special pairs in the reaction centers, where multistep electron transfer reactions proceed to separate electron-hole pairs far apart
across the lipid bilayer thylakoid membrane. The well-organized and asymmetric molecular arrangement across the membrane play an important role
in the charge separation of the photosynthetic reaction center. In a
series of studies, the simulation of the light harvesting and the succeeding charge separation processes with Langmuir-Blodgett (LB) film molecular
assemblies were examined. These molecular devices can be used for the
photoelectric conversion and are called molecular photodiodes. The electrical energy created can be used further to reduce carbon dioxide and to
oxidize water. Our research on artificial photosyntheses with LB films
will be reviewed.

1. INTRODUCTION

In biosystems molecules organize themselves into complex functional entities
with cooperating components of molecular dimensions. For example, well-organized molecular assemblies in lipid bilayer membranes play an important role in
photosynthetic processes of plants and bacteria[1]. In 1984, the atomic structure of the reaction center, i.e. the basic machinery for the beginning events
of photosynthesis, of a purple bacterium R. viridis was determined with x-ray
diffraction[2]. All reaction centers are complexes containing protein subunits
and donor-acceptor molecules. These centers span an inner membrane in the
plant or bacterial cell, and their donor-acceptor complexes perform charge
separation that creates a potential gradient across the membrane. Before the
charge separation can take place at any reaction center, solar energy must be
harvested by light-absorbing antenna pigments and transmitted to the center.
In green plants, the charge separation occurs in two different reaction centers
working in series. The result of these primary processes is that reduced
products (NADPH) appear on one side of the thylakoid membrane and oxygen appears on the other. The four-electron oxidation of water to molecular oxygen
is catalyzed by the oxygen-evolving complex containing manganese. NADPH is
used further in a series of dark reactions by which CO_2 is converted into
useful fuels.

To design artificial photosynthetic molecular systems[3-5] for solar energy
conversion, it is of great interest to mimic the elaborate molecular machinery

for the light harvesting, the charge separation, and the catalytic multi-electron oxidation of water and reduction of CO_2. The consideration of the structure and function of the asymmetric spatial arrangement of electron donors and
acceptors in the charge separation unit across the thylakoid is most essential.
The Langmuir-Blodgett (LB) film is one of the most appropriate artificial
material by which the spatial arrangement of the various functional moieties
across the film can be constructed readily at atomic dimensions. In a series
of studies, we have attempted to simulate the elemental processes of the photosynthesis by taking advantage of LB monolayer assemblies.

In this paper, recent developments from our group will be reviewed with
emphasis on the following subjects:

a) molecular photodiodes of complex LB films with heterostructure consisting
   of an electron acceptor (A), a sensitizer (S), and an electron donor (D)
   amphiphiles where viologen, pyrene (or porphyrin), and ferrocene moieties
   act as A, S, and D units, respectively;

b) molecular photodiodes consisting of unidirectionally oriented amphiphilic
   folded type S-A-D triads;

c) photoelectric conversion by a monolayer of highly oriented amphiphilic
   linear type A-S-D triads;

d) kinetics of photoinduced electron transfer in LB films and the effect of an
   electrical double layer;

e) simulation of the primary process of the photosynthetic reaction center by
   mixed monolayer with triad and antenna molecules; and

f) catalytic reduction of CO_2 with metal complexes.

2. MOLECULAR PHOTODIODE WITH HETEROGENEOUS A/S/D LB FILMS

The first molecular photodiode was fabricated with a molecularly ordered
film on a gold optically transparent electrode (AuOTE) prepared by the LB
method[6] as shown in Figure 1(a), where hydrophilic parts and hydrophobic units
are indicated by circles and squares, respectively. With their amphiphilic
properties, three functional compounds tend to orient regularly in the heterogeneous LB films. Another interesting and fascinating application of LB films
is their use as controlled-thickness spacers or "distance keepers". Therefore,
the distances between the three functional moieties, i.e. A, S, and D, can be
closely controlled at known values. The electron transfer process in such
molecular assemblies is free from any complication due to diffusion. Kuhn and
Möbius[3,4] have previously studied the distance dependence of the rate of photoinduced electron transfer in LB films. Their proposed dependence agrees with
the experimental[7-9] and theoretical[10] results for non-adiabatic electron transfers, which are described by the following equations[10]:

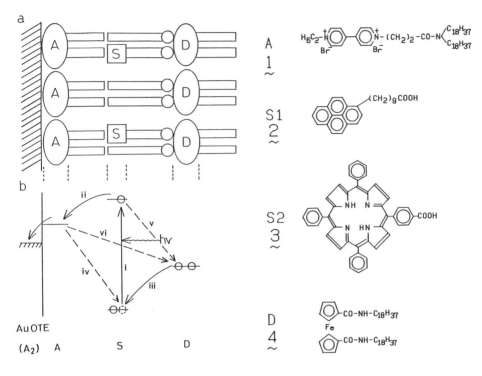

FIGURE 1
Molecular photodiode with heterogeneous A/S/D LB film on AuOTE: a, structure; b, energy diagram.

FIGURE 2
Structural formulae of A, S and D amphiphiles used to construct heterogeneous A/S/D LB films.

$$k = k(r)\exp(-\Delta G^*/RT) \tag{1}$$

$$k(r) = k_o\exp\{-\beta(r - r_o)\} \tag{2}$$

In addition to the distance dependence, the effect of the standard free energy difference $\Delta G°$ for the electron transfer, i.e. the difference between the energy levels of the excited (or ground state) donor and the ground state (or excited) acceptor[10-12], is another important factor in determining the rate of electron transfer. Relationship (3) was first introduced by Marcus where the free energy barrier for the reaction, i.e. ΔG^* in eq.(1), is given in terms of the reorganization energy λ[10]:

$$\Delta G^* = \lambda/4(1 + \Delta G°/\lambda)^2 \tag{3}$$

The changes in bond lengths of the reactants and the changes in solvent orientation coordinates in the electron transfer are related to λ[10]. In Marcus's original theory, the motion of the nuclei was treated classically. There have been several attempts to treat the nuclear coordinates quantum mechanically and to modify the equation for the energy gap $\Delta G°$[10,12]. In connection with the design of the proper energy diagram for the molecular photodiode, the inverted region, where the rate decreases with an increase in a large excess of $-\Delta G°$, predicted by eq.(3) is most important. The presence of the inverted region has been confirmed experimentally by the use of internal electron transfer systems with rigid spacers[13].

Keeping the distance and the $\Delta G°$ dependence in mind, we considered how to design a better molecular photodiode. In Figure 1(b), the energy diagram of the A/S/D molecular photodiode is depicted as a function of distance across the LB film. If the forward processes indicated by arrows with solid lines are accelerated and the backward processes with dashed lines are retarded by setting the distances and the energy levels appropriately, the photoinduced vectorial flow of electrons can be achieved. Namely, the acceleration by setting $-\Delta G°$ equal to λ is assumed for the forward electron transfer processes ii and iii, while the retardation, as a consequence of the inverted region, is assumed for the back-electron transfer processes iv and v. Once an electron-hole pair is separated successfully, the recombination of the pair across the large separation by LB film (process vi) is hindered.

The three kinds of functional amphiphilic derivatives used for the first A/S/D type molecular photodiode[6] are shown in Figure 2 together with porphyrin sensitizer 3 used later. By depositing these three amphiphiles on AuOTE, as shown in Figure 1(a), and by use of the resulting electrode as a working electrode in a photoelectrochemical cell[7], the photoinitiated vectorial flow of electrons was achieved and detected as photocurrents. The AuOTE is a metal electrode and hence does not by itself possess a rectifying ability as does a semiconductor electrode[7]. In spite of the inability of the substrate electrode to rectify, the photocurrent had opposing directions depending on the spatial arrangement of A/S/D or D/S/A. The direction was in accordance with the energy level profile across LB films in Figure 1(b).

Not only three-layer systems, but also stacks of multilayers of each component, e.g. in the form of A,A,A/S,S,S/D,D,D and D,D,D/S,S,S/A,A,A[6,14] were tried. Much higher photocurrents than those for A/S/D and D/S/A were observed. The direction of the photocurrent also agreed as expected for these multilayered systems. In addition to pyrene as a sensitizer, an amphiphilic porphyrin derivative[14] was also examined. The photocurrent spectrum due to porphyrin was observed under visible light illumination.

In such heterogeneous LB films, the long alkyl chains intervened between the A and S and between the S and D moieties. As a result, the electron transfer decreased to the point that part of the excited sensitizers were deactivated by the emission of photons. To cope with this problem, polyimide LB films consisting of A, S, and D units were used for constructing more efficient molecu-

FIGURE 3

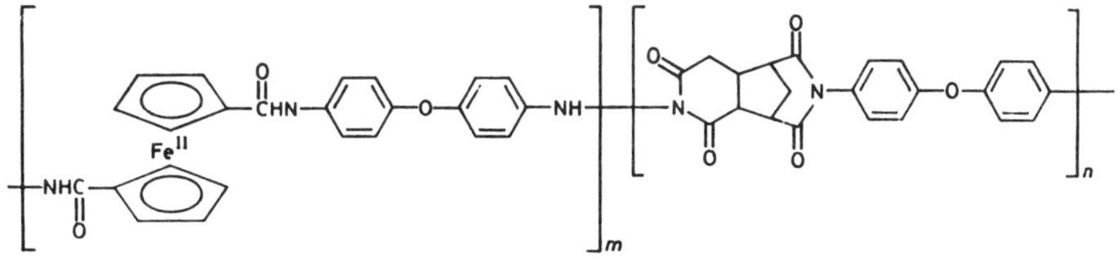

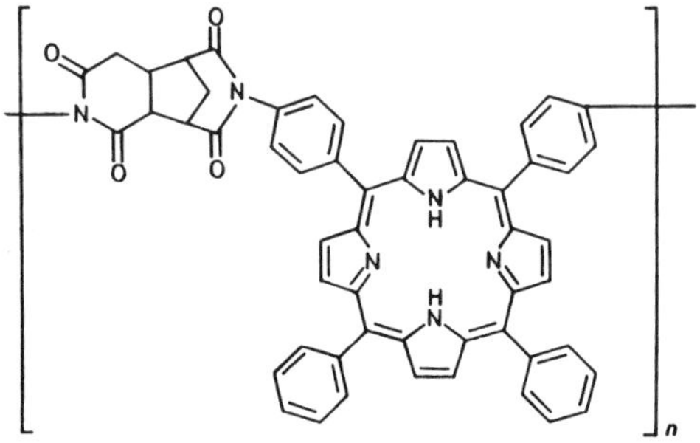

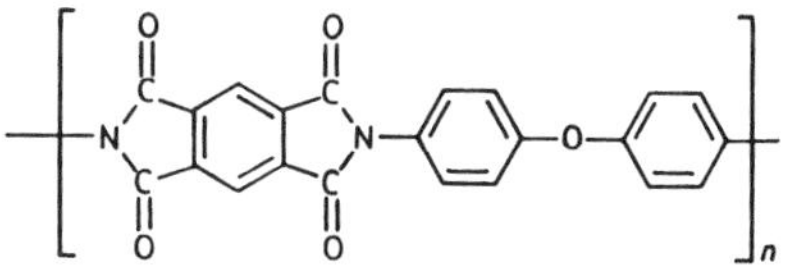

Structural formulae of D, S and A components of polyimide heterogeneous LB film molecular photodiode.

lar photodiodes in collaboration with Kakimoto, Imai, and their co-workers[15].
They have reported the preparation and properties of polyimide LB films[16].
Since polyimide LB films have no long alkyl spacer between the layers (monolayer thickness 0.4 - 0.6 nm), electrons should be more readily transferred. The structures of A, S, and D components are shown in Figure 3. For example, sub-μA order of photocurrents were observed for AuOTE's which were coated with six layers of A, two layers of S, and six layers of D. These magnitudes are ca. 10 times larger than those for photodiodes with conventional LB films.

3. MOLECULAR PHOTODIODE CONSISTING OF FOLDED TYPE S-A-D TRIADS

Another approach to shortening the distances between the functional moieties is the use of unidirectionally oriented amphiphilic triad monolayers[6,17]. Each triad contained an A, S, and D moiety as its functional subunits. Other groups studied also the two-step photodriven charge separation and back electron transfer reactions of triad molecules of the S-A_1-A_2[18] or A-S-D[19,20] types. They succeeded in retarding charge recombination in homogeneous solution, but did not attempt to orient these triad molecules in one direction and thus to conduct a direct photoelectric conversion. In our first amphiphilic triad 8 (SAD C11)[6], A, S, and D corresponded to viologen, pyrene, and ferrocene moie-

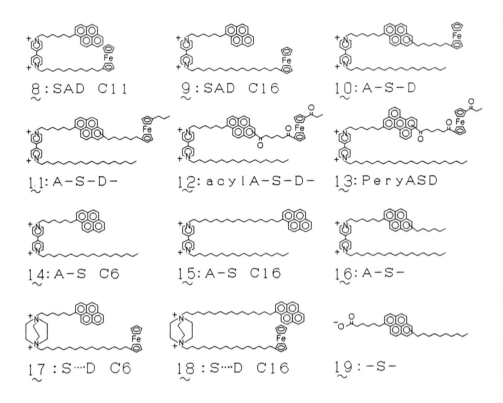

FIGURE 4

Structural formulae of folded type S-A-D and linear type A-S-D triads and their reference A-S, S-D and -S- compounds.

ties, respectively. The viologen moiety is hydrophilic, whereas the pyrene and ferrocene moieties are hyrophobic. Subunits A and S were linked together with a C_6 alkyl chain, while subunit D was linked to subunit A with a longer C_{11} alkyl chain. Later, a modified triad 9 (SAD C16) was synthesized in which the subunit D was linked to A with a much longer C_{16} alkyl chain[17] in order to improve the balance of the two distances between A and S and between S and D. The structures of these two S-A-D triads are shown in Figure 4 together with the linear type A-S-D triads and their reference compounds, A-S, S-D, and -S-.

Due to their amphiphilic properties, the A, S, and D moieties were considered to be arranged spatially in this order, owing to the difference in length of two alkyl chains, from the electrode to the electrolyte solution perpendicularly as shown in Figure 5(b). This arrangement was confirmed by observing anodic photocurrents, whose direction was in accordance with the photoinduced vectorial flow of electrons expected from the energy diagram of the oriented triad, as depicted in Figure 5(c). From the shape illustrated in Figure 5(b),

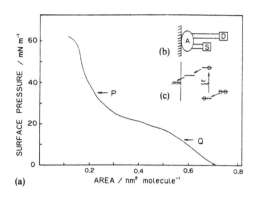

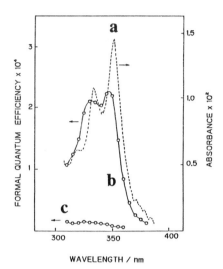

$\underset{\sim}{20}$ $C_{19}H_{39}$ COOH   AA

FIGURE 5

(a) Surface pressure - area isotherm of a mixed monolayer of triad 8 with AA (20); (b) schematic representation of oriented triad 8; (c) corresponding energy diagram.

FIGURE 6

Absorption spectrum (spectrum a) and photocurrent spectra (spectra b and c for films deposited at 35 and 15 mN/m, respectively) of mixed monolayer of 8 and AA (1:2).

S-A-D is called "folded" type triad. The photodiode function of these folded type S-A-D triads was studied in detail[17] in terms of i) the wavelength of the incident photons, ii) the surface pressure applied for monolayer deposition, iii) the applied electrode potentials, iv) the distances between A and S and between S and D, and v) the role of the D moiety.

In Figure 6(a) is shown an absorption spectrum for two layers of the mixed monolayer of triad 8 and arachidic acid 20 (AA) (1:2) which was deposited on a quartz plate. The photocurrent spectra b and c correspond to the mixed mono-layers which were deposited on AuOTE's at the two different surface pressures indicated by P and Q, respectively, in Figure 5(a). The formal quantum efficiency, plotted on the ordinate, for the action spectra was defined as the number of photoelectrons flowing per number of incident photons[7]. The action spectrum was similar to the absorption spectrum of the pyrene moiety, indicating that the photocurrent was induced by excitation of the sensitizer.

The most important and interesting point shown in Figure 6 is that the photocurrent increased by a factor of ca. 20 when the surface pressure for film deposition was changed only from 15 to 35 mN m^{-1}. The increase in the surface concentration by compression is insufficient to explain such a dramatic in-crease in the photocurrent. Rather, the non-oriented structure of triad 8 and

the unfavorable relative location of the A, S, and D moieties in the monolayer
at the low surface pressure might be responsible for the small photocurrent.
By contrast, the much larger photocurrents observed at the high surface pressure support the assumption that the folded type S-A-D triad molecule is oriented in the more favorable structure as shown in Figure 5(b).

The potential dependence of anodic photocurrents, i.e. an i - V characteristic of the molecular photodiode was what we expected from its energy diagram
shown in Figure 5(c).

A higher efficiency of triad **9** than that of triad **8** was also observed, which
might be attributed to a better matching in concurrent electron transfer reactions between A and S and between S and D owing to an improved balance between
the A-S and S-D distances.

Another amphiphilic compound **14** (A-S C6) without the D moiety in Figure 4
confirmed a positive contribution of the D moieties in triads **8** and **9**.

4. MOLECULAR PHOTODIODE CONSISTING OF LINEAR TYPE A-S-D TRIADS

In order to improve the orientation of triad molecules in the monolayer,
"linear"-type A-S-D triad compounds, whose structures are also shown in Figure
4, were synthesized[21]. The mixed monolayer of a linear-type A-S-D triad **10**
with behenic acid exhibited a much higher photocurrent than did that of the
folded-type S-A-D triads. This indicates that a more ideal spatial arrangement
of the A, S, and D moieties was attained for the linear triad molecules in the
mixed monolayer. A favorable orientation of the linear triad molecule was also
accomplished at a relatively low surface pressure of 15 mN m^{-1} in contrast with
the inactive orientation of the folded-type triad molecules deposited at this
surface pressure. The importance of the presence of the D moiety was again
confirmed by comparing the efficiency of the linear triad with that of a reference A-S- compound **16**.

5. KINETICS OF PHOTO-INDUCED ELECTRON TRANSFER AND THE EFFECT OF ELECTRICAL DOUBLE LAYER

As described in section 2, kinetics of photoinduced multi-step electron
transfer plays a crucial role in efficiencies of the molecular photodiodes. To
clarify the energy gap and the distance dependence of the photoinduced electron
transfer in heterogeneous LB films or triad monolayers, nanosecond and picosecond laser photolyses were carried out.

As a reference in photoinduced multi-step electron transfer in the A/S/D and
D/S/A LB films, the luminescence decay curves were recorded for sensitizers
such as pyrene and Ru(bpy)$_3^{2+}$ derivatives confined as one monolayer in LB
films. These films also contained other monolayers of acceptor or donor amphi-

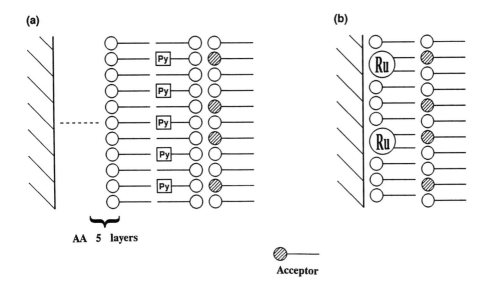

(a)

AA 5 layers

(b)

Acceptor

FIGURE 7
Heterogeneous LB films used for kinetic study of photo-induced electron transfer: a, PDA (2) and A (or D); b, RuC19 (21) and A (or D).

philes which were deposited apart from the sensitizer monolayer by a fixed
distance as shown in Figure 7. The structures of the sensitizers and those of
amphiphilic acceptors and donors are shown in Figures 8 and 9, respectively.
In addition to the difference of 1.0 V in the oxidation potentials between the
excited pyrene and the $Ru(bpy)_3^{2+}$ derivatives, the redox potentials of four
types acceptors ranged widely up to 1.8 V. This enabled us to examine the

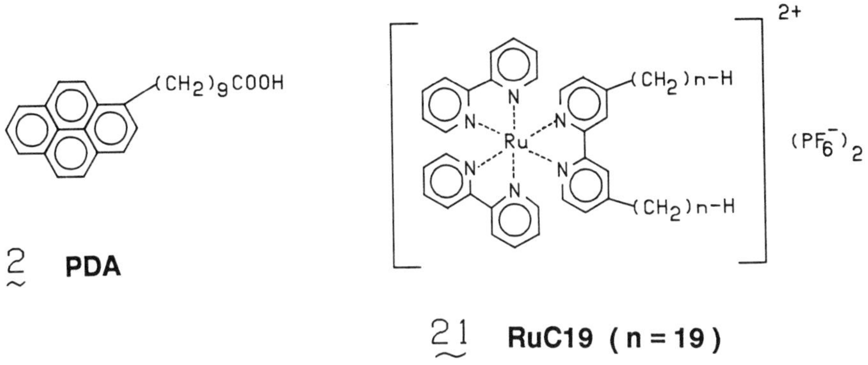

2 **PDA**

21 **RuC19 (n = 19)**

FIGURE 8
Structural formulae of the sensitizers used in the kinetic study: a, pyrene-
decanoic acid 2 (PDA); b, RuC19 (n = 19) 21.

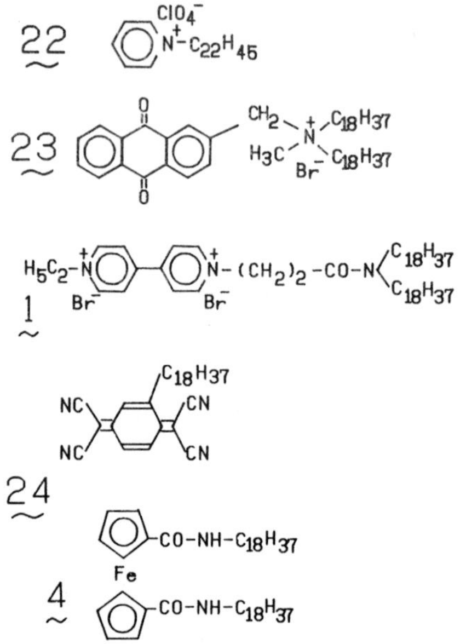

FIGURE 9
Structural formulae of the acceptor and donor amphiphilic derivatives used in the kinetic study.

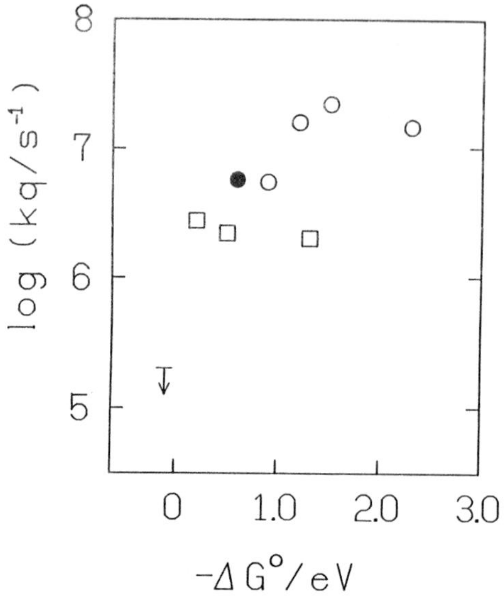

FIGURE 10
Rate constants of electron transfer quenching as a function of standard free-energy $\Delta G°$ of the quenching for PDA (circles) and RuC19 (squares).

energy gap $\Delta G°$ dependence of the photoinduced electron transfer rate of the A/S LB films[22]. The results are shown in Figure 10. It may be concluded from the plots in Figure 10 that the possibility of a Marcus-type inverted region exists at highly negative $\Delta G°$, although it is not definite yet because of the limited data.

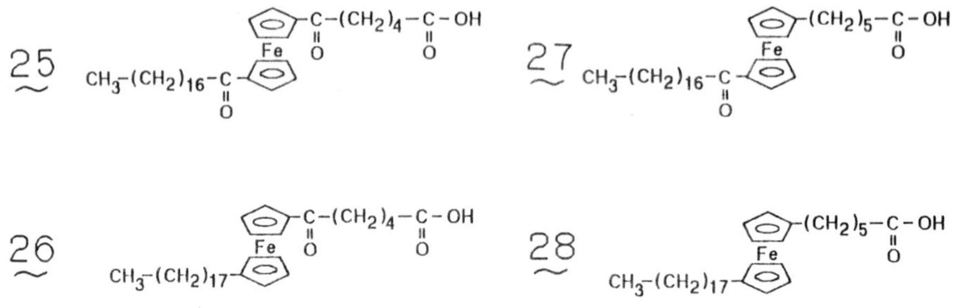

FIGURE 11
Structural formulae of amphiphilic ferrocene derivatives used as electron donor.

To examine the energy gap law for S/D LB films, four kinds of amphiphilic ferrocene derivatives (25-28 in Figure 11) as electron donors with different standard redox potentials $E°$'s and with the same alkyl chain spacer were synthesized[23]. The comparison of the energy gap $\Delta G°$ dependence of the reductive quenching of the $Ru(bpy)_3^{2+}$ derivative 21 (RuC19) with the ferrocene derivatives for three different systems, i.e. the LB films, the micellar, and the solution systems, implies that the electrical potential difference between the hydrophilic head-groups in LB films has to be taken into account for estimation of the effective energy gap $\Delta G°$. Namely, an efficient photoinduced electron transfer quenching happens even in a S/D LB film in which the reaction was expected to be up-hill ($\Delta G° > 0$) on the basis of half-wave potentials and thus too slow to be detected. The result contradicted also with that of the corresponding solution system.

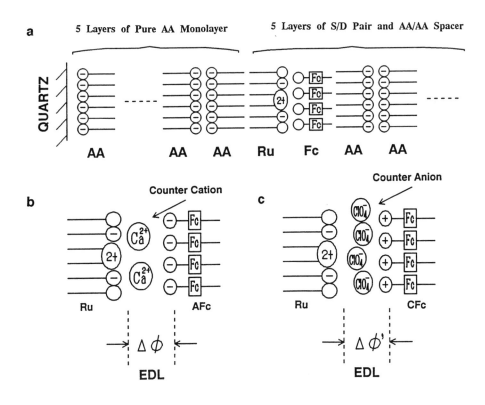

FIGURE 12
Structure of heterogeneous LB film (a) used to study the effect of electrical double layer on kinetics of electron transfer quenching and supposed structures of the electrical double layers for anionic (b) and cationic (c) head groups of amphiphilic ferrocene derivatives 25 and 29, respectively.

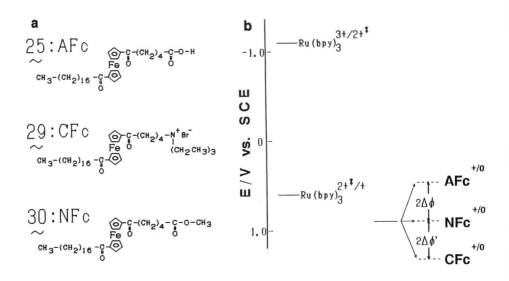

FIGURE 13
Structural formulae of ferrocene derivatives (a) used to study the effect of
head-group charge and supposed change in energy diagram (b) due to the change
in the inner potential difference in the electrical double layer.

The effect of the inner potential difference in the electrical double layer
was further studied by using three kinds of amphiphilic ferrocene derivatives,
AFc (25), CFc (29), and NFc (30), with an anionic, a cationic, and a nonionic
head-group, respectively[24] (Figure 13(a)). All the redox potentials of three
ferrocene derivatives were ca. 0.9 V vs. SCE and more positive than the reduction potential of excited $Ru(bpy)_3^{2+}$ of 0.6 V vs. SCE. The luminescence decay
curves for RuC19 in three types of S/D systems were recorded together with that
in a reference LB film in which the pure monolayer of AA was deposited in place
of the ferrocene donor layer. Only the decay curve for S/AFc showed a fast
decay component, while the other two curves are very similar to that for the
reference LB film and showed almost a single exponential decay without appreciable quenching.

The change in sign of the head-group charge was expected to vary signs of
the electrical double layer and thus the effective energy gap $\Delta G°$. As shown in
Figure 12, electron transfer quenching is expected to occur in region of electrical double layer where the hydrophilic head-groups of the RuC19 and the
ferrocene amphiphiles are facing each other. As shown in the schematic diagram
of the LB film structure in Figures 12(b) and (c), it seems to be quite reasonable to assume that i) the positive charges of $Ru(bpy)_3^{2+}$ are neutralized by
the carboxylate anions present in the same plane in the mixed monolayer; ii)

the negative charges of the head-group of the anionic ferrocene derivative are in contact with their counter Ca^{2+} cations; and iii) the positive charge of the cationic donor in contact with their counter ClO_4^- anions. The potential difference will readily become a few hundreds millivolts as observed for mono- layers at the air-water interface[25] and for micellar surfaces[26]. If this potential difference is taken into account, the energy diagram has to be shift- ed as shown by the dashed lines in Figure 13(b). Thus, appreciable electron transfer quenching, observed specifically in the RuC19/the anionic ferrocene derivative system, is rationalized. It is noteworthy that the electrostatic potential effect was proposed to explain why the photochemical reaction follows the L pathway in the reaction center with almost C_2 symmetry[27].

In order to increase the response time and efficiency of the molecular photodiode consisting of folded type S-A-D triad molecules, it is important to know the distance and orientation dependence of the electron transfer quenching in A-S and S-D monolayers. For this purpose, A-S (Figure 14(a)) and folded type S-D (e.g. **17**, **18** in Figure 4) amphiphiles with alkyl chain spacer of different carbon numbers were prepared. The luminescence decay of these LB films were measured with the picosecond laser photolysis[28]. The linear

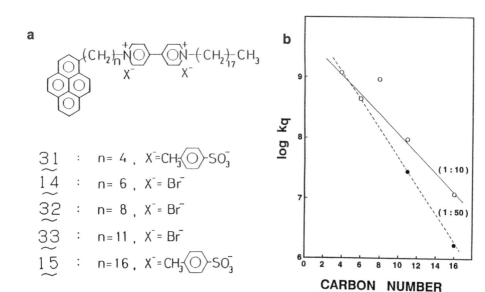

FIGURE 14
Structural formulae of A-S compounds (a) used to study the effect of alkyl chain length between A and S moieties on kinetics of electron transfer quench- ing in mixed monolayers of A-S and AA.

decrease in log k_q with an increase in carbon number was observed for the A-S monolayers as shown in Figure 14(b), while the regular dependence of log k_q on the alkyl chain lengths could not be obtained for folded type S-D systems. This result indicates that the alkyl chains in the A-S diad molecules are extended in their monolayers and therefore the distance between A and S can be controlled by changing the carbon number, while this is not the case for the folded type S-D diads. Folding of the alkyl chain in the S-D diad in the monolayer may account for the non-ideal result.

6. SIMULATION OF THE PRIMARY PROCESS IN THE PHOTOSYNTHETIC REACTION CENTER BY MIXED MONOLAYER WITH TRIAD AND ANTENNA MOLECULES

In the next step, we simulated the light harvesting and succeeding charge separation processes by a monolayer assembly consisting of synthetic antenna pigments and triad molecules, as illustrated in Figure 15[5,29]. For the light-harvesting (H) antenna pigments, an amphiphilic pyrene derivative was used[30]. For the amphiphilic linear triad molecule, a perylene moiety, as the S unit, and viologen and ferrocene moieties, as the A and the D units, respectively, were used. The structures of the antenna 19 and the triad 13 molecule are also shown in Figure 4. Because of the overlap of the emission spectrum of the antenna pyrene and the absorption spectrum of the sensitizer perylene moiety of

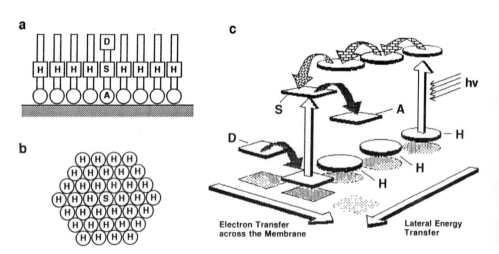

FIGURE 15

Schematic representation of the artificial photosynthetic reaction center by a monolayer assembly of Pery A-S-D triad 13 and antenna -S- 19 molecules for light harvesting (H), energy migration and transfer, and charge separation via multistep electron transfer: a, side view of monolayer assembly; b, top view of a triad surrounded by antenna molecules; c, energy diagram for photoelectric conversion in a monolayer assembly.

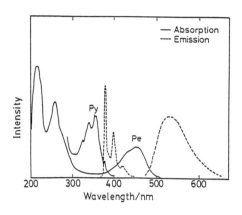

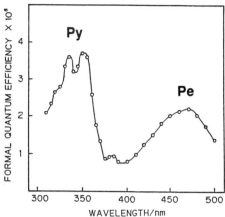

FIGURE 16

UV and visible absorption and emission spectra of antenna pyrene **19** and perylene triad **13** derivatives in ethanol.

FIGURE 17

Photocurrent spectrum of the mixed monolayer of triad **13** and antenna **19** with a molar ratio of 1:4 deposited on AuOTE.

the triad (Figure 16), light energies harvested by the antenna molecules were efficiently transferred to the sensitizer moiety of the triad. Thus, the excitation energy of the perylene moiety should be converted to electrical energies via multistep electron transfer across the monolayer as described above.

Figure 17 shows the photocurrent spectrum of the mixed monolayer of the triad and the antenna with a molar ration of 1:4. Maxima of anodic photocurrents at ca. 350 and 470 nm are found. These correspond to the adsorption maxima of pyrene and acylated perylene. The anodic direction of the photocurrent agrees with the energy diagram and the orientation of the triad shown in Figure 15. In contrast, negligible photoresponse was obtained with the pure antenna monolayer. The result indicates that charge separation in the triad molecules was initiated by light absorption both with perylene sensitizer itself and with the pyrene antennas followed by the energy transfer.

7. CATALYTIC REDUCTION OF CO_2 WITH METAL COMPLEXES

The object in this section is to construct an artificial photosynthetic system using a complex LB film consisting of two or three types of LB film stacked in series as shown in Figure 18. One of them has a photoinduced charge separation function as described above, and the others possess catalytic activities for oxidation and reduction reactions achieved by the separated holes and

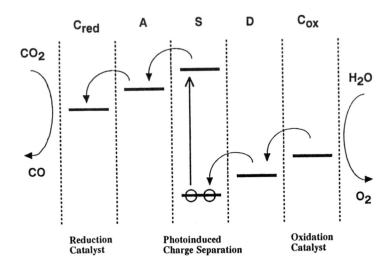

FIGURE 18

Artificial photosynthetic system using a complex LB film consisting of a charge separation A/S/D (or A-S-D) layer sandwiched between two redox catalysis monolayers.

electrons, respectively. The electrocatalytic activity of LB films of amphiphilic nickel cyclams[31] was examined for the reduction of CO_2, since nickel cyclam is known to possess high catalytic activity and selectivity to CO[32].

Two amphiphilic nickel(II) complexes with the long-alkyl-substituted 1,4,8,11-tetraazacyclotetradecane (cyclam) shown in Figure 19(a) were synthesized, i.e. nickel(II) tetrakis(N-hexadecyl)cyclam 34 (Ni-R_4-Cyclam) and nickel(II) N-hexdecylcyclam 35 (Ni-R_1-Cyclam). Glassy carbon disk electrodes were modified with LB films of these amphiphilic nickel complexes and the electrocatalytic activities of these electrodes were examined by cyclic voltammetry (CV) in aqueous solutions saturated with N_2 or CO_2 (Figure 19(b)). The redox waves of the LB films of the alkyl derivatives of nickel(II) cyclam observed under N_2 were shifted in the positive direction relative to that of nickel(II) cyclam in a homogeneous solution. Under CO_2, the increase in the reductive current of these electrodes compared with that under N_2 was observed at more negative potentials than -1.3 V vs. SCE in an aqueous solution with a constant pH value. This result shows that the LB films of these complexes possess sufficient electrocatalytic activity to reduce CO_2 despite their minimal amounts.

In addition to the electrochemical study of the LB modified electrodes, the electrocatalytic reduction of CO_2 on mercury with nickel(II) cyclam itself was studied in detail[33] by CV, polarography, and electrocapillarity in aqueous

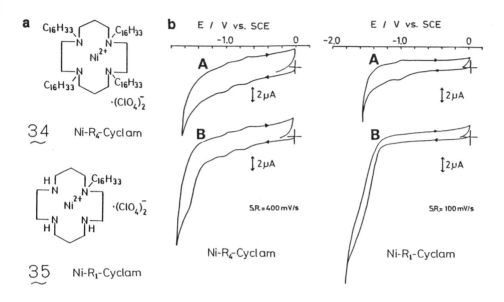

FIGURE 19

Structural formulae (a) of amphiphilic Ni(II) complexes with N-long alkyl substituted cyclams and their catalytic activities for carbon dioxide electroreduction (b) studied by cyclic voltammetry under (A) N_2 and (B) CO_2.

solutions, with and without the catalyst under N_2, CO, and CO_2. The effect to the catalytic activity and the stability of the nickel complexes of the short-alkylated cyclams as determined by the position and the number of the introduced alkyl groups to cyclam were also investigated[34,35]. The products of the electrocatalytic reduction of CO_2 with nickel(II) complexes of cyclam and the alkylated cyclams were analyzed *in situ* during their CV by the modified Differential Electrochemical Mass Spectroscopy (DEMS) originally developed by Heitbaum and co-workers[36]. In our DEMS apparatus, an amalgamated gold mesh electrode is placed in contact with a nonwetting polytetrafluoroethylene inlet membrane, one side of which contacts the solution being electrolyzed and the other side contacts the high-vacuum system of a mass spectrometer[37]. CO and H_2 were shown to be the major products with these nickel(II) cyclams at low and high negative potentials, respectively. Derivatization of ruthenium complexes for catalytic CO_2 reduction with long-alkyl chains is now being attempted.

8. CONCLUSION

With the near completion of the study on the reduction of CO_2 with LB monolayer catalysts, we started the study of the amphiphilic catalysts which can oxidize water to O_2 by positive holes generated with the LB assembly for charge separation. Our final goal is the construction of LB film molecular assemblies for artificial photosynthesis as schematically shown in Figure 20. We hope to attain this goal in the near future.

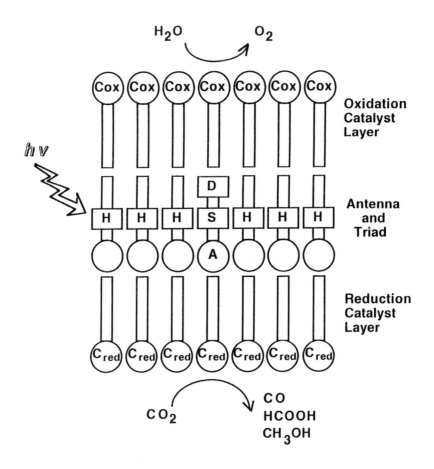

FIGURE 20
Schematic structure of LB film molecular assembly for artificial photosynthesis.

ACKNOWLEDGEMENTS

I would like to thank all my colleagues who collaborated in this research. This research was supported by a Grant-in-Aid for Scientific Research 61470076 and those on Priority Areas 63604534 and 63612506 from the Ministry of Education, Science, and Culture and by a Grant from the Nissan Science Foundation.

REFERENCES

1) L. Stryer, Biochenistry, 3rd ed. (Freeman, New York, 1988).

2) J. Deisenhofer, O. Epp, K. Miki, R. Huber and H. Michel, J. Mol. Biol. 180 (1984) 385; Nature 318 (1985) 618.

3) H. Kuhn, J. Photochem. 10 (1979) 111; Thin Solid Films 99 (1983) 1.

4) D. Möbius, Ber. Bunsenges. Phys. Chem. 82 (1978) 848; Acc. Chem. Res. 14 (1981) 63.

5) M. Fujihira, Mol. Cryst. Liq. Cryst. 183 (1990) 59.

6) M. Fujihira, K. Nishiyama and H. Yamada, Thin Solid Films 132 (1985) 77.

7) T. Osa and M. Fujihira, Nature 264 (1976) 349; M. Fujihira, N. Ohishi and T. Osa, Nature 268 (1977) 226; M. Fujihira, T. Kubota and T. Osa, J. Electroanal. Chem. 119 (1981) 379.

8) J. W. Verhoeven, Pure Appl. Chem. 58 (1986) 1285.

9) J. R. Miller, J. V. Beitz and R. K. Huddleston, J. Am. Chem. Soc. 106 (1984) 5057; J. R. Miller, Nouv. J. Chim. 11 (1987) 83.

10) R. A. Marcus and N. Sutin, Biochim. Biophys. Acta 811 (1985) 265.

11) R. A. Marcus, J. Chem. Phys. 43 (1965) 2654; P. Siders and R. A. Marcus, J. Am. Chem. Soc. 103 (1981) 748.

12) J. Ulstrup and J. Jortner, J. Chem. Phys. 63 (1975) 4358.

13) J. R. Miller, L. T. Calcaterra and G. L. Closs, J. Am. Chem. Soc. 106 (1984) 3047.

14) M. Fujihira, K. Nishiyama and H. Yoneyama, J. Chem. Soc. Jpn. (1987) 2119.

15) Y. Nishikata, A. Morikawa, M. Kakimotao, Y. Imai, Y. Hirata, K. Nishiyama and M. Fujihira, J. Chem. Soc., Chem. Commun. (1989) 1772.

16) M. Suzuki, M. Kakimota, T. Konishi, Y. Imai, M. Iwamoto and T. Hino, Chem. Lett. (1986) 395; M. Kakimoto, M. Suzuki, T. Konishi, Y. Imai, M. Iwamoto and T. Hino, Chem. Lett. (1986) 823; Y. Nishikata, M. Kakimoto, A. Morikawa, I. Kobayashi, Y. Imai, Y. Hirata, K. Nishiyama and M. Fujihira, Chem. Lett. (1989) 861.

17) M. Fujihira and H. Yamada, Thin Solid Films 160 (1988) 125.

18) S. Nishitani, N. Kurata, Y. Sakata, S. Misumi, A. Karen, T. Okada and N. Mataga, J. Am. Chem. Soc. 105 (1983) 7771; N. Mataga, A. Karen, T. Okada, S. Nishitani, N. Kurata, Y. Skata and S. Misumi, J. Phys. Chem. 88 (1984) 5138.

19) T. A. Moore, D. Gust, P. Mathis, J. C. Mialoeq, C. Chachaty, R. V. Bensasson, E. J. Land, J. C. Doizi, P. A. Liddel, W. R. Lehman, G. A. Nemeth and A. L. Moore, Nature, 307 (1984) 630.

20) M. R. Wasielewski, M. P. Niemczyk, W. A. Svec and E. B. Pewitt, J. Am. Chem. Soc. 107 (1985) 5562.

21) M. Fujihira and M. Sakomura, Thin Solid Films 179 (1989) 471.

22) M. Fujihira, K. Nishiyama and K. Aoki, Thin Solid Films 160 (1988) 317.

23) T. Kondo, H. Yamada, K. Nishiyama, K. Suga and M. Fujihira, Thin Solid Films 179 (1989) 463; T. Kondo and M. Fujihira, Kobunshi Ronbunshu 47 (1990) 921.

24) T. Kondo and M. Fujihira, Chem. Lett. (1991) 191; T. Kondo, M. Yanagisawa and M. Fujihira, Electrochim. Acta submitted.

25) V. Vogel and D. Möbius, Thin Solid Films 132 (1985) 205; 159 (1988) 73; J. Colloid Interface Sci. 126 (1988) 408.

26) C. Wolff and M. Grätzel, Chem. Phys. Lett. 52 (1977) 542; M. Grätzel, Heterogeneous Photochemical Electron Transfer (CRC Press, Florida, 1989).

27) J. R. Norris and M. Schiffer, C&EN News July 30 (1990) 22.

28) H. Yamada, H. Ohtani and M. Fujihira, to be submitted.

29) M. Fujihira, 2nd International Symposium on Bioelectronic and Molecular Electronic Devices R&D Association for Future Electron Devices, Dec.12-14, 1988, Fujiyoshida, Japan, pp. 35-38; M. Fujihira, M. Sakomura and T. Kamei, Thin Solid Films 180 (1989) 43.

30) M. Fujihira, T. Kamei, M. Sakomura, Y. Tatsu and Y. Kato, Thin Solid Films 179 (1989) 485.

31) Y. Hirata, K. Suga and M. Fujihira, Thin Solid Films, 179 (1989) 95.

32) M. Beley, J. P. Collin, R. Ruppert and J. P. Sauvage, J. Am. Chem. Soc. 108 (1986) 7461.

33) M. Fujihira, Y, Hirata and K. Suga, J. Electroanal. Chem. 292 (1990) 199.

34) U. Akiba, Y. Nakamura, Y. Hirata, K. Suga and M. Fujihira, J. Chem. Soc., Chem. Commun. submitted.

35) Y. Hirata, Y. Nakamura, U. Akiba, K. Suga and M. Fujihira, submitted.

36) O. Wolter and J. Heitbaum, Ber. Bunsenges. Phys. Chem. 88 (1984) 2.

37) Y. Hirata, K. Suga and M. Fujihira, Chem. Lett. (1990) 1155.

Photochemical Processes in Organized Molecular Systems
K. Honda (Editor-in-Chief)
© Elsevier Science Publishers B.V., 1991

FRACTALS AND EXCITATION TRANSFER IN MOLECULAR ASSEMBLIES

Iwao YAMAZAKI, Naoto TAMAI* and Tomoko YAMAZAKI

Department of Chemical Engineering, Faculty of Engineering, Hokkaido
University, Sapporo 060, Japan

Electronic excitation transfer has been studied with several kinds of molecular assemblies such as vesicles, Langmuir-Blodgett films. The morphology
of molecular assembly frameworks affects the dynamics of energy relaxation
processes, the fractal behaviors are found in spatial distribution of chromophores or in site-energy distribution. A molecular optical device capable
of spatially-controlled photoexcitation transport and switching is proposed.

1. INTRODUCTION

Transport and trapping of electronic excitation energy have been the subject
of extensive theoretical and experimental works. Special attention has recently been paid to dynamics of the Förster-type excitation energy transfer in
molecular assemblies of restricted molecular geometries, for examples, vesicles,
Langmuir-Blodgett (LB) films and photosynthetic light-harvesting antenna pigment systems. Several examples with which we are concerned in this paper are
shown schematically in Figure 1. One can expect to observe new aspects of the
excitation transfer quite different from systems in which acceptor molecules
are randomly and uniformly distributed in the three-dimensional (3D) rigid or
fluid medium.

The present paper will review our recent experimental results on the excitation transfer in vesicles and LB films. Several types of fluorescence decay
kinetics for the excitation transfer will be presented in relation to their
assembly geometries and fractal configurations.

2. EXCITATION TRANSFER IN 2D SYSTEM: DYES ADSORBED ON VESICLE SURFACES[1,2]

Vesicles are static colloidal particles which consist of surfactant molecules of two long hydrocarbon chains connected to the polar head groups (Figure
1a). Single compartment vesicles contain 2000-10000 monomers per vesicle, and

---

* Present address: Masuhara Microconversion Project, ERATO, JRDC, Kyoto
  Research Park, Chudoji-minami, Simogyo-ku, Kyoto 600, Japan.

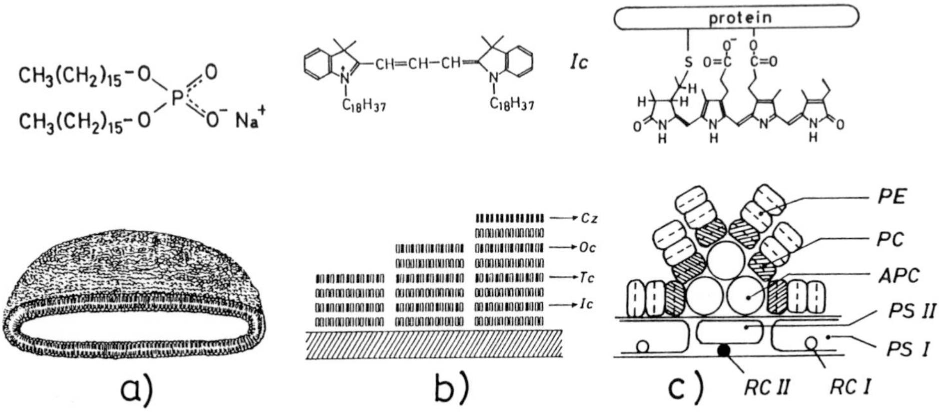

FIGURE 1

Schematic illustration of organized molecular assemblies: (a) vesicles, (b) LB multilayer films, and (c) biological antenna of algae.

their surface area is usually as large as $10^5 - 10^6$ Å^2. When donor and acceptor molecules are adsorbed on the surface, one can examine kinetics of the 2D energy transfer. We have investigated the excitation energy transfer between rhodamine 6G (donor) and malachite green (acceptor) and between rhodamine B (donor) and malachite green (acceptor) adsorbed on anionic vesicles consisting of dihexadecylphosphate (Figure 1a). Figure 2 shows the donor fluorescence decay curves together with the results of analyses based on the fractal theory.

The fractal denotes a self-similar structure of dilatational symmetry which will have great potential to describe a multitude of irregular structures. Klafter and Blumen[3] proposed a fluorescence decay function of the donor in a 2D plane where the acceptors are distributed in a space of the fractal dimension:

$$\rho(t) = \exp[\, -t/\tau_D - \gamma_A(t/\tau_D)\,]\ , \qquad (1)$$

$$\text{where}\quad \beta = \bar{d}/s, \quad \text{and}\quad \gamma_A = x_A(d/\bar{d})V_d R_0^{\bar{d}}\,(1-\beta). \qquad (2)$$

τ_D is the lifetime of donor, and s is the order of the multipolar interaction; x_A is the fraction of fractal sites occupied by acceptors; d and $\bar{d}$ are the Euclidian and fractal dimensions, respectively; and V_d is the volume of unit sphere in d dimension.

The fluorescence decay curves can be analyzed with eq 1 by the conventional curve-fitting method. By varying γ_A and β values with the τ_D value being fixed, the best-fitting curves were obtained for different surface densities of acceptors. The results are shown in Figure 2. On increasing acceptor concen-

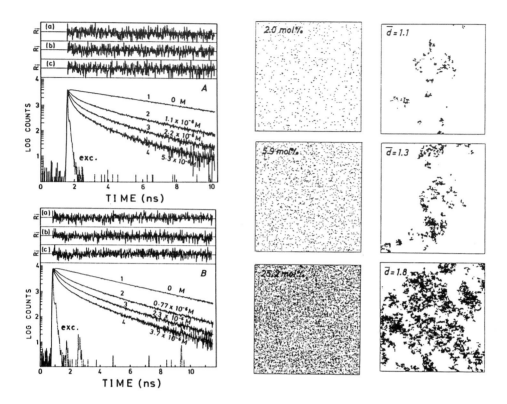

FIGURE 2
Fluorescence decay curves of donor, rhodamine 6G (A) and rhodamine B (B), in the presence of acceptor, malachite green. The best-fit curves are based on the fractal dimension d = 1.3.

FIGURE 3
Computer-generated plots of the distribution of dye molecules adsorbed on vesicle surfaces under the assumption of a square lattice (200 x 200). The right column is the fractal distribution, and the left is the random and uniform distribution.

tration, γ_A increases linearly with the concentration, whereas β is almost constant. The fractal dimension d can be derived directly from β values through eq 2; $\bar{d}$ = 1.31 ± 0.078 for rhodamine 6G and $\bar{d}$ = 1.32 ± 0.049 for rhodamine B.[1] A model picture was obtained through the computer graphics for the fractal distribution of dye molecules on the surface under an assumption of Levy's dust model. In this model, we assume that, in each step of molecular adsorption, an adsorption site is grown at a distance r apart from the previous site with the probability P(r) expressed as $P(r) \propto r^{-\bar{d}}$, where $\bar{d}$ is the fractal dimension associated with a random-walk of step distance r.

3. EXCITATION TRANSFER IN 2D SYSTEM: LB MONOLAYER FILMS[4]

The LB film is typical of the 2D molecular assembly. The excitation energy relaxation of dyes in LB monolayers was investigated with rhodamine B (Figure 4) in the concentration range of 0.01–30 mol% at 80 K and 295 k. The time-resolved fluorescence spectra exhibit a dynamic Stokes shift decaying logarithmically during 30 ps in higher concentrations. The fluorescence decays can be analyzed by an equation including stretched exponential function:

$$\rho(t) = A_1 \exp[-t/\tau_D - \gamma(t/\tau_D)^\beta] + A_2 \exp(-t/\tau_D) \qquad (3)$$

The mechanisms of singlet excitation energy relaxation were interpreted in terms of (1) energy migration among energetically disordered monomer sites represented by the ultrametric space (hierarchical energy distribution or Gaussian distribution of the density of excited state) and (2) energy trapping at aggregate sites spread in a monolayer. It was found that the dynamic Stokes shift and the fluorescence decay are related to the diffusion length of the excitation energy transport[4] as is shown in Figure 4.

4. SEQUENTIAL EXCITATION TRANSPORT IN LB MULTILAYERS[2]

With stacking multilayered LB films, the sequential excitation transfer from an outer layer to an inner layer was studied. We have prepared three types of the LB films consisting of sequences of donor (D) and acceptors (A_1, A_2 and A_3); namely, D–A_1 (hereafter referred to as 2L), D–A_1–A_2 (3L), D–A_1–A_2–A_3 (4L).

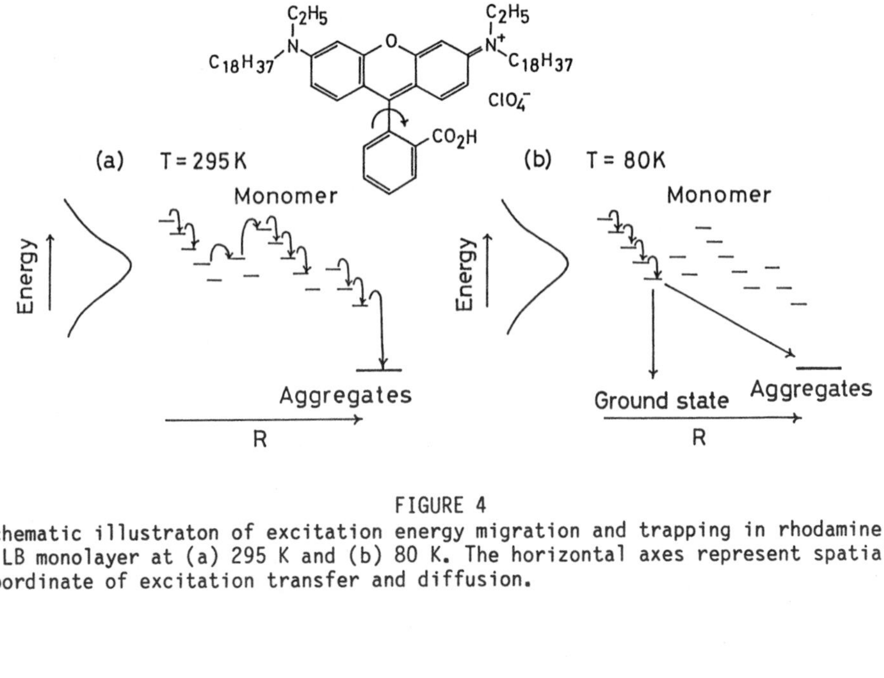

FIGURE 4
Schematic illustraton of excitation energy migration and trapping in rhodamine B LB monolayer at (a) 295 K and (b) 80 K. The horizontal axes represent spatial coordinate of excitation transfer and diffusion.

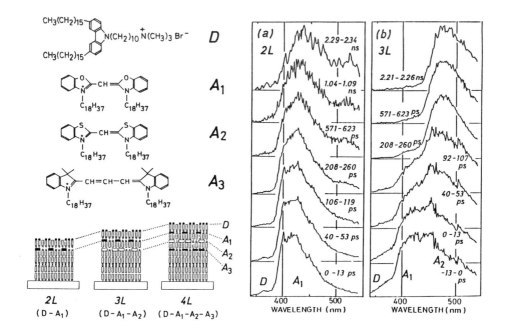

FIGURE 5
Schematic illustration of stacking structures of the LB multilayer films and molecular formulas of the pigment chromophores.

FIGURE 6
Time-resolved fluorescence spectra of the LB multilayer films: (a) 2L, D-A$_1$ and (b) 2L, D-A$_1$-A$_2$. The excitation wavelength is 295 nm.

The structures of these multilayer films are illustrated schematically in Figure 5. Six to eight layers were deposited on a quartz plate in the following order: (1) five layers of palmitic acid – cadmium salt, (2) monolayer(s) consisting of palmitic acid and small amounts of dyes (A$_1$, A$_2$ and A$_3$) and (3) monolayer of carbazole (D) and (4) a monolayer of palmitic acid. An outer layer of palmitic acid prevents the multilayered structure from being destroyed. The concentration of pigment molecules was 10 mol% in each layer.

Figure 6 shows the time-resolved fluorescence spectra of the LB multilayers of 2L and 3L. Each spectrum is normalized to the maximum intensity. It is seen that the spectrum changes significantly with time in the picosecond time range. In 2L, following excitation of the D layer at 295 nm, a fluorescence band of D appears weakly at 350 nm and A$_1$ band at 420 nm. In 3L, the fluorescence bands of D and A$_1$ appear in the initial time region, and then A$_2$ band appears at 470 nm after 40 ps. Similar spectral change can be seen in 4L. It is seen that the fluorescence from the inner layer rises more slowly than those of the outer layers, and that the transfer rate depends on the interlayer chromophore distance. These sequential time behaviors can be fitted approximately with the decay kinetics of $\exp(-2kt^{1/2})$ type. The values of the rate constants fall in between 0.02 to 0.2 $ps^{-1/2}$. These results show that the energy transfer takes place sequentially from the outer to the inner layers, similarly to the photosynthetic light-harvesting antenna pigment system.

5. PHOTOCHROMIC LB FILMS AS AN OPTICAL SWITCHING DEVICE[5]

Spyropiran (SP) and its derivatives are known to show a photochromic reaction. Using their derivatives with a long alkyl chain, we can prepare a monolayer or multilayer LB film containing photochromic chromophores. In this study, the LB multilayers consisting of thiacyanine (TC), SP and indocarbocyanine (IC) were prepared (Figure 7). When the photochromic LB film is irradiated with UV light (363 nm), SP is converted to merocyanine (MC). This sequence of chromophores TC-MC-IC gives substantial spectral overlaps between absorption of acceptors and fluorescence of donors, then the Förster excitation transfer takes place in this pathway. On the other hand, when the LB film is irradiated with green light (545 nm), MC is converted to SP, and the sequence of the energy transport is closed. It follows that the fluorescence emissions from particular layers can be switched by irradiation of UV or green light as is shown in Figure 8: Under the UV irradiation, i.e., the sequence of the excitation transport is opened, the fluorescence is emitted from IC at 725 nm, but disap-

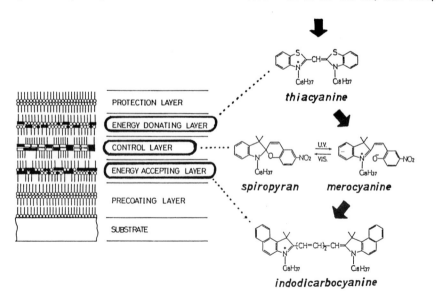

FIGURE 7
Stacking structures of the photochromic LB films.

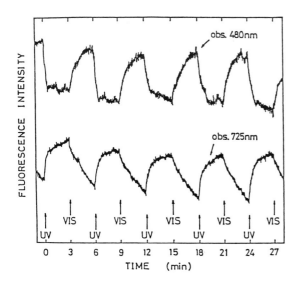

FIGURE 8
Photoswitching of the fluorescence from the donating
and accepting layers monitored at 480 nm and 725 nm,
respectively. The fluorescence intensities are regu-
lated by UV or green light
irradiation.

pears in TC at 480 nm; under the green light irradiation, i.e., the sequence is
closed, the fluorescence disappears in IC, but emits from TC. It is seen from
Figure 8 that the fluorescence intensities are switched in an out-of-phase between the donor layer and the acceptor layer. This type of LB films may function as a 2D-pattern logical device.

6. CONCLUDING REMARKS
   Our future aims and goals are the fabrication of molecular optical devices

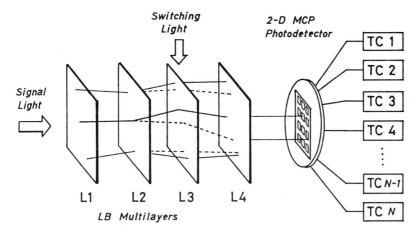

FIGURE 9
Model of an excitation transfer network and optical switching device equipped
with a 2D photon-counting detector as spatial- and time-correlators (TC).

capable of spatially controlling and switching photoexcitation transport. The net processes are illustrated schematically in Figure 9. To achieve this, improvement and development are needed in sample materials and optical detection techniques. In a sophisticated molecular assembly, LB multilayer film, there are several techniques that are able to control the chromophore distribution and orientation specifically, by introducing cyclodextrin LB films and polymer LB films. As for the detection system, a 2D photon-counting detector is neces sary for investigating spatial- and time-correlation of photonic energy transport in the multilayered molecular network (Figure 9).

ACKNOWLEDGEMENT

The authors acknowledge Professor Shigeo Tazuke for his helpful discussion and continued encouragement on this study.

REFERENCES

1) N. Tamai, T. Yamazaki and I. Yamazaki, Fractal Behaviors in Two-Dimensional Excitation Energy Transfer on Vesicle Surfaces, in: Ultrafast Phenomena V, Springer Ser, Chem. Phys. Vol. 46, eds. G.R. Fleming and A.E. Siegman (Springer, Berlin, 1986) pp. 449-453.

2) I. Yamazaki, N. Tamai and T. Yamazaki, J. Phys. Chem. 94 (1990) 516.

3) J. Klafter and A. Blumen, J. Chem. Phys. 80 (1984) 875.

4) N. Tamai, T. Yamazaki and I. Yamazaki, Can. J. Phys. 68 (1990) 1013.

5) T. Minami, I. Yamazaki, T. Yamanaka and N. Tamai, Photonic Energy Transport and Switching in Photoresponsive LB Films, in: Proceedings of the MRS International Meeting on Advanced Materials, Vol. 12, eds. M. Doyama, S. Somiya, R.P.H. Chang and S. Tazuke (MRS, Pittsburgh, 1989) pp. 267-272.

Photochemical Processes in Organized Molecular Systems
K. Honda (Editor-in-Chief)
© Elsevier Science Publishers B.V., 1991

MICROPHOTOCONVERSION:
EXPLORATORY CHEMISTRY BY LASER AND MICROFABRICATION

Hiroshi MASUHARA[†]

Microphotoconversion Project, Exploratory Research for Advanced Technology
(ERATO) Program, Research Development Corporation of Japan, 15 Morimoto-
cho, Shimogamo, Sakyo-ku, Kyoto 606, Japan

A future possibility of spatially resolved photochemistry by laser and an importance
of micrometer dimension are pointed out and discussed. What are the subjects and
how to do chemistry in micrometer-size volume are proposed, and preparation of
micrometer reaction field is considered. As practical tools, micrometer space-
resolved and picosecond time-resolved spectroscopy and laser manipulation have
been developed and applied to polymer systems. As an ultimate goal of the present
work, we explore an integrated chemical system where micrometer reaction fields
are arranged spatially, each of them corresponds to one reaction step, and
reactions are energized, interrogated, and controlled by laser pulse.

1. INTRODUCTION

Photochemistry is now recognized as a superior approach to thermochemistry for
elucidating reaction mechanism. Chemical reaction consists of many steps and is
understandable as a path from reactants to products. Reaction starts from a particular
electronic, vibronic, as well as rotational energy levels of reactants and leads to
products with specific energy levels. If we excite molecules with narrow energy
resolution, we can specify electronic structures of both initial and final species. This
state-to-state chemistry has been studied in the gas phase with high resolution
spectroscopy. On the other hand, chemical reaction processes can be directly followed
by fast spectroscopic method, when the reaction is triggered with a very short laser
pulse. This chemistry of ultrafast phenomena is another exciting field recently
developed. It is well known that these sophisticated chemistry has been realized in
photochemistry, particularly, by introducing lasers with high energy and temporal
resolution as excitation and monitoring light sources. In Table 1 characteristics of
photochemistry and thermochemistry are summarized and compared with each other[1].

Most of the studies on photochemistry have been performed in solution and gas
phase, so that spatial resolution, another important characteristic of lasers, has not
received a lot of attention. In the cases of molecular assemblies and organic solids,
however, photochemistry should have a spatial meaning. Some minute areas can be

---

[†] Kyoto Institute of Technology, Matsugasaki, Kyoto 606, Japan

H. Masuhara

TABLE 1. Characteristics of photochemistry and thermochemistry

	Photochemistry	Thermochemistry
energetic resolution	10^{-5} eV	Boltzmann distribution
temporal resolution	10^{-14} sec	> sec
spatial resolution	sub μm	cm
energy density	TW/cm^2 sec	KW/cm^2 sec
temperature	low temperature (room)	high temperature
mode	non-contact	contact

TABLE 2. Physical quantities in μm and cm dimensions

	μm	cm
Number of Molecules		
Gas	2.7×10^7 /μm^3	2.7×10^{19}/cm^3
Liquid	6×10^9 /μm^3	6×10^{21}/cm^3
Weight	$\sim 10^{-12}$ g	~ 1 g
Surface Area	6 μm^2	6 cm^2
Electric Field	1 MV/cm	10^2 V/cm
Atmospheric Pressure	10^{13} Pa/cm^2	10^5 Pa/cm^2
Photon Density	10^2 KJ/cm^2	1 mJ/cm^2
Diffusion Time	100 μs/μm	10 hr/cm
Mean degree of Polymerization	5×10^3	5×10^7

These quantities are calculated under normal conditions such as room temperature and atmospheric pressure. The same field, pressure, and photon numbers are focused into μm and cm dimensions.

excited selectively by focused laser beam, while the neighborhoods are still remained undisturbed. It is also possible to drive and interrogate chemical reaction as a function of position of the sample.

In the above, we have described photochemistry and spectroscopy from energetic and temporal viewpoints, namely, laser has been used as a light source for reactions and interrogations. On the other hand, photochemistry is now extended to control chemical reaction by changing conditions of reaction field under photoillumination. Polarity, pH, and viscosity of solvent are adjusted, adsorption/desorption equilibrium and hydrophobic/hydrophilic property of surface are controlled, and solubility, conductivity, and morphology are changed by illuminating photoresponsive molecules. If key molecules, which isomerize or ionize upon excitation, are connected to polymer chain, their structural and chemical changes are amplified to bulk physical properties[2]. These are very interesting phenomena which can be used for controlling chemical reactions.

We expect that a new photochemistry will involve energetic, temporal, and spatial resolutions and laser will be used for energizing, interrogating, and controlling reactions. From this idea, we propose an exploratory study on a new integrated chemical system where minute reaction fields are arranged spatially and the reaction step in each field is driven, measured, and controlled with laser. In this system temporal and spatial arrangements of the minute reaction sites are the most important. Microfabrication techniques such as microlithography, laser ablation, chemical vapor deposition, and scanning electrochemical microscope are fruitful for preparing the microchemical reaction sites. Namely, we explore a new chemistry by utilizing laser and microfabrication.

In this review we point out the importance of spatial resolution in chemistry, summarize our ideas, describe the new tools of space- and time-resolved spectroscopy and manipulation, and propose Microphotoconversion.

2. IMPORTANCE OF SPATIAL RESOLUTION IN CHEMISTRY

In the history of chemistry, it has been believed that μm is just the bulk. The dimension of molecules is extremely smaller than μm. Every chemist has wished to see molecules directly in nm order, which is now possible by scanning tunneling microscope. Langmuir-Blodgett films where molecules are oriented in nm dimension are now widely used. Lipid, protein, and well-designed, synthetic molecules undergo self-organization, when dispersed in water, and form organized system with nm resolution. Then, we have to ask whether μm is just the same as bulk or not in chemistry. Before discussing our idea, the basic parameters and properties in μm and cm (the dimension of the bulk) are summarized in Table 2.

Weight and surface area are, of course, diminished from cm to μm. The number of molecules contained in cubic μm cell reduced from cubic cm cell by many orders of magnitude. It is worth noting that the number of molecules which can be excited by light is limited in μm dimension, so that the total number of molecules in the excited state becomes saturated. This is a sever condition for getting nice S/N value in spectroscopic measurements. It is considered that direct mesurements in ns and ps time domains require at least μm small dimension.

Photon density in μm minute area can be quite easily increased up to MJ/cm^2 even by using conventional lasers. As the dimension is reduced from cm to μm, electric field and pressure are enhanced by 10^4 and 10^8 times, respectively. Namely, strong perturbations upon chemical reaction are applicable in μm-size volume. This means that a new chemical behavior is expected, and a new way to control chemical reactions is possible.

It is basically important that the dimension of polymers and molecular assemblies are sometimes close to μm. A high polymer having 5×10^3 mean degrees of polymerization may have a length in the order of μm if it is completely extended. In cm-order this polymer is still deemed to be a point and the polymer with comparable length to cm may have 5×10^7 degrees of polymerization. Molecular assemblies such as vesicles, membranes, colloids, micelles, and so on may have a scale close to μm. Therefore, the conformation of polymers and the structure of molecular assemblies can be strongly influenced by the wall of a μm-size cuvette.

In the case of small molecules diffusion motion amplifies the nm dimension up to μm. Diffusion time in solution can be estimated by assuming the same diffusion constant as that of the cm bulk. Using $x = \sqrt{2Dt}$, where x is the distance, D is the diffusion constant of molecules, and t is the time during which the molecule crosses

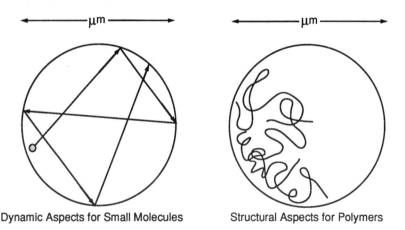

Dynamic Aspects for Small Molecules   Structural Aspects for Polymers

Figure 1
Size effect upon small molecules and polymers

over x, the value of t for μm and cm are calculated. As shown in Table 2, it takes only 100 μs for μm while hours for 1 cm, using D ~ 10^{-4} cm^2/sec. This indicates that even small molecules are affected dynamically by the wall in a μm cubic cell. This is just the same as the subject of dynamics in confined geometry[3]. These conditions are illustrated in Figure 1.

We believe that μm is a specific dimension for analyzing dynamic structure and chemical reaction of various molecular systems. Another important reason why μm is crucial dimension is that diffraction limit of laser light is just the order of wavelength, namely, sub μm. Concerning the time resolution, it is well known that electron as well as energy transfer, proton transfer, isomerization, and dissociation in condensed phase take place in the time region of ps. Vibrational relaxation and solvation dynamics can be followed with the sub ps resolution. Sophisticated lasers and detection methods using multi-channel diode array, single photon counting electronics, streak camera, and frequency up-conversion are now available, so that ps time resolution should be satisfied even in μm-order small volume.

We know that μm and ps themselves are not an advanced measure for spatial and temporal resolution, respectively. At the present stage of investigation, scanning tunneling microscope makes it possible to see molecules in nm order, and femtosecond laser is available. Nevertheless, we believe that it is very important and reasonable to satisfy μm and ps resolutions simultaneously, because ps dynamic analysis in nm-order area is impossible. We have to remind that flash and laser photolyses have made great contributions for elucidating chemical reaction mechanisms, giving spectroscopic and kinetic information of transient species[4]. Namely, step-by-step following of a reaction is indispensable for understanding dynamics, and of course this is also true for the studies in minute small areas. Furthermore, chemical modification of solid surface, preparation of reaction field, and manipulation of particles and liquid droplets are possible in the μm-order dimension. We consider now that μm-ps spectroscopy and chemistry will be a fundamental basis for the studies on dynamic structure and reaction of various molecular systems.

3. WHAT AND HOW IN MICROMETER CHEMISTRY

3.1. What Chemistry in Micrometer-size Volume

Generally it is believed that liquid properties such as viscosity, polarity, and temperature are uniform in solution. However, this is questioned if we look at the solution with small dimension. When we change the liquid property only at some positions in the three-dimensional space, there arises a nonequilibrium array of minute reaction volumes with different characters. Near the wall, this is already realized without any perturbations, which is confirmed for porous Vycol glass and polymer gel[3]. The most powerful method for analyzing these liquid properties is to monitor fluorescence dynamics. Polarity and viscosity near the wall of the cell and at the liquid/liquid interface will be particular, and adsorption/desorption of molecules may change their effective concentration in the minute volume. Fine microdroplet or capsule is one of such systems. Surface tension might induce a special pressure effect upon the dynamics in the small liquid droplet. We have started our research by studying elementary processes such as fluorescence depolarization and excimer as well as exciplex formation in μm-size volume. Dynamics of rotational diffusion and association as well as dissociation of molecules is well elucidated in the bulk solution, and their analysis in μm-size reaction volume is expected to show the characteristics[5,6].

Ionic species like contact ionpair, solvent-separated ionpair, and free ion radicles are expected to show characteristic behavior and will be another nice probe for elucidating chemical reactions. Particularly, ion radicals are the most fruitful candidate, since they have long lifetime and are very sensitive to interfacial effect This possibility will be confirmed if transient absorption spectral measurement in μm-size dimension will be introduced

Chemical reaction in solution can be controlled by changing viscosity, polarity, pH, reactant concentration and so forth with laser excitation. Irie published a series of reports concerning development of photoresponsive molecules in solution and their introduction to polymers[2]. Cis-trans isomerization of stilben and azobenzene, zwitter ion formation, radical formation, ring-formation and ring-cleavage, photoionization of leuco marachite green, and inclusion or exclusion of metal ions in crown ether derivatives are key molecular reactions for controlling solution properties upon photoexcitation. pH and reactant concentration are changed by dissociating H^+ and metal ions from mother molecules under photoexcitation. Micropolarity is also photochemically controlled. This will be a unique method for controlling chemical reactions in the μm-size volume. Particularly, it is quite challenging to follow the reaction by fast kinetic spectroscopy.

One more interesting thing is to synthesize polymer gel or film containing photoresponsive molecule and to change their volume by photo-illumination. Dynamic aspects of photoinduced volume change can be studied in small dimension[7].

It is now strongly expected that many interesting behavior will be seen in μm-order volume, and detailed spectroscopic analysis should been done. Such a study has never been given as far as we know, which means that we have many subjects to study dynamics in the μm-size dimension at the present stage of investigation. The nature of dynamic structure and chemical reaction in small volume is a scientifically important problem.

3.2. How to do Chemistry in Micrometer-size volume

In order to measure and to conduct reaction in the μm-size volume, we need some methods with spatial and temporal resolution. Furthermore, some tools to choose small materials as well as reagents, to transfer them to a certain point with µm resolution, and to induce reactions are required. We have started to develop these in order to realize the chemistry we described in the above section.

The first is dynamic microspectroscopy which enables us to measure fluorescence and transient absorption in the small geometry. Time resolution should be in ps-order, because elementary processes occur in this time region. Namely, 3-dimensional space- and time-resolved fluorescence spectroscopy, using confocal laser scanning microscope and time-correlated single photon counting, has been already reported[8] and described below. Picosecond microphotolysis, giving transient absorption spectra in the µm-size reaction volume, is also possible under a confocal laser scanning microscope[9]. Surface, interface, and their thin layers are also very important minute areas for space-resolved spectroscopy. Variable-angle time-resolved total internal reflection fluorescence spectroscopy makes it possible to analyze dynamic structure and chemical reaction in the interface layers[10,11]. A depth profile of chemical reaction will be shown. Some nonlinear spectroscopy methods are also a fruitful candidate, for example, surface second harmonic generation is well known and utilized for many purposes. These spectroscopic techniques will be extended to a new methodology to control chemical reaction at minute confined geometry.

The second tool is a manipulation method like hand. It is impossible to conduct in chemical reaction in µm-size dimension by conventional glass, pipette, beaker, spatula, while laser can do everything. By utilizing laser trapping phenomena, it is now possible to choose various kinds of particles such as liquid droplets, latex particles, catalyses, beads, and biocells, and to fix them at a certain position in the three dimensional space[12]. Therefore, laser manipulation will receive a lot of attention from chemical viewpoint.

Some miniature devices of valve and cock are also needed to conduct chemical reactions in µm-size volume. In order to switch on or off the device, photocontrol of gels is again a fruitful candidate. It is now possible to prepare photoresponsive gels which change shape and expand in µm order upon excitation[5,7]. If the quick response is attained, we will be able to construct a system where photoresponsive gels are parts of chemical conversion system. Innovations of these microchemical machines coincide with those of micromachine, microoptics, microelectronics, microsurgery, and so forth.

Dynamic microspectroscopy and laser manipulation are indispensable tools for chemistry in µm-size volume. Their development and application have been started already in this Project, and some of the results will be described later.

3.3. How to Prepare Micrometer-size Reaction Field

Chemical reactions proceed in reaction fields and are very sensitive to surrounding environmental conditions. In general reaction rates and yields are determined/controlled by reaction fields, which is clearly demonstrated by solvent effects upon reactions. Therefore, we consider that preparation of μm-size reaction field is critical for conducting chemistry in minute spot.

The first trial for preparing the reaction field is photophysical modification of glass, polymer, semiconductor, membrane, and other various kinds of solids. Using lasers, it is possible to modify small surface, since laser light can be focused to the wavelength order. Laser annealing is the most simple method which can change aggregation and orientation of molecules/atoms in the solid. The new physical properties are expected only in the irradiated region.

More chemical approach than structural change is to plant molecules on the surface. Photochemical modification of materials surface is very promising since only minute region irradiated can be modified[5,13]. When the photofunctional molecules are introduced in some small areas, physical and chemical properties of those parts can be controlled by laser pulse.

Another approach is to fabricate the materials surface and form a chemically meaningful pattern. Conventional microlithographic technique is the most general method to prepare a special pattern on the solid surface. Recently scanning tunneling microscope is used not only to observe the surface structure but also to prepare minute line or circuit. This will be a new attractive tool for creating minute reaction field.

Further sophisticated approach is based on simultaneous fabrication and modification of organic solid. Laser ablation resulting μm-order pattern is now

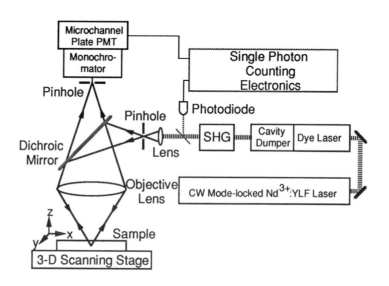

Figure 2
Dynamic fluorescence microspectroscopy system

possible. The bond scission is induced densely, which means that concentrated radicals, reactive with foreign molecules, are formed on fabricated surface. Therefore, when ablation is performed in solution containing some reagents or under reactive gas atmosphere, molecules are introduced exclusively onto the fabricated surface[5,14,15].

Microlithography, scanning tunneling microscope, laser ablation, and related techniques have been used mainly for electronics engineering, however, chemical application will be very fruitful. Preparation of minute reaction field will be the first step for constructing an integrated chemical system where minute reaction fields are arranged spatially, matching to temporal behavior of chemical reaction.

4. DYNAMIC FLUORESCENCE MICROSPECTROSCOPY SYSTEM

This system is based on two techniques[8]. One is a confocal laser scanning microscope recently proposed and widely used in the field of optics. Since two pinholes are set in the microscope, this microscope does not suffer from interference effect of laser beam, giving good S/N value. Another advantage is that a temporal response function of ps laser and fluorescence does not become diverse after passing the optics. This is an important factor for getting high time-resolution. The third advantage is that this microscope enables us to measure fluorescence as a function of the 3-dimensional position. The other technique is a time-correlated single photon counting which is well-known for photochemists.

A schematic diagram of our system is shown in Figure 2. The second harmonic of a cavity dumped dye laser (Coherent 702-1 dye laser and 7220 cavity dumper), synchronously pumped by a CW mode-locked Nd:YLF laser (Quantronix 4217 ML), was used as an excitation light source. The pulse width and repetition rate of the 290 nm light were 2 ps and 3.8 MHz, respectively. The laser beam was condensed by a

TABLE 3. Performance of dynamic fluorescence microspectroscopy system

Spatial resolution	
depth	0.5 μm
lateral	0.3 μm
Temporal resolution (instrumental response)	2 ps (33 ps)
Wavelength resolution (wavelength region)	1 nm (300-1000 nm)

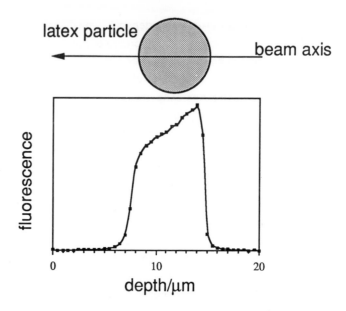

Figure 3
Fluorescence profile of pyrene-doped,
single lattex particle

lens and led to an entrance pinhole of a confocal microscope (Zeiss UMSP-50). This
lens is a zoom one which matches the beam diameter with a numerical aperture of the
microscope. The beam is reflected and focused onto a sample. Fluorescence from the
sample is collected by the same objective lens and led to the other pinhole in front of a
detector. Fluorescence from defocused areas of the sample can not form an image on
the pinhole, so that it is mostly cut off. Thus, depth information is obtainable in addition
to lateral resolution. Moving the microscope stage 3-dimensionally, we can measure
ps fluorescence rise and decay curves as a function of position. The performance of
this system, confirmed experimentally by analyzing fluorescence data of model
samples, is summarized in Table 3.

By using the fluorescence microspectroscopy system, it is possible to estimate the
chromophore concentration in the micrometer-size volume. Fluorescence intensity is,
of course, proportional to excitation intensity, so that depth profile of fluorescence
intensity shows how excitation light is absorbed along the depth. Assuming the
Lambert-Beer equation it is very simple to calculate the chromophore concentration in
the small volume. In Figure 3 fluorescence profile of pyrene doped in a poly(methyl
methacrylate) latex particle is shown. The pyrene concentration was calculated to be 5
x 10^{-3} M. The microscopy system has a depth resolution, which means that any

scattering and absorbing materials in front of or behind the focussed particle does not affect the measurement. Namely, absorption distribution in any spot in the 3-dimensional space can be measured.

Another application is to study excimer kinetics in a small liquid droplet, which is described in the preceding chapter[5].

5. LASER MANIPULATION IN MICROMETER-SIZE VOLUME

It was first demonstrated by Ashkin that a μm-size particle could be trapped by a focused laser beam[12]. The conditions are (i) the refractive index of the particle is larger than that of the surrounding environments, and (ii) the laser beam should be focused into a small spot with a lens with large NA value. This laser trapping phenomenon can be explained with a schematic diagram shown in Figure 4. At the interface the laser beam refracted, namely, transverse momentum changes its direction by $\Delta P = P_1 - P_0$. According to conservation rule of momentum, the particle should receive the momentum change $-\Delta P$. The sum of these vector changes results in the power, moving the particle to the focus point.

In order to develop the technique as a chemical manipulation tool, we introduce the second laser beam for inducing chemical reaction and for fabricating the particles[16]. A schematic diagram of our system is shown in Figure 5. A 1064 nm laser beam from a CW Nd:YAG laser (Spectron SL-903U) was introduced to an optical microscope (Nikon, Optiphoto XF, objective lens X 100, focused into ca. 1μm spot) as a trapping light source. The third harmonic of a Q-switched Nd:YAG laser (Qunta-Ray, DCR-II, fwhm=7 nm, 355 nm) was coaxially led to the microscope and used as an excitation pulse. Time-resolved fluorescence spectra are recorded by a photodiode array (Princeton Instruments, DSIDA). Fluorescence rise and decay curves in ps time region can be measured by a microchannel plate photomultiplier connected to single photon counting electronics.

Laser trapping behavior of a poly(methyl methacrylate) latex particle in water is shown in Figure 6(a). The trapping laser beam was introduced perpendicularly to the plane of the photograph. Since the photograph in Figure 6 was taken with the XY stage being moved along the X direction, untrapped particles were transferred along this direction while the trapped one was fixed at the same position. Similarly, it was demonstrated that the particle was trapped and manipulated along the Y and Z directions as well. It was confirmed that various kinds of particle were trapped by the present system. Polymer latex particles, organic liquid droplets, oil droplets, glass beads, TiO_2 particles, and salmonella germs in water are such samples, while water droplets could not be captured in organic solvent[16,17].

Time-resolved fluorescence spectroscopic measurement of the trapped particle is now very easy, since the 3-dimensionally space- and ps time-resolved fluorescence

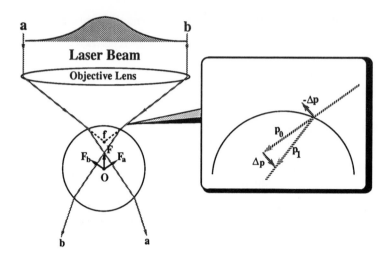

Figure 4
Schematic diagram of momentum transfer in laser trapping

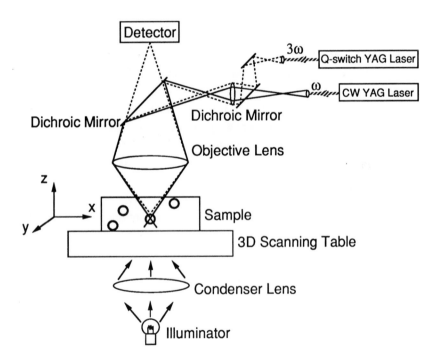

Figure 5
Laser trapping-spectroscopy-ablation system

(a)

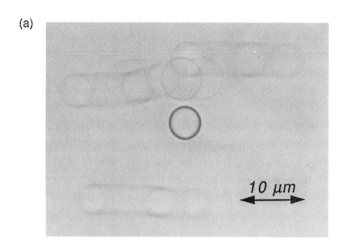

(b)

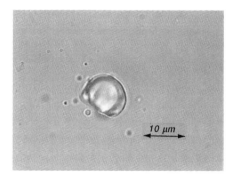

Figure 6
Laser trapping and ablation behavior of small particles
(a) latex particles, (b) an ablated microcapsule and
toluene droplet poured out from the capsule

measurement is available. Adjusting the optical condition of the excitation laser beam, a certain position of the particle, surface or inner part, can be chosen and its fluorescence can be examined. This means that photochemical modification of the surface is possible and photoinduced chemical reaction can be conducted in any position in the trapped particle.

Increasing excitation intensity, the irradiated part of the trapped particle is easily ablated, a hole being formed, and fragmented materials are ejected into the bulk water. Laser ablation of the trapped microcapsule containing toluene solution of pyrene is shown as an example in Figure 6(b). Just upon ablation the microcapsule was broken and the toluene solution flowed out. We will be able to transfer a single microcapsule to a certain position in the 3-dimensional space, to pour out inner solution, and to modify chemically the position with the solution. Under the trapping condition we can observe the whole behavior of laser ablation of particles undergoing Brownian motion and follow the reaction.

The present system composed of two lasers, a microscope, and optics are named as a laser trapping-spectroscopy-ablation system. Our laser manipulation, using this system, has made it possible to obtain molecular as well as electronic information on the trapped particle, to induce photochemical reaction, and to fabricate it. This means chemical manipulation in μm-size volume is now realized.

6.   MICROPHOTOCONVERSION: OUR ULTIMATE GOAL

Chemistry in μm order dimension have not received a lot of attention. This is because typical subject in chemistry concerns with molecular design, preparation of molecular assemblies like LB film, high-order structure of polymers, and so on, and all ideas are based on nm dimension. It is worth noting, however, that dynamic aspects of chemical reaction needs a large volume where molecules can diffuse, encounter with each other, and separate from the reaction partner. We believe that μm is a very important spatial dimension for conducting chemical reactions.

Now it is possible to prepare minute reaction fields by using microfabrication techniques. Characterization of molecules and detailed analysis of chemical reactions can be performed in the μm-size volume, and the ps time-resolution is attained just like in the bulk kinetics. Photofunctional molecules can be planted onto small area of materials surface, which will be used in controlling chemical reaction by laser light. Now we consider that μm-ps chemistry, including spectroscopy, fabrication and reaction, will be an indispensable basis for the studies on dynamic structure and chemical reaction.

In general, chemical reaction consists of several elementary processes in combination. For each process the most suitable environmental condition could be more easily adjusted in the case of a minute reaction field compared to the bulk.

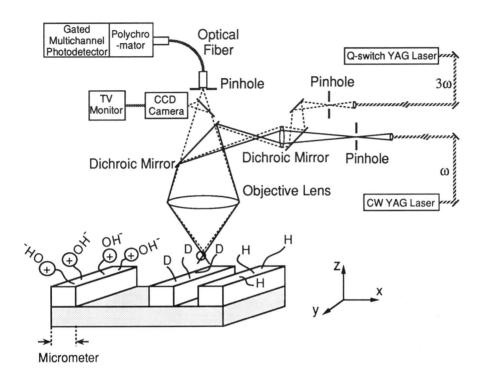

Figure 7
Schematic diagram of Microphotoconversion system,
a kind of micro factory

Therefore, the selectivity and high efficiency will be attained. One exploratory idea we proposed is to create an integrated chemical system where one reaction field corresponds to one reaction step and reactions are controlled by laser light. A schematic diagram of this system is shown in Figure 7. Minute reaction fields are spatially arranged where reaction proceeds sequentially along the array of the reaction fields. One can imagine biological cells where functional organs are embedded purposively. In our case every reaction field is prepared artificially, reaction step characteristic of μm-size volume is conducted, and reactions are energized, manipulated as well as controlled completely by laser light. This novel molecular and materials conversion system, named Microphotoconversion system, will be a prototype of micro factory, which is our ultimate goal. Furthermore, the present research will also contribute to various kinds of technology related to lasers and materials in addition to chemistry.

ACKNOWLEDGEMENT

The author wishes to express his science thanks to Drs. Kitamura, Misawa, and Sasaki for their invaluable discussion.

REFERENCES
1) T. Matsuura, News Letter of the Special Research Project "Frontiers of Highly Efficient Photochemical Processes" supported by the Grant-in-Aid from the Ministry of Education, Science, and Culture, Japan 1 (1986) 1.

2) M. Irie, Advances in Polym. Sci. 94 (1990) 28.

3) J. Klafter and J. M. Drake, Molecular Dynamics in Restricted Geometries (John Wiley & Sons, NY, 1989).

4) S. Claesson, Fast Reactions and Primary Processes in Chemical Kinetics (Almqvist & Wiksell, Uppsala, 1967); H. Masuhara, Application of lasers to transient absorption spectroscopy and nonlinear photochemical behavior of polymer systems, in: Lasers in Polymer Science and Technology: Applications vol. II, eds. J. P. Fouassier and J. F. Rabek (CRC Press, Boca Raton, 1990) pp. 235-260.

5) N. Kitamura and H. Masuhara, Micrometer photochemical dynamics in organized molecular systems, this volume.

6) H. Misawa, M. Koshioka, K. Sasaki, N. Kitamura, and H. Masuhara, Abstract of the Symposium on Photochemistry Jpn. (Kyoto, 1990) p 7.

7) M. Ishikawa, N. Kitamura, and H. Masuhara, Abstract of the 13th IUPAC Symposium on Photochemistry (Coventry UK, 1990) p 285.

8) K. Sasaki, M. Koshioka, and H. Masuhara, Abstract of the 37th Symposium on Appl. Phys. Jpn. (Saitama, 1998) p 819; Technical Digest of the 17th Int'l Quantum Electronics Conf. (Anaheim, 1990) p 170.

9) M. Koshioka, K. Sasaki, and H. Masuhara, Abstract of the Symposium on Photochemistry Jpn. (Kyoto, 1990) p 307.

10) M. Yanagimachi, M. Toriumi, and H. Masuhara, Abstract of the 13th IUPAC Symposium on Photochemistry (Coventry UK, 1990) p 289.

11) H. Masuhara, N. Mataga, S. Tazuke, T. Murao, and I. Yamazaki, Chem. Phys. Lett. 100 (1983) 415; H. Masuhara, S. Tazuke, N. Tamai, and I. Yamazaki, J. Phys. Chem. 90 (1986) 5830.

12) A. Ashkin, Phys. Rev. Lett. 24 (1970) 156; Science 210 (1980) 1081; A. Ashkin, J. M. Diedzic, J. E. Bjorklom, and S. Chu, Opt. Lett. 11 (1986) 288.

13) N. Ichinose, N. Kitamura, and H. Masuhara, Chem. Lett. (1990) 1945.

14) N. Shimo, T. Uchida, and H. Masuhara, Foreign gas effect upon excimer laser ablation of polymer, in: Laser Ablation for Materials Synthesis, Mat. Res. Soc. Symp. Proc. 191, eds. D. C. Paine and J. C. Bravman (Materials Research Society, Pittsburgh, 1990) p 91.

15) T. Uchida, N. Shimo, and H. Masuhara, Abstract of the 13th IUPAC Symposium on Photochemistry (Coventry UK, 1990) p 290.

16) H. Misawa, M. Koshioka, K. Sasaki, N. Kitamura, and H. Masuhara, Chem. Lett. (1990) 1479.

17) H. Misawa, N. Kitamura, and H. Masuhara, Polym. Preprint 39 (1990) 3106.

Photochemical Processes in Organized Molecular Systems
K. Honda (Editor-in-Chief)
© Elsevier Science Publishers B.V., 1991

MICROMETER PHOTOCHEMICAL DYNAMICS IN ORGANIZED MOLECULAR SYSTEMS

Noboru KITAMURA and Hiroshi MASUHARA[†]

Microphotoconversion Project, Exploratory Research for Advanced Technology (ERATO) Program, Research Development Corporation of Japan, 15 Morimoto-cho, Shimogamo, Sakyo-ku, Kyoto 606, Japan

Chemical and physical processes characteristic to μm-order dimensions were demonstrated. As μm-size effects, the photoresponse time of dilation of microgels and the efficiency of pyrene excimer formation in liquid droplets dispersed in water were shown to decrease with reducing their size to μm. Simultaneous fabrication and modificaiton of materials surface was explored to prepare micropatterns on the surface with photofunctional molecules. Besides bulk materials, microfabrication of individual small particles dispersed in solution was also demonstrated to be possible by the currently developed laser micromanipulation technique for the first time. Importance of μm-ps chemistry is discussed on the basis of the present results.

1. INTRODUCTION

Chemical and physical properties of organized molecular systems such as polymers, Langmuir-Blodgett films, and so forth are characterized by their microstructures (i.e., microphase separation, three-dimensional structures, interpenetration networks, etc.) and such inhomogeneity is common for various materials. Nevertheless, the modes of chemical reactions and/or molecular interactions occurring in microstructures of these molecular systems are generally believed to be the same with those in bulk. Such an assumption may not be warranted and, indeed, the phenomena characteristic to μm domains have been sometimes reported. For example, it is well known that the conformation of a polymer chain in solution is largely influenced by a vessel wall in μm-order dimensions.[1] Electronic conductivity of polypyrrole or poly(3-methylthiophene) fibers has been reported to increase with decreasing the diameter of the fibers to sub-μm.[2] Analogous μm-size effects are expected in various systems, in particular, for dynamic processes in which short lived transient species are likely to be influenced by μm-order structures.[3] From a

---

[†]Permanent address: Kyoto Institute of Technology, Matsugasaki, Sakyo-ku, Kyoto 606.

synthetic view point, furthermore, elucidation of characteristic properties in μm domains will lead to future design of a series of μm functional sites on materials surface as well as to control of chemical processes occurring in μm reaction sites.

For basic understanding of the characteristic features in μm dimensions space- and time-resolved spectroscopy is clearly indispensable. As discussed in the preceding chapter, laser can be focused onto a μm spot with a short pulse width of ns~subps, so that only photochemical/photophysical techniques can be applied to these studies.[3] Furthermore, lasers provide potential means to prepare μm structures by using lithographic or ablation technique.[4,5] We expect that these techniques will open a new research field of μm-ps chemistry, including fabrication of a series of spatially arranged μm functional sites with a variety of photoresponsive molecules. Along the strategy, we are exploring i) μm-size effects on photochemical processes, ii) μm photochemical patterning and modification of polymer surface, and iii) microfabrication of individual small particles. In this review, we describe our recent results on these subjects and demonstrate the importance of space- and time-resolved chemistry.

2. MICROMETER SIZE EFFECTS ON PHOTOCHEMICAL PROCESSES

In μm-order dimension, chemical and physical processes are expected to be different from those in the bulk as discussed in the preceding chapter[3] and in the introduction of the review. To study μm-size effects on photochemical processes, however, conventional techniques are not applicable and every experiment should be conducted under a microscope. Our space- and time-resolved spectroscopy[3,6] and laser micromanipulation techniques[3,7-9] are indispensable to demonstrate μm-size effects as described below.

2.1. Photoinduced Volume Expansion of Polymer Gels[10]

Polyacrylamide (AA) gels containing triphenylmethane leuco derivatives (TPMX where X is OH or CN) are known to show a large photoinduced volume change; dilation.[11] For mm-order disk-shaped gels, the time necessary to reach the equilibrium gel volume upon photoillumination has been reported to be the order of hour.[11,12] Recently, we showed that the photoresponse time was very much improved by reducing the size of rod-shaped AA-TPMX gels (100:1 molar ratio) to μm.

Photoirradiation of the AA-TPMCN gels in pure water and monitoring of the volume change were performed under an optical microscope equipped with a CCD camera and a TV monitor. The photoresponse time (α) and the equilibrium volume ratio, V_∞/V_0, where V_0 and V_∞ are the volume of the microgel at t (irradiation time) = 0 and ∞, respectively, are summarized in Table 1. The most important finding is that α is faster for the smaller size of the gel with V_∞/V_0 being almost constant around 5.6 ~ 6.0 irrespective of the initial radius (R_0) of the microgel. For the microgel with R_0 = 11 μm,

Table 1. Size Effects on Photoinduced Volume Change of AA-TPMCN
Microgels in Water

Ro (μm)	V_∞/V_0	α (sec)
180	6.0	1080
90	5.8	250
47	5.6	110
10	5.6	11

α was 11 sec, while that with R_0 = 180 μm was 1080 sec. It is apparent that microgels are advantageous with respect to the fast response to the volume change upon photoirradiation.

Photoinduced dilation of the present rod-shaped microgels was analyzed by the following equations.

$$[R_\infty - R(t)] = (R_\infty - R_0)(6/\pi^2)\sum_{n=1}^{\infty} n^{-2} \exp(-n^2 t/\alpha) \qquad (1)$$

$$\alpha = R^2_\infty/\pi^2 D \qquad (2)$$

R_∞ and $R(t)$ are the radii of the microgel at $t = \infty$ and $t = t'$, respectively. α is related to the diffusion coefficient of the gel network, D. These equations have been proposed by Tanaka and Fillmore to analyze the dilation process of spherical polyacrylamide gels in water.[13] When $t/\alpha > 0.25$, eq. 1 is simplified to eq. 3 where A is constant under the present assumption.

$$[R_\infty - R(t)]/(R_\infty - R_0) = A \exp(-t/\alpha) \qquad (3)$$

Plotting $[R_\infty - R(t)]$ and t for the present data, we obtained α from the slope as summarized in Figure 1. The relationship between α and R_∞ is explicable by eqs. 2 and 3, indicating that the photoinduced volume change of the microgel is governed by the diffusion of the gel network. The diffusion coefficient, D, was determined to be 1.2 x 10^{-7} cm²/sec, which was slightly smaller than the reported value for polyacrylamide gel (D = 2.4 ~ 4.0 x 10^{-7} cm²/sec).[14-16]

The change in microstructures of the gels during photoinduced dilation is of primary importance to understand the diffusion of the gel network as well as the μm-size effects. The fluorescence lifetime of TPM cation is governed by the rotational relaxation of the phenyl ring in the molecule and therefore, by the microviscosity around the dye. We

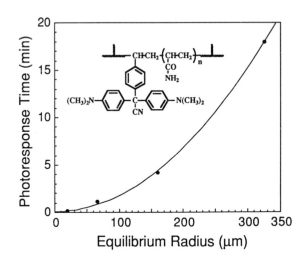

FIGURE 1
Photoinduced volume change of AA-TPMCN microgels in water

studied fluorescence dynamics of thin film (20 μm thickness) AA-TPMCN (100:1 molar ratio) gels based on time-correlated single photon counting measurements.[17] Fluorescence from TPM cation in the gels showed non-single exponential decay, so that the average fluorescence lifetime (τ_{av}) estimated by integration of a whole decay curve with time was measured as a function of V_t/V_0, where V_t is the volume of the microgel at t = t'.

In the initial stage of dilation, τ_{av} was estimated to be ~30 ps. Knowing that the fluorescence lifetime of malachite green dye, a model compound of TPM cation in the gels, is ~ 1 ps in water, the rotation of the phenyl ring of TPM cation is highly hindered in the present gels. With increasing V_t/V_0 upon photoillumination, τ_{av} sharply decreases. The results manifest that dilation accompanies expansion of the gel network and uptake of water molecules, leading to relatively free rotation of the phenyl ring. Further photoillumination, however, had almost no effect on τ_{av} and gave ~ 10 ps at $V_t/V_0 > 4$. This corresponds to the viscosity around TPM cation to be ~ 7 cP as estimated from the viscosity dependence of the fluorescence lifetime of malachite green in various alcohols. Such local viscosity, ~ 7 cP, in the gel around $V_t/V_0 = 4 ~ 6$ seems to be very high (i.e., equilibrium volume $V_\infty/V_0 = 5 ~ 6$, Table 1). One of the possible reasons will be the fact that the rotational relaxation of the phenyl ring of TPM cation is restricted to some extent by the acrylamide main chain of the gels. The microstructures of water in the vicinity of TPM in the gels may be a possible reason as well. Since the gels are inhomogeneous in microstructures, we expect that time-resolved fluorescence spectroscopy with subμm spatial resolution will provide further information on dynamic structure and μm-size effect of the present AA-TPMCN gels.

2.2. Excimer Formation Dynamics in Liquid Paraffin Droplets[18]

μm-size effects were also observed for pyrene excimer formation in liquid paraffin droplets. To study the excimer formation dynamics in an individual droplet, we applied the currently developed laser trapping-picosecond spectroscopy system[3,7,8]. An idividual micrometer liquid paraffin/pyrene (1.1 x 10⁻² M) droplet dispersed in water was manipulated freely by focusing a 1053 nm laser beam from a CW mode-locked Nd:YLF laser on a droplet under an optical microscope (x 100, NA = 1.30). Various size of the droplets were selected by laser trapping technique and pyrene in a droplet was excited by the second harmonics (290 nm, 2ps, 3.8 MHz) of a cavity dumped dye laser, synchronously pumped by a CW mode-locked Nd:YLF laser. Fluorescence was detected by a microchannel plate photomultiplier and single photon timing electronics as described elsewhere.[6]

Fluorescence spectra in Figure 2 reveals the efficient pyrene excimer formation in a μm-order liquid paraffin droplet. However, the efficiency of the excimer formation is dependent on the size of the microdroplet. The μm-size effects on I_E/I_M, where I_E and I_M are the fluorescence intensity of excimer (470 nm) and monomer (384 nm), respectively, are summarized in Figure 3. The I_E/I_M value decreases with decreasing the diameter of the microdroplet below 10 μm while the value tends to level off above 10 μm. Analogous results were also observed for pyrene/silicon oil droplets.

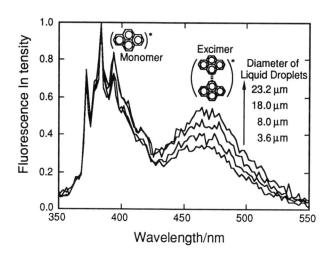

FIGURE 2
Pyrene excimer formation in various size of liquid paraffin droplet in water

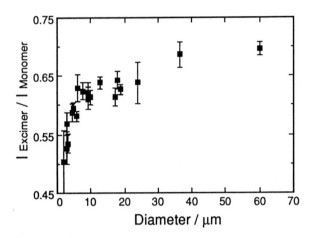

FIGURE 3
Micrometer size effect on pyrene excimer formation in liquid
paraffin droplets in water

One possible origin of the present results will be the μm-size effect on the viscosity in liquid droplets. To estimate the solution viscosity in individual droplets, we performed dynamic fluorescence anisotropy measurements on p-terphenyl/liquid paraffin droplets in water.[18] Preliminary results indicate that the rotational relaxation time of p-terphenyl increases with decreasing the diameter of the droplet, which supports the μm-size effects on the solution viscosity. We suppose that mutual association and/or orientation of liquid paraffin and water at the water/oil interphase will be reflected on the solution viscosity. Namely, the smaller the diameter of a droplet, the larger the contributions of such surface association/orientation to the solution viscosity in a droplet.

Another explanation is to assume that the solubility of pyrene in liquid paraffin droplets is dependent on the size of the droplet. Namely, for the liquid paraffin droplets with their diameter of subμm or less, we suspect that partition of water molecules into the organic layer may result in the decrease in the solubility of pyrene in droplets, rendering a decrease in I_E/I_M with decreasing the droplet diameter below 10 μm.

Although the solution properties and/or the concentration of molecules dissolved in microdroplets are generally believed to be same with each other, this is not correct as demonstrated by spectroscopic measurements of individual particles.[19] The factor not considered in bulk chemistry plays a dominant role in μm chemistry and elucidation of these factors will become exceedingly important.

3. MICROMETER PHOTOCHEMICAL FABRICATION AND MODIFICATION OF POLYMER SURFACE

The μm size-effects on dynamic structure and chemical reactions are very important. From practical view point, on the other hand, how to prepare a series of μm reaction sites is critical to realize analogous size effects. Photochemical patterning and modification of materials surface by photofunctional molecules (S) is a possible preparation method and, we expect that this will create the integrated chemical system in which reactions proceed sequentially along spatially arranged μm reaction sites. Patterning and modification of materials surface should be accompanied by, i) high spatial resolution, ii) applicability of the technique to various S, and iii) well-defined chemical bonding of S to the surface. In particular, well-defined chemical structure of the modified surface is quite important to study photochemical and photophysical processes occurring at the surface. Therefore, conventional methods of photoresist[20] and surface graft polymerization[21] are not applicable to the present purpose owing to complexity in chemical structure of the modified surface. As the first attempt, we explored μm patterning and modification of polymer surface by photochemical techniques.

3.1. Photochemical Surface Modification of Polymer Films by Aromatic Molecules[22]

Photochemical polar addition of alcohol to aromatic olefin[23] was applied to surface functionalization of poly(2-hydroxyethyl methacrylate) (PHEMA) films by pyrene or phenanthrene; eq. 4. A PHEMA film (20 nm thickness) prepared on a quartz plate was irradiated at 313 nm for 30 ~ 60 min with the film being in contact with acetonitrile solution of aromatic olefin (1 or 2; 2 ~ 20 x 10^{-3} M) and dimethyl terephthalate (DMTP, 5 x 10^{-3} M). After the photoreaction, the film was immersed in acetonitrile over 1 day to remove the unreacted molecules.

The absorption and fluorescence spectra of the pyrene-modified PHEMA film are shown in Figure 4 together with those of 1 in acetonitrile. In Figure 4 , absorption by a quartz plate and PHEMA itself is subtracted from the observed spectrum, so that the

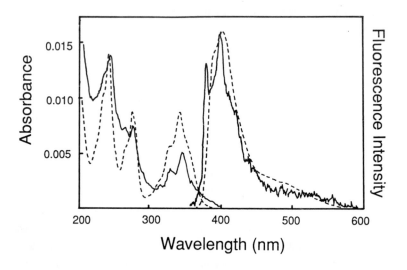

FIGURE 4

Absorption and fluorescence spectra of pyrene-modified PHEMA film (solid line) and 1 (broken line) in acetonitrile. Absorbance and fluorescence intensity of the spectra are normalized at 248 nm and the maximum intensity, respectively

characteristic absorption around 240, 270, and 350 nm is ascribable to the pyrenyl chromophore introduced into the film by the photoreaction. The fluorescence spectrum of the modified film showed the vibrational structure and a broad band around 377 and 530 nm, respectively, which could not be seen in the spectrum of **1** (Figure 4). Furthermore, since the spectrum of **1** adsorbed in a PHEMA film was almost identical with that in acetonitrile, the changes in the spectrum before and after the photoreaction are indicative of the reaction of the double bond in **1** and a -OH group of the film; eq. 4.

The photoreaction and fluorescence of **1** in acetonitorile were efficiently quenched by triethylamine and DMTP, respectively. Photochemical surface modification was successful when **1** or DMTP was replaced by **2** or 1-cyanonaphthalene, respectively, while the photoreaction did not proceed in the absence of DMTP or 1-cyanonaphthalene as an electron acceptor. These results strongly indicate that the radical cation of **1** (or **2**) produced by photoinduced electron transfer between the excited state of **1** (or **2**) and DMTP (or 1-cyanonaphthalene) is an intermediate of the photoreaction and, the pyrenyl group could be covalently introduced to the PHEMA film. As a characteristic feature of the present surface functionalization method, the amount of S introduced to the film can be controlled by the concentration of aromatic olefin as well as by photoirradiation time at a given concentration of the olefin.

Characterization of the pyrene-modified PHEMA film was performed by both X-ray photoelectron spectroscopy (XPS) and secondary ion mass spectroscopy (SIMS). XPS spectra demonstrate that the C-H band intensity at 288 eV increases upon the photoreaction while the C-O (290 eV) and C=O (297 eV) band intensity remains almost constant before and after surface modification, indicating the introduction of the pyrenyl group to the film. SIMS spectra in Figure 5 also prove that the PHEMA film is modified by pyrene in the surface layer of 7 nm as revealed by the larger C/O or H/O intensity ratio of the pyrene-modified film as compared with that of a PHEMA film. We suppose that the functionalized depth will be controllable by the affinity of solvent to the film as in the case of surface photograft polymerization in solution.[24]

Spatially-resolved surface functionalization of the film by pyrene was successful based on photoirradiation of the film through a photomask. Surface patterning of the film with ~50 μm lateral resolution has been achieved. The resolution will be improved by contact exposure of a film with a photomask or by the use of a laser as a light source. Besides aromatic olefin, the cyclopropane derivatives of aromatic hydrocarbon were also usable for the present photochemical surface modification. We believe that μm reaction sites with a variety of photofunctional molecules will be spatially arranged by the present technique and, photochemical/photophysical processes occurring at the modified surface will be elucidated by three-dimensional space- and time-resolved fluorescence and absorption spectroscopy.

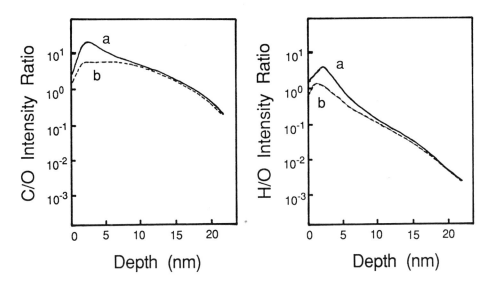

FIGURE 5
SIMS depth profiles of C/O and H/O intensity ratios for the pyrene-modified (a) and non-modified (b) PHEMA films

3.2. Fixation of Dyes on Laser Ablated Patterns[25]

Laser ablation is a possible candidate for the technique applicable to simultaneous micropatterning and modification of materials. Various reactive species are supposed to be produced on material surface during laser ablation, so that functional molecules will be chemically fixed to the reactive species. To demonstrate such possibility, we studied excimer laser (KrF, 248 nm) ablation of polymer films.

Laser ablation is strongly dependent on environmental conditions.[26] For example, the etching rate of a spin-coated poly(methyl methacrylate), PMMA, film was accerelated or suppressed with increasing in oxygen or argon gas pressure, respectively, as compared with that in vacuo. For poly(ethylene terephthalate) films, laser ablation in the presence of reactive gaseous molecules such as ammonia or hexamethyldisilane resulted in implantation of nitrogen or silicon atoms to the film, respectively, as revealed by XPS spectra of the ablated films. These results clearly prove that the ablated surface is highly reactive.

If reactive species produced during laser ablation are long lived, the ablated surface could be functionalized by various methods with a variety of molecules. Indeed, the ablated PMMA shows a strong ESR signal even one hour after laser ablation. Another important aspect is the formation of emissive species on the ablated surface of PMMA as shown in Figure 6 while no emission was observed before ablation. In the initial stage of ablation (360 pulse irradiation), broad emission was observed around 600 nm. Further pulse irradiation (4000 pulses) resulted in the appearance of new emission around 440 nm.

The emission was quenched by immersing the ablated PMMA into an aqueous dye solution. In particular, when we use rhodamine 101, the emission quenching accompanies the fixation of the dye onto the PMMA film as typically seen in Figure 7 and the dye cannot be removed even after prolonged sonication of the dye-PMMA film in pure water.

It is very interesting to note, furthermore, that this specific dye fixation strongly depends on the laser fluence for ablation. Namely, when laser ablation of a PMMA plate is conducted with the laser fluence below the ablation threshold (40 mJ/cm^2; threshold was estimated to be ~500 mJ/cm^2), the dye is fixed on the ablated surface (Figure 7a). Contrary, ablation with the laser power above threshold (i.e., 570 mJ/cm^2) leads to the selective dye fixation on non-ablated surface as clearly seen in Figure 7b. One possible explanation is that, predominant chemistry is oxidation of the surface at low laser power ablation below 500 mJ/cm^2, while crosslinking of reactive polymeric species will be a major path at high laser power ablation (> 500 mJ/cm^2). It has been reported that photodegradation of PMMA produces the reactive intermediate radicals and these radicals are known to undergo oxidation or crosslinking.[27] The formation of crosslinked polymers at high laser power ablation will be a reasonable consequence

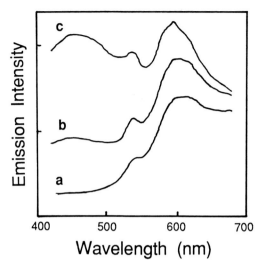

FIGURE 6
Emission spectra of the ablated surface of PMMA excited at 320-380 nm.
Laser fluence: 40 mJ/cm^2. The number of the laser pulse; a) 360, b) 1440,
c) 4000

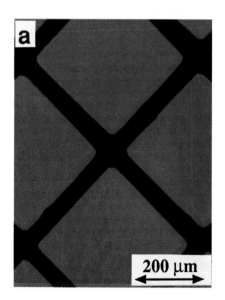 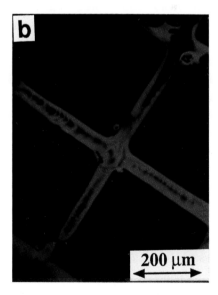

FIGURE 7
Fluorescence patterns on the laser ablated PMMA plates. Ablated by a)
40 mJ/cm^2, 4000 pulses, b) 570 mJ/cm^2, 700 pulses in air. Ablated plates
were soaked in aqueous solution of rhodamine 101

since the concentration of the intermediate radicals should be higher than that ablated below the threshold.

At < 500 mJ/cm^2, dye fixation will be ascribed to oxidation of the PMMA surface (Figure 7a) while efficient crosslinking on the surface at > 500 mJ/cm^2 disfavors the dye fixation. Inspection of Figure 7b reveals, indeed, that the dye is not fixed on the nonablated surface but on the close vicinity of the ablated pattern. It has been well known that the morphological changes of polymers take place not only on the irradiated part but on the neighborhood of the ablated part.[28] The morphological changes including partial oxidation of PMMA below a photomask and the dye fixation on such part will be the primary reason for the results in Figure 7b.

Variation of the mode of the dye fixation is quite interesting since a positive or negative dye pattern can be fabricated on materials surface depending on the laser fluence (Figure 7). The relationship between chemical nature of the ablated surface and the dye pattern mentioned above suggests that, beside rhodamine 101, a variety of molecules can be introduced onto the surface. Other techniques such as surface graft polymerization, affinity labelling, and so forth will be also applicable to surface functionalization. Simultaneous microfabrication and modification of materials by laser ablation could play an important role for preparation of integrated chemical systems.

4. MICROFABRICATION OF INDIVIDUAL SMALL PARTICLE[7,8],

Laser trapping technique[29,30] was applied to microfabrication of an individual small particle dispersed in solution for the first time. When an intense pulsed laser (355 nm, 7 ns fwhm) was irradiated on an optically trapped, pyrene-doped, PMMA latex particle, laser ablation of the particle was observed even in water. Laser ablation of the particle was depended on, i) the fluence of the pulsed laser, ii) the pulse repetition rate, as well as on iii) the nature of the ablated materials. For pyrene-doped PMMA latex particles, indeed, complete ablative decomposition, partial ablation, ablative penetration by a small hole (< 1 μm, Figure 8), or ablative fusion of two particles was shown to be possible depending on the experimental conditions mentioned above. These results indicate a possibility of controlled fabrication/modification of individual μm-order particles in solution.

Further important features of the laser trapping-ablation are as follows. Firstly, when an untrapped particle was irradiated by the pulsed laser, we could not confirm laser ablation since the untrapped particle disappeared from the ocular field of the microscope. Laser trapping is thus necessary for precise microfabrication of a particle. Secondly, the diameter of the small hole fabricated by laser ablation on a particle (Figure 8) is much smaller than that expected from the aperture angle depicted by the numerical aperture of the objective lens as well as from the refractive indices of a

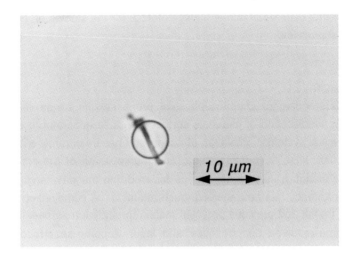

FIGURE 8
Laser ablation of an optically trapped, pyrene-doped PMMA particle in water

particle and a medium. The results will be explained by both nonlinear optical and photochemical effects.

Laser ablation has been frequently discussed based on nonlinear multi-photon absorption of dyes doped in materials. It has been reported that the higher singlet excited state of pyrene generated by multi-photon absorption induces bond scission of PMMA.[31,32] Multi-photon absorption is proportionally related to the square of the photon number, so that the bond scission takes place at the central part of the gaussian pulsed laser beam, rendering a minute hole on a pyrene-doped PMMA particle. The multi-photon absorption will be further accelerated by self-focusing effects of the laser beams.[33] An intense laser beam propagating in a medium induces an increase in the refractive index of the medium depending on the square of the laser beam intensity. Average laser intensity is around few tens MW and several hundreds MW for the trapping and pulsed laser beams, respectively, so that self-focusing effect of the laser beams will be a probable reason for the production of a minute hole (< 1μm) on a particle.

Besides pyrene-doped PMMA particles, laser trapping-ablation was successful for polystyrene latex particles, melamine resin wall microcapsules,[34] and so forth. Furthermore, since characterization of an optically trapped particle can be made spectroscopically in ns-ps time regime,[3,8,9] selection-characterization-microfabrication of an individual small particle dispersed in solution is achieved simultaneously by our currently developed laser trapping-spectroscopy-ablation technique. Clearly, the technique enables us to tweezer a single μm particle three-dimensionally, transfer the particle to any desired spatial position and to fabricate it in controlled fashion (Figure 8). We expect that such approach is highly potential for the studies on chemistry and physics of small particles.

5. CONCLUSION

Basic understanding of chemical/physical processes in μm-ps resolution and microfabrication/modification of materials surface are quite important to construct μm integrated chemical systems. However, μm chemistry has been rarely explored due to experimental difficulties in measurements and manipulation of μm-order samples. Furthermore, although a laser beam can be focused into the wavelength order spot, such spatial coherency has not received much attention. A combination of laser and microscope is fruitful and provides potential means to study μm-ps chemistry such as space- and time-resolved spectroscopy and laser trapping-spectroscopy-ablation techniques. Based on these techniques, it has been shown for the first time that μm-ps chemistry involves characteristic features which are not expected from the knowledge of bulk chemistry. Furthermore, we demonstrated that μm-order reaction sites could be prepared by photochemical methods combined with surface modification techniques.

The present approach of the study will contribute to various fields of science. For example, μm-ps spectroscopic technique could be applied to elucidate chemical/physical processes occurring in various organized molecular systems, biological tissues, catalyses, and so forth. μm laser trapping-ablation-ps spectroscopy will play an important role for inspection and modification of LSI circuits or microelectronic devices as well. Furthermore, elucidation of characteristics of μm chemistry contributes to future design of microsensors as well as to application of micromachines to biological systems in vivo. Clearly technical developments of more sophisticated spectroscopy/manipulation methods and further understanding of μm-ps chemistry will open a new field of science.

ACKNOWLEDGEMENT

N.K. was a previous laboratory staff member and H.M. was a collaborator of the late Prof. Shigeo Tazuke. They would like to present their sincere condolences and to express their heartfelt thanks for his continuous encouragement to scientific activities of N.K. and H.M.

The works described in this review were done by M. Ishikawa and N. Tamai (2.1), H. Misawa, K. Sasaki, and M. Koshioka (2.2 and 4), N. Ichinose (3.1), and T. Uchida and N. Shimo (3.2). The authors are greatly indebted to these Project members for their collaboration and thanks are also due to S. Hitomi for her help.

REFERENCES
1) F. Rondelez, D. Ausserre, and H. Hervet, Ann. Rev. Phys. Chem. 38 (1987) 317.

2) Z. Cai and C.R. Martin, J. Am. Chem. Soc. 111 (1989) 4138.

3) H. Masuhara, Microphotoconversion: Exploratory chemistry by laser and microfabrication, this volume.

4) Laser and particle-beam chemical processing for microelectronics, Materials Research Society Symposium Proceedings, Vol. 101, eds. D.J. Ehrlich, G.S. Higashi, and M.M. Oprysko (Materials Research Society, Pittsburgh, 1988).

5) R. Srinivasan and B. Braren, Chem. Rev. 89 (1989) 1303.

6) K. Sasaki, M. Koshioka, and H. Masuhara, XVII International Quantum Electronics Conference (Anaheim, 1990) Technical Digest p170

7) H. Misawa, M. Koshioka, K. Sasaki, N. Kitamura, and H. Masuhara, XVII International Quantum Electronics Conference (Anaheim, 1990) Technical Digest p170.

8) H. Misawa, M. Koshioka, K. Sasaki, N. Kitamura, and H. Masuhara, Chem. Lett. (1990) 1479.

9) H. Misawa, M. Koshioka, K. Sasaki, N. Kitamura, and H. Masuhara, Chem. Lett. (1991) 469.

10) M. Ishikawa, N. Kitamura, and H. Masuhara, XIII IUPAC Symposium on Photochemistry (Coventry, 1990) Abstracts p285.

11) M. Irie, Adv. Polym. Sci. 94 (1990) 27.

12) M. Irie and D. Kungwatchakan, Macromolecules 19 (1986) 2476.

13) T. Tanaka and D.J. Fillmore, J. Chem. Phys. 70 (1979) 1214.

14) Y. Li and T. Tanaka, J. Chem. Phys. 92 (1990) 1365.

15) T. Tanaka, L. Hocker, G. Benedek, J. Chem. Phys. 59 (1973) 5151.

16) A. Peters and S.T. Candau, Macromolecules 19 (1986) 1952.

17) N. Tamai, M. Ishikawa, N. Kitamura, and H. Masuhara, Symposium on Photochemistry, Japan (Kyoto, 1990) Abstract p9.

18) H. Misawa, M. Koshioka, K. Sasaki, N. Kitamura, and H. Masuhara, Dynamics in Small Confining Systems, eds. J.M. Drake, J. Klafter, and R. Kopelman (Materials Research Society, 1990) Extended Abstracts p141.

19) H. Misawa, M. Koshioka, K. Sasaki, N. Kitamura, and H. Masuhara, XIII IUPAC Symposium on Photochemistry (Coventry, 1990) Abstracts p288.

20) T. Sugawara and T. Matsuda, Polym. Preprints Japan 39 (1990) 597.

21) S. Tazuke, T. Matoba, H. Kimura, and T. Okada, A novel modification of polymer surface by photografting method, in: Modification of Polymers, American Chemical Society Symposium Series, Vol. 121 (American Chemical Society, 1980) p217.

22) N. Ichinose, N. Kitamura, and H. Masuhara, Chem. Lett. (1990) 1945.

23) R.A. Neunteufel and D.R. Arnold, J. Am. Chem. Soc. 95 (1973) 4080.

24) S. Tazuke and R. Takasaki, J. Polym. Sci. Polym. Chem. Ed. 21 (1983) 1517.

25) T. Uchida, N. Shimo, and H. Masuhara, XIII IUPAC Symposium on Photochemistry (Coventry, 1990) Abstracts P290.

26) N. Shimo, T. Uchida, and H. Masuhara, Foreign gas effect upon excimer laser ablation of polymer, in: Laser Ablation for Materials Synthesis, Materials Research Society Symposium Proceedings Vol. 191, eds. D.C. Paine and J.C. Bravman (Materials Research Society, Pittsburgh,1990) p.91.

27) H.H.G. Jellinek, Aspects of Degradation and Stabilization of Polymers (Elsevier, Amsterdam, 1978).

28) A. Itaya, A. Kurahashi, H. Masuhara, Y. Taniguchi, and M. Kiguchi, J. Appl. Phys. 67 (1990) 2240 .

29) A. Ashkin, Phys. Rev. Lett. 24 (1970) 156.

30) A. Ashkin, Science 210 (1980) 1081.

31) H. Masuhara, H. Hiraoka, and E.E. Martinero, Chem. Phys. Lett. 135 (1987) 103.

32) H. Masuhara and H. Fukumura, Polym. News in press (1990).

33) Y.R. Shen, The Principles of Nonlinear Optics (Wiley, New York, 1984) Chap. 17.

34) H. Misawa, N. Kitamura, and H. Masuhara, Polym. Preprints Japan, 39 (1990) p.1245.

CONTRIBUTED POSTERS

The editors of this volume wish to acknowledge the contributions of the following scientists to poster session of the Memorial Conference. These poster titles are grouped into the same categories as the book chapters themselves.

I: Photoinduced electron and energy transfer processes

Fluorescence Decay Characteristics of 9,9'-Bianthryl in Nonpolar versus Polar Polymer Matrices at Room Temperature.
   K.A. Al-Hassan

ESR Studies on Photodissociation of 2,4,6-Trimethylbenzoyl Phosphine Oxide.
   M. Kamachi, K. Kuwata, W. Schnabel

Time-Resolved EPR Studies on Photoinduced Electron Transfer Reactions from Metalloporphyrins to Quinones.
   S. Yamauchi, M. Satoh, K. Akiyama, S. Tero-Kubota, M. Iwaizumi

The Emitting Species of Azaaromatics Adsorbed on Silica Gel, Alumina, and Silica-Alumina.
   S. Suzuki, H. Murai, H. Kawaguchi, K. Ohashi

Preparation of Novel Organometallic Ions As Studied by Laser Ablation-Molecular Beam Method.
   H. Sato

Phototautomerization of 7-Hydroxycoumarins Due to Intramolecular Proton-Transfer Reactions in Hydroxylic and Nonhydroxylic Solvents.
   T. Moriya, T. Hiraga

Excited State Properties of Molecular Crystals under High Pressure.
   T. Hiraga, T. Moriya

Photoionization Study of Shorter Diphenylpolyenes in Polar Solvents by Using Transient Resonance CARS Technique.
   T. Dudev, T. Kamisuki, N. Akamatsu, C. Hirose

Electronic Transition between Different Ion-Core States of Carbon Monoxide.
   S. Sekine, C. Hirose

Conformational Change and Energy Transfer in Rat Liver Phenylalanine Hydroxylase.
   F. Tanaka, N. Tamai, I. Yamazaki , N. Kaneda, T. Nagatsu

Femtosecond-Picosecond Dynamics of the Excited 9,9'-Bianthryl.
   T. Okada, S. Nishikawa, K. Kanaji, N. Mataga

Time-resolved Observation of Excitation Hopping between Two Anthryl Moieties Attached to Both Ends of Methylene and Polystyrene Chains.
  T. Ikeda, B. Lee , H. Ushiki , K. Horie

Femtosecond Laser Photolysis Studies on the Proton Transfer Process in the Excited Hydrogen Bonding Complexes.
  H. Miyasaka, K. Wada, A. Tabata, N. Mataga

Picosecond Fluorescence and Absorption Microspectroscopy for Inhomogeneous Photochemical Processes.
  N. Kitamura, K. Sasaki, M. Koshioka, H. Masuhara

Pressure Dependence of the Electric Field in the Hollow Cathode Discharge.
  A. Wada, H. Hondo, C. Hirose

Photodissociation of Highly-Excited Triplet State of Benzophenone Studied by a Time-resolved Thermal Lensing Technique.
  Y. Kajii, T. Suzuki, Y. Takatori, K. Shibuya, K. Obi

II: Photoredox reactions in solution

Synthesis and Photoreduction of Viologen Derivatives Having Benzophenone Moiety.
  C. Tanaka, Y. Nambu, T. Endo

Structural Dependence on Photoaddition of Alcohols to Arylalkenes. Solvent and Additive Effects on Photoinduced Electron Transfer Reactions.
  K. Mizuno, I. Nakanishi, N. Ichinose, M. Sawasaki, Y. Otsuji

Salt-induced Photoamination of Arenes by Aliphatic Amines via Arene/p-Dicyanobenzene Exciplexes.
  K. Shima, M. Yasuda

Photolysis of 1,1,1-Triarylalkanes, Alkenes, and Alkynes.
  M. Shi, Y. Okamoto, S. Takamuku

Photochemical Fixation of Carbon Dioxide in Oxoglutaric Acid Using Isocitrate Dehydrogenase and Cadmium Sulfide.
  H. Inoue, Y. Kubo, H. Yoneyama

Magnetic Field Effects upon Photoreaction of Bifunctional Chain Molecules.
  Y. Tanimoto

Methylene Chain Length and External Magnetic Field Effects on Bichromophoric Photochemistry.
  R. Nakagaki, K. Mutai, S. Nagakura

Wavelength-depending Photofading of Phthalocyanin with an Electron Donor.
  Y. Kaneko, Y. Nishimura, T. Arai, H. Sakuragi, K. Tokumaru, M. Kiten, S. Yamamura, D. Matsunaga

Effects of Organic Solvents to Enhance the Charge Separation Efficiency in Xanthene Dyes-sensitized Reduction of Viologens in Aqueous Solution.
  N. Takane, Y. Nishimura, Y. Kuriyama, H. Sakuragi, K. Tokumaru

G-G Specific DNA Strand Scission by Hydroperoxynaphthalimide upon Photoirradiation.
  S. Matsugo, I. Saito

Multi-electron Photoredox Process by the Use of Reduced Viologen in Hydrophobic Microenvironment.
    T. Matsuo, H. Nakamura, H. Iwata, A. Yokoo

Photochemistry of μ-Peroxodicobalt (III) Complexes with Polyamine Ligands: Oxygenation of Photoinduced Co (II) Complexes in Aqueous Solutions.
    S. Matsufuji, N. Shinohara

Intramolecular Energy/Electron Transfer in Covalently-linked Heterobinuclear Systems Consisting of Polypyridine-Ru(II)/Os(II) and -Ru(II)/Rh(III) Complexes.
    M. Furue, M. Hirata, M. Naiki, T. Oguni, S. Kinoshita, T. Kushida, M. Kamachi

No-Way, One-Way, Intermediate-Way, and Two-Way Photochemical Isomerization of Aromatic Olefin.
    T. Arai, M. Tsuchiya, Y. Ogawa, Y. Kikuchi, H. Okamoto, H. Furuuchi, T. Karatsu, H. Sakuragi, K. Tokumaru

Eu^{3+}/Eu^{2+} Photoredox Reaction - One Electron Reduction or Oxidation of Substrates.
    A. Ishida, S. Takamuku

III: Photochemistry in organized molecular systems

Assembly of Oriented Porphyrins in Synthetic Bilayer Membrane and Their Photochemistry.
    I. Hamachi, H. Iwasaki, Y. Ishikawa, T. Kunitake

Homogeneous Alignment of Liquid Crystals Modulated by Polarization-Photochromism of Surface Azobenzenes.
    Y. Kawanishi, Y. Suzuki, T. Tamaki, K. Ichimura

Analysys of Fluctuation in the Structure of Cyclodextrin Inclusion Complexses Based on Fluorescence Spectroscopy.
    A. Nakamura, K. Saito, Y.-Q. Du, F. Toda

A Photophysical Study of Copolymers Containing Carbazole Chromophore in Monolayer and LB Multilayers.
    T. Miyashita, T. Yatsue, M. Matsuda, M. Van der Auweraer, F.C. De Schryver

Permeability Control of Phospholipid Bilayers Containing Dye Molecules Capable of Photoisomerization.
    Y. Yonezawa, T. Sato, H. Fujiwara, M. Kijima, H. Hada

Time-resolved Observation of Isothermal Phase Transition of Liquid Crystals Induced by Photoisomerization of Azobenzene Dopant.
    S. Kurihara, T. Sasaki, H.-B. Kim, T. Ikeda, S. Tazuke

Photochemistry of Polysilanes: Transient Intermediates Produced by the Photolysis of Polysilanes.
    A. Watanabe, Y. Tsutsumi, M. Matsuda

Photoconductivity of Conjugated Polymer Containing Anthracene and Sulfur in Backbone.
    E. Kobayashi, Y. Kakinuma, J. Jiang, S. Aoshima, J. Furukawa

Fluorescence Studies of Alternating Carbazole Polymers.
    Y. Itoh, M. Nakada, A. Hachimori, S.E. Webber

Bacteriorhodopsin Orientation in Polymer Film.
    S. Kunugi

Classification Based on Structure for Macromolecular Chain Dynamics.
    H. Ushiki, F. Tsunomori, S. Kashimori

Organic Photochemistry of Highly Organized Molecules. Highly Sequence-Selective Photoreaction of 5-Bromouracil-containing Hexanucleotides.
    I. Saito, H. Sugiyama, Y. Tsutsumi

Control of Photochemical and Photophysical Processes in LB Films by Polyion Complexation.
    K. Nishiyama, M. Kurihara, M. Fujihira

Novel Surface Polyfluorinated Molecular Assemblies and Their Photochemical Characterization.
    H. Inoue, M. Matsuda, M. Okada, Y. Kameo, M. Hida

Reversible Change in Optical Rotatory Power of Cholesteric Liquid Crystalline Polymers Induced by Photochromism.
    Y. Suzuki, K. Ichimura, N. Tamaoki,, K. Ozawa, T. Matsumoto

Photochemistry of Amphiphilic Stilbazoles in Ordered Molecular Assemblies.
    H. Ihara, A. Ohmori, J. Nakazono, M. Shimomura

Photoregulation of Permeability across a Membrane from a Graft Copolymer Containing a Photoresponsive Polypeptide Branch.
    S. Inoue

Excited Triplet States and Triplet Energy Migration in Concentrated Polymer Matrices.
    S. Ito, H. Katayama, T. Tawa, M. Yamamoto

Folding Processes for Intramacromolecular Reaction Studied by Polymer Chain Dynamics Methods.
    T. Torii, H. Ushiki

Microviscosity and Photoresponse of Ionic Gels: Picosecond Time Resolved Fluorescence Studies.
    N. Tamai, M. Ishikawa, N. Kitamura, H. Masuhara

IV: Toward integrated molecular systems

Novel Photoelectrochromic Monolayer Assemblies.
    T. Nagamura

Photochemical Surface Modification of Polymer Films by Photofunctional Molecules.
    N. Ichinose, N. Kitamura, H. Masuhara

Polyoxometaloeuropates As Molecular Analogue of Eu^{3+}-doped Metal Oxide Phosphors.
    H. Naruke, T. Yamase

Photochemical Hole Burning in a Focused Laser Spot.
    N. Kishii, N. Asai, S. Tamura, N. Matsuzawa, J. Seto

Decomposition Process of Fulgide Films in Write-Erase Cycles.
T. Yoshida, A. Morinaka, K. Murase

Photochemical Hole Burning of Ionic Porphins and the Deuterated Analogues.
K. Sakoda, M. Maeda, K. Kominami, M. Iwamoto

Photo-amplified Storage Optical Memory by Pyrene LB Films.
M. Fujihira, T. Kamei

Area Modulative Liquid Crystal Light Shutter Arrays.
S. Sugihara

Photoelectrical Properties of Electrochemically-doped Polymer-Dye System.
Y. Shirota, N. Noma, K. Namba

Fluorescent Micro-Pattern Formation on Polymer Surface by Laser Ablation Method.
T. Uchida, H. Sugimura, K. Kemnitz, N. Shimo, H. Masuhara

Photochromism in Polymer Matrices at Cryogenic Conditions.
Y. Munakata, T. Tsutsui, S. Saito

Electroreflectance Study of Electrode Reaction of Adsorbed Molecules on Electrode Surfaces.
T. Sagara, K. Niki

In situ Photoelectric Monitoring of Phthalocyanine Thin Film Growth.
N. Minami, M. Asai

Photocatalytic Reaction over Ion-exchangeable Layered Niobates.
K. Domen, T. Onishi

Photocontrol of Antigen-Antibody Reactions.
M. Harada, M. Sisido, J. Hirose, M. Nakanishi

Laser Manipulation-Spectroscopy-Ablation of a Single Particle in Solution.
H. Misawa, M. Koshioka, K. Sasaki, N. Kitamura, H. Masuhara

Multi-mode Chemical Signal Transducer and Its Conjugated Function.
T. Iyoda, T. Saika, K. Honda, T. Shimidzu

Structure-dependent Photoelectric Properties in Porphyrin LB Films.
M. Yoneyama, T. Nagao, H. Tanaka, T. Murayama

Photon-gated Photochemical Hole Burning in Tetraphenylporphine Derivatives.
S. Tamura, N. Kishii, N. Asai, N. Matsuzawa, J. Seto

Electron Spin Resonance Study of Positive Holes in J-Aggregates of a Cyanine Dye with Variation of Size Formed on AgBr Microcrystals.
T. Tani, Y. Sano

Emission Spectra in Efficient Formation of Metal Oxide Particles.
M. Fujita, N. Shimo

Photochemical Hole Burning of Quinizarin-Cyclodextrin Inclusion Compound in LB Film.
Y. Sakakibara, H. Takahashi, T. Tani

AUTHOR INDEX

SUBJECT INDEX